AA002939

MATERIAUX 2010

IOP Conference Series: Materials Science and Engineering
Volume 28

Mahdia, Tunisia
4 - 7 November 2010

ISBN: 978-1-63439-839-8
ISSN: 1757-8981

Printed from e-media with permission by:

Curran Associates, Inc.
57 Morehouse Lane
Red Hook, NY 12571

Some format issues inherent in the e-media version may also appear in this print version.

Copyright© by the Institute of Physics
All rights reserved.

Printed by Curran Associates, Inc. (2015)

For permission requests, please contact the Institute of Physics
at the address below.

Institute of Physics
Dirac House, Temple Back
Bristol BS1 6BE UK

Phone: 44 1 17 929 7481
Fax: 44 1 17 920 0979

techtracking@iop.org

Additional copies of this publication are available from:

Curran Associates, Inc.
57 Morehouse Lane
Red Hook, NY 12571 USA
Phone: 845-758-0400
Fax: 845-758-2634
Email: curran@proceedings.com
Web: www.proceedings.com

Table of contents

Volume 28

MATERIAUX 2010
4–7 November 2010, Mahdia, Tunisia

Accepted papers received: 8 December 2011
Published online: 7 February 2012

Papers

012001

Photopyroelectric back detection configuration for thermal diffusivity measurement of undoped and doped GaSb semiconductors I Abdellaziz, I Mellouki, S Abroug and N Yacoubi pg. 1

012002

Cellulose fibers obtained by organosolv process from date palm rachis (Phoenix dactylifera L.) H Ammar, M Abid and S Abid pg. 7

012003

Chemical durability and Structural approach of the glass series (40-y) Na_2O-yFe_2O_3-5Al_2O_3-55P_2O_5-by IR, X-ray diffraction and Mössbauer Spectroscopy S Aqdim, El H Sayouty, B Elouad and J M Greneche pg. 12

012004

Electronic band structure calculations for $Ga_xIn_{1-x}AS_yP_{1-y}$ alloys lattice matched to InP A Bechiri, F Benmakhlouf, H Allouache, S Bacha and N Bouarissa pg. 20

012005

Study of the Resistance of the Composite Columns under Axial Compression R Belounis and M Mimoune pg. 27

012006

First-principle study on electronic structure property of GaN/AlN N Benayad, M Dine El Hanani and M Djermouni pg. 35

012007

Optical glass surfaces polishing by cerium oxide particles D Bouzid, N Belkhie and T Aliouane pg. 46

012008

Use of argillaceous materials containing sediments of dam in the retention of heavy metals Z Dali-Youcef pg. 52

012009

Elasticity modulus evaluation for a porous material by acoustic methods S Debboub, Y Boumaiza and A Boudour pg. 57

012010

Dispersion of the carbon nanotubes (DWNTC) by the cationic surfactants imidazolinium type A Zafour, D Abdelkader, E Flahaut, N Rebizi and A Skender pg. 65

012011

Corrosion evolution of the aluminum alloys used in overhead transmission lines E Rhaiem, T Bouraoui and F El Halouani pg. 71

012012

Sound velocity variation as function of polarization state in Lead Zirconate Titanate (PZT) Ceramics W Essolaani and N Farhat pg. 77

012013

Low polarization switching in Lead Zirconate Titanate (PZT) Ceramics N Farhat and W Essolaani pg. 83

012014

Structural and magnetic properties of Pr_2Co_7 powders R Fersi, N Mliki and L Bessais pg. 92

012015

Material Choice for spindle of machine tools S Gouasmi, B Merzoug, G Abba and L Kherredine pg. 96

012016

Poly [(pyrrole-2, 5-diyl) co(4-Nitro benzylidene)]: A new synthetic approach and electronic properties H Gherras, A Hachemaoui, A Yahiaoui, M Belbachir and A Belfedal pg. 101

012017

Synthesis and crystal structure of $Sr_2As_2O_7$ from single-crystal data F Edhokkar, A Hadrich, M Graia and T Mhiri pg. 109

012018

Study of the valorization of phosphogypsum in the region of Gafsaas filler in paper S Hassen, Z Anna, E Elaloui E, M N Belgacem and E Mauret pg. 116

012019

The porous nature of ZnO thin films deposited by sol–gel Spin-Coating technique M Karyaoui, A Ben Jaballah, R Mechiak and R Chtourou pg. 125

012020

Structural and magnetic investigation on tetragonal R-Fe alloy with 1:12 stoechiometry S Khazzan, N Mliki and L Bessais pg. 133

012021

Evaluation and measurement of the damping properties of laminated CFRP composite plates L Kherredine, S Gouasmi, R Laissaoui and N E Zeghib pg. 137

012022

Study by AES, EELS Spectroscopy of electron Irradiation on InP and $InPO_4$/InP in comparison with Monte Carlo simulation Z Lounis, M Bouslama, K Hamaida, C Jardin, A Abdellaoui, A Ouerdane, M Ghaffour and N Berrouachedi pg. 146

012023

Synthesis of Plaster for moulers's elaboration of imprint denture M Hamiane, N Rabahi, M Saidi and M Salhi pg. 154

012024

AES and EELS tools associated to TRIM simulation methods to study nanostructures on III-V semiconductor surfaces A Ouerdane, M Bouslama, M Ghaffour, A Abdellaoui, A Nouri, K Hamaida and Y Monteuil pg. 160

012025

A study on Fluoroanhydrite; a solid waste of the chemical industry: Characterization and valorization attempts F Rais, A Kamoun, A Jelidi and M Chaabouni pg. 171

012026

TEM and stress relaxation analysis of 99.1% aluminum processed by Equal Channel Angular Extrusion A Rebhi, T Makhlouf, Y Champion, J P Couzinié and N Njah pg. 177

012027

Granulometric distribution of natural and flux calcined chert from Ypresian phosphatic series of Gafsa-Metlaoui basin compared to diatomite filter aid R Saidi, A Tlili, A Fourati, N Ammar, A Ounis and F Jamoussi pg. 183

012028

Mechanical properties of tricalcium phosphate–alumina composites S Sakka, F Ben Ayed and J Bouaziz pg. 191

012029

Effect of fluorapatite additive on the mechanical properties of tricalcium phosphate-zirconia composites I Sallemi, F Ben Ayed and J Bouaziz pg. 201

012030

The damage of the cutting tools out of carbide metallic during the turning of a soaked and not hardened steel XC38 M Seghouani, A Tafraoui and S Lebaili pg. 209

012031

First-principales calculations on the origins of the gap bowing in $Pb_{1-x}Ca_xS$, $Pb_{1-x}Ca_xSe$ and $Pb_{1-x}Ca_xTe$ alloys C Sifi, M Slimani and H Meradji pg. 219

012032

Grafting of 4-aminomethylbenzensulfonamide-lipoic acid conjugate on gold nanoparticles M Stiti, H Bouzit, M Abdaoui and J Y Winum pg. 227

012033

A Novel poly {(2,5-diyl thiophene) (2-Thiophenyl Methine)} from the Acid-Catalyzed Polycondensation of thiophene and thiophene-2-carbanaldehyde A Belmokhtar, A Yahiaoui, A Hachemaoui, N Sahli, A Benyoucef, A Belfedal and M Belbachir pg. 235

012034

Optimization of the synthesis based on Resorcine and formaldehyde to obtain carbon nanoxérogels A Zweg, E Elaloui, S Nouri and Y Moussaoui pg. 244

012035

Synthesis, crystal structure and spectroscopic investigations of a new tetrachloridozincate $NH_3(C_6H_4)_2NH_3ZnCl_4$ W Wamani, M Belhouchet and T Mhiri pg. 252

012036

Quantification of surface state effects in GaAs MESFETs Z Hadjoub, K Cheikh, A Zouyed, A Khoualdia and A Doghmane pg. 259

012037

Acoustic microscopy evaluation of anisotropy effects on elastic properties of Na_2CoGeO_4 single crystals A Doghmane, I Al-Suraihy, I Hadjoub and Z Hadjoub pg. 266

012038

Bulk metallic glass matrix composite for good biocompatibility F Hadjoub, W Metiri, A Doghmane and Z Hadjoub pg. 272

012039

Evolution of microstructural and mechanical properties of an industrial 2017 Aluminium alloy processed by equal channel angular extrusion M A Rekik, T Makhlouf and N Njah pg. 277

012040

Crystal structure and spectroscopic investigations of a new benzidinuim dichloride $NH_3(C_6H_4)_2NH_3Cl_2$ W Wamani, M Belhouchet and T Mhiri pg. 284

012041

Synthesis, Calorimetric and X ray diffraction studies of the solid Solution $[Rb_{0.92}(NH_4)_{0.08}]_2TeCl_4Br_2$ W Ben Aribia, M Abdelmouleh, A Van Der Lee and A Kabadou pg. 291

012042

An investigation into the effects of conventional heat treatments on mechanical characteristics of new hot working tool steel M L Fares, M Athmani, Y Khelfaoui and A Khettache pg. 297

012043

X-ray studies of structure defects in recycled aluminum alloy during equal channel angular extrusion: Dislocation density and stored energy T Makhlouf and N Njah pg. 307

012044

A study of the phase transition behaviour of $[(NH_4)_{0.63}Li_{0.37}]_2TeBr_6$ R Karray, D Linda, A Van Der Lee, A Ben Salah and A Kabadou pg. 314

012045

Oxidations of organic matter present in the phosphoric acid 54% by the ozone: characterization of groups carbonyls upstream and downstream of the ozonation D Linda, A Louati, C Chtara and A Kabadou pg. 323

012046

Decomposition behavior of the $NaH_2\ (PO_4)_{0.48}(AsO_4)_{0.52}\ H_2O$ compound above room temperature shown by the study of Raman X- ray powder and ac-conductivity Kh Jarraya, N Guiblin, N E Ghermani and T Mhiri pg. 333

012047

A Theoretical study of Propagation of Rayleigh surface waves in functionally graded piezoelectric (FGPM) half-space I Ben Salah, A Njeh and et M H Ben Ghozlen pg. 340

012048

Magnetic and structural properties of nanocrystalline $PrCo_3$ K Younsi, L Bessais, V Russier and J C Crivello pg. 350

012049

Magnetic and magnetocaloric properties of monovalent substituted $La_{0.5}Ca_{0.45}A_{0.05}MnO_3$ (A = Na, Ag, K) perovskite manganites A Mehri, W Cheikh-Rouhou Koubaa, M Koubaa and A Cheikh-Rouhou pg. 354

012050

Effect of strontium substitution on the physical properties of $Nd_{0.5}Ca_{0.5-x}Sr_xMnO_3(0.0\leq x\leq 0.5)$ manganites R M'nassri, W Cheikhrouhou-Koubaa, M Koubaa and A Cheikhrouhou pg. 360

012051

Magnetocaloric properties of $Nd_{0.5}Ca_{0.5-x}Sr_xMnO_3$ $(0.0\leq x\leq 0.5)$ manganites R M'nassri, W Cheikhrouhou-Koubaa, M Koubaa and A Cheikhrouhou pg. 368

012052

Structure, magnetic and magnetocloric properties of $Pr_{0.7}Ca_{0.3}Mn_{1-y}Cr_yO_3$ A Selmi, W Cheikhrouhou-Koubaa, M Koubaa and A Cheikhrouhou pg. 373

012053

Effect of Ga doping on the physical properties of $La_{0.7}(CaSr)_{0.3}Mn_{1-x}Ga_xO_3$ (x = 0, 0.025, 0.05, 0.075 and 0.1) S Othmani and M Balli pg. 380

012054

Effect of monovalent doping on the physical properties of $La_{0.7}Sr_{0.3}MnO_3$ compound synthesized using sol-gel technique F Ayadi, F Saadaoui, W Cheikhrouhou-Koubaa, M Koubaa, A Cheikhrouhou, L Sicard and S Ammar pg. 386

012055

Effect of sodium substitution on the structural, magnetic and magnetocaloric properties of $Nd_{0.5}Sr_{0.5}MnO_3$ perovskite manganites A Mehri, W Cheikhrouhou Koubaa, M Koubaa and A Cheikhrouhou pg. 394

PREFACE

The national conference 'MATERIAUX 2010', which took place in Mahdia (Tunisia), from 4–7 November 2010 was organized by The 'Tunisian Materials Research Society: Tu-MRS' in collaboration with the Materials Physics Laboratory of Sfax 'LPM', Faculty of Sciences, Sfax University and the Research Unit ' Physique, Informatique et Mathématiques ', Faculty of Sciences, Gafsa University

The First National Conference on Materials 'MATERIAUX 2006' was organized in Douz (Tunisia) in December 2006. This was followed by 'MATERIAUX 2007' held in Hammamet (Tunisia) in April 2007 and the National Conference 'MATERIAUX 2009' organized jointly with the Second 'Journées Internationales de la Physique des Matériaux et Applications: JIPMA 2009' in Gafsa (Tunisia) from 20–24 December 2009.

'MATERIAUX 2010' is intended to provide an excellent opportunity for National, Maghreb and International researchers to make their own work on materials known to a wider audience and to have discussions with other participants. This conference will also be an opportunity to exchange experiences, create and consolidate cooperation between different research structures in the Maghreb countries and also the countries around the Mediterranean.

This conference will equally promote research development, contribution to collaboration between universities and the socio-economical milieu.

More than 300 senior researchers, Professors, PhD and Masters students attended this conference from Tunisia, Algeria, Morocco, France, Spain and Canada. Several researchers, engineers and managers from industrial firms also attended.

The conference consisted of plenary talks (8), oral contributions (40) and poster presentations (250). The topics of the Conference were:

- Nano-materials, nano-systems, thin films, surfaces and interfaces, applications
- Multifunctionnal materials, magnetic materials, dielectric materials, superconducting materials, applications,...
- Materials for electronics, informatics and communication
- Ceramics, glasses, polymers
- Natural materials (phosphates, clay,...)
- Metallic materials, alloys, metallurgy,...
- Others (materials and environment, materials and energy, biomaterials,...)

I want to thank the scientific committee, the organizing committee, the local committee and everyone who contributed to the organization of this meeting for their invaluable efforts in order to guarantee the complete success of this conference.

Abdelwaheb Cheikhrouhou
President of 'Tu-MRS'
Chairman of the Conference 'MATERIAUX 2010'

Committies

Organizing Committee

Chairman
CHEIKHROUHOU Abdelwaheb (Faculté des Sciences de Sfax)

Members
ALAYA Sahbi (Faculté des Sciences de Gabès)
BENNACEUR Raouf (Faculté des Sciences de Tunis)
BEN SALEM Mohamed (Faculté des Sciences de Bizerte)
CHEIKHROUHOU-KOUBAA Wissem (Faculté des Sciences de Sfax)
EL JANI Belgacem (Faculté des Sciences de Monastir)
EZZAOUIA Hatem (Centre de Recherches et de Technologies de l'Energie, Technopole de Borj Cédria)
LAMLOUMI Jilani (Ecole Supérieure des Sciences et Techniques de Tunis)
REZIG Bahri (Ecole Nationale d'Ingénieurs de Tunis)

Local Committee

Chairman
CHEIKHROUHOU Abdelwaheb (Faculté des Sciences de Sfax)

Members
CHEIKHROUHOU-KOUBAA Wissem (Faculté des Sciences de Sfax)
KOUBAA Mohamed (Institut Supérieur de Biotechnologie de Sfax)
NJEH Anwar (Institut Préparatoire aux Etudes d'Ingénieurs de Sfax)
BEN SALAH Issam (Faculté des Sciences de Sfax)
TAKKALI Férid (Faculté des Sciences de Sfax)
REGAIEG Yassin (Faculté des Sciences de Sfax)
OTHMANI Safa (Faculté des Sciences de Sfax)
MNASSRI Rafik (Faculté des Sciences de Sfax)

Secretariat
BEN GHOZLEN Afifa (Faculté des Sciences de Sfax)
BOUGHARIOU Sana (Faculté des Sciences de Sfax)

Scientific Committee
M. ADDOU, Faculté des Sciences de Kénitra (Morocco)
N. AMDOUNI, Faculté des Sciences de Tunis (Tunisia)
M. BACCOUCHE, Faculté des Sciences d'Annaba (Algeria)
H. BATIS, Faculté des Sciences de Tunis (Tunisia)
A. BELAFHAL, Faculté des Sciences d'El Jadida (Morocco)
M.H. BEN GHOZLEN, Faculté des Sciences de Sfax (Tunisia)
R. BENNACEUR, Faculté des Sciences de Tunis (Tunisia)
B. BENYOUCEF, Université Abou Bakr Belkaid, Tlemcen (Algeria)
M. BEN SALEM, Faculté des Sciences de Bizerte (Tunisia)
B. BESSAIES, Centre de Recherches et Technologies de l'Energie (Tunisia)
H. BOUCHRIHA, Faculté des Sciences de Tunis (Tunisia)
W. BOUJELBEN, Faculté des Sciences de Sfax (Tunisia)
A. CHEIKHROUHOU, Faculté des Sciences de Sfax (Tunisia)
W. CHEIKHROUHOU-KOUBAA, Faculté des Sciences de Sfax (Tunisia)
R. CHTOUROU, Centre de Recherches et Technologies de l'Energie (Tunisia)
M. DEBBABI, Ecole Nationale d'Ingénieurs de Monastir (Tunisia)
A. DAKHLAOUI, Faculté des Sciences de Bizerte (Tunisia)
A. DINIA, Université de Strasbourg (France)
B. ELJANI, Faculté des Sciences de Monastir (Tunisia)
A. ELJAZOULI, Faculté des Sciences Ben Msik, Casablanca (Morocco)
Z. FAKHFAKH, Faculté des Sciences de Sfax (Tunisia)
A. GASMI, Faculté des Sciences d'Annaba (Algeria)
A. GHARBI, Faculté des Sciences de Tunis (Tunisia)
R. GHARBI, Faculté des Sciences de Sfax (Tunisia)
K. GUIDARA, Faculté des Sciences de Sfax (Tunisia)
H. GUERMAZI, Institut Préparatoire aux Etudes d'Ingénieurs de Sfax (Tunisia)

S. GUERMAZI, Faculté des Sciences de Sfax (Tunisia)
M. HADDAD, Faculté des Sciences de Meknès (Morocco)
A. HAJ AMARA, Faculté des Sciences de Bizerte (Tunisia)
D. HAMANA, Faculté des Sciences de Constantine (Algeria)
N. KAMOUN, Faculté des Sciences de Tunis (Tunisia)
S. KADDOUR-CHARFI, Faculté des Sciences de Tunis (Tunisia)
M. KADDOUR, Faculté des Sciences de Sfax (Tunisia)
M. KHITOUNI, Faculté des Sciences de Sfax (Tunisia)
T. MHIRI, Faculté des Sciences de Sfax (Tunisia)
Y. MLIK, Institut Préparatoire aux Etudes Scientifiques et Techniques (Tunisia)
N. MLIKI, Faculté des Sciences de Tunis (Tunisia)
A. NJAH, Faculté des Sciences de Gafsa (Tunisia)
M. OUESLATI, Faculté des Sciences de Tunis (Tunisia)
K. ZELLAMA, Faculté des Sciences d'Amiens (France)

Invited Speakers

AMMAR-MERIAH Souad, ITODYS, Université Paris Diderot (France)
BEN SALEM Mohamed, Faculté des Sciences de Bizerte, Université du 7 Novembre á Carthage (Tunisia)
CHEIKHROUHOU Abdelwaheb, Faculté des Sciences de Sfax, Université de Sfax (Tunisia)
DAKHLAOUI Amel, Faculté des Sciences de Bizerte, Université du 7 Novembre á Carthage (Tunisia)
DJABBAR Ahmed, Université des Sciences et des Technologies de Lille (France)
DURASTANTI Félix, Centre d'Etudes et Recherche en Thermique, Environnement et Systèmes (C.E.R.T.E.S.), Université Paris Est- Créteil (France)
FERY-FORGUES Suzanne, Université Paul Sabatier, Toulouse (France)
GIRAUD Romain, Laboratoire de Photonique et de Nanostructures, CNRS/LPN, Marcoussis (France)

IOP Conf. Series: Materials Science and Engineering 28 (2012) 012001 doi:10.1088/1757-899X/28/1/012001

Photopyroelectric back detection configuration for thermal diffusivity measurement of undoped and doped GaSb semiconductors

I. Abdellaziz[1], I. Mellouki, S. Abroug, N. Yacoubi
Photothermal Laboratory, IPEIN, BP 62, Merazaka 8000, Nabeul, Tunisia.

E-mail: ines.abdellaziz.haouala@gmail.com; yacoubi20042003@yahoo.fr; mellouki_imene@voila.fr

Abstract. Photopyroelectric technique in back detection configuration (BPPE) was used to investigate thermal diffusivity measurement of undoped and doped GaSb bulk samples. A theoretical model was developed and specific design of pyroelectric sensor based on PVDF film of 25μm thickness was constructed. The methodology is based on a designed back photopyroelectric detection cell and thermal diffusivity value is obtained directly as a result of a scan of the BPPE phase and amplitude signal as a function of square root modulation frequency.
This method involves the thermally thick limit of the pyroelectric signal in the standard BPPE. To avoid the conventional requirement for set-up transfer function a signal normalization procedure is implemented.
Obtained thermal diffusivity coefficients by means of this technique are $16.7 \pm 0.5*10^{-6}$ m^2s^{-1}, $6.3 \pm 0.3 *10^{-6}$ m^2s^{-1}, and $7.3 \pm 0.2 *10^{-6}$ m^2s^{-1} respectively for undoped GaSb, Te- doped GaSb and Zn-doped GaSb semiconductors.
By studying samples of GaSb with different types of doping (doping with Te and with Zn) we notice a change in the thermal diffusivity function of doping.

1. Introduction

Photothermal techniques have proved to be useful experimental methodologies for carrying out thermal characterizations in the liquid and solid state [1, 2]. Several photothermal methodologies, involving Mirage Effect [3] and photopyroelectric detection [4, 5] have been reported for measuring thermal diffusivity.

During the past few years, application of photopyroelectric technique which have the advantage of being non destructive has been considerably involved.
Recently, many useful applications of photopyroelectric technique have been reported in studying thermal properties such as specific heat(c), thermal conductivity (K), thermal diffusivity (D) and effusivity (e) using frequency scan of photopyroelectric signal [6, 9].
The basic mechanism of photopyroelectric effect consisted in changing temperature on one side of pyroelectric transducer consequence of an absorption modulated radiation by a sample placed in good thermal contact with the other pyroelectric transducer side.

[1] To whom any correspondence should be addressed.

As consequence, this will induce charge on pyroelectric tranducer electrodes and appear as a photopyroelectric voltage signal when connected to an external circuit [10].

In this paper, the back detection configuration utility of photopyroelectric technique for performing precision measurement of thermal diffusivity is shown.

This procedure is illustrated through measurements of thermal diffusivities for GaSb undoped and doped semiconductor samples.

Gallium antimonide is a direct gap semiconductor having good carrier's mobility and significant electro-optical potential in the near IR range.

Due to its potential applications in low loss optical fibers, high speed opto-electronic devices, high efficiency solar cells, for long wavelength detectors (λ> 1.5μ m) and IR laser diode, researches on GaSb have been enhanced [11].

High quality GaSb substrates with definite types of impurities, witch play an important role in electrical and optical properties of GaSb and free carrier concentration; in addition to low level dislocation densities are required for the epitaxial growth of these diodes [12].

2. Theory

In Figure.1, 1-D geometry for the photopyroelectric technique in a back detection configuration is shown [6]. Back photopyroelectric structure cell used in this study consists of air (g), an opaque pyroelectric sensor (p) and an opaque sample (s).

Sample is irradiated with modulated monochromatic light beam of angular frequency ω given by $\omega = 2\pi f$ where f is the modulation frequency. Light with intensity I_0 is impigning on the upper surface of sample, which absorbs light on its surface with optical absorption coefficient β.

Detected signal is proportional to pyroelectric coefficient p of detector and to temperature distribution along the detector thickness.

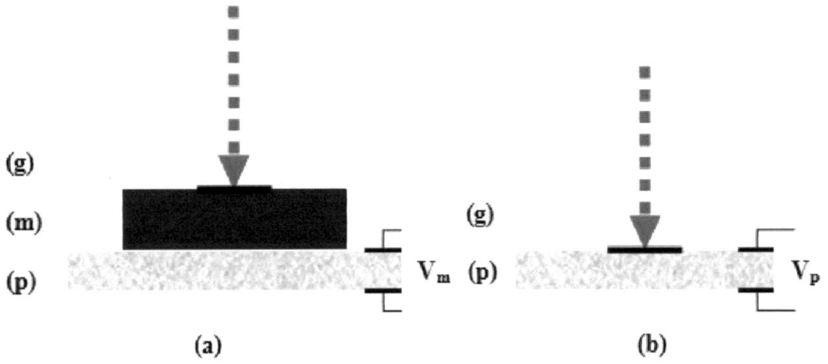

Figure 1. Schematic representation of Back PPE cell with (a) and without sample (b).

The Photopyroelectric voltage is related to spatially averaged temperature field in the bulk of pyroelectric film by the following relationship [5]:

$$V_p(\omega,t) = \frac{pL_p}{\varepsilon\varepsilon_0} < T_p(\omega) > \exp^{j\omega t} \quad (1)$$

Where:

p: pyroelectric coefficient,

ε : relative dielectric constant of the detector

ε_0 : Vacuum dielectric permittivity

L_p : detector thickness,

$<T_p(\omega)>$: Spatially averaged temperature field in pyroelectric film bulk

Temperature distribution through PVDF film $T_p(x,\omega)$, is obtained by solving one-dimensional thermal diffusions equation in each medium consisting the PPE cell.
Heat propagation across the whole chamber is governed by heat diffusion equations of each medium coupled via boundary conditions at the interfaces $T_i = T_j$ and $K_i \frac{dT_i}{dx} = K_j \frac{dT_j}{dx}$, *i* and *j* representing consecutive media.
General expression of photopyroelectric voltage is established by A.Mandelis and [13] and M.Chirtoc [14].
An analytical expression of a 3-layer model cell can be obtained by suitable particularizations of more general photopyroelectric systems [6].
In this study, the detector is considered optically opaque and sample is thermally thick and optically opaque. Under such circumstances, normalized voltage signal expression is significantly simplified [6].
Photopyroelectric amplitude and phase of a sample (m) can be expressed respectively as:

$$|V_m| = \frac{2 * V_0 e^{-\sqrt{\frac{\pi f}{D_m}} L_m}}{\left(\frac{e_m}{e_p}+1\right)\left(\frac{e_g}{e_m}+1\right)} \quad (1) \qquad \text{and } \varphi_m = \varphi_0 - \sqrt{\frac{\pi f}{D_m}} L_m \quad (2)$$

V_m obtained by integrating the diffusion equations is normalized by the ratio V_p, where V_p is the signal measured directly over the detector covered with very thin layer of graphite.
Assuming this case, normalized PPE amplitude and phase expressions can be expressed respectively as:

$$|V_N| = \frac{|V_m|}{|V_p|} = \frac{2 * e^{-\sqrt{\frac{\pi f}{D_m}}}}{\left(\frac{e_m}{e_p}+1\right)\left(\frac{e_g}{e_m}+1\right)} \quad (3) \qquad \text{and } \varphi_N = \varphi_m - \varphi_p = -\sqrt{\frac{\pi f}{D_m}} L_m \quad (4)$$

In preview equations, following symbols are used:
L_m : sample thickness,
D_m : sample thermal diffusivity
e_m : sample thermal effusivity
e_p : detector thermal effusivity
e_g: gas thermal effusivity
V_0 : experimental set up transfert function

$\mu_i = \left(\frac{D_i}{\pi f}\right)^{1/2}$: Thermal diffusion length in medium i.

3. Experimental setup

The photopyroelectric experimental setup system is shown in Fig 2. It comprises an optical probe (light source and chopper), the custom-made pyroelectric cell and measuring system. Beam of 25 mW He-Ne laser (632.8 nm) modulated by a variable frequency mechanical chopper of type SR540 is impigned into sample surface which is in thermal contact with pyroelectric detector.

Pyroelectric detector was a 25µm thick PVDF film with 1.5 µm thick Al electrodes evaporated on both surfaces for electrical contacts. Thermal and electrical PVDF parameters values (as provided by manufacturer Good fellow) are gathered in table 1.

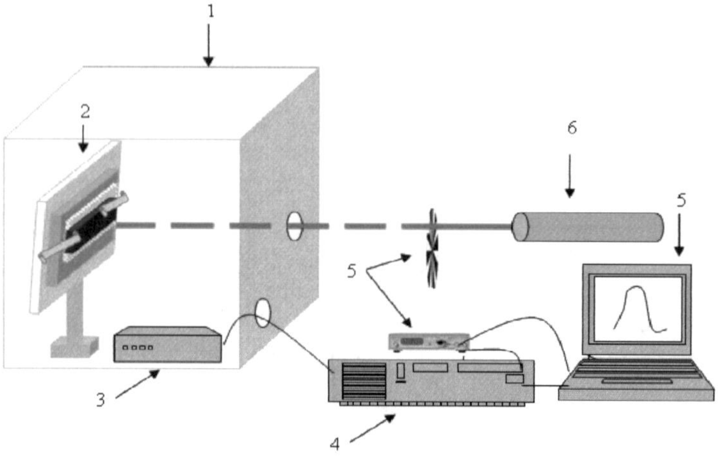

Figure.2. Schematic diagram of PPE experimental setup: 1-PPE Cell, 2-Pyroelectric sensor, 3-current voltage converter, 4- look-in amplifier, 5- Mechanical Chopper, 6-Fixed Laser Source

To ensure good thermal contact between bulk sample and PVDF detector, thermal grease is employed. A very thin layer of graphite of 1 μm thickness is deposited directly by pulverization on both sample surface and PVDF respectively.

The BPPE signal (amplitude and phase) was processed with a Standford Research System EG& G 5210 lock-in amplifier. The data acquisition, processing and analysis were performed with adequate software.

Table 1. Thermal and electrical parameters of PVDF.

	Thermal conductivity $Wm^{-1}K^{-1}$	Thermal diffusivity m^2s^{-1}	relative dielectric constant
PVDF	0.14	7,4 10^{-8}	8.4

4. Results and discussion

Photopyroelectric intensity signal and its phase were independently measured as a function of chopping frequency for a 632.8 nm wavelength.

At the same frequency of modulation, PPE amplitude and phase are respectively measured with (Figure.1 (a)) and without (Figure.1 (b)) sample.

Typical plots (theoretical (line) and experimental (point)) of the behaviour of the normalized phase (radians) and natural logarithm normalized amplitude of the BPPE signal as a function of square root chopping frequency for respectively undoped, Te-doped and Zn-doped GaSb samples are displayed in figures 3, 4 and 5.

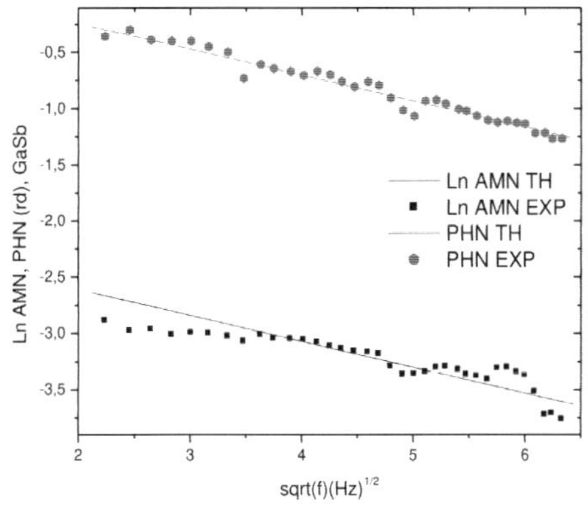

Figure 3. . Experimental and theoretical variation of the natural logarithm of the normalized amplitude and normalized phase (radian) of photopyroelectric signal V_N with the square root of the modulation frequency for thermal diffusivity of 16,7 10^{-6} m^2s^{-1} for undoped GaSb sample

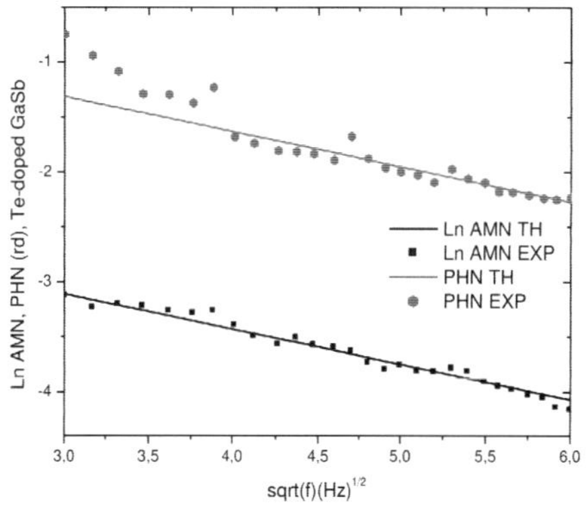

Figure 4. Experimental and theoretical variation of the natural logarithm of the normalized amplitude and normalized phase (radian) of photopyroelectric signal V_N with the square root of the modulation frequency for thermal diffusivity of 6,3 10^{-6} m^2s^{-1} for Te-doped bulk GaSb semiconductor.

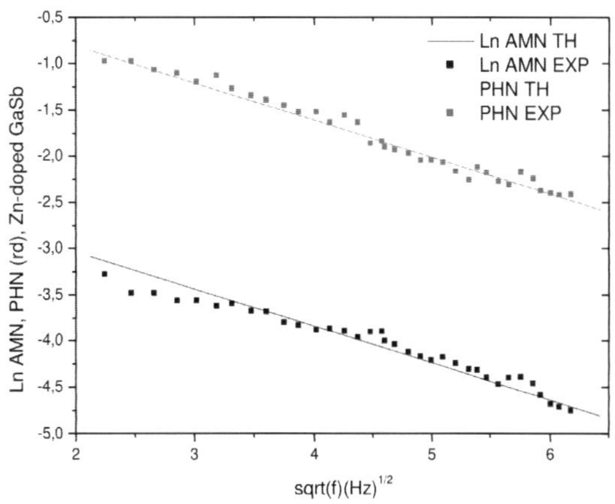

Figure 5. Experimental and theoretical variation of the natural logarithm of the normalized amplitude and normalized phase (radian) of photopyroelectric signal V_N with the square root of the modulation frequency for thermal diffusivity of 7,3 10^{-6} m^2s^{-1} for Zn-doped bulk GaSb semiconductor.

Photopyroelectric amplitude and phase measured for PVDF detector only are used for normalisation. This normalalization involves pyroelectric signal ratio with and without sample. Equations for both amplitude (3) and phase (4) of PPE signal were used to fit experimental results. Figures 3, 4 and 5 contain the best fit performed with theses equations on the experimental data.

The overall methodology is based on linear behaviour of natural logarithm normalized amplitude and normalized phase simultaneously as function of square root modulation frequency.

A common slope for this normalized signal (natural logarithm of normalized amplitude and normalized phase (radians)), which depends on the thermal diffusivity sample, is obtained for each sample and it is shown to be directly related to sample thermal diffusivity.
This slope is measured from normalized amplitude and phase signals and used for evaluating thermal diffusivity sample through equation (3) and (4).
From these fits we obtained from slope, thermal diffusivity coefficient for each sample. Results are summarized in table 2.

Table 2. Thermal diffusivities coefficients of undoped and doped bulk GaSb semiconductors

Sample	Thickness (µm)	Thermal diffusivity ($*10^{-6}m^2s^{-1}$)
GaSb	515	16.7 ±0.52
Te-doped GaSb	495	6.3 ± 0.28
Zn-doped GaSb	610	7.3 ± 0.25

Noise at low frequency (under 3, 5 $Hz^{1/2}$) showed by the gap between experimental data and corresponding theoretical plots is due to mechanical chopper.
Concerning the accuracy of the method, absolute uncertainty type for thermal diffusivity measurement is obtained by statistical method and is gathered in Table 2.
As presented in Table 2, the results obtained on several bulk GaSb samples shows a variation of thermal diffusivity coefficient as function of doping type (Te, Zn).

5. Conclusion

In conclusion, in this paper we have presented the application of a particular normalisation procedure based on photopyroelectric technique, in thermal diffusivity measurement.

Using a very thin layer of graphite deposited on the sample, establish the optically opaque condition which simplify the general expression of photopyroelectric signal.
By studying samples of GaSb with different types of doping (doping with Te and with Zn) we notice a change in the thermal diffusivity function of doping.

Photopyroelectric experiments have proven to be a valuable tool to measure directly the thermal diffusivity coefficient of bulk semi-conductors and it shows additionally sensitivity to doping.
A major advantage of the photopyroelectric technique over other reported photothermal techniques lies in its extreme sensitivity, simplicity and non destructive probing ability; also it consists of the possibility to carry out fast estimates of the thermal diffusivity of bulk semiconductors, by measuring the photopyroelectric signals with and without sample.

References

[1] Boccara A. C. A 1980 *Appl. Phys. Lett.* **36** 213-233.
[2] Balderas -López J. A. A 2003 *Rev. Sci. Inst.* **74** 700-702.
[3] Boccara A. C. A 1980 *Appl. Phys. Lett.* 36 130.
[4] Zakaria A. A 2006 *Sci. Asia.* **32** 47-52.
[5] Albuquerque J.E. A 2005 *Appl. Phy. Lett.* 87.
[6] Chirtoc M. A 2009, chap5 (Thermal wave physics and related photothermal techniques: basic principles and recent developments) Editor E.Marin, transworld research network.
[7] Hadj Sahraoui A. A 2002 *Rev. Sc. Inst.***73** 2766-2772.
[8] Balderaz-Lopez J. A. A 2001 *Anal. Sci.* **17** 519-522.
[9] Pittois S. A 2001 *Anal. Sci.* **17** 110-113.
[10] Blackburn H. A 1970 *Inf. Phys.* **10** 191.
[11] Krier A. A 1991 *Semi. Sci. and Tech.* **6** 1066.
[12] Meyer J. R. A 2001 *App. Phys.* **89**. 5815.
[13] Mandelis A. A 1985 *J. Appl. Phys.* **57** 4421.
[14] Chirtoc M. A 1989 *Phys. Rev.* B **40** 14.

IOP Conf. Series: Materials Science and Engineering **28** (2012) 012002 doi:10.1088/1757-899X/28/1/012002

Cellulose fibers obtained by organosolv process from date palm rachis (Phoenix dactylifera L.)

H. Ammar[1], M. Abid, S. Abid
Laboratoire de Chimie Appliquée: Hétérocycles, Corps Gras et Polymères, Université de Sfax, Faculté des Sciences Sfax, 3000, Tunisie

E-mail: houcine_fss@yahoo.fr

Abstract. In this preliminary study, the chemical composition of Tunisian DPR was established and discussed. The main characteristic of this agri-residue was its high lignin content in comparison with that of alfa plant. CIMV process was used to selectively separate cellulose fibres, hemicelluloses and lignin at atmospheric pressure. The obtained unbleached pulp was analysed in accordance with Kappa index and degree of polymerisation and then bleached by treating successively with peroxyacids and hydrogen peroxide in basic media.

1. Introduction

In northern African countries and others allover the world where forest is not developed at all, the creation of sustainable resources of non-wood raw material for paper industry is extremely needed. Straws, grass, bagasse and various agri-residues can be utilized to obtain non-wood pulps for making paper. In Tunisia paper industry uses an annual plant Stipa tenacissima commonly called alfa and imports pulps in order to compensate the lack of woody fibres. This plant is extensively cultivated in 300 000 ha. Furthermore Tunisia has more than four millions date palms which occupy 23 000 ha and after the date fruit harvesting important quantities of date palm rachis and leaves wastes are accumulated every year. This agricultural residue can be viewed as source of reinforcing fibres for composite materials. This way of valorisation has been the subject of some studies recently reported [1-3].

The use of date palm rachis and leaves was also envisaged in the paper making context but few studies were published on the pulp production from this vegetable matter. Khristova et all. reported that pulping of date palm rachis and leaves from Sudan leads to pulps with variable yields and mechanical properties depending on experimental conditions [4]. The date palm rachis (DPR) gave best results when using the alkaline sulphite-antraquinone process while the leaves were best pulped with the soda method. Ghosh and Nag described the production of good quality paper of desirable mechanical properties from date palm leaves using mechanical pulping process [5]. More recently Khiari et all. prepared and characterized several chemical pulps from Tunisian DPR using the classical soda-antraquinone cooking [6]. These pulps gave paper sheets with good properties without the need of refining operations. The classical chemical process present however some serious shortcomings consisting in (i) the lignin and hemicelluloses degradation and then their utilization in low added value applications such as producing process energy, (ii) the air and water pollution and (iii) the high investment costs.

---

[1] To whom any correspondence should be addressed.

We have showed in our recent communication that CIMV process was very efficient to separate selectively cellulose fibres hemicelluloses and lignin from Tunisian alfa grass [7]. The numerous advantages of this procedure compared with conventional alkali pulping led us to examine the possibility of applying this process to destructuring DPR. To the best of our knowledge no study has been reported on separating cellulose fibres, lignin and sugars from this vegetable matter by an organosolv process. The present study deals with a first investigation to examine the potential use of Tunisian DPR and to evolve improved procedure for extracting cellulose fibres without sacrificing lignin and hemicelluloses.

2. Experimental

2.1. Raw materials

DPR was collected randomly from five 10-15 year-old trees in the south of Tunisia. The rachis length was about 2 m and with 3-5 cm diameter. Air dried DPR was chopped to 2-4 cm in size. The acetic acid (99 to 100%), formic acid (98 to 100%) and hydrogen peroxide (~ 50% by mass) used were commercial products (Merck Eurolab).

2.2. DPR chemical analysis

Small pieces of DPR were defibrated and then analyzed by the method of Van Soest and Wine known as NDF-ADF method (Neutral Detergent Fiber-Acid Detergent Fiber). Cellulose, lignin and hemicellulose contents were determined by treating DPR sample successively with NDF, ADF and potassium permanganate reagents [8].

NDF reagent composition: 150 g sodium lauryl sulfate, 93.05 g of ethylene diamine tetra-sodium, 22.8 g of disodium phosphate, 34.05 g of sodium borate, 10 ml of water and 50 mL of ethyl cellulose in 5 liters of water.

ADF reagent composition: 100g cetyltrimethylammonium bromide (CTAB) in 5 liters of 1N sulfuric acid.

Potassium permanganate reagent composition: 2 volumes of an aqueous solution of $KMnO_4$ (0.3 M) to 1 volume of a buffer solution containing: 6 g $Fe(NO_3)_3$ $9H_2O$, 0.15 g $AgNO_3$ in 100 ml of water, 500ml of glacial acetic acid, 5g of potassium acetate and 400 ml of tertiary butyl alcohol.

Ash was determined in accordance with the following standards (TAPPI 211 om-93). The dry sample of vegetable matter is heated in an oven at 550 °C for 3 hours. (Ash % = weight of ash / weight of the initial dry vegetable matter x 100).

2.3. Cooking

Cooking was carried out in a one-litre glass reactor at atmospheric pressure. DPR (30g), cut into fragments of an average length of 2-4 cm, was first impregnated with the cooking liquor (300mL formic acid / acetic acid / water: 30/50/20: v/v/v) at 60°C for 60 min. After impregnation, the reactional medium is brought up to 107°C for 3 hours. The obtained pulp was filtered, pressed and washed twice acetic acid solution, then with water. It was finally dried and analyzed. The used acids were recycled by simple distillation. The addition of water to the obtained residue after evaporation of the acids, led to the precipitation of the lignin. These were filtered, then washed until neutral pH with distilled water and dried. The solubilized sugars in the water were obtained in the form of syrup after concentration.

2.4. Bleaching

The unbleached pulp obtained after DPR cooking, is bleached by means of a PaEP sequence. The operating conditions were as follow:

Delignification sequence Pa: The unbleached pulp (10 g) was introduced in reactor at atmospheric pressure and then 80mL of formic acid / acetic acid mixture (25/75:v/v) and hydrogen peroxide (1.8g)

were added. The reactional medium is heated at 60°C after 1 hour the delignified pulp was filtered, pressed, washed with distilled water and dried.
Bleaching sequence ***EP:*** The delignified pulp (10 g) was introduced in 100 mL of sodium hydroxide solution (0.4g) containing hydrogen peroxide (0.6g) and then the mixture (pH = 11) was heated for 1 h at 90°C. The pulp thus bleached was first filtered, washed several times with distilled water and then pressed.

2.5. Analysis of pulps
Chemical characteristics are measured in accordance with the following standards: Kappa number (AFNOR NF T 12-018 and T 12-019), DPv (degree of polymerisation of cellulose chain obtained by viscosity measure) (AFNOR NF T 12-005).

3. Results and discussion

3.1. Chemical composition
The chemical composition of DPR used in this work is shown in Table 1 and compared to those of some agricultural residues and annual plants [7, 9-13]. DPR chemical compositions from other works related to DPR are also given in Table 1.

Table 1. Chemical constituents of Tunisian DPR and some vegetable matters

Vegetable matters	cellulose	hemicelluloses	lignin	ash (SiO_2)
Alfa [7]	55	31	9	3
DPR (This work)	41,2	28,5	27,3	3
DPR [6]	45	29,8	27,2	5
DPR [4]	43,1	n. d	23,8	5,6
DPR [13]	44	28	14	2,5
Banana stem [9]	43,6	38,3	11	7,1
Rice straw [10]	37	22	18,3	14,6 (7)
Sorghum bagasse [11-12]	48,8	33,1	16,1	1,4 (0,6)
Sugarcane bagasse [11-13]	41,2	28,4	19,6	2,9 (1,3)

The main remark concerns the high amount of lignin in DPR (27 %) in comparison to those reported for annual plants such as alfa (9 %) or agri-residues which vary from 11 to 20 %. This relatively high content in lignin which is comparable to those collected from literature from other works [4, 6, 14] apart that found by Bendahou et al. [15] could negatively impact the cellulose fibres delignification and thus could constitute a serious drawback to pulping process. On the opposite, the relatively low content in ash (3%) which can be considered as very close to those known for some non-wood sources such as alfa is mainly constituted by Ca, K, Na and Cl with a very small amount of silicon which is profitable for using DPR in pulping. Finally, it is worth noting that DPR contains, as already reported in literature [4, 6, 15], lower amount of cellulose (45 %) than alfa (55 %). Nevertheless, this cellulose content allows envisaging the valorisation of this agri-residue for papermaking applications.

3.2. Pulping and bleaching
The selective separation of cellulose, hemicelluloses and lignin of DPR is achieved at atmospheric pressure using a mixture of formic acid / acetic acid / water according to the protocol previously used for alfa grass. The formic acid essentially played the role of proton donor which has the effect of hydrolyzing both lignin and hemicelluloses while the acetic acid served as solvent for the fragments of these two by-products. The water intervened at the level of the dissociation of the organic acids and participated in hemicelluloses hydrolysis. These cooking agents were easily recyclable without combustion of the cooking liquor.

The chemical characteristics of the DPR unbleached pulp in terms of Kappa Index (KI), viscosity and degree of polymerization (DP_v) as well as yield and rejects are reported in Table 2 which also includes the results taken from our previous published work related to alfa cooking. It clearly appeared that DPR pulping was not sufficient when alfa optimized cooking conditions were used as indicated by the moderate unbleached pulp yield (35 %), the high value of KI (41.5) and the important fraction of rejects (7 %). This difference between the two vegetable matters behaviors when treated in the same experimental conditions can be attributed probably to the DPR lignin content which is very higher than that of alfa. On the opposite, viscosity and DP_v values, which can be considered as close to those from alfa, indicated limited degradation of the cellulose.

Table 2. Chemical CIMV Cooking of DPR and Alfa

Unbleached Pulp	DPR	Alfa	DPR
Yield (% of dry matter)	31	49.06	9
Rejects (% of dry matter)	7	-	3.5
Kappa Index	41.5	28.5	33
Viscosity (cm^3/g)	724	617	690
DP_v	1084	909	997

Cooking conditions
[a] Impregnation: 1h at 60°C; cooking liquor fa/aa/water: (30/50/20: v/v/v); L/M ratio: 10/1; cooking time: 3h.
[b]Impregnation: 1h at 60°C; cooking liquor fa/aa/water: (35/45/20: v/v/v); L/M ratio: 15/1; cooking time: 5h.

Some preliminary experiments showed that increasing the cooking time as well as the cooking liquor composition improved DPR delignification with a progressive decrease in rejects and KI. Controlling key pulping variables is therefore very important to determine the optimized conditions for DPR cooking. The percentage of formic acid and the cooking time seem to be the main factors that should be controlled when cooking DPR in formic acid / acetic acid / water environment since unbleached pulp yield and viscosity depended predominantly on these factors.

In order to improve its delignification, the DPR unbleached pulp was in first treated with a solution of peroxyacids initially synthesized by adding hydrogen peroxide in a mixture of formic acid / acetic acid and then bleached by hydrogen peroxide in basic media. This simple and innovative way of refining DPR pulp proved to be perfectly complementary to the cooking process in a carboxylic acid medium. DPR pulp was finally obtained with relatively low KI (6) and high DP_v (911) but the pulp yield with respect to dried raw material was limited (31).

4. Conclusion

This preliminary study shows that DPR components can be selectively separated at atmospheric pressure by using a mixture of formic acid / acetic acid / water as cooking reagent. This organosolve process permits the production of cellulose fibres in moderate yield without significant degradation but the lignin content remains relatively high. Work is in progress to select with care the best cooking parameters as well as bleaching sequences which produce cellulose fibres exhibiting suitably chemical, mechanical and optical properties that could be acceptable in the manufacture of printing and writing grades of paper.

Acknowledgment
The authors acknowledge the Ministry of Higher Education, Scientific Research and Technology in Tunisia for their financial support.

References

[1] Al-Sulaiman F. A 2002 *App. Comp. Mat.* **9** (6) 377–396.
[2] Abou-Shark B and Hamid H 2004 *Polymer Degradation and Stability* **85** (3) 967–973.
[3] Taha I, Steurnagel L and Ziegmann G 2007 *Composite Interfaces* **14** (7–9) 669–684.
[4] Khristova P, Kordsachia O and Khider T 2005 *Bioresource Technology* **96** 79–85.
[5] Ghosh S. K and Nag D 2009 *Ippta J.* **21** (2) 95.
[6] Khiari R, Mhenni M. F, Belgacem M. N and Mauret E 2010 *Bioresource Technology* 101 775–780.
[7] Ammar H, Mallek H, Abdelkefi F, Benjelloun-Mlayeh B and El Gharbi R 2009 *Journal de la Société Chimique de Tunisie* **11** (1) 69-75.
[8] Van Soest P. J and Wine R. H 1968 *J. Assoc. Off. Anal. Chem.* **51** 780-785.
[9] Astrid M. M, Benjelloun M. B, Delmas M and Bravo R 2005 *Appita Journal* **58** 393.
[10] Lam H. Q, Le Bigot Y, Delmas M and Avignon G 2003 *Appita Journal* **54** 2.
[11] Lam H. Q 2000 Séparation sélective de la cellulose, des hémicelluloses et des lignines par le système catalyseur/solvant: acide formique/acide acétique, de matières végétales à teneur variable en silicium. *Thesis, INPT, France.*
[12] Panagiotopoulos I. A, Bakker R. R, De Vrije T, Koukios E. G and Claassen P. A M 2010 *International Journal of Hydrogen Energy* **35** 7738-7747.
[13] Inyang M, Gao B, Pullammanappallil P, Ding W and Zimmerman A. R 2010 *Bioresource Technology* **101** 8868–8872.
[14] El Morsy M. M. S 1980 *Fibre Science and Technology* **13** (4) 317–321.
[15] Bendahou A, Dufresne A, Kaddamai H and Habibi Y 2007 *Carbohydrate Polymers* **68** 601–608.

Chemical durability and structural approach of the glass series (40-y) Na_2O-yFe_2O_3-5Al_2O_3-55P_2O_5- by IR, X-ray diffraction and Mössbauer Spectroscopy

S. Aqdim[a,b] [1], El. H. Sayouty[a], B. Elouad[c], J. M. Greneche[d]
[a]Laboratoire de Physique de l'Etat Condensé, Université Hassan II Ain Chock, Faculté des Sciences Casablanca B.P. 5366 Maârif, km 8 route d'El Jadida, Casablanca-Morocco.
[b]Laboratoire de Chimie Minérale, Département de Chimie, Université Hassan II Ain Chock, Faculté des Sciences-Casablanca B.P. 5366 Maârif, km 8 route d'El Jadida, Casablanca-Morocco.
[c]Laboratoire d'Elaboration, Analyse Chimique et Ingénierie des Matériaux (LEACIM), Université de la Rochelle, avenue Michel Crépeau, 17042 La Rochelle Cedex 01, France.
[d]Laboratoire de Physique de L'Etat Condensé, UMR CNRS 6087, 72085 Le Mans Cedex 9. France

E-mail: said_aq@yahoo.fr; hassayout@yahoo.fr; belouadi@univ-lr.fr; jean.marc.greneche@univ-lemans.fr

Abstract. The relationship between the composition and structure for the glasses of general composition (40-y)Na_2O-yFe_2O_3-5Al_2O_3-55P_2O_5 ($5 \leq y \leq 20$), has been studied. The chemical durability and density of these glasses increase with increasing Fe_2O_3 content. The dissolution rate (D_R), calculated from the weight loss in distilled water at 90°C for up to 20 days was ≈ 3.10^{-9} g/cm^2/mn which is 30 times lower than that of window glass. The structure and valence states of the iron ions in the glasses were investigated using, X-ray diffraction, 57Fe Mössbauer spectrometry, potentiometric analysis, and infrared spectroscopy. Both Mössbauer spectrometry and potentiometric analysis allow to estimate both Fe^{2+} and Fe^{3+} contents in all these glasses. X-ray diffraction indicates that the local structure of iron phosphate glasses is related to the short range structures of $NaFeP_2O_7$. Infrared spectra indicate the formation of P–O–Fe bonds in the pyrophosphate glasses that replace P–O–Na bonds. The presence of a small content of Al_2O_3 in the glass seems to play a role as a network modifier. The addition of Fe_2O_3 to Al_2O_3 in phosphate glasses favours the enhancement of the formation of pyrophosphate units because iron ions have stronger effect on the depolymerization of metaphosphate chains than the aluminium ions. Finally, the I.R spectra indicate that the presence of P-O-Fe bands of these glasses containing more than 15 mol%Fe_2O_3 is consistent with their good chemical durability.

[1] To whom any correspondence should be addressed.

1. Introduction

Phosphate glasses have recently gained great interest due to their potential application in various technological fields: optics, vitrification of high level nuclear wastes, etc [1-20]. However this class of materials should face a serious limitation resulting from their high sensitivity to the ambient moisture [21-23]. Since our early work [22] on iron phosphate glasses we have recognized the significant wet corrosion resistance of some compositions, as confirmed by many authors [2, 10, 22]. Various studies have shown that the combination of P_2O_5 with oxide like ZnO, Al_2O_3 and more specifically Fe_2O_3 can give rise to glass compositions with high chemical durability [1, 11-17]. Such corrosion resistance was attributed to the change of chemical bonds in the vitreous state, as the bridging oxygen P-O-P bonds tend to be replaced by P-O-M bonds (with M= Zn, Al, Fe, etc). The presence of Fe originates two possible valence states, Fe^{2+} and Fe^{3+} the equilibrium of which in phosphate glasses depends upon the melting atmosphere. In addition, the chemical durability has been found to independent of the Fe^{2+}/Fe^{3+} ratio in some phosphate glasses [14]. It is thus important to follow the evolution of the concentration of Fe valence states which can be easily and accurately estimated by using transmission ^{57}Fe Mössbauer spectrometry.

The aim of the present study is to investigate the structural change and the durability modification versus composition as the ratio of Fe_2O_3 is varied along the series of phosphate glasses (45-y)Na_2O-yFe_2O_3-5Al_2O_3-55P_2O_5 with $5 \leq y \leq 20$.

2. Experimental section

In the absence of any redox phenomenon, the investigated glasses of the (40-y)Na_2O- yFe_2O_3-5Al_2O_3-55P_2O_5 series belong to quaternary system Na_2O-Fe_2O_3-Al_2O_3-P_2O_5. As it is shown in Fig. 1, these materials can be located on one line of pseudo ternary system 9Na_2O.11P_2O_5-9Al_2O_3. 11P_2O_5-9Fe_2O_3.11P_2O_5 which represents the planar section of the Na_2O-Fe_2O_3-Al_2O_3-P_2O_5 tetrahedral diagram, corresponding to a constant molar fraction of P_2O_5 $\chi(P_2O_5)$ =0,55. The chemical compositions of the present glasses are listed in Table 1. The glass samples were prepared from the appropriate mixture of the starting compounds Na_2CO_3, Al_2O_3, Fe_2O_3 and $(NH_4)_2HPO_4$. The weighted mixture of about 10 g were thoroughly in an agate mortar before to be submitted to moderate heat treatments between 300°C and 500°C in order to get an homogeneous system before the glass preparation. The melt was achieved for about 15 min at 1070 ±10°C while the quenching down to room temperature allows to obtain glassy pellets with approximate sizes of 5-10mm diameter and 1-3mm thickness. The vitreous state was first evidenced from the shiny and transparency aspect and then confirmed from XRD patterns. For the chemical durability analysis, pieces of 0.9x0.9x0.3 cm^3 were prepared after polishing using a 400-grit polishing paper and then finished with SiC paper; they were then immersed into a flask filled with 100ml of distilled water at 90°C for 20 days. The dissolution rate (D_R) was then determined from the weight loss during the aqueous treatment at 90°C. The specific mass (density) was determined using the picnometry with both powdered and monolith glasses.

^{57}Fe Mössbauer spectra were recorded at 300 K and at 77 K using a conventional constant acceleration transmission spectrometer with a $^{57}Co(Rh)$ source and a bath cryostat. The Mössbauer sample consists of powdered glass containing 5mg Fe/cm2. An α-Fe foil was used as calibration sample while the values of isomer shift (IS) are quoted to that of α-Fe at 300K. The Mössbauer spectra were fitted by means of the MOSFIT program, assuming the superposition of quadrupolar doublets with lorentzian lines. Indeed, the interest of using Mössbauer spectrometry lies on the distinction of Fe valence states and consequently the estimation of the iron redox ratio $Fe^{2+}/(Fe^{2+}+Fe^{3+})$ from the spectral area. The values of this ratio were then compared to those obtained by potentiometric method in dosing both iron Fe^{2+} and iron Fe^{3+} with dichromate solution after complete digestion of the glass sample employing concentrated H_2SO_4. The experimental error in the value of $Fe^{2+}/(Fe^{2+}+Fe^{3+})$ was found about ± 2. Infrared (IR) spectra were measured for each glass between 400 and 1400 cm^{-1} using mX-1 and NIC-3600 FTIR spectrometers. The analysed samples were prepared by pressing a mixture of about 2mg of glass powder with 100 mg of anhydrous KBr powder.

3. Results and discussion

3.1. Structural approach by Infrared spectroscopy

The infrared recorded spectra for glass series (40-y) Na_2O-5Al_2O_3-yFe_2O_3-55P_2O_5 are illustrated in Fig 1. The analysis of these spectra shows a clear evolution from the metaphosphate to the pyrophosphate modes. As a matter of fact with increasing of iron (III) oxide, the vibrational bonds of the pyrophosphate tend to be the dominant feature of the spectrum. As can be evidenced in Fig 2, the band (νs PO_3) at 1050cm^{-1} appears as the only vibrational mode in the glass. This band is characteristic in the pyrophosphate glasses since it is related to νs PO_3. However in the region of 500-700cm^{-1}, it is still possible to observe the vibrational mode related to P-O-P bonds the number of which tends to decrease when the glass composition is changed from the metaphosphate to the pyrophosphate domains. Such behaviour results from the depolymerisation of the metaphosphate into pyrophosphate compounds. Indeed the depolymerization phenomenon corresponds to the increase break of the P-O-P bonds.

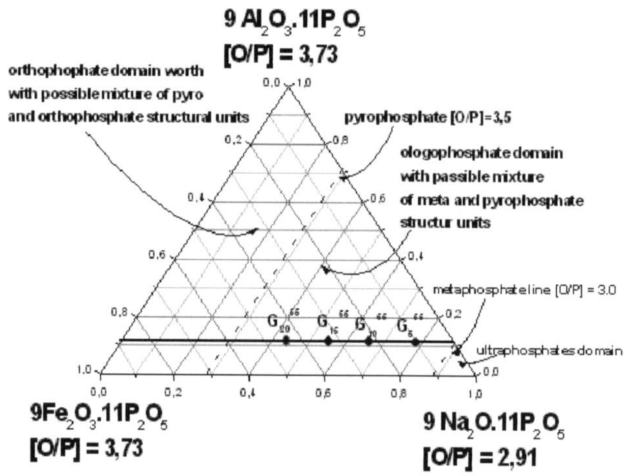

Figure 1. Fig.1: Localization of the investigated glass compositions G_5^{55}, G_{10}^{55}, G_{15}^{55} and G_{20}^{55} within the ternary diagram (9Na_2O.11P_2O_5)-(9Al_2O_3.11P_2O_5)-(9Fe_2O_3.11P_2O_5). Table 1 gives the corresponding compositions within the quaternary system Na_2O-Fe_2O_3-Al_2O_3-P_2O_5

The IR spectrum of G_5^{55} is very similar to that of the metaphosphate composition [10, 14, 25] as it exhibits all the modes characteristic of a metaphosphate family (δ(PO_2) = 550cm^{-1}, νs (P-O-P) = 725cm^{-1}, and νas (PO_2) = 1198 cm^{-1}) in addition to the diphosphate modes νas (PO_3) = 1070 cm^{-1}; νs(PO_2) and νs (P-O-P) =755 cm^{-1} [26, 27]. Furthermore, the structure medication deduced from the vibrational spectroscopy is totally compatible with localization of the analyses compounds (G_5^{55}, G_{10}^{55}, G‶$_{15}^{55}$, G_{20}^{55}) inside the ternary diagram given in Fig. 1. As a matter of fact, G_5^{55} is located in the olygophosphate domain close to the metaphosphate line. Therefore it is normal that its structure contain mainly the metaphaphate units. Indeed when the molar fraction of Fe_2O_3 is increased, the composition (Fig 1. and Table 1) of the glass becomes closed to the pyrophosphate line. For G_{20}^{55} for example, the dominant structure units are expected to be those of pyrophosphate. This is the reason why its infrared spectrum contain mainly a large vibration band assigned to νs (PO_3) = 1060cm^{-1}. Since the number of P-O-P has decreased between G_5^{55} and G_{20}^{55}, it is normal to observe a decrease of the corresponding intensity νs (P-O-P) ≅ 700cm^{-1}.

Table 1. Glass starting compositions in terms of basic oxides (see Fig 1 for the position inside the ternary system with constant P_2O_5 ratio)

Sample	Starting glass Composition (mol %)				ratio$[Fe^{2+}]/[Fe^{2+}+Fe^{3+}]$,%	[O/P] ratio *	(D_R) ($g/cm^2/mn$)	ρ (g/cm^3)
	Na_2O	Fe_2O_3	Al_2O_3	P_2O_5	± 1		20 days	± 0.02
G_5^{55}	35	5	5	55	6,5	3.09	(5± 0,2) x10^{-7}	2.33
G_{10}^{55}	30	10	5	55	6,6	3.18	(9± 0,2)x10^{-8}	2.63
G_{15}^{55}	25	15	5	55	8,2	3.27	(2.5± 0,2)x10^{-8}	2.72
G_{20}^{55}	20	20	5	55	7,5	3.36	(3.± 0,2)x10^{-9}	2.81

3.2. ^{57}Fe Mössbauer Spectrometry and iron Redox Phenomenon

The 300K Mössbauer spectra of some Na-Al-Fe phosphate glasses are illustrated in Fig. 3. As those obtained at 77K, they exhibit a central prevailing quadrupolar doublet with broadened lines and a small line at higher velocities. The first part has to be well described by means of at least 4 quadrupolar doublets which can be unambiguously attributed to high spin state Fe^{3+} ions in octahedral sites. Nevertheless, it is important to notice that a quadrupolar splitting distribution can be also successfully used, that is consistent with a disordered local structure as expected in a glassy system. The second part of the Mössbauer spectra (Fig. 2) has to be associated to a minor quadrupolar component, well described by two doublets, unambiguously ascribed to the presence of Fe^{2+} ions. The refined values of mean Mössbauer hyperfine parameters, as isomer shift (δ), quadrupole splitting (ΔE_Q), and fraction of Fe^{2+} are listed in Table 2.

Table 2. Mössbauer characteristics of the glass series $(40-y)Na_2O-5Al_2O_3-yFe_2O_3-55P_2O_5$

Sample	$[Fe^{2+}]/[Fe^{3+}+Fe^{2+}]$; %	Molar	Stage 1 Chi Sqr. (*N*=15, *df*=1)	*p*	Stage 2 Chi Sqr. (*N*=15, *df*=1)
$G_5^{55}{}_5$	6	0.43	1.07	0.64	2.51
G_{10}^{55}	7	0.42	1.00	0.73	2.55
G_{15}^{55}	8	0.41	0.97	0.63	2.61
G_{20}^{55}	7	0.40	0.93	0.88	2.66

The values of quadrupolar splitting suggest Fe^{2+} and Fe^{3+} ions with distorted octahedral coordination [24, 25]. This description allows thus to determine the iron redox ratio $Fe^{2+}/(Fe^{2+}+Fe^{3+})$ from the corresponding spectral area of each component, assuming the same value of f-Lamb Mössbauer factor. It can be concluded that the increasing Fe_2O_3 content favours a reduction of Fe^{3+} into Fe^{2+} ions during melting in air. It seems thus that the significant factor affecting the redox state of the iron in sodium-aluminium-iron phosphate glasses could be the presence of ammonium phosphate $(NH_4)_2HPO_4$ in a batch composition as a source of P_2O_5. As a matter of fact, reducing conditions were created during the melting process which favours the reduction of Fe^{3+} into Fe^{2+} [24]. However this reduction does not exceed 10%. Some authors have suggested that the $Fe^{2+}/(Fe^{3+}+Fe^{2+})$ ratio is more related to the melting temperature than to the glass composition of [8, 16]. The molar fractions of $Fe^{2+}$and Fe^{3+}in the title of series of glasses were also determined from the potentiometric technique. As it can be seen in Fig 4, the obtained values are slightly higher than those determined from the Mössbauer study. However all the values lie in the same range of about $[Fe^{2+}]/[Fe^{2+}]+[Fe^{3+}]$ = 6-10%.

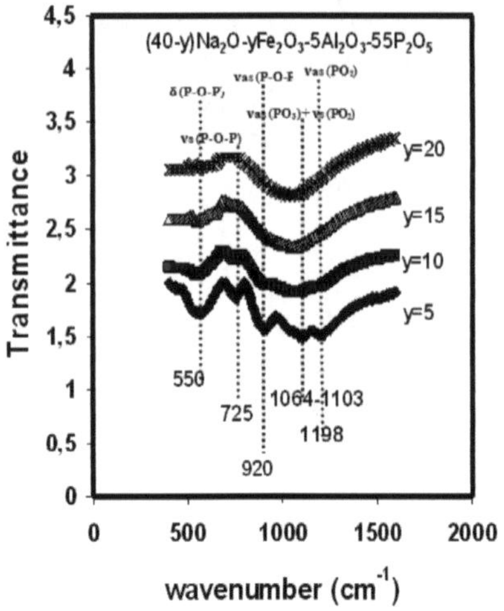

Figure 2. IR spectra of the series of (40-y)Na_2O -yFe_2O_3-5Al_2O_3-55P_2O_5 glasses, y= 5, 10, 15 and 20

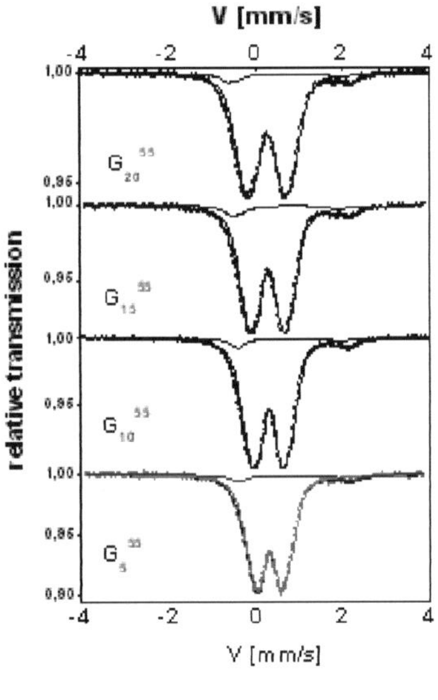

Figure 3. Room temperature Mossbauer spectra of (40-y)Na_2O -yFe_2O_3-5Al_2O_3-55P_2O_5 glasses

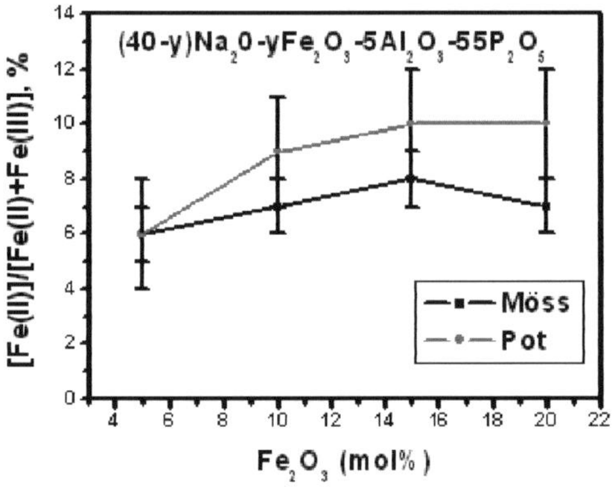

Figure 4. Variation of relative % of Fe^{2+} versus initial molar fraction in the glass series (40-y)Na_2O-5Al_2O_3-yFe_2O_3-55P_2O_5

3.3. Density and Molar Volumes

The specific mass (Density) of vitrified phosphates is increasing with increasing molar fraction along the series (40-y)Na_2O-yFe_2O_3-5Al_2O_3-55P_2O_5. As it is shown in Fig. 5, from density measurements, it was possible to deduce the value of the molar volume and oxygen radius, calculated from the approximate hypothesis of close packing of oxygen anions O^{2-}, having r(O^{2-})cal recapitulated for each composition in Table 3.

Table 3. Variation of the density and related molar volume in (40-y)Na_2O-yFe_2O_3-5Al_2O_3-55P_2O_5. system (* M = molar mass; ρ = density; N_A = Avogadro number; N_0 = number of oxygen atoms in the molar formula)

Sample	Molar formula (number of O/mol)	(Molar mass)	Density ρ(g/cm³)	Molar volume In (nm)³ $V=M/\rho N_A \times N_o{}^*$	Calculated oxygen radius (nm) $r_{cal}(O^{2-})$
G_5^{55}	35Na_2O.5Fe_2O_3.5Al_2O_3.55P_2O_5 (340)	11288	2,33	0,0238	0,144
G_{10}^{55}	30Na_2O.10Fe_2O_3.5Al_2O_3.55P_2O_5 (350)	11776	2,63	0,0213	0,139
G_{15}^{55}	25Na_2O.15Fe_2O_3.5Al_2O_3.55P_2O_5 (360)	12264	2,72	0,0209	0,138
G_{20}^{55}	20Na_2O.20Fe_2O_3.5Al_2O_3.55P_2O_5 (370)	12752	2,81	0,0204	0,137

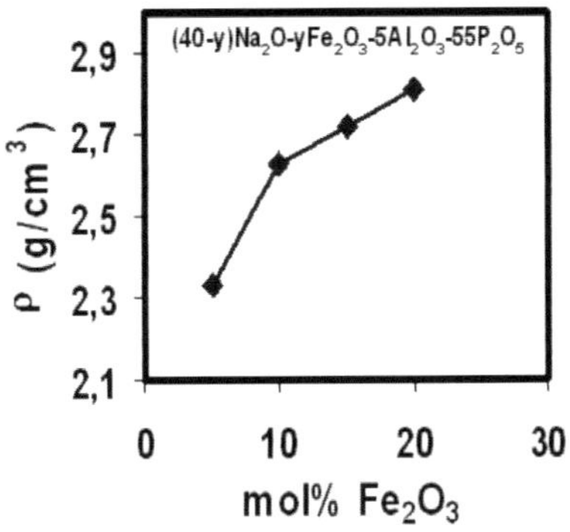

Figure 5. Variation of the specific mass (density) versus mol% in (40-y)Na_2O -yFe_2O_3-5Al_2O_3-55P_2O_5 system

3.4. Analysis of the Chemical Durability of the Glass Series (40-y)Na_2O-yFe_2O_3-5Al_2O_3 55P_2O_5

The chemical durability of the series of glasses (40-y)Na_2O-yFe_2O_3-5Al_2O_3-55P_2O_5 was approached using the measurement of the dissolution (D_R) rate which was defined as the weight loss of the glass expressed in terms of $g.cm^{-2}.mn^{-1}$. The values of D_R reported in Table 1, show a dissolution decrease versus the Fe_2O_3 content of our samples after their immersion in 100ml of distilled water, heated at 90°C for 20 consecutive days (Fig 6). There was no detectable difference in the visual appearance of the glass samples G_{10}^{55}, G_{15}^{55}, and G_{20}^{55} immersed in water at 90°C during the attack's period. However the glass sample G_5^{55} was weakly corroded. The glass containing 20 mol % Fe_2O_3 had a (D_R) 30 times less than the D_R for window glass and ~ 100 times less than the D_R for BABAL glass of composition 27BaO_2-45B_2O_3-18Al_2O_3-10Fe_2O_3 which has been studied as alternative materials for nuclear waste immobilization [9, 18].

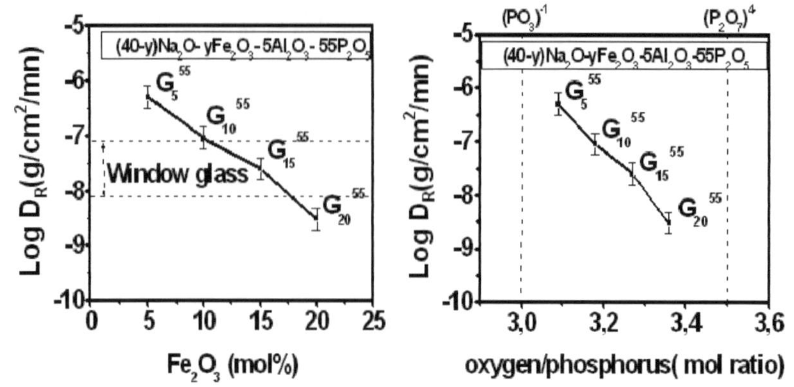

a b

Figure 6. Dissolution rates (D_R) of the glass series (40-y)Na_2O-yFe_2O_3-5Al_2O_3-55P_2O_5 versus: a)Fe_2O_3(mol%); b) ratio [O/P].

4. Correlation between the structure and the high durability of Iron Phosphate Glasses

The IR techniques have confirmed the structural evolution of the glass network along series (40-y) Na_2O-yFe_2O_3-5Al_2O_3-55P_2O_5, from a mixture of pyrophosphate and metaphosphate 3 < [O/P] < 3,5 towards the pyrophosphate line with [O/P] = 3,5 (Fig. 1). The determination of the molar volume (Table 3) tends to confirm the increase of the covalent character of the chemical bond with increasing Fe_2O_3 content. The increase of this covalent character may explain the chemical strength of glass network which becomes more resistant to the chemical attack in an aggressive aqueous medium at 90°C. The presence of a small content of Al_2O_3 in the glass seems to play a role as a network modifier when Fe_2O_3 content increases (y≥15) [10]. The chemical resistance of Aluminium-iron phosphate glasses (regarding aqueous attackat 90°C) is attributed to the increasing number of Fe-O-P bonds in the glass [19-24]. Such bonds are expected to be more water resistant than the P-O-P ones, which constitute the majority of bonds in the low iron concentrated glasses.

5. Conclusion

The structural and proprieties physic-chemical of sodium-aluminium-iron-phosphate glasses have been studies by using several techniques. The structural approach of these glasses by I.R spectroscopy show a radical change of structure when the Fe_2O_3 content increases. This change lead to the formation of short units groups of pyrophosphates. Results of spectroscopy Mössbauer indicate the proportions of Fe^{2+} and Fe^{3+} in the glass. The parameters hyperfins show that the last occupied the merely octahedral sites more or less distorted. The structure glass can be considered as pyrophosphate units connected with ferric and ferrous iron in octahedral or distorted octahedral coordination. The improved chemical durability is attributed to the replacement of the easily hydrated Na-O-P and P-O-P bonds by corrosion resistant Fe-O-P bands. The O/P ratio is also an important factor to aqueous chemical durability. These glasses have a dissolution rate (D_R) 30 times less than the D_R for window glass and ~ 100 times less than the D_R for BABAL glass which has been considered as alternative materials for the immobilization of nuclear waste substance.

References

[1] Aqdim S, Sayouty E. H and Elouadi B 2008 *Eurasian chem. Tech. Journal* **10** 9.
[2] Brow R. K, Arens C. M, Yu X and Day D. E 1994 *Phys. Chem. Glasses* **35** 132.
[3] Brow R. K 2000 *J. Non-Cryst. Solids* **1** 263-264 1.
[4] Mogus-Milankovic A, Pivac B, Furic K and Day D. E 1997 *Phys. Chem. Glasses* **38** 74.

IOP Conf. Series: Materials Science and Engineering **27** (2012) 012003 doi:10.1088/1757-899X/27/1/012003

[5] Mogus-Milankovic A, Rajic M, Drasner A, Trojko R and. Day D. E 1998 *Phys. Chem. Glasses* **39** 70.
[6] Day D. E, Wu Z, Ray C. S, Hrma P 1998 *J. Non-Cryst. Solids* **241** 1.
[7] Sales B. C, Boatner L. A 1984 Science **226** 45.
[8] Fang X, Ray C. S, Mogus-Milankovic A and Day D. E 2001 *J. Non-Cryst. Solids* **283** 162.
[9] Fang X, Ray C. S, Marasinghe G. K and Day D. E 2000 *J. Non-Cryst. Solids* **293** 263-264.
[10] Mogus-Milankovic A, Gajovic A, Santic A and Day D. E 2001 *J. Non-Cryst. Solids* **289** 204.
[11] Karabulut M, Marasinghe G. K, Ray C. S, Waddill G. D, Day D. E, Bady Y. S, Saboungi L, Shastri S, Haeffner D 2000 *J. Appl. Phys.* **87** 2185.
[12] Karabulut M, Marasinghe G. K, Ray C. S, Day D. E, Ozturk O, Waddill G. D 1999 J Non-Cryst. Solids **249** 105.
[13] Booth C. H, Gallen P, Bucher J. J, Edelstein N. M, Shuh D. K, Marasinghe G. K, Karabulut M, Ray C. S, Day D. E 1999 *J. Mater. Res.* **14** 2628.
[14] Ray C. S, Fang X, Karabulut M, Marasinghe G. K, Day D. E 1999 *J. Non-Cryst. Solids* **249** 1.
[15] Sales B. C, Abraham M. M, Bates J. B and Boatner L. A 1985 *J. Non-Cryst. Solids* **71** 103.
[16] Brooks J. S, Williams G. L, and Allen D. W 1992 *Phys. Chem. Glasses* **33** 171.
[17] Kurkjian C. R, and Sigety E. A 1968 *Phys. Chem. Glasses* **9** 73.
[18] Yu X, Day D. E, Long G. J and Brow R. K 1997 *J. Non-Cryst. Solids* **215** 21.
[19] Kumar B and Lin S 1991 *J. Am. Ceram. Soc.* **74** 226.
[20] Reis S. T, Karabulut M, Day D. E 2001 *J. Non-Cryst. Solids* **292** 150.
[21] Ouchetto M 1993 Thèse de Doctorat d'Etat es Sciences, Faculty of Sciences Rabat, Morocco.
[22] Aqdim S 1990 Diplôme des Etudes Supérieures de 3ème Cycle de Spécialité, Faculty of Sciences Rabat, Morocco.
[23] Arbib E. H 1998 Thèse de Doctorat d'Etat es Sciences, Faculty of Sciences Rabat, Morocco.
[24] Tallant D. R, Nelson C 1986 *Phys. Chem. Glasses* **27** 75.
[25] Nelson N. B and Exarhos G. J 1979 *J. Chem. Phys.* **71** 2739.
[26] Brow R. K, Kirkpatrick R. J and Turner G. L 1993 *J. Am. Ceram. Soc.* **76** 919.
[27] Fang X, Ray C. S, Mogus-Milankovic A, Day D. E 2001 *Non-Cryst. Solids* **283** 162.

IOP Conf. Series: Materials Science and Engineering **28** (2012) 012004 doi:10.1088/1757-899X/28/1/012004

Electronic band structure calculations for $Ga_xIn_{1-x}As_yP_{1-y}$ alloys lattice matched to InP

A. Bechiri[a1], F. Benmakhlouf[a], H. Allouache[b], S. Bacha[b] and N. Bouarissa[c]
[a] Physics Department, Faculty of Science, University of Tebessa, 12002 Tebessa, Algeria,
[b] Department of Physics, Faculty of Science, 23000 Annaba University, Algeria
[c] Department of Physics, Faculty of Science, King Khalid University, Abha, P.O.Box 9004, Saudi Arabia

E-mail: bechiri@yahoo.fr

Abstract. A pseudopotential formalism coupled with the virtual crystal approximation are applied to study the effect of compositional disorder upon electronic band structure of cubic $Ga_xIn_{1-x}As_yP_{1-y}$ quarternary alloys lattice matched to InP. The effects of compositional variations are properly included in the calculations. Very good agreement is obtained between the calculated values and the available experimental data for the lattice–matched alloy to InP. The absorption at the fundamental optical gaps is found to be direct within a whole range of the y composition whatever the lattice-matching to the substrate of interest. The alloy system $Ga_xIn_{1-x}As_yP_{1-y}$ lattice matched to InP is suggested to be suitable for an efficient light emitting device (ELED) material.

1. Introduction

The III-V quaternary semiconductor alloys are promising candidates for many device applications. The capabilities of III-V quantum well devices can frequently be expanded by introducing quaternary layers to the design [1]. $Ga_xIn_{1-x}As_yP_{1-y}$ is one of the major materials for the fabrication of optoelectronic devices, such as edge or surface emitting long wave-length semiconductor lasers [2-4]. Information on the energy band gap and the lattice matching condition to available substrates is a prerequisite for many practical applications. Generally speaking, a given quaternary comprises a vast two-dimensional space of compositions [x, y]. In practice, however, most experiments have focused on the one-dimensional subsets, [x, y(x)] that are (approximately) lattice matched to one of the common binary substrate materials [1,5,6] (which are InP for the quaternary alloy being studied here). The band gap of a quaternary alloy can change widely while maintaining the lattice match completely to a binary crystal used as a substrate.

There has been little motivation to pursue the band structure of GaInAsP lattice matched to GaAs, since its direct band gap spans roughly the same range as direct-gap AlGaAs, that is considerably

[1] To whom any correspondence should be addressed.

easier to grow [1]. However, lattice matched to InP, GaInAsP has been the subject of numerous investigations [6-12] and it is currently employed in commercial optoelectronic and electronic devices. In the present paper, the energy band-gaps for the $Ga_xIn_{1-x}As_yP_{1-y}$ alloys lattice matched to substrate, namely InP are calculated over the entire composition range of x and y by using the empirical pseudopotential method (EPM) under the virtual crystal approximation (VCA) which found to be reliable for quaternary semiconductor alloys when the compositional variations are involved [13].

2. Method of calculation

The electronic structure of the quaternary alloy system of interest is calculated using the empirical pseudopotential method (EPM) under the virtual crystal approximation (VCA) in which the effects of composition disorder are involved. Together with the VCA that takes into account the alloy disorder, the EPM offers the most effective means of obtaining the overall band structure of quaternary alloys [13]. In the EPM, the actual atomic potential is replaced by a pseudopotential and a set of atomic form factors $V_a(G)$ are treated as adjustable parameters that can be fit to available experimental data. G are reciprocal lattice vectors. The pseudo-wave functions $\psi_{n,k}(r)$ and the band energy values $E_n(k)$ are a solution of

$$\left[\frac{p^2}{2m}+V(r)\right]\psi_{n,k}(r) = E_n(k)\psi_{n,k}(r) \tag{1}$$

where V(r) is given in the reciprocal lattice as,

$$V(r) = \sum_G V_a(G)S(G)e^{iGr} \tag{2}$$

and S(G) is the structure factor. The $\psi_{n,k}(r)$ have the Bloch form, and they can be expanded in a set of plane waves.

First starting $V(G)$'s for the elements involved are chosen; a model potential based on atomic term values is a good starting point. Next the structure is included via S(G), and the Schrödinger equation is solved for the energy eigenvalues and wavefunctions. The empirical pseudopotential parameters are optimized using the non-linear least-squares method [14,15]. Our non-linear least-squares method requires that the root-mean-square deviation of the calculated level spacing (LS`s) from the experimental ones defined by,

$$\delta = \left[\frac{\sum_{(i,j)}^{m}\left\{\Delta E^{(i,j)}\right\}^2}{(m-N)}\right]^{\frac{1}{2}} \tag{3}$$

should be minimum. In Eq.(3), $\Delta E^{(i,j)} = E_{exp}^{(i,j)} - E_{cal}^{(i,j)}$, where $E_{exp}^{(i,j)}$ and $E_{cal}^{(i,j)}$ are the observed and calculated LS`s between the ith state at the wave vector $k = k_I$ and the jth at $k = k_J$, respectively, in the m chosen pairs (i,j). N is the number of the empirical pseudopotential parameters. The experimental band-gap energies for GaAs, InAs, GaP, and InP pure binary compounds at Γ, X and L high-symmetry points used in the fitting procedure are given in Table 1. The calculated energies given by solving the empirical pseudopotential secular equation depend nonlinearly on the empirical pseudopotential parameters. The starting values of the parameters are improved step by step by iterations until δ is minimized.

Table 1. Band-gap energies of GaAs, InAs, GaP and InP.

Compound	$E_{\Gamma\Gamma}$ (eV)		$E_{\Gamma X}$ (eV)		$E_{\Gamma L}$ (eV)	
	Exp.[10]	Cal.	Exp.[16]	Calc.	Exp.[16]	Calc.
GaAs	1.42	1.42	1.91	1.90	1.73	1.71
InAs	0.36	0.35	1.37	1.37	1.07	1.08
GaP	2.74	2.78	2.26	2.26	2.63	2.60
InP	1.35	1.34	2.21	2.19	2.05	1.93

The dimension of our eigenvalue problem is a (136x136) matrix. Usually, 59 plane waves lead to a good convergence within a few percent between theory and experiment for the energy band gaps as shown in Table 1. Generally, for zinc-blende structures, just three form factors are sufficient to determine each atomic potential. The adjusted local symmetric and antisymmetric pseudopotential form factors for all of the binary compounds of interest are listed in Table 2. This table contains also the used lattice constants for these binary compounds.

Table 2. Pseudopotential parameters for GaAs, InAs, GaP and InP

Compound		GaAs	InAs	GaP	InP
Form factors (Ryd)	$V_S(3)$	-0.23539	-0.21401	-0.22528	-0.20031
	$V_S(8)$	0.0126	0.00	0.00	0.00
	$V_S(11)$	0.057018	0.046505	0.109755	0.068396
	$V_A(3)$	0.073883	0.045385	0.084005	0.106112
	$V_A(4)$	0.05	0.045	0.06	0.06
	$V_A(11)$	0.01	0.01	0.03	0.03
Lattice constant (Å)		5.6533	6.0583	5.4505	5.8687

For calculating the potential of the ternary and quaternary alloys under load, we have adopted a similar methodology to that we used previously [13,17] and in which compositional disorder was incorporated. Thus, for a mixed ternary crystal AB_xC_{1-x}, the potential is expressed as,

$$V_{alloy}(r) = V_{VCA}(r) + V_{dis}(r) \tag{4}$$

where $V_{VCA}(r)$ is the periodic virtual crystal potential and $V_{dis}(r)$ is the non-periodic potential due to the compositional disorder,

$$V_{VCA}(r) = xV_{AB}(r) + (1-x)V_{AC}(r) \tag{5}$$

and

$$V_{dis}(r) = -p[x(1-x)]^{\frac{1}{2}} \sum_j \Delta(r - R_j) \tag{6}$$

where j indicates that the summation over j is over the AB and AC molecular sites,

$$p = \frac{\left[\sum_{i=1}^{n} N_i \Delta(d_i)^2\right]^{\frac{1}{2}}}{\sum_{i=1}^{n} N_i \Delta(d_i)} \tag{7}$$

where d is the nearest-neighbor distance and N the number of nearest-neighbor sites, which is 12 for the FCC structure, *i* indicates the *i*th nearest-neighbor sites and n can be extended to include the whole

crystal. Although N_i becomes large and is approximately proportional to r^2, the potential $\Delta(r)$ decreases more rapidly owing to screening, thus ensuring convergence of p.
For a quaternary system of the form $A_xB_{1-x}C_yD_{1-y}$, the form factors are obtained within the VCA as,

$$V_{alloy}(G) = V_{VCA}(G) = xyV_{AC}(G) + (1-x)yV_{BC}(G) + x(1-y)V_{AD}(G) + (1-x)(1-y)V_{BD}(G) \quad (8)$$

However, when the effects of compositional variations are included, $V_{alloy}(G)$ becomes,

$$V_{alloy}(G) = V_{VCA}(G) - x(1-x)\left[yC_{ABC} + (1-y)C_{ABD}\right] - y(1-y)\left[xC_{ACD} + (1-x)C_{BCD}\right] \quad (9)$$

where $C_{ABC}, C_{ABD}, C_{ACD}$ and C_{BCD} are the bowing parameters of pseudopotential form factors of ternary alloys $A_xB_{1-x}C, A_xB_{1-x}D, AC_yD_{1-y}$ and BC_yD_{1-y}, respectively.

The lattice constant of both ternary and quaternary alloys is assumed to follow Vegard's rule [18]. The lattice matching conditions for $Ga_xIn_{1-x}As_yP_{1-y}$ quaternary system on InP substrate are as follows :

$$x = \frac{0.1896y}{0.4182 - 0.0132y} \qquad (0 \le y \le 1) \quad (10)$$

3. Results

The band structure of $Ga_xIn_{1-x}As_yP_{1-y}$ lattice matched to InP was calculated throughout the high-symmetry points Γ, X , and L in the Brillouin zone. The As composition energy variations of the conduction band edges at Γ, X, and L with respect to the top of the valence band are obtained. The results are quantitatively presented in Table 3 and displayed in Fig. 1. In view of Table 3, one can note that the numerically calculated results agree with the experimental data extremely well. The curves in Fig. 1 are the quadratic lest-squares fits to our results according to the analytical expressions,

$$E_{\Gamma\Gamma}(eV) = 1.34 - 0.76y + 0.12y^2 \quad (11)$$

$$E_{\Gamma X}(eV) = 2.19 - 1.37y + 0.67y^2 \quad (12)$$

$$E_{\Gamma L}(eV) = 1.93 - 0.9y + 0.22y^2 \quad (13)$$

Of particular importance are the non-linear coefficients values which represent the optical bowing parameters for the band gaps at Γ, X, and L high-symmetry points. These parameters are induced by composition disorder. Basic properties of $Ga_xIn_{1-x}As_yP_{1-y}$ compositions lattice-matched to InP reported an optical bowing parameter for the band gap at Γ of 0.138 eV [12]. The measurement of Nahory et *al.* [21] reported a bowing coefficient at Γ of 0.12 eV. Furthermore Vurgaftman et *al.* [1] have recommended in their recent review a bowing parameter of 0.13 eV. All these values are in very good agreement with the value found by the present calculations (see Eq. 11).

Table 3: Direct and indirect band-gap energies in $Ga_xIn_{1-x}As_yP_{1-y}$ lattice matched to InP substrate for some compositions y, compared to available experimental data

Material	$E_{\Gamma\Gamma}$ (eV)		$E_{\Gamma X}$ (eV)		$E_{\Gamma L}$ (eV)	
	Cal.	Exp.	Cal	Exp.	Cal	Exp.
$Ga_{0.11}In_{0.89}As_{0.24}P_{0.76}$ / InP	1.17	1.18[19]	1.89	_	1.74	1.83[19]
$Ga_{0.28}In_{0.72}As_{0.60}P_{0.40}$ / InP	0.93	_	1.61	_	1.47	_
$Ga_{0.47}In_{0.53}As$ / InP	0.71	0.721[20]	1.48	_	1.26	1.2[19]

The extent of the direct-to indirect–band–gap transition is examined by our calculations. Our results (see Fig. 1) imply a direct band gap for the $Ga_xIn_{1-x}As_yP_{1-y}$ alloy at all compositions y whatever the lattice-matching to the InP substrate of interest. However, the alloy of interest has a direct gap $E_{\Gamma\Gamma}$ being a considerable distance from the $E_{\Gamma X}$ and $E_{\Gamma L}$ band gaps, so that it is suggested that $Ga_xIn_{1-x}As_yP_{1-y}$ lattice matched to InP froms a potential efficient light emitting device (ELED) material. Using the universal tight binding model based on a modified pseudocell, Shim and Rabitz [6] have reported that for lattice matching to InP, the absorption at the fundamental optical gap in $In_{1-x}Ga_xAs_{1-y}P_y$ is direct ($\Gamma \rightarrow \Gamma$) within a whole range of the y composition ($0 \le y \le 1$) which agrees with our predictions.

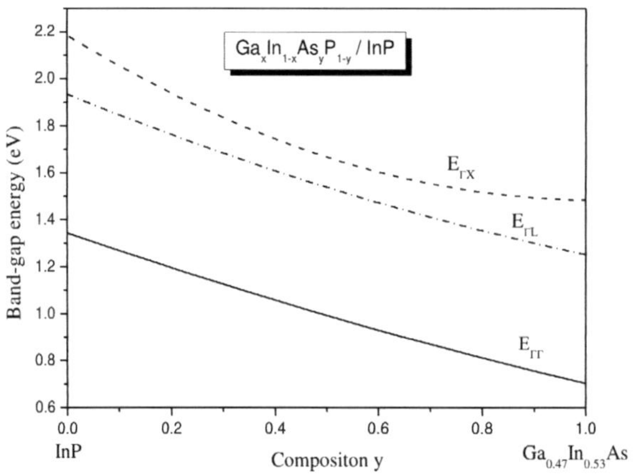

Figure 1. Direct and indirect band-gap energies in $Ga_xIn_{1-x}As_yP_{1-y}$/InP as a function of composition y.

The dependence of the band-gap $E_{\Gamma\Gamma}$ on the composition x and y using a 3D surface and a contour plot are presented in Fig. 2(a) and Fig. 2(b), respectively. The composition-dependent band-gap variation of $Ga_xIn_{1-x}As_yP_{1-y}$ is somewhat parabolic (i.e., the bowing effect). While the 3D plot shows clearly how the energy gap changes versus the composition parameters x and y, the contour plot provides a convenient way to read numerical values from the graph. The energy gap decreases as y increases while it is an increasing function of x. It approaches the energy band-gaps of the binary constituents when x and y approaches 0 and/or 1.

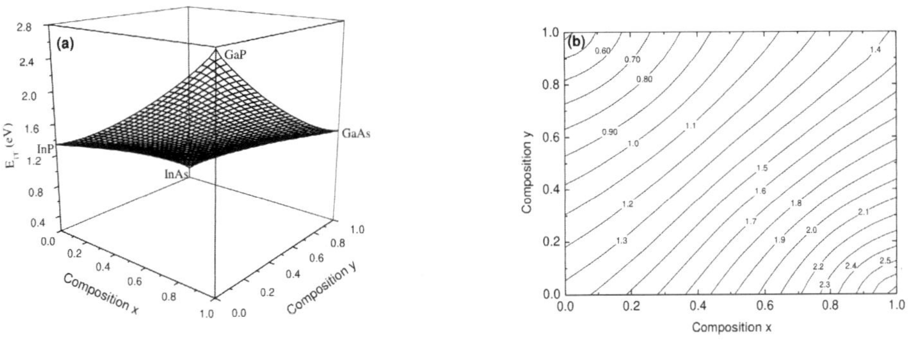

Figure 2. Direct band-gap energy ($E_{\Gamma\Gamma}$) in $Ga_xIn_{1-x}As_yP_{1-y}$ as a function of composition (x and y) using (a) 3D surface and (b) contour map.

In Fig. 3, we show the variation of the $E_{\Gamma\Gamma}$ band-gap energy as a function of the composition y over the range 0-1 for the quaternary alloy system $Ga_xIn_{1-x}As_yP_{1-y}$ lattice matched to InP. The $E_{\Gamma\Gamma}$ optical band-gap energy ranges from 0.71 to 1.34 eV due to the direct gap induced by the Γ-conduction minimum band over the entire range of y. The $E_{\Gamma\Gamma}$ band-gap calculated by Shim and Rabitz [6] cover 0.72 eV < $E_{\Gamma\Gamma}$ < 1.42 eV, while the experimental band gap has been reported from 0.721 to 1.35 eV [10,20]. Note that, while the experimental results differ somewhat from those of ref.[6], they agree very well with the present calculations. However, we do believe that the satisfaction of our results for the alloy under investigation lattice-matched to InP support the quality of the EPM calculations. It is worth mentioning that the $E_{\Gamma\Gamma}$ band gap ranges show significant differences according to the particular lattice-matched substrates. This implies that the alloy of interest could provide more opportunities to obtain a desired band gap depending on the lattice-matched substrates and the composition y. This in turn may opens up the possibility of developing new electronic devices with special features and properties.

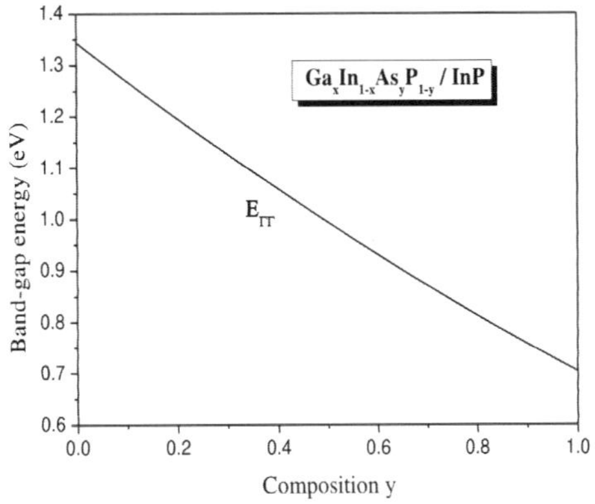

Figure 3. Direct band-gap energy ($E_{\Gamma\Gamma}$) in $Ga_xIn_{1-x}As_yP_{1-y}$ lattice matched to InP as a function of composition y

4. Conclusion

This work reports on an EPM within an improved VCA (in which the effects of composition disorder are taken into account) determination of the electronic band structure of the $Ga_xIn_{1-x}As_yP_{1-y}$ quaternary alloys lattice matched to InP substrate for $0 \leq y \leq 1$. A summary of the key findings follows: (i). Very good agreement was obtained between the band gap predictions and available experimental data for the alloy $Ga_xIn_{1-x}As_yP_{1-y}$ lattice-matched to InP (ii). Our calculation predicted that the alloy $Ga_xIn_{1-x}As_yP_{1-y}$ lattice-matched to InP exhibits a $(\Gamma \rightarrow \Gamma)$ direct band gap in the full range of y composition. (iii). $Ga_xIn_{1-x}As_yP_{1-y}$ could provide more diverse opportunities to obtain a desired band gaps which opens up the possibility of discovering new electronic devices with special features and properties. (iv). Considerable differences according to the particular lattice matched substrate were found for $E_{\Gamma\Gamma}$ band gap ranges in the alloy of interest which could provide additional opportunities to obtain a desired band gap.

References

[1] Vurgaftman I, Meyer J. R, Ram-Mohan L. R 2001 *J. Appl. Phys.* **89** 5815.
[2] Iga K, Kinoshita S 1996 *Process Technology for Semiconductor Lasers* Springer, Berlin.
[3] Quillec M 1996 *Materials for Optoelectronics*, Kluwer Academic Pub., Boston.
[4] Shim K, Rabitz H 2000 *J. Korean Phys. Soc.* **37** 124.
[5] Bouarissa N 1999 *Infrared. Phys. Technol* **40** 423.
[6] Shim K, Rabitz H 1998 *Phys. Rev. B* **57** 12874.
[7] Adachi S 2009, Properties of Semiconductor Alloys: Group-IV, III–V and II–VI Semiconductors, John Wiley & Sons, New York.
[8] Swaminathan V and Macrander A. T 1991 Materials Aspects of GaAs and InP Based Structures, Prentice Hall, Englewood Cliffs, NJ.
[9] Swaminathan V 1992 in Indium Phosphide and related Materials: Processing, Technology, and Devices, edited by A.Katz, Artech House, Boston.
[10] Adachi S 1982 *J. Appl. Phys.* **53** 8775.
[11] Pearsall T. P 1982 *GaInAsP alloy Semiconductors*, John Wiley and Sons, New York.
[12] Levinshtein M, Rumyantsev S, Shur M 1999, Handbook Series on Semiconductor Parameters, Vol.2, World Scientific, Singapore, and references cited therein.
[13] Bouarissa N 1999 *Superlattice Microstrutures* **26** 279.
[14] Kobayasi T, Nara H 1993 Bull.Coll.Med.Sci, Tohoku Univ.**2**.
[15] Bouarissa N, Aourag H 1995 *Mater.Sci.Eng.B* **34**, 58.
[16] Adachi S 1989 *Phys.Rev.B* **39** 12612.
[17] Bouarissa N 1998 *Phys. Lett. A* **245**, 285.
[18] Reported in: Al-Douri Y 2003 *Mat. Chem. Phys.*, ***78*** 625.
[19] Adachi S 1989 *J. Appl. Phys.* **66** 6030.
[20] Perea E. H, Mendez E. E, Fonstad C. G 1980 *Appl. Phys. Lett.* **36** 978.
[21] Nahory R. E, Pollack M. A, Johnston W. D and Barns R. L 1978 *Appl.Phys.Lett.* **33** 659.

IOP Conf. Series: Materials Science and Engineering **28** (2012) 012005 doi:10.1088/1757-899X/28/1/012005

Study of the Resistance of the Composite Columns under Axial Compression

R. Belounis [a1], **M. Mimoune** [b]

[a] Department of Civil Engineering, University of Constantine, Algeria.

[b] Department of Civil Engineering, Professor and Director of the Laboratory of Soil Mechanics and Structure LMSS University of Constantine, Algeria.

E-mail: malinerad@yahoo.fr; mmimoune@gmail.com

Abstract. The principal goal of this work is to determine the resistance of the tubular steel columns filled with concrete. In order to carry out this, of experimental work and the non-linear analyses by finite elements were employed to compare them with the theoretical resistance of the composite columns of numerical calculation's results by the developed program. In general they can become a powerful tool in gaining a better understanding of the resistance of composite columns. The principal part of work is designed to clarify the performance of the steel-concrete composite columns: how materials act as an integral body, giving a higher resistance than the material composition, each in its field.

1. Introduction

Tubular composite columns filled with concrete have become increasingly popular in structural applications all over the world. This is due to their high strength, as composite material and many other properties such as high ductility and load capacity. Although the risk of a major earthquake in a few countries is small, this type of columns can offer many other benefits, such as speed of construction, safety aspects, and possible use of simple standardized assemblies. An indication that, composite column provides the economies of a concrete column with the rapid construction of a steel one, resulting in significant savings in the overall structure of a construction project.

2. Experimental work

Many researchers tried to find an approach between theory and test results. Thus in 1807, Thomas Young showed that the compression bars is influenced by the actual geometric imperfections such as initial curvature, centering defects and heterogeneity of the composite material [1].

The composite columns made from hollow steel filled with concrete have become the most widely used in building construction of large heights such as skyscrapers, and bridges, this is due to the confinement of concrete applied by the section of steel tube which gives a very high compressive strength [2] even if the column is very slim compared to the one of reinforced concrete [3] [4].
Several studies around the world have shown that there are factors of major influences on the behavior of composite columns such as:

[1] To whom any correspondence should be addressed.

-Ratio of wall thickness D / t or B / t, and the slenderness ratio of column D / L or B / L, respectively, which explain the phenomena of local buckling of the tube walls and the buckling of the column.
-The characteristic strength of concrete filling: high strength concrete, lightweight concrete. [5] [6] [7] that have shown great utility in terms of resistance of the cross section of the composite column.
-Yield strength structural steels (steel tube of high resistance) [8], the use of resistance above 355 MPa, have forced the researcher to change the conditions of Eurocode 4: (1) and (2) so they can be applied when using high strength steels [9].
We even made comparative studies on the accuracy of different computer codes such as EC4, BS, ACI, LRFD, and the Chinese Code, with the experimental results, and we found that we must adjust the safety factor given by the EC4 [10].
We consider the local buckling of the walls as a major risk, because it can lead to rupture of the composite columns, it is expressed by the ratios D / t, and B / t, which must meet the following condition for column's safety:

$D/t \leq 52\sqrt{\frac{235}{fy}}$ (1) for circular sections and $B/t \leq 90\sqrt{\frac{235}{fy}}$ (2) for square and rectangular sections.

It's the same for buckling which is conditioned by the slenderness λ with:

$$\lambda = \sqrt{\frac{Nplrd}{Ncr}} \text{ [10]} \qquad (3)$$

3. Composite columns made from hollow steel Section filled with concrete
Concrete added to the profile can distinguish several types of composite columns filled with concrete figure1:

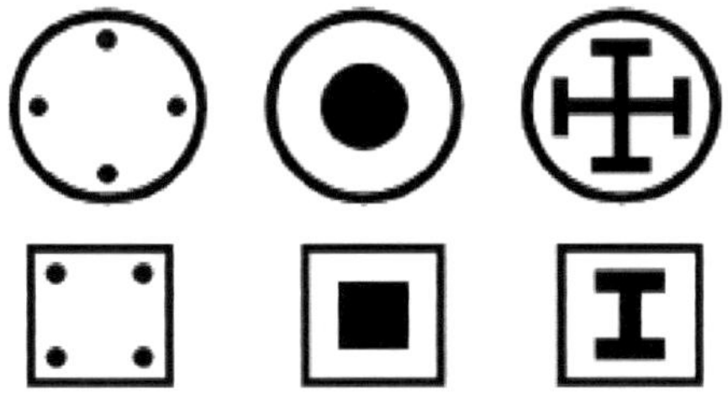

Figure 1. Different types of composite columns filled with concrete

3.1. Methods of calculation
For the calculation of composite columns we chose the simplified calculation method of EC4 to determine the normal force of plastic compressive strength *Nplrd* for the rectangular and square sections [11].

$$N_{plrd}=A_{a}.f_{y}+A_{c}.f_{ck}+A_{s}.f_{sk} \qquad (4)$$

where : *Aa, Ac, As:* they are respectively cross sections of steel, concrete and reinforcement, and *fy, fck, fsk* its respective strengths of steel, concrete and reinforcement
In the absence of transverse reinforcement the third term is zero.

A composite column has sufficient strength if: $N_{sd} \leq \chi . N_{plrd}$ (5)

Where : $\chi = \dfrac{1}{\phi + \left[\phi^2 - \bar{\lambda}^2\right]} \leq 1$ (6)

$$\phi = 0.5\left[1 + \alpha\left(\bar{\lambda} - 0.2\right) + \bar{\lambda}^2\right] \qquad (7)$$

and: α=0.21.

$$\bar{\lambda} = \sqrt{\frac{N_{plrd}}{N_{cr}}} \tag{8}$$

Where the critical load of a composite column is:

$$N_{cr} = \frac{\pi^2 (EI)_e}{l^2} \tag{9}$$

With: $(EI)_e$ is the stiffness of the column.
And l: is the buckling length of the column assumed to be rigid and isolated ($l = L$).
For columns subjected to composed bending are determined interaction points *A*, *B*, *C* and *D*.
Point *A*: compressive strength alone:

$N_A = N_{pl.Rd} \quad M_A = 0$

Point *B*: flexural strength alone:

$N_B = 0 \quad M_B = M_{pl.Rd}$

Point *C*: flexural strength similar to that associated to the position of point *B*, but with a resultant non-zero in compression:

$$N_C = N_{pm.Rd} = A_C.0{,}85\frac{f_{ck}}{\gamma_C} \quad (section\ enrobée)$$
$$M_C = M_{pl.Rd}$$
$$= A_C\frac{f_{ck}}{\gamma_C}(section\ creuse\ remplie\ de\ béton)$$

Note: *fck* must eventually be assigned a factor $[1+\eta_c\frac{t}{d}\frac{f_y}{f_{ck}}]$ if it is a hollow circular section.

Point *D*: maximum moment resistance

$$N_D = \frac{1}{2}N_{pm.Rd} = \frac{1}{2}A_c\ 0{,}85\frac{f_{ck}}{\gamma_c}(section\ enrobée)$$
$$= \frac{1}{2}A_c\frac{f_{ck}}{\gamma_c}(section\ creuse\ remplie\ de\ béton)$$

3.2 Calculation of composite columns subjected to compression and composed bending

In our work we developed a program in FORTRAN 90 by calculating the resistance of hollow section composite columns filled with concrete to determine the influence of various parameters such as, the geometric characteristics of the cross section of the column, and the mechanical characteristics of used materials.

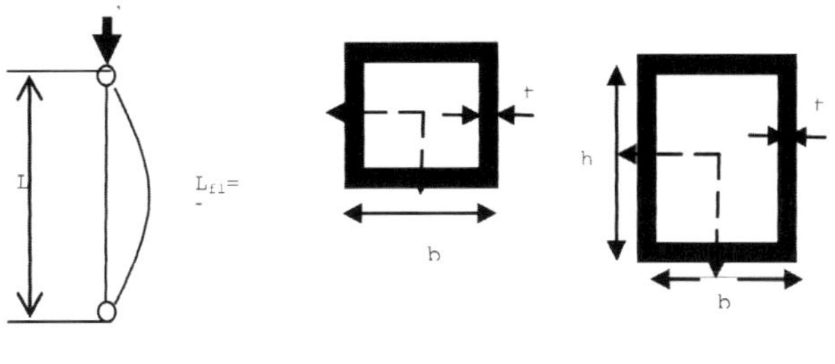

Figure 2. Sections of composite columns studied.

The program for calculating the resistance of composite columns filled with concrete was checked first with leading a confrontation between the results of theoretical calculations and experimental results obtained from the literature.

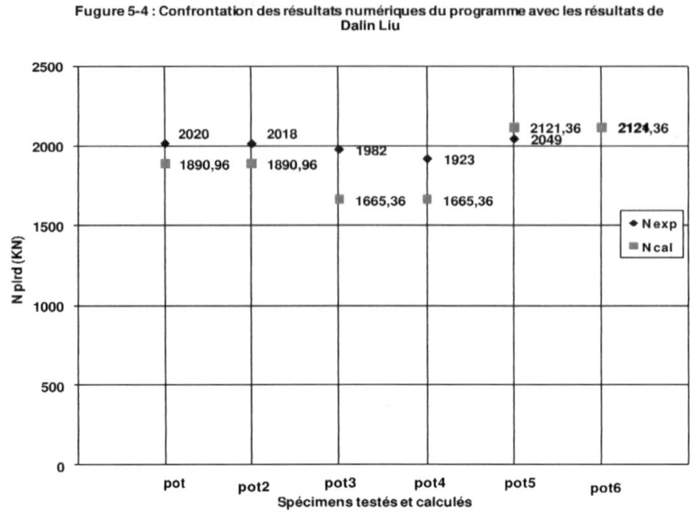

Figure 3. Comparison of numerical results with the results of Liu Dalin.

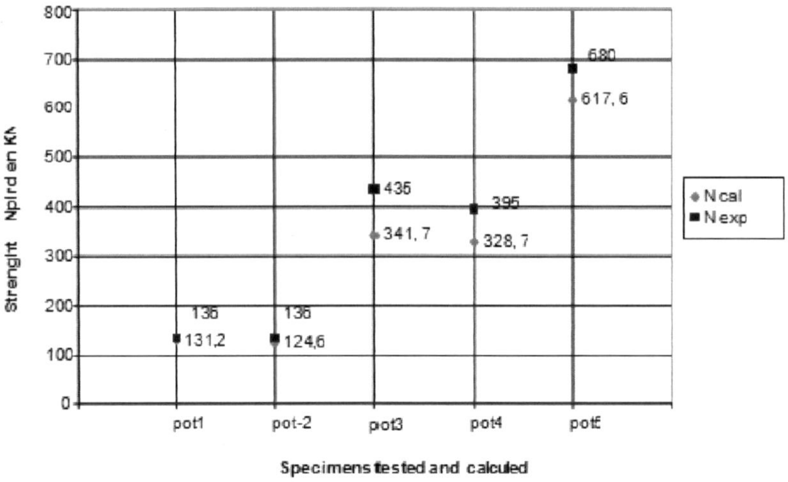

Figure 4. Comparison of numerical results with the results of Khandakar M, Anwar H

It is noted for Figures 3 and 4 that the results of numerical calculation are close to experimental [11] Liu Dalin, and Khandakar M, H and Anwar [12].

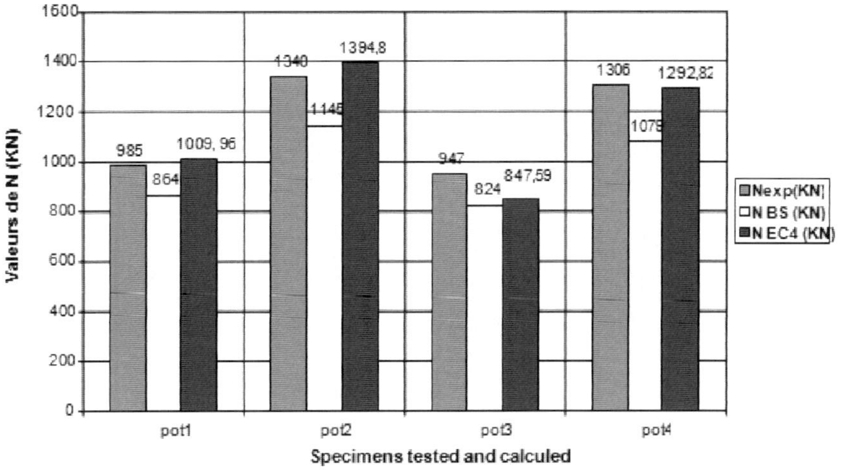

Figure 5. Comparison between the tests, the British code BS, and EC4

In Figure 5 we see that the two computer codes gave values more or less close, with some accuracy of Eurocode 4 compared to the British Standard BS 5100.

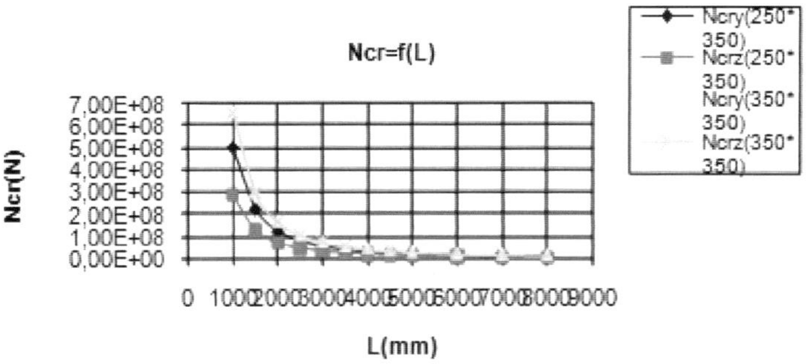

Figure 6. The influence of buckling length on the critical normal force according to column cross section

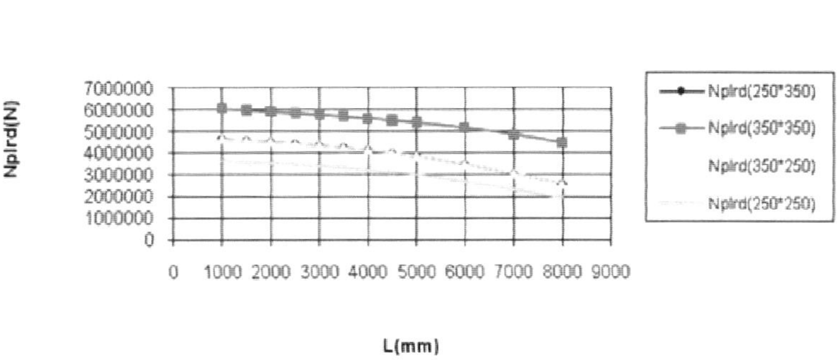

Figure 7. The influence of the dimensions of the column section.

IOP Conf. Series: Materials Science and Engineering **28** (2012) 012005 doi:10.1088/1757-899X/28/1/012005

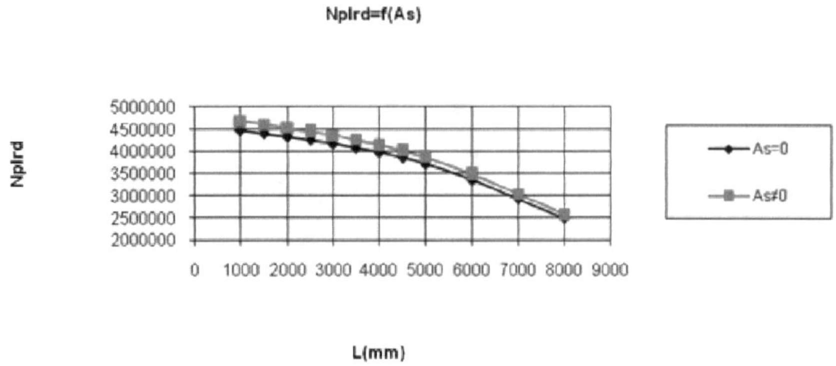

Figure 8. The influence of longitudinal reinforcement.

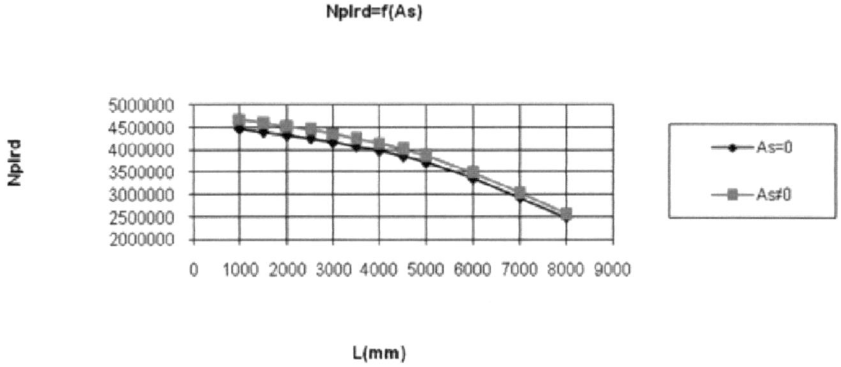

Figure 9. The influence of steel grade.

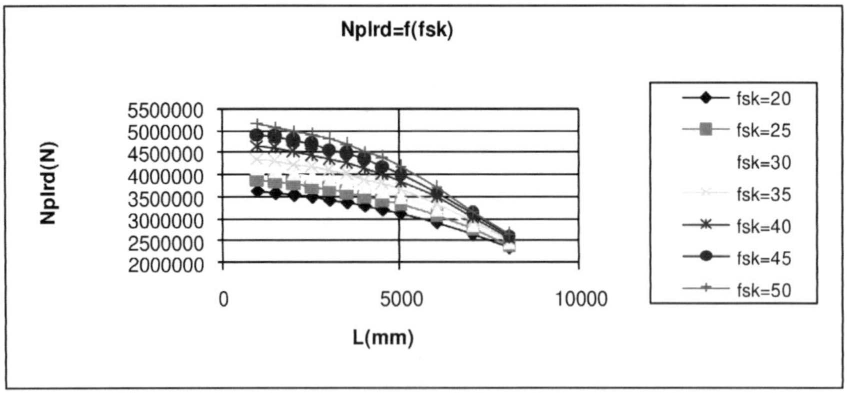

Figure 10. The influence of the strength class of concrete

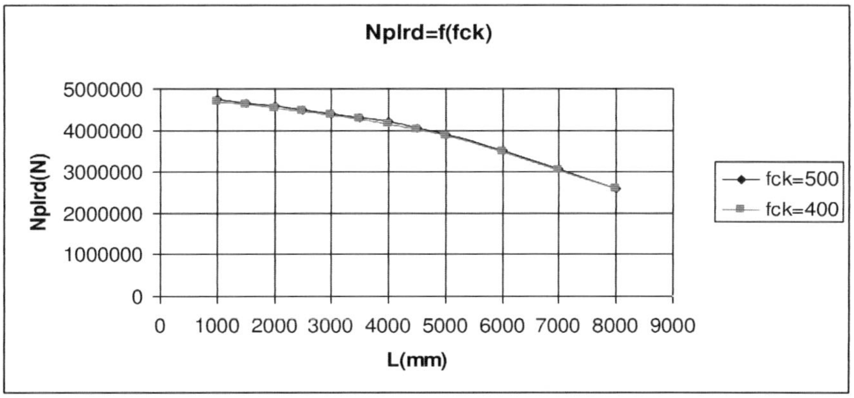

Figure 11. The influence of the elastic limit of steel tube.

4. Discussion of results

Noticing the figures from 6 to 11 we can see that the strength of tubular composite columns filled with concrete is inversely proportional to the length of the buckling which is taken equal to the real length. First we have studied the effect of the buckling length as a function of the cross section of columns on the critical normal force we found difference of 40% between *Ncry* and *Ncrz* of the rectangular section and 0% of the square section, the column strength was reduced by varying the length of buckling 1000-8000 mm without changing the dimensions of the cross section, and whatever the material used and its resistance, then we must always look for the buckling length with the appropriate cross section to better resist external loading. It was also noticed that column strength is large enough in the presence of longitudinal reinforcement *As*, the grade of steel has also played an important role, as well as the characteristic strength of concrete used for filling the steel tube *fsk*, and tensile strength of steel tubes *fck* that have shown an increase in resistance when increasing if the values of these parameters.

On the other hand, the variation of resistance of composite columns by varying the parameters mentioned above is included between 2300 kN-2500kN for the curve that shows the effect of reinforcement and 3000kN for curve which has the effect of steel grade, 2500KN for curves showing the effect of resistance of concrete and steel, so the variation is very important in varying shades of steel, and less important for other parameters. A convergence of results is by increasing the length of buckling by varying all parameters and for all curves except for curve which treats the influence of the cross section when we note that the difference between the curves is fixed along the axis buckling lengths.

5. Conclusions

This article was based on the study of the influence of some parameters on the resistance of composite columns hollow section filled with concrete under axial loading, the results obtained have shown that these parameters have great influence on the resistance force calculated by the method of EC4 of composite columns and we must always maintain a certain ratio between the buckling length of the column and the dimensions of the cross section.

The results of the study have mounted that the resistance increases in square sections compared to the rectangular ones and the increase was approximately 25%. It is also concluded that the resistance of composite columns decreases with increasing their length whatever are loading conditions, materials used, or the cross section itself.

References

[1] Fellah F, Seridi A 2007 Séminaire national de Génie civil, Annaba.
[2] Shanmugan N. E, Lakshmi B 2001 Journal of Constructional Steel. Research **57** 1041–1080.

IOP Conf. Series: Materials Science and Engineering **28** (2012) 012005 doi:10.1088/1757-899X/28/1/012005

[3] Kefeng.T, John M. N, Xincheng P 2003 *ACI Journal* **22**.
[4] Wang Y.C 1999 *Journal of Constructional Steel Research* **49** 25–41.
[5] Giakoumelis G and Lam D 2004 *Journal of Constructional Steel Research* **60** 1049-1068.
[6] Baig M. N, Jiansheng F., Jianguo N 2006 *ISSN* **11** 657-666.
[7] M.Mouli, H.Khelafi. Engineering Structures (2006), doi:10.1016/j.engstruct.2006.10.00.
[8] Mohanad Mursi, Brian Uy .Journal of Constructional Steel Research 60 (2004) 1825–1848
[9] Mimoune M, Mimoune F. Z, Tahi A 13th International Conference for Building and Construction. 15 – 19 Juin 2006 Cairo – Egypt.
[10] Eurocode 4 –conception et dimensionnement des structures mixtes acier –béton. Partie 1-1: règles générales et règle pour les bâtiments
[11] Dalin L. 2005 *Journal of Constructional Steel. Research* **61** 902–911.
[12] Khandaker M, Anwra H 2003 *Composites: part B* **34** 715-725.

IOP Conf. Series: Materials Science and Engineering **28** (2012) 012006 doi:10.1088/1757-899X/28/1/012006

First-principle study on electronic structure property of GaN/AlN

N. Benayad[a1], M. Dine El Hanani[a], M. Djermouni[b]
[a]Laboratoire des Matériaux Magnétiques, Département de Physique, Faculté des Sciences, Université Djilali Liabes de Sidi Bel Abbés, Sidi Bel Abbés 22000, Algérie
[b]Laboratoire de Modélisation et Simulation en Sciences des Matériaux, Faculté des Sciences, Université Djilali Liabes de Sidi Bel Abbés, Sidi Bel Abbés 22000, Algérie

E-mail: dr.benayad@yahoo.fr

Abstract. Ferromagnetism of GaN and AlN based magnetic semiconductors was investigated by ab initio calculations based on the local density approximation. In a system of Mn atom doped GaN and AlN, the ferromagnetic ordering of Mn magnetic moments was induced by hole doping. It was also found that *3d* transition metal atoms of Fe and Co showed the ferromagnetic ordering of their magnetic moments in GaN and AlN without any additional carrier doping treatments. Appearance of the ferromagnetism in these systems suggests possibility for a fabrication of a transparent ferromagnetic which will have great impact on industrial applications in magneto optical devices.

1. Introduction

III-V nitrides are important in electronics, microelectronics and materials science. In particular GaN/AlN semiconductors have recently attracted extensive attention because of their importance to both science and technology for optoelectronic applications in the short-wave length range as well as for high-temperature, high-power and high-frequency electronic devices. This group III-V nitrides is considered as non-magnetic semiconductors. We can obtain by alloying them a new class: DMS (semiconductor magnetic diluted), with a sizable amount (a few percents or more) of magnetic elements, such as manganese Mn. The studies of DMS based on III-V and their heterosructures have offered a wide variety of materials and structures, making it possible to explore further the effect of the exchange interaction in semiconductors. Most remarkably Mn doping in GaN and AlN leads to ferromagnetism [1]. This behavior is generally called carrier-induced ferromagnetism because hole carriers introduced into the system mediate the ferromagnetic coupling between the Mn ions [2], although its microscopic mechanism has been controversial. The key to understanding the mechanism of the carrier-induced ferromagnetism is to elucidate the nature of the doped hole carriers as well as the exchange interaction between the holes in the host valence band and the localized d orbitals of magnetic ions, so-called p-d exchange interaction. Recently, there have been a few first principles electronic structure calculations of Mn doped GaN systems. Fong and al [3] performed electronic structure calculations of Fe and Mn doped GaN using the Tight-binding linearized muffin tin orbital (TB-LMTO) method. Kronik and al [4] considered (Mn, Ga) N in wurtzite structure and performed

[1] To whom any correspondence should be addressed.

electronic calculations using a plane-wave pseudo-potential method. Sanyal and al [5] studied the exchange interaction in (Mn, Ga) N and compared it with that in (Mn, Ga) As using the ab initio plane wave method within the local spin density approximation LSDA and (LSDA+U) calculations. The aim of this paper is to show that by using the full potential linear muffin tin orbital (FP-LMTO) within the LDA, the results of structural and electronic properties calculations are accurate compared to others works. The crystal is divided into two regions: non- overlapping muffin tin spheres surrounding every atom and the interstitial region between the spheres. Within the spheres, the potential is expanded in spherical harmonics, while the interstitial regions it is expanded in terms Fourier series. Provides the best accuracy at the price of increasing the computational time.

2. Results and discussion

2.1. Structural optimization

The calculations were first carried out assuming ideal zinc blend (ZB) and wurtzite (WZ) structures. For wurtzite structures, the axis ratio c/a and the internal parameter u for both GaN and AlN are c/a = 1.636, 1.616, u = 0.375, 0.376 respectively were used. The calculated total energies and pressures for several lattice constants were fitted with the empirical Birch equation of state to obtain equiliberium lattice constants. For zincblende and wurtzite structures, the forces acting on atoms and the elements of the stress tensor have converged within 10^{-4} Ry/(a.u)^3. Table I summarizes the results of our calculations for lattice parameters.

Table I. Calculated lattice constants a_0 and c_0, bulk modulus, pressure derivative (B'_0) and internal parameter *u* compared to experiment and other works of GaN and AlN zinc blend (B_3) and wurtzite (B_4) phases.

Materials	References	a_0 (A^0)	c_0 (A^0)	B_0 (GPa)	B'_0	c/a	u (A^0)
	Present	4.47	-	194.58	3.97	-	-
GaN (B_3)	Experiment	4.50[a]	-	185[a]	-	-	-
	Other works	4.46[b]	-	202[b]	4.32[b]	-	-
	Present	3.158	5.166	194.2	4.1	1.636	0.375
GaN (B_4)	Experiment	3.192[c]	5.183[c]	188[c]	3.2[c]	1.624[c]	0.375[c]
	Other works	3.18[d]	5.151[d]	-	-	1.62[d]	-
	Present	4.34	-	201	4.002	-	-
AlN (B_3)	Experiment	4.37[a]	-	215[a]	-	-	-
	Other works	4.349[b]	-	211[b]	3.90[b]	-	-
	Present	3.07	4.961	205	3.996	1.616	0.376
AlN (B_4)	Experiment	3.11[c]	4.979[c]	185[c]	5.7[c]	1.601[c]	0.385[c]
	Other works	3.11[d]	4.979[d]	-	-	1.601[d]	0.382[d]

[a]Ref [6] [b]Ref [7] [c]Ref [8, 9, 10,11] [d]Ref [12]

For zinc blend (B_3) and wurtzite (B_4), the lattice constants from LDA calculations are slightly smaller than the experimental values. The equilibrium lattice constants to obtain from LDA calculations are in good agreement with those calculated by first principles method within different approximations. Our calculated bulks are slightly larger than experimental and other theoretical works except for AlN zinc blend (B_3). This is essentially due to the effect of the LDA on the calculations. We note an anomaly in comparing our results of pressure derivative of the bulk modulus of AlN wurtzite (B_4).

2.2. The relative stability and phase transition

In order to calculate the ground state properties of GaN and AlN, the total energies are calculated in GaN/AlN (NaCl B_1, Zinc blend B_3, Wurtzite B_4) structures for different volumes around the equilibrium cell volume V_0. Figure 1 show the fitted E_{tot} versus V curves of the three phases of GaN and AlN considered, calculated by using LDA calculation. The pressure is given by:

$$P = \left(\frac{B_0}{B_0'}\right)\left[\left(\frac{V_0}{V}\right)^{B_0'} - 1\right]$$

From this equation we deduce the transition volume:

$$V(P) = V_0\left[1 + P\left(\frac{B_0'}{B_0}\right)\right]^{-1/B_0'}$$

Where B_0 is the bulk modulus at the equilibrium V_0 and B_0' is its pressure derivative.

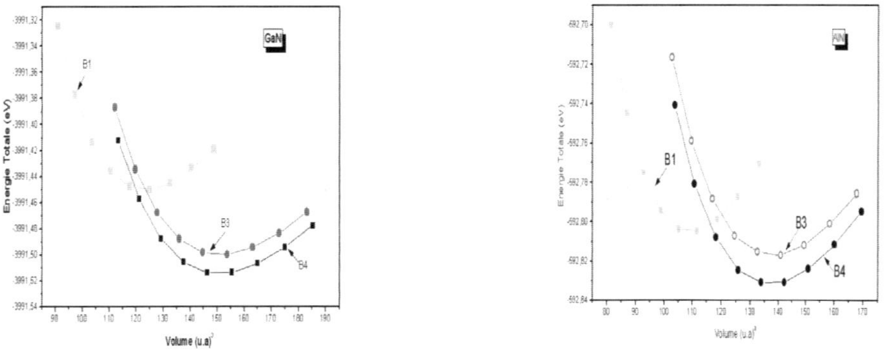

Figure 1. Calculated total energies as a function of volume for zinc blend, wurtzite and rocksalt GaN and AlN with LDA calculation.

Table II: transition pressures P_t for GaN and AlN.

		P_t	$V_t(B_3)$	$V_t(B_1)$	$V_t(B_3)/V_0$	$V_t(B_1)/V_0$	$\Delta V\%$
GaN	Present	33.9	132.46	110.34	0.87	0.72	14.6
Zinc blend - Rocksalt	Experiment	-	-	-	-	-	-
	Other Works	38.15[a]	-	-	-	-	-
GaN	Present	47.20	127.39	107.43	0.84	0.71	13.2
Wurtzite - Rocksalt	Experiment	47-50[b]	-	-	-	-	-
	Other Works	42.9[c]	-	-	-	-	-
AlN	Present	6.48	133.87	106.81	0.97	0.77	19.6
Zinc blend - Rocksalt	Experiment	-	-	-	-	-	-
	Other work	7.1[b]	-	-	-	-	-
AlN	Present	15.83	128.311	103.689	0.75	0.93	17.9
Wurtzite - Rocksalt	Experiment	14-16.6[d]	-	-	-	-	-
	Other Works	9.1[c]	-	-	-	-	-

[a]Ref [14] [b]Ref [15] [c]Ref [16] [d]Ref [17]

It's clear from the E(V) curve, that the wurtzite (B_4) structure is the most stable at room condition, which is consistent with experimental results and other theoretical works. The structure phase stability is determined by calculating Gibbs free energy for both phases ($G = E_{tot}+PV-TS$) since the theoretical calculations are performed at 0 K, Gibbs free energy becomes equal to enthalpy ($H= E_{tot}+PV$) [13]. The transition pressure (P_t) of the zinc blend to rocksalt and wurtzite to rocksalt of GaN and AlN are determined from the constraint of equal static lattice enthalpy. The results for the pressure transition P_t obtained are listed in table II, compared with other avaible theoretical results and experimental data. The present results agree with the theoretical values and experimental data. Our results are similarity between the zinc blend and wurtzite and show that the gradient corrections to the LDA have very small effects on P_t of GaN/AlN.

2.3 Elastic constants

In the following, we study the GaN and AlN compounds in their metastable zinc blend phase because there are little experimental data and theorical works under strain and stress effect, behind this we will also study it in the wurtzite phase.

The elastic properties define the properties of material that undergoes stress, deforms and then recovers and returns to its original shape after stress ceases. These properties play an important part in providing valuable information about the binding characteristic between adjacent atomic planes, anisotropic character of binding and structural stability. Hence, to study the stability of these compounds in zinc blend (B_3) and wurtzite (B_4) structures, we have calculated the elastic constants at equilibrium lattice parameter. The elastic moduli require knowledge of the derivative of the energy as a function of the lattice strain. It is possible to choose this strain in such a way that the volume of the unit cell is preserved. In the case of cubic system, there are only three independent elastic constants namely C_{11}, C_{12} and C_{44}. Thus for the calculation of them, we have used the Mehl method [18, 19], which state the use of the volume conserving orthorhombic and monoclinic tensors in determining the shear constants C_{11}-C_{12} and C_{44} respectively. So set of three equations means to determinate all the constants, which means that three types of strain must be applied to the starting crystal.

The first type involves calculating the elastic modulus $(C_{11}+2C_{12})$, which are related to the bulk modulus B: $B = 1/3\,(C_{11}+ 2C_{12})$.

The second type involves calculating the elastic modulus $(C_{11}-C_{12})$, which are related to the equilibrium volume: $P/V_0 = C_{11}-C_{12}$.

The third type involves calculating $C_{44} = 2P/V_0$.

Table III: Elastic constants of zinc blend GaN and AlN.

		C_{11}	C_{12}	C_{44}
GaN	Present	279	152	271
	calcul[a]	285	159	155
	calcul[b]	296	154	206
AlN	Present	309	147	283
	calcul[a]	313	160	192
	calcul[b]	304	152	199

[a]Ref [20] [b]Ref [21]

For zinc blend -GaN and zinc blend -AlN, there are no experimental data and there are several other calculations. Our calculations are all in good agreement with the other works.

In the case of wurtzite (B_4) phase, there are five independent elastic constants namely C_{11}, C_{12}, C_{13} and C_{44}. These elastic constants are related to the bulk modulus B_0 and four shear constants $(C_{11}+C_{12}$, C_{11}-C_{12}, $1/2\ C_{33}$ and $2C_{44})$ [22]. Hence, a set of five equations is needed to determine all the constants.

The first equation involves calculating the bulk modulus B which related to the elastic constants by: $2/9(C_{11}+C_{12}+2\,C_{13}+1/2\,C_{33})$

The second and third ones calculating the shear constants $(C_{11}+C_{12})$ and C_{33} by applying the uniform in plane strain tensor. $\begin{pmatrix} \delta & 0 & 0 \\ 0 & \delta & 0 \\ 0 & 0 & \delta \end{pmatrix}$ And the inter- layer strain tensor $\begin{pmatrix} 0 & 0 & 0 \\ 0 & 0 & 0 \\ 0 & 0 & \delta \end{pmatrix}$ Respectively.

Application of these strains changes the total energy from its unstrained values as following:

$$\Delta E\,(V_0) = V_0\,(C_{11}+C_{12})\,\delta^2 \text{ and } \Delta E\,(V_0) = 1/2\ V_0\, C_{33}\ \delta^2.$$

The fourth equation involves applying volume conserving monoclinic strain tensor $\begin{pmatrix} \delta & 0 & 0 \\ 0 & -\delta & 0 \\ 0 & 0 & 0 \end{pmatrix}$

Which transform the energy to: $\Delta E\ (V_0) = V_0\ (C_{11}\text{-}C_{12})\,\delta^2$

Finally, for the last type of deformation, we used the volume conserving triclinic strain tensor given by: $\begin{pmatrix} 0 & 0 & \delta \\ 0 & 0 & 0 \\ \delta & 0 & 0 \end{pmatrix}$

This change the total energy to $\Delta E\,(V_0) = 2V_0\, C_{44}\,\delta^2$

The four shear constants are determined by a polynomial fit of the calculated total energy as a function of the shear strains curves. The calculated values of elastic constants in wurtzite-B_4 are listed in table IV. The positive values of the elastic constants, indicating that B_4 phase is elastically stable.

Table IV: Elastic constants of wurtzite GaN and AlN.

		C_{11}	C_{12}	C_{13}	C_{33}	C_{44}
GaN	Present	114	241	177	449	384
	calcul[a]	365	135	114	381	109
AlN	Present	150	128	383	40	144
	calcul[a]	396	137	108	373	116

[a]Ref [23]

Our results for both GaN and AlN are note in good agreement with others calculations obtained with ab initio calculations.

2.4 electronic properties

The electronic band structures of zinc blend and wurtzite in both GaN and AlN along symmetry lines are displayed in figures 2 and 3 for LDA calculation. The calculated band energies at normal symmetry points in the Brillouin zone are given in Table V and VI. The band structures are calculated at the appropriate theoretical equilibrium lattice constants for LDA.

In zinc blend-B_3 and wurtzite-B_4 for GaN, we find the valence band maximum (VBM) at Γ point and the conduction band minimum (CBM) at Γ point; predicting that this compound is a direct gap material (Γ-Γ) for both phases 1.52, 2.3 (eV) respectively. In the absence of spin-orbit interaction, the valence band below the Fermi levelare triply and doubly degenerate at Γ point.

In zinc blend phase the calculated (Γ-Γ) band gap value is in agreement with the others results [24, 25, 26] and the experiment results [27], but for others gaps (Γ-X), (Γ-L) we have either theoretical or experimental values to compare with them.

In wurtzite-phase the calculated (Γ-Γ), (Γ-K) and (Γ-M) band gaps values are in agreement with the others calculations. For the zinc blend/ wurtzite AlN, the zero energy reference is the valence band maximum. It occurs at the Γ point, whereas the conduction band minimum occurs at the X point. Therefore, the band gap of zinc blend Al N is (Γ-X) indirect.

The calculated (Γ-Γ), (Γ-X) and (Γ-L) band gaps values are in agreement with others predictions calculations [28], [29] (see the table V for comparison).

In wurtzite AlN, the band gap is direct at Γ point, this is in close agreement with the results of [30], [31].

(a) Zinc blend phase (b) wurtzite phase

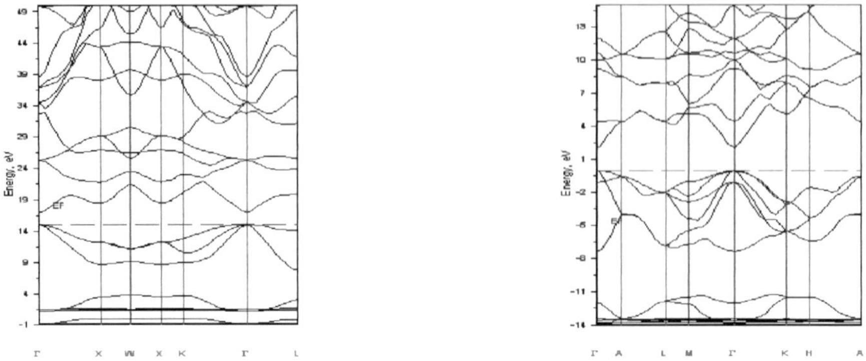

Figure 2. LDA band structures of GaN along the principle high-symmetry directions in the Brillouin zone in (a) zinc blend and (b) wurtzite phases.

(a) Zinc blend phase (b) wurtzite phase

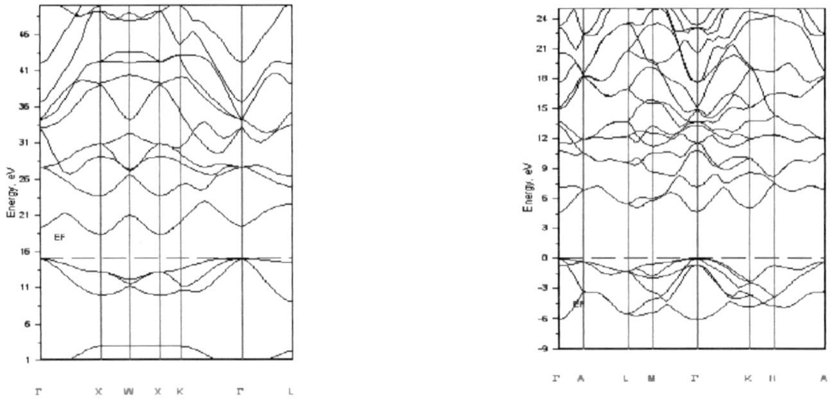

Figure 3. LDA band structures of AlN along the principle high-symmetry directions in the Brillouin zone in (a) zinc blend and (b) wurtzite phases.

2.5. Density of state

(a)

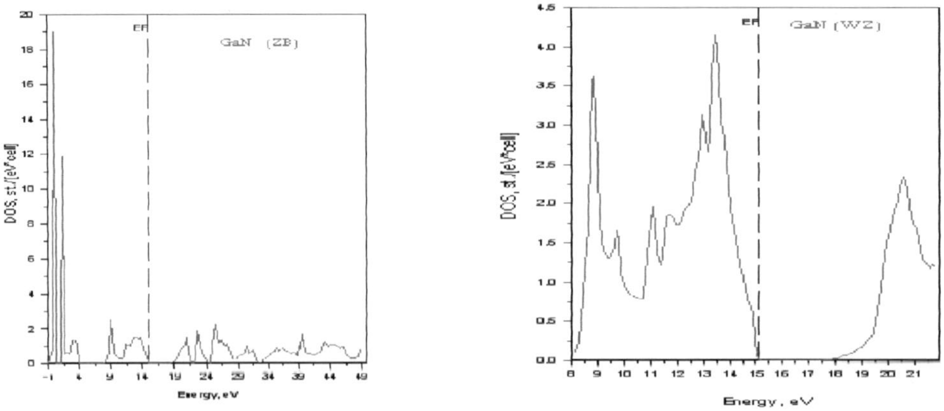

(b)

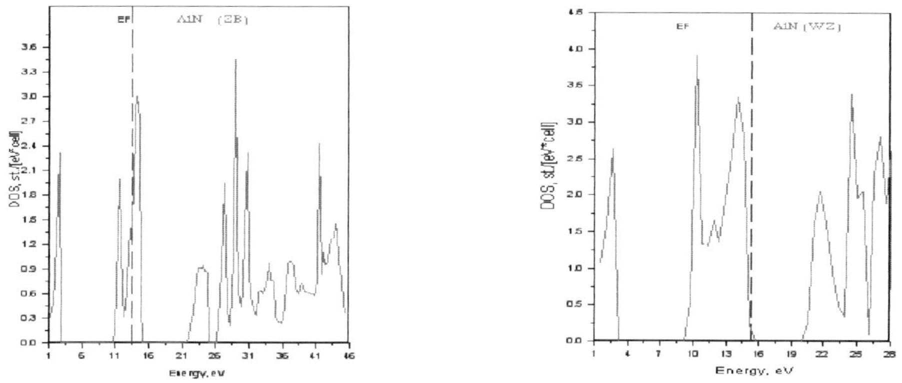

Figure 4. Calculated total density of states in (a) GaN (zinc blend- wurtzite) and (b) AlN (zinc blend- wurtzite) respectively.

In order to check the accuracy of our band structures results. We have calculated the total density also the partial density of state (PDOS) is calculated but not shown. In this case we present the calculated total density of states (TDOS) of GaN/AlN in zinc blend and wurtzite structures.
For both zinc blend-GaN and wurtzite-GaN, the total DOS presents three regions, the lower part of the valence bands are dominated by N-*2s* states and the upper part by N-*2p* and Ga-*4p* states.

Figure 4 shows the total density of states (TDOS) for both zinc blend AlN and wurtzite AlN structures. While not much difference is expected between DOSs obtained for the tow structures. The total DOS presents three regions: the lower part of valence bands is dominated by N-*2s* sates and the upper part by N-*2p* and Al-*3p* states. The Al-*3s* states contribute to the lower valence bands.

2.6. Band gaps pressure coefficients

In order to investigate the pressure effect on zinc blend GaN and AlN energy gaps, we examine the band energies at the selected symmetry points as a function of the pressure. The results of our calculation for the direct and indirect band gaps $E_g(\Gamma)$, $E_g(X)$ and $E_g(L)$ versus the pressure are shown in figures 5, table VI summarize the results of our calculated linear and sub-linear coefficients of band gaps compounds compared to the avaible theoretical works [32,33,34]. We notice that our results are in good agreement with the other prediction calculations. Our results for first pressure derivative coefficients (α) of band gaps agree well with FP-LMTO calculations of Kim and al [35] and FP-LAPW calculations of Wei and Zunger [34]. Our results for second pressure derivative coefficients (β) of band gaps are also in good agreement with LMTO calculations of Christensen and Gorczyca [32]. Since the linear pressure coefficients of $E_g(\Gamma)$, $E_g(X)$ and $E_g(L)$ are positive these main band gaps are also increased under pressure. However the increase of E_g (X) is less important than that of Eg (Γ) and $E_g(L)$.

We also study the behavior of the energy band gaps versus the volume for GaN and AlN in zinc blend structure (see figures 5).

From figure 5, we notice that at equilibrium volume the GaN has an direct gap (Γ-Γ), and the AlN has an indirect gap (Γ-X). The slope dE_g/dV of the lines are given in table VII, the negative linear pressure coefficients indicate that the indirect band gaps (Γ -X) decreases with increase of the pressure.

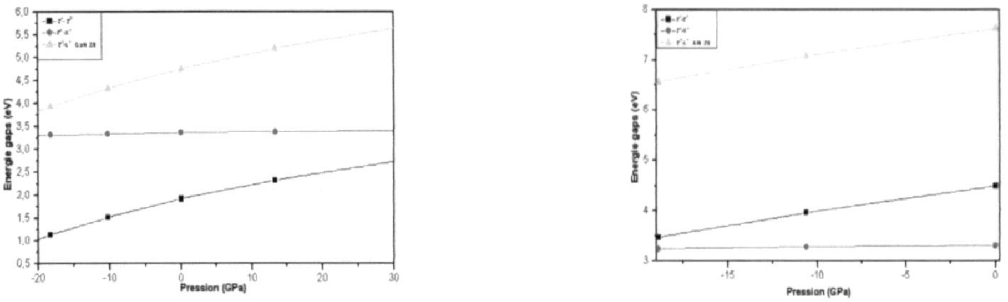

Figure 5. Direct and indirect band gap energies versus presure in zinc blend (a) GaN (b) AlN respectively.

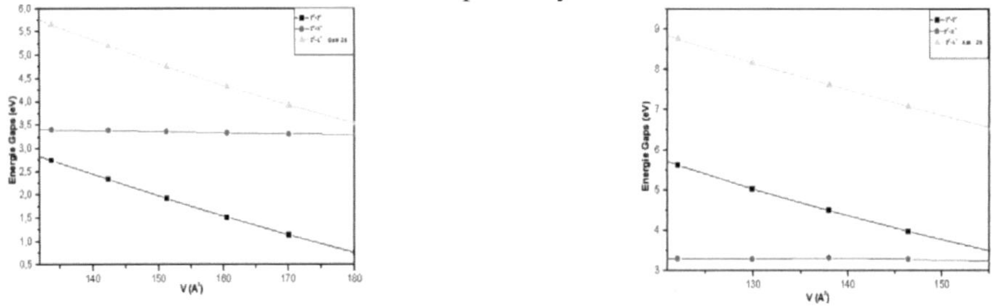

Figure 6. Calculated volume coefficients of energy gap in zinc blend structure of GaN and AlN respectively

Table VI: pressure coefficients related to $E_g(P) = E(a_0) + \alpha P + \beta P^2$ of the calculated values of the energy gaps for zinc blend GaN/AlN compared with other calculations respectively.

GaN	α (meV/GPa)		β (meV/GPa)	
	Present work	Other calculations	Present work	Other calculations
$E(\Gamma^v - \Gamma^c)$	37.4	31[a], 32[b], 40[c]	-0.363	-0.38[c]
$E(\Gamma^v - X^c)$	2.16	1.7[a]	-0.024	-0.03[c]
$E(\Gamma^v - L^c)$	35.6	32.1[a], 42[c]	-0.356	-0.38[c]
AlN	α (meV/GPa)		β (meV/GPa)	
	Present work	Other calculations	Present work	Other calculations
$E(\Gamma^v - \Gamma^c)$	42.9	42[a], 31[c]	-0.59	-0.34[c]
$E(\Gamma^v - X^c)$	1.21	1.9[a], 1.9[b], 1.7[c]	-0.013	-0.03[c]
$E(\Gamma^v - L^c)$	44.05	41.1[a], 44[c]	-0.067	-0.38[c]

Table VII: calculated volume coefficients of energy gap in zinc blend structure of GaN and AlN respectively.

GaN	λ (eV/A^3)		γ (eV/A^3)	
	Present work	Other calc	Present work	Other calc
$E(\Gamma^v - \Gamma^c)$	-0.092	-	1.58	-
$E(\Gamma^v - X^c)$	-0.0025	-	3.74	-
$E(\Gamma^v - L^c)$	-0.113	-	2.17	-
AlN	λ (eV/A^3)		γ (eV/A^3)	
	Present work	Other calc	Present work	Other calc
$E(\Gamma^v - \Gamma^c)$	-0.13	-	2.51	-
$E(\Gamma^v - X^c)$	-0.16	-	3.61	-
$E(\Gamma^v - L^c)$	-0.15	-	3.05	-

2.7. Ternary alloys

The calculated spin-polarized band structures for $Ga_{0.75}Mn_{0.25}N$ and $Al_{0.75}Mn_{0.25}N$ alloys at predicted equilibrium lattice constants are depicted in figures 7 and 8, which are drawn along the symmetry directions in the first Brillouin zone. We define the majority-spin component to be the one that contains the largest number of electron. The states in the energy range from -8.3 to -3.1 eV for $Ga_{0.75}Mn_{0.25}N$ (10.1 to 14.1 eV for $Al_{0.75}Mn_{0.25}N$) come mainly from N-*p* states with minor contributions from Mn-*d* Ga (Al)-*s*, *p* states.

The Fermi level is running through the impurity band, the $Ga_{0.75}Mn_{0.25}N$ and $Al_{0.75}Mn_{0.25}N$ alloys are half metallic. In this case, they possess a Fermi surface for the majority spin.
In this figures we see that both majority and minority spin components display a band gap. This indicates that the introduction of Mn does not destroy the semiconducting nature of these materials.

The calculated total of states (DOS) curves at the predicted equilibrium lattice constants for ferromagnetic GaN: Mn and AlN: Mn are shown in figure 9. In comparison with Ga and Al, Mn has extra four valence electrons, which fill spin-up e_g and t_{2g} band. The doubly degenerated band is fully occupied, while the triply degenerated band is only two thirds filled so the Fermi level falls just into the latter 100% spin polarized band.

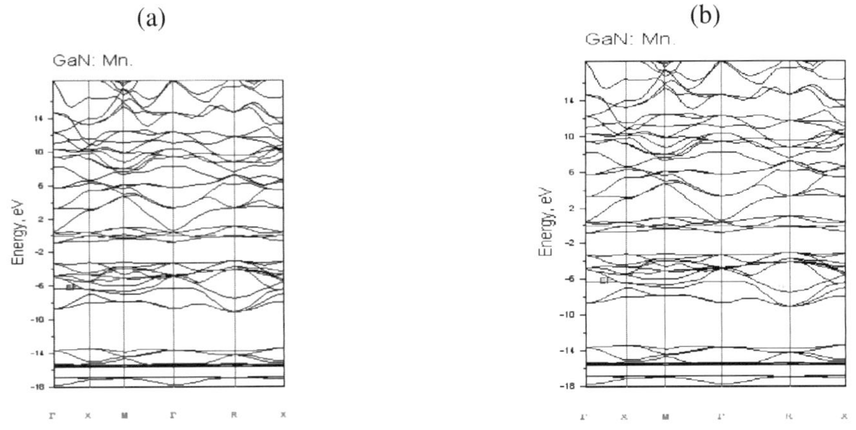

Figure 7. Spin-polarized band structure of zinc blend $Ga_{0.75}Mn_{0.25}N$: (a) minority spin (b) majority spin.

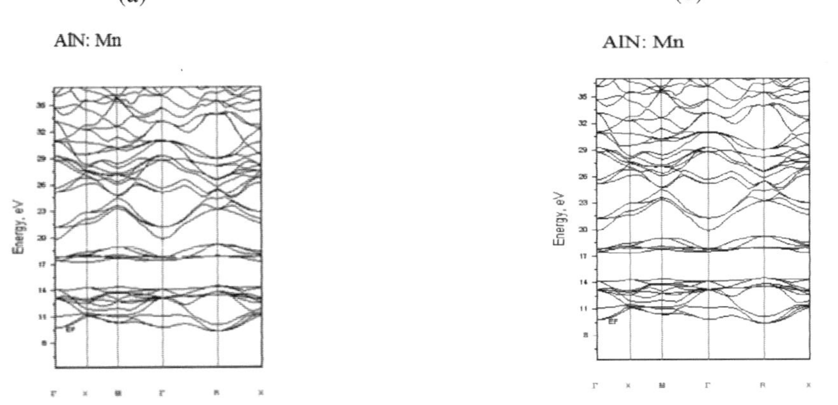

Figure 8. Spin-polarized band structure of zinc blend $Al_{0.75}Mn_{0.25}N$: (a) minority spin (b) majority spin.

(a) zinc blende $Ga_{0.75}Mn_{0.25}N$ (b) zinc blende $Al_{0.75}Mn_{0.25}N$

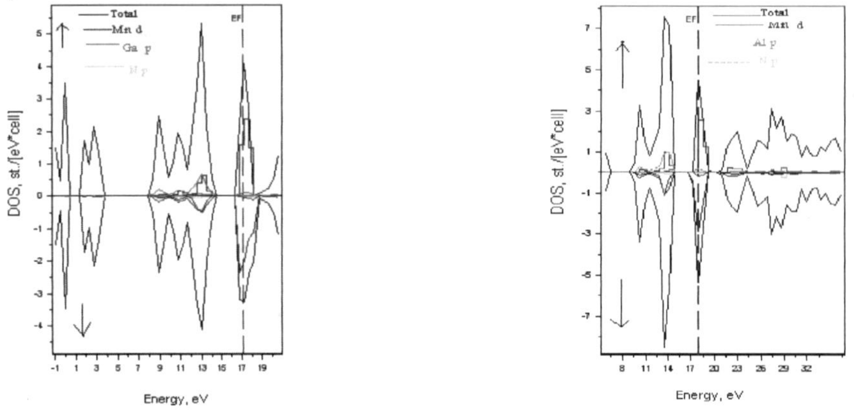

Figure 9. Total and partial spin polarized DOS in ferromagnetic states.

3. Conclusion

In this study we have presented a complete theoretical analysis of the structural, elastic and high pressure properties of GaN and AlN by using FP/LMTO method. The use of LDA for the exchange correlation potential permitted us to obtain good structural parameters. The results regarding the high pressure structural phase transition are in excellent agreement with the experimental data and agree with the other theoretical works. Plus we have performed a first principles FP/LMTO calculation of the electronic structures and magnetic properties of ordered ferromagnetic GaMnN and AlMnN with the zinc blend lattice. Optimal bulk geometries have been calculated by total energy minimization. The Ga-*3d* semi-core electrons have been explicitly treated as valence electrons. We found that both materials possess a band of well defined spin, which is primarily due to hybridization of Mn-*3d* and N-*2p* orbitals.

References

[1] Perdew J. P and Wang Y 1992 *Phys. Rev. B* **45**, 13 244.
[2] Schwarz K and Blaha P 2003 *Comput. Mter. Sci.* **28** 259.
[3] Kim K, Lambrecht W. R. L and Segall B 1996 *Phys. Rev. B* **53** 16310.
[4] Van Schilfgaarde M, Sher A and Chen A. B 1997 *J. Cryst. Growth.* **178** 8.
[5] Shimada K, Sota T, and Suzuki K 1998 *J Appl. Phys.* **84** 4951.
[6] Trampert A, Brandt O and Ploog K. H 1998 in crystal structure of group III Nitrides, edited by Pankove J. I and Moustakas T. D, Semiconductors and Semimetals **50.**
[7] Kanoun M. B, Goumri-Said S, Merad A. E, Merad G, Cibert J and Aourag H 2004 *Semicond. Sci. Technol.* **19** 1220.
[8] Xia H, Xia Q, and Ruoff A. L 1993 *Phys. Rev. B* 47 **12** 925.
[9] Edgar J. H 1994 Electronic Materials Information Service (EMIS) Datareviews Series 'Institution of Electrical Engineers, London.
[10] Perlin P, Jauberthie-Carillon C, Ti L, Miguel A. S, Grzegory I, Polian A 1991 High Press. Res. **7** 96.
[11] Perlin P, Jauberthie-Carillon C, Ti L, Miguel A. S, Grzegory I, Polian A 1992 *Phys. Rev .B* **45** 83.
[12] Savrasov S. Y 1996 *Phys. Rev. B* **54** 16470.
[13] Wettling W, Windscheif J 1984 *Solide State Commun.* **50** 33.
[14] Abu-Jafara M, Al-Sharifb A. I, Qteish A 2000 *Solid Stat .Comm.* **116** 389.
[15] Serrano J, Rubio A, Ernandez E. H, Munoz A and Mujica A, *Phys. Rev. B* **62** 16612.
[16] Xia H, Xia Q, and Ruoff A. L 1993 *Phys. Rev. B* 47 **12** 925.

[17] Vollstadt H, Ito E, Akaishi M, Akimoto S, and Fukunaga O 1990 *Proc. Jpn. Acad ,Ser. B: Phys. Biol. Sc.* **66** 7.
[18] Mehl M. J, Osburn J. E, Papaconstantopoulos D. A and Klein B. M 1990 *Phys. Rev. B* **41** 10311.
[19] Mehl M. J, Klein B. M, Papaconstantopoulos D. A 1995 in: J. H. Westbrook, R.L. Fleisher (Eds.) Principles, Intermetallic Compounds **1**.
[20] Fast L, Wills J. M, Johansson B, Eriksson O 1995 *Phys. Rev. B* **51** 17431.
[21] Yamaguchi M. Y, Yagi T, Azuhata T, Sota T, Suzuki K, Chichibu S and Nakamura S 1997 *J. Phys. :Condens. Matter* **9** 241.
[22] Vogel D, Kruger P, and Pollmann J 1997 *Phys. Rev. B* **55** 12836.
[23] Stampft C and Van de Walle C. G 1999 *Phys. Rev. B* **58** 5521.
[24] Wright A. F and Nelson J. S 1994 1995 *Phy. Rev. B* **50** 2159 **51** 7866.
[25] Lambrecht W. R. L and Segall B, Strite S, Martin G, Agarwal A, Moroko H and Rockett A 1994 *Phy. Rev. B* **50** 14155.
[26] Christensen N. E and Corczyca I 1994 *Phys. Rev. B* **50** 4397.
[27] Wei S. H and Zunger A 1999 *Phys. Rev. B* **60** 5404.

IOP Conf. Series: Materials Science and Engineering **28** (2012) 012007 doi:10.1088/1757-899X/28/1/012007

Optical glass surfaces polishing by cerium oxide particles

D. Bouzid[1], N. Belkhie, T. Aliouane
Laboratoire d'Optique Appliquée, Institut d'Optique et de Mécanique de Précision, Université Ferhat Abbas Sétif 19000 Algérie

E-mail: ja_bouzid@yahoo.fr;

Abstract. The use of powders in metallic oxides as means of grinding and polishing of the optical glass components have seen recently a large application in optical industry.
In fact, cerium oxide abrasive is more used in the optical glass polishing. It is used as grains abrasive in suspension or fixed abrasive (pellets); these pellets are manufactured from a mixture made of cerium oxide abrasive and a organic binder.
The cerium oxide used in the experiments is made by (Logitech USA) of 99 % purity, the average grain size of the particle is 300 nm, the density being 6,74 g /cm^3 and the specific surface is 3,3042 m^2/g.
In this study, we are interested in the surfaces quality of the optical glass borosilicate crown (BK7) polished by particles in cerium oxide bounded by epoxy. The surfaces of the optical glass treated are characterized by the roughness, the flatness by using the microscope Zygo and the SEM.

1. Introduction
Several methods such as polishing abrasive in suspension, float polishing [1], magnetic field assisted fine finishing [2], elastic emission polishing [3] ,laser polishing [4] are used to finish the optical surfaces.

The chemical mechanical polishing (CMP) is mostly used in the optical industry, this technique is influenced by a great number of technological parameters. The two most important are the type of pad polishing and the polishing abrasive [5-6].

Fixed abrasive polishing is particularly suitable for the mass production. The advantages of this technique are that only pure water is used as a lubricant and unlike in free abrasive polishing no chemically unreliable slurry is needed [7]. In this technique, the abrasive grains are bound by epoxy, therefore, the orientation of the grains towards the sample surface is constant. In this case, the pressure and the weight loss also are constant.

2. Experimental procedure
The manufacture of the pellet-tools was carried out by uniaxial compression at ambient temperature using a hyrdaulic press with a load of 95 MPa.(see figure 1) the mashes consisting of the cerium oxide abrasives (produced by logitech Inc USA) of 300 nm grain size and a specific surface of 3,3042 m^2/g, as shown in Figure 2 and an epoxy. The polishing tool is made of 16 pellets tied together on a metallic support with a resin. It rotates at a speed of 132 rpm. The glass sample BK7, also rotates at a speed of

[1] To whom any correspondence should be addressed.

450rpm. Its relative movements makes it oscillate in the plane of the pellets with an amplitude of 10 mm. The sample is continuously sprayed with distilled water at a rate of 500ml/ min. Figure3 illustrates the polishing process.

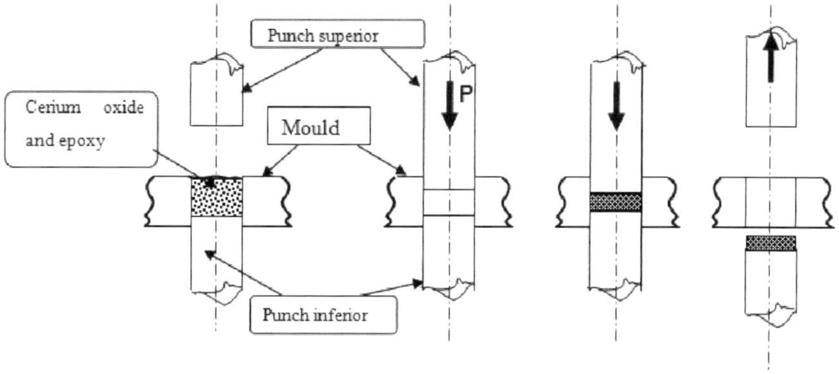

Figure 1. Manufacture of cerium oxide pellet

Figure 2. SEM photography of the cerium oxide abrasive grains

The surface of the glass is attacked by water at the same time it is subjected to intense mechanical action. The mechanism of material removal is the result of a combination of both mechanical and chemical actions [8], the mechanical action of polishing agent particles controls the rate of material removal whereas chemical action causes breakage of network connections Si-O-Si. The action of two elements (water and abrasive grains) allows the formation of the layer of silica gel with mechanical properties that differ from the sample. This layer is easily removed by the abrasive grains.

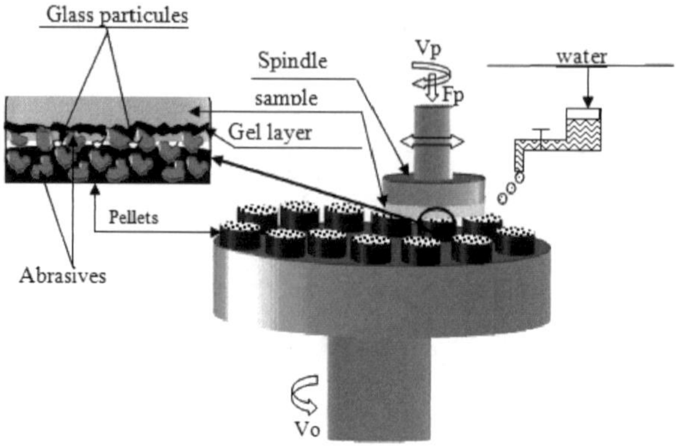

Figure 3. Method of polishing optical glass by cerium oxide pellets polishing process

3. Results and discussion

3.1. Characterization of the glass surface

Polished surfaces are characterized by the Zygo microscope (New vew 100), Figures4 a-b show respectively the surface of the glass ground by the fraction F7 and polished during 10 minutes. These figures show a marked improvement in the surface, the ridges are abraded by the abrasive particles gradually as the polishing time increases until the surface becomes very smooth, and if the polishing continues, the grains begin to scratch the surface[6][9]. The evolution of the surface during polishing was observed by scanning electron microscope. Figure 5.a shows the glass surface lapped by F7. The surface is completely covered with defects. Figure 5b shows the surface after 2 minutes polishing. At this stage, the surface begins to improve by a sharp decrease in defects.

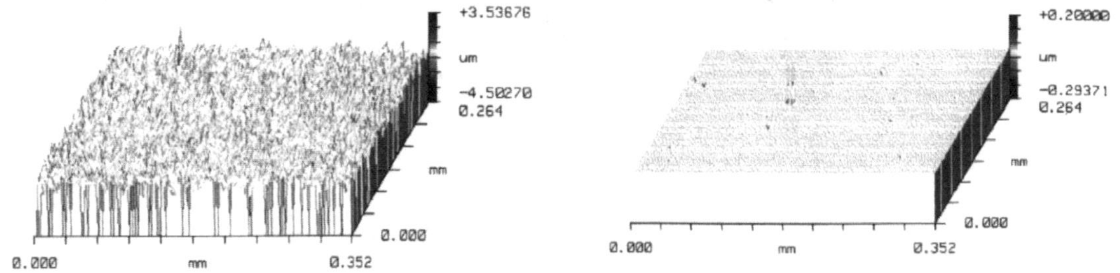

(a) (b)

Figure 4. Microstructure of the optical glass by Zygo
(a)0min polishing surface; (b) after 10 min polishing surface)

After 10 minutes polishing (Figure 5c) there is again a small defect.

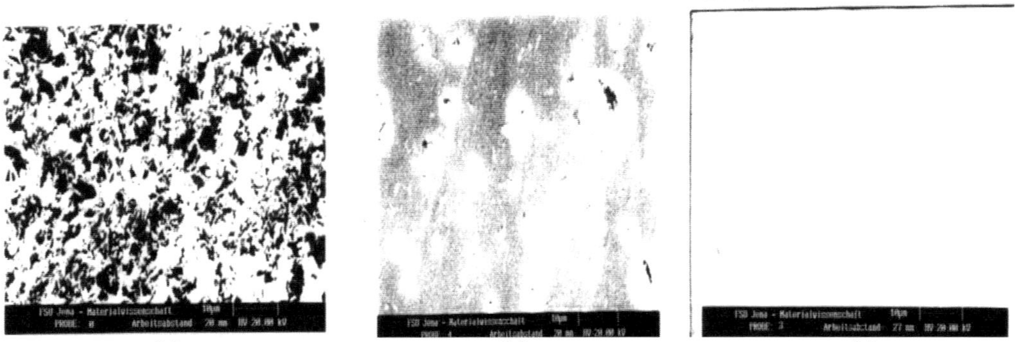

(a) (b) (c)

Figure 5. Evolution of the surface by SEM
(a)lapped surface by F7; (b) after 2 min polishing; (c) after 10min polishing.

Figures 6 a-b illustrate the distribution of the peak and the valley of the lapped glass surface with the F7 and polished by the cerium oxide during 10 minutes. These figures show that before polishing, the number of cavities at a depth of less than 1 μm is very large and the largest number of peak to a height of less than 1 micron, which proves that the fine grinding is done in good conditions. After 10 minutes of polishing, we find that the depth and height of the valley decreased significantly (the distance between the trough and crest of 0.002 μm) which shows that the polished surface is closer to the ideal surface.

Figures 7a-b shows respectively the topography of the glass surface before and after 10 minutes of polishing. These figures clearly illustrate the changes undergone by the surface during polishing; the improvement of the glass surface is characterized by reduced defects found on the surface of smooth glass.

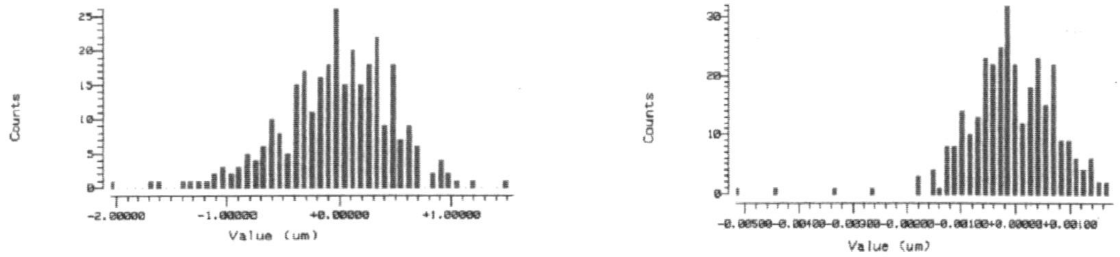

(a) (b)

Figure 6. Distribution of the peak and the valley
(a) Lapped surface with F7; (b) after 10 min polishing

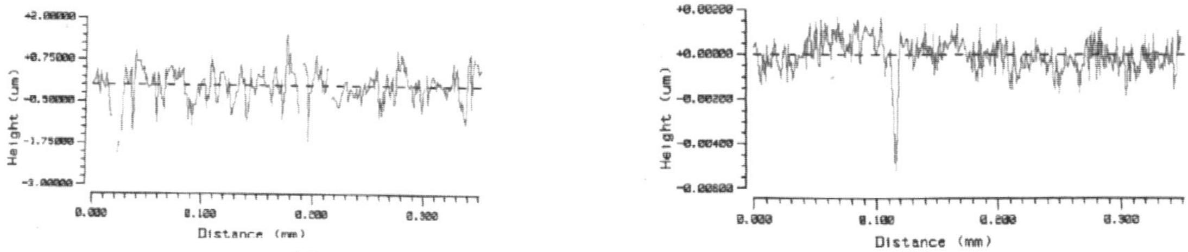

(a) (b)

Figure 7. Surface profiles
(a) before polishing; (b) after 10 min polishing

3.2. Surface glass flatness

We are interested in changing the overall shape of the glass surface during polishing, we found that it changes shape during polishing, it is either slightly concave or slightly convex. To obtain more stable production conditions, it is necessary that the shape of the polished surface tends more towards the concavity rather than convexity. This defect is characterized by the degree of flatness or shape deviation. To better illustrate these changes we have made measurements and observations using a Talyserf and an interferometer phase. Figures 8 shows the shape of the surface during three minutes polishing and the degree of concavity is characterized by the number of rings. The shape deviation on the polished surface for 10 minutes is 68.4 nm.

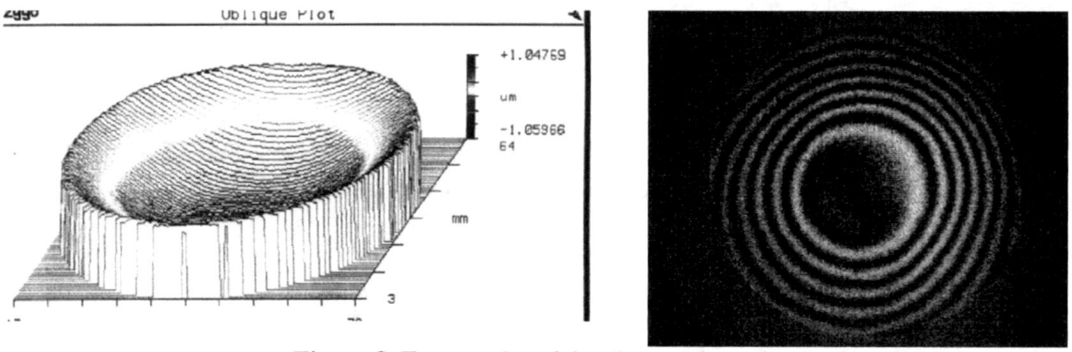

Figure 8. Topography of the glass surface after 3 min polishing

3.3. Pellet surface

The maximum surface roughness of the tool varies from 32.046 to 7.886 µm. This explains why the pellet surface is also polished by the workpiece; this confirms the results obtained [8] whereby SiO_2 particles are localized on the tool surface. At the beginning of polishing, the glass surface is rough (peak to valley = 8.039 µm), having undergone a fine grinding with the fraction F7, while the roughness (peak to valley) of the pellet surface is also important (32.046 µm). This means that before polishing (tp = 0 min), the two contacting surfaces are rough, during polishing, there is a presence of a mutual polishing and the most important peaks on both surfaces are uniformly abraded. The abrasive grains having sharp edges become, progressively as the polishing goes on rounded, before leaving the binding matrix. The lifetime of the pellet is estimated at 40 hours polishing, which means a capacity of about 240 polishing surfaces, which makes this technology commercially very profitable compared to the traditional technique, while the speed of material removal of the glass is 4.150 10-3 mm^3 / s.

4. Conclusion

Fixed abrasive polishing can be used to obtain high-quality surfaces in a very short time (within 5 minutes of polishing, the roughness rms achieved is 10 nm). It is very profitable since it only uses abrasive tools that are technically simple to perform. Compared with non-conventional techniques such as float polishing and polishing by a magnetic field, the fixed abrasive polishing has very satisfactory results on the roughness and the polishing rates. For the free abrasives polishing, the time required is higher than the technique of fixed abrasive polishing. Economically, this method is very profitable compared to other techniques.

References

[1] Bennet M, Schaffer J. J, Shibano Y, Namba Y 1987 *Appl. Opt.* **26** (4) 696.
[2] Kurobe T, Imanaka K 1987 *Precisin Engneering* **9** (1) 31 43.
[3] Mori Y, Yamanachi K, Endo F 1987 *6th INT. Conf .on Production Engineering Osaka.*
[4] Vega F, Lupon N, Lebrian J. A, Laguarta F 1998 *Opt. Eng* **37** (1) 272 279.
[5] Bouzid D, Bouzid S, Bouafia M, Jungstand U, Herold V *Glass technology*, Vol. Fev. 97 18 22.

[6] Aliouane T, Bouzid D, Belkhir N, Bouzid S, Herold V 2005 *J. phys. IV* **124** 123 128.
[7] Bouzid D, Zegadi R, Jungstand U, Herold V 2001 *Glass Technology Glass* **42** (2) 60.
[8] Koucky J, Matusek M 1984 *Glass Technology* **25** (5) 240 243.
[9] Belkhir N, Bouzid D, Herold V 2009 *Appl. Surf. Science* **255** 7951 7958

Use of argillaceous materials containing sediments of dam in the retention of heavy metals

Z. Dali-Youcef [1]
Laboratoire des matériaux LABMAT, ENSET, BP 1523, El M'naouer, Oran, Algeria.
Tel/Fax: 213 041 41 98 06

E-mail: zohra_dali 2002@yahoo.fr

Abstract. The interaction of heavy metals like Fe (II), and Cu (II) ions from aqueous solutions is studied using sediments of dam activated by impregnation with ammonium chloride (VNE) and thermally activated at 300°C (V300). This study of retention of two metals allowed us to notice shows that: the rate of retention of metals using "V300" (thermal activation) is better then using "VNE". The retention of metals is governed by the model of adsorption of Langmuir for iron and of Freundlich for copper. Because of high surface area per unit area, activated mud is the most effective adsorbent and exhibits high capacity of adsorption of iron and copper. Mud of dam activated by impregnation with ammonium chloride (VNE) and thermally activated at 300°C has a considerable potential for the removal of heavy metal cationic species from aqueous solution.

1. Introduction

Heavy metals are toxic pollutants which essentially result from industry. They are also the common contaminants of underground water. Numerous processes done to eliminate dissolved heavy metals, including ionic exchange, precipitation, ultrafiltration, osmosis inversion, électrodialysis [1-3] as well as other methods using clay in natural shape [4] or having a treatment of physical activation (thermal) [5] or chemical activation [6]. Nevertheless, some of these approaches can be expensive and unprofitable in developing countries. Consequently, the need exists for a strategy of treatment which is simple and strong, which uses the resources and local constraints. In this respect dam sediments are chosen materials for retention of heavy metals. Algeria knows besides the problem of pollution, the envasement of these dam. The mud removed after dredgings of dams, aroused our attention. We report here a study on the ability of an activated mud to uptake iron and copper. Dam sediments underwent acid activation by impregnation with ammonium chloride and thermal activation by heating up to 300°C.

2. Materials and methods

2.1. Materials

Dam sediments (mud)) were obtained by the breach of the dam of Fergoug (West Algerian). The mud is crushed and sieved to 2 mm diameter to eliminate pebbles and other solids. The powder was

[1] To whom any correspondence should be addressed.

characterized without any preliminary purification. The specific surface of mud was calculated using method of adsorption and désorption of NH_3 gas (B.E.T) by using a Microméritics Gemini 2375; it is equal to 54,58 m^2/g and surface of micropores is 29,644 m^2/g. The mud was characterized by diffraction of the X-rays using a generator of the X-rays (Philips typify PW 3710) with an anticathode of copper. The spectrum of diffraction of the X-rays confirms that the mud constitutes bédeilite (1), quartz (2), saponite (3) and dolomite (4) (figure 1)

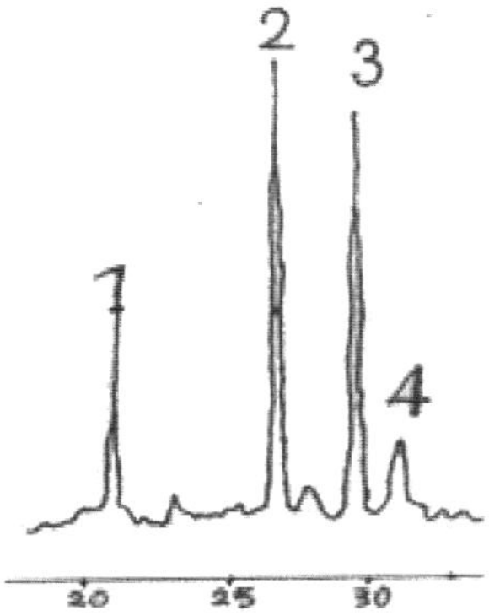

Figure 1. Spectrum of X-ray of mud.

2.2. Activation of materials

The mud underwent an acid activation: 100g of mud were impregned in 100 mL solution of ammonium chloride (1 N) and put in a mitten oven at 150°C during 24 hours. The mud so obtained is called VNE. The analysis of X-rays diffraction (figure2), shows that the activated mud (VNE) is one mixture of vermiculite (1,4), chlorite (2), saponite (3,6) and quartz (5). The specific surface increases is 271,68 m^2/g and surface of micropores is 78,18 m^2/g [7-10]. The water used in the preparation of these solutions is a water distilled and demineralized its conductivity equal 1,05 µS/ cm.

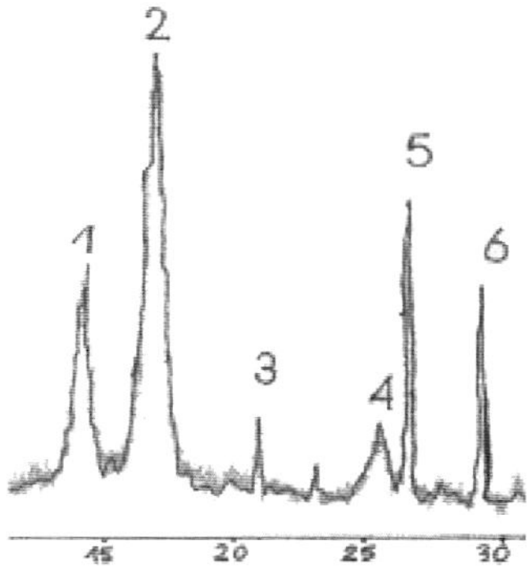

Figure 2. Spectrum of X-ray of VNE.

The mud underwent a treatment of thermal activation: the heating of the mud was done in a mitten oven (Nabertherm D-2804) provided with a system of control and regulation of the temperature (Tmax. = 1200°C, P = 9KW). The heating was up to 300°C during 24 hours. The mud so treated is noted V300. The specific surface increases to 202,14 m^2/g and surface of micropores (86,32 m^2/g). The X-rays diffraction analysis (figure 3) shows a high rate in quartz (2) (not purification of mud before heating), bédeilite (1), saponite (3) and dolomite. No change of structure with regard to the raw mud is observed.

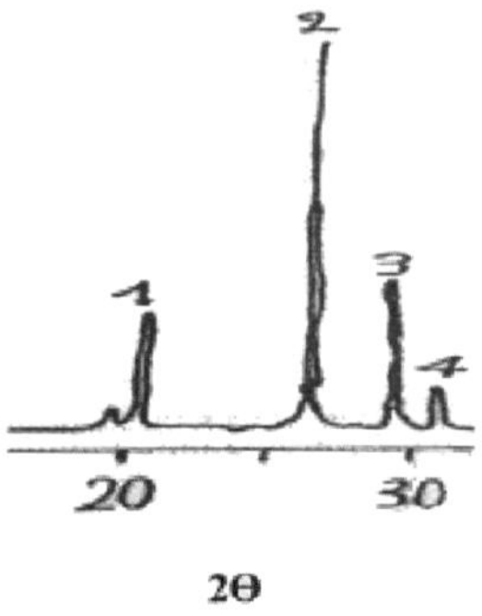

Figure 3. Spectrum of X-ray of V300.

2.3. Adsorption of heavy metals

Metals studded are: copper (II), and iron (II) used as their coloured salts, $CuSO4.5H_2O$ and $FeSO_4$. The materials used are mud activated by impregnation with ammonium chloride (VNE) and mud activated thermically at 300°C (V300). A mass m (g) of material mixed to a volume V (ml) solution of metal of initial concentration Ci (mg/l) rest during a fixed time (mn), then centrifuged and the supernatant is analyzed with a spectrophotometer of the type Perkin Lamda 25

2.4. Influence of material type on the retention of heavy metals

We disperse a mass of 1g of materials "VNE" and "V300" in a 100 ml volume of metal solution of initial concentration Ci (70 mg/L). The suspension is leaved to rest at 25°C, then centrifuged and the supernatant is analyzed by spectrophotometry. Then we obtained concentration at equilibrium (Ce) and reduction ratio (R% = (Ci-Ce)/Ci).

Ce (mg/L) is metal concentration in the supernatant.

The concentration of metal at surface of material (x/m) is defined according to Vandamme [11]:

$$x/m = (Ci - Ce)\,V/m \qquad (1)$$

Where the following parameters represent:

x/m: quantity of metal per unit of mass of mud activated (mg/g)

Ci: initial concentration (mg/l)

Ce: residual concentrations at equilibrium of the liquid phase (mg/l) V: volume of adsorbed (L)

m: mass adsorbent (g)

The evolution of the retention rate of metal according to the time of the activated mud is given in figures 2, and 3.

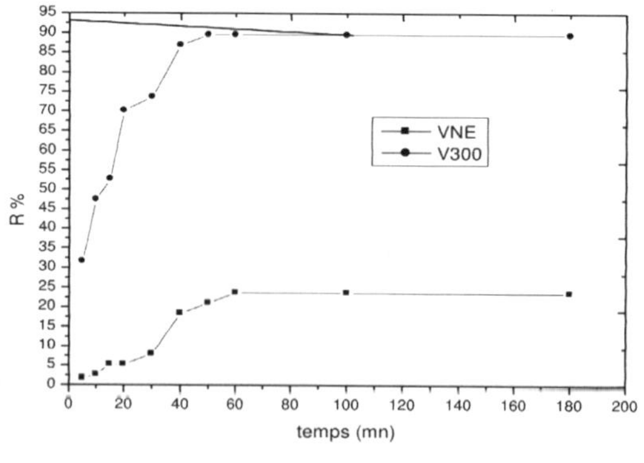

Figure 4. Evolution of the rate of retention of copper (II) according to the solid for an initial concentration of 70 mg/L.

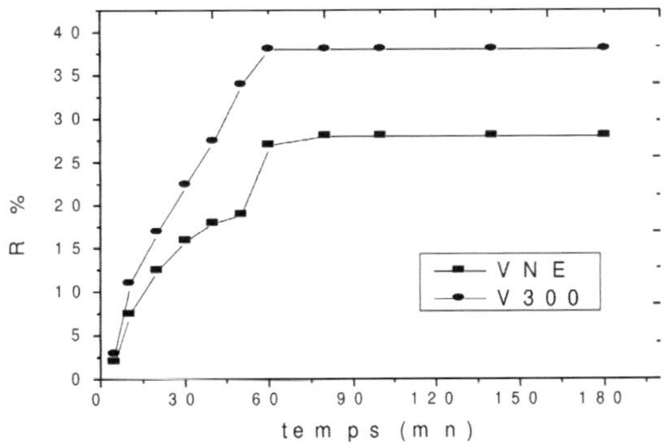

Figure 5. Evolution of the rate of retention of iron (II) according to the solid for an initial concentration of 70 mg/L.

The results observed in figures 4 and 5, show that time adsorption equilibrium of metal on activated mud chemically and thermically is approximately 60 minutes. Both materials show a better affinity to copper (R% = 90%). These results also state that the process of sorption can be regarded as being fast because of the saturation of the solid reached at the end of 60 mn. The curves represented by the two figures arrive to the same conclusion: the thermically activated mud (at 300°C) adsorbs better metals studied: iron, and copper. The rate of retention of two metals by the intermediary of the solid V300 (thermal activation) is higher than that obtained by the use of material VNE (chemical activation), this result is in agreement with the literature [12]. The heat treatment probably causes some modifications in the structure of clay. The departure of water of adsorption (interfoliaceous water) starting from 220°C (dehydration) releases from the sites and increases in the specific surface [13-15] of material, as well too specific surface of material "V300" (202,14 m2/g) as well as the surface of the micropores (86,32 m2/g) are higher than those of the material "VNE" (specific surface 183,07 m2/g and surfaces it micropores 78,18 m2/g), consequently the capacity of adsorption of treated material is improved by the increase of specific surface and by the increase of surface of the micropores, which are essential parameters of a good adsorbent. The thermally treated mud having a specific surface higher than that of the mud treated chemically, show a better affinity and capacity of adsorption with metals (iron, and copper).

2.5. Isotherm of fixing of heavy metals on materials

Suspensions containing 1g material in 100ml of metal solution with various concentrations, are put under constant agitation during 24 hours at the temperature of 25°C; then the suspensions are centrifuged and the supernatants analyzed by spectrophotometry. The initial concentrations vary from 10 to 600mg/L of iron and copper. The results of Ce (mg/L) obtained enable us to calculate parameters of Langmuir and Freundlich. Table 1 show parameters of Langmuir and Freundlich. The adsorption of iron obeys to Langmuir model. Adsorption of copper on "V300" obeys to model of Freundlich.

Table 1. Results of the parameters of the isotherm of fixing of iron and copper on VNE and V300 at 25°C.

Model	Langmuir			Freundlich		
Parameters	R^2	b	Qo (mg/g)	R^2	n	K
Elimination of iron on VNE	0,9691	0,062	2,18	0,7151	0,79	0,01
Retention of iron on V300	0,9791	0,017	2,30	0,8066	1,87	0,11
Retention of copper on V300	0,9728	$3,4\ 10^{-4}$	1428,57	0,9906	1,52	1,96

3. Conclusion

As the first development of sediments of Algerian dam, the availability and low cost of this mineral make it a good candidate for adsorption processes. For these reasons activated mud was selected in all the tests reported herein.

Copper and iron are adsorbed by the use of materials containing sediments of dam. The isotherms of adsorption show a high affinity adsorbate adsorbing, particularly in the case of copper. In the light of the results obtained, we note that the heat treatment of the mud showed an increase in specific surface compared to the acid treatment by ammonium chloride. Among studied metals, copper is fixed much more easily to the activated mud. Our study of sorption of two metals enabled us to note that:

- The rate of retention of metals by the intermediary of the solid "V300" (thermal activation) is definitely higher than that obtained by the use of the material "VNE" (chemical activation).
- The retention of copper with two materials elaborated is better than retention of iron: rate of retention of copper is 90%.
- The retention of metals is governed by the model of adsorption of Langmuir for iron and Freundlich for copper.

Consequently, muds activated chemically by ammonium chloride and thermally at 300°C are successfully used as adsorbing agent and has a high adsorption capacity for Fe (II), and Cu (II) ions. The high selectivity in the bonding of Cu(II) suggests "V300" and "VNE" be useful for removal of these toxic heavy metal ions from waste waters. This ability can be explored in treatment technologies since muds activated chemically by ammonium chloride and thermally at 300°C are a cheap, an abundant, and locally available resource.

References

[1] Ayla Ö, Dursun Ö, Ibrahim H. E 2004 *Adsorption* **10** 317 326.
[2] Wilczak A, Keinath T. M 1993 *Water Environ.Res.* **65** 238 244.
[3] Ömer Y, Yalcin A, Fuat G, 2003 *Water Research* **37** 948 952.
[4] Brigatti M F, Corradini F, Carlo Franchini G, Mazzoni S, Medici L, Poppi L 1995 *Applied Clay Science* **9** 383 395
[5] Chu K H, Hashim M A 2000 *J. of Chem. Tech. and Biotech.* **75** 1054 1060
[6] Vengris T, Binkiene R, Sveikauskaite A 2001 *Applied Clay Science* **18** 183 190.
[7] Gates W. P, Anderson J. S, Raen M. D, Churchman G. J 2002 *Applied Clay Science* **20** 189 197.
[8] Falaras P, Lezou F, Seiragakis G, Petrakis D 2000 *Clays and Clay Minerals* **48** 549 556.
[9] Srasra E, Trabelsi-Ayedi M 2000 *Applied clay Science* **17** 71 84.
[10] Mokaya R, Jones W, Davies M. E, Whittle M. E 1993 *J. of the American Oil Chemists' Society* **70** 241 244.
[10] Mokaya R, Jones W, Davies M. E 1993 *Oil Chem.* **70** 241.
[11] Vandamme H, Bergaya F 1987 *J. de Chimie Physique* **84** 1075 1082.
[12] Sabah E, Turan M, Celik M. S 2002 *Water-Res.* **36** 3957 3964.
[13] Sarikaya Y, Onal M, Baran B, Alemdaroglu T 2000 *Clays and Clay Minerals* **18** 557 562.
[14] Bojemueller E, Nennemann A, Lagaly G 2001 *Applied Clay Science* **18** 277 284.
[15] Do D. D 1998 Adsorption Analysis: Equilibria and Kinetics, Imperial College Press, Queenland.

IOP Conf. Series: Materials Science and Engineering **28** (2012) 012009 doi:10.1088/1757-899X/28/1/012009

Elasticity modulus evaluation for a porous material by acoustic methods

S. Debboub[1], Y. Boumaiza and A. Boudour
Faculty of sciences, Department of Physics, Laboratory LEAM, Badji Mokhtar University-Annaba B.P.12 (23000) Annaba-Algeria

E-mail: deb_salima@yahoo.fr

Abstract. The main scope of this study is the elastic properties characterization for porous silicon due to its importance in the electronics manufacturing by micro acoustic methods. This is why high frequency ultrasound and acoustic microscopy (acoustic signature V(z)) can be used. The interpretation of experimental results for the studied material can be led with modelling the acoustic signature V(z) that is merely based on certain aspects of the wave propagation theory in porous media. We can highlight the relation between porosity and elastic properties and the velocities of surface and longitudinal waves.

1. Introduction
Since the porous material contains heterogeneities it is highly complex and consists of two media. In this case we referred to the experimental results [1] determined from the acoustic microscope and micro ultrasound to adapt the Biot theory at higher frequencies to gigahertz range by means of different parameters to model the reflection coefficient $R(\theta)$ and the acoustic signature V(z). Acoustic microscopy is a technique that uses ultrasound in a frequency range that varies from megahertz to 1 gigahertz [2]. One target to highlight the role of porous materials is very important to the industry such as silicon for its luminescent properties. Based on the acoustic signatures of the material and its reflection coefficient, we can determine the elastic parameters (E, G, υ) against their porosity. Porous silicon is obtained by HF treatment and the pore size is in range of a nanometer.

2. Biot Theory
The theory of Biot (1956) provides a general framework to study sound propagation in porous materials [3,4]. Originally this model was developed to describe the behavior of porous rocks and enabled a macroscopic description of the coupled propagation of small amplitude perturbations in a two-phase elastic medium. Biot's formalism was applied in case of acoustic porous materials saturated with air. Taking One consider the deformable frame vibration in equations for porous media lead to the introduction of two additional mechanical parameters that are intrinsic to the elastic structure of medium: the Young's modulus E and Poisson's coefficient ν. We shortly clarify herein the assumptions required for the implementation of this theory.

The wavelength is assumed larger in comparison to the pore radius.
- The mechanical displacements of both phases liquid/ solid are supposed smaller.
- The porosity is supposed opened (the pores are assumed to be interrelated).
- We assume that the matrix is elastic and isotropic.

[1] To whom any correspondence should be addressed.

2.1. Constraints and deformations in a porous medium for an elastic structure

On the basis of constraints present in an elastic solid or a fluid, stresses in a porous medium are defined as forces acting on the deformable frame or the air inside the pores per area unit of porous material. Stress-strain relations are expressed for any porous medium as beneath:

In the solid phase:

$$\sigma^{s}_{i,j}=[(P-2N)\theta^{f}+Q\theta^{f}]\delta_{i,j}+2Ne^{s}_{ij} \tag{1}$$

So in the fluid phase:

$$\sigma^{f}_{i,j}=-\Phi p=Q\theta^{s}+R\theta^{f} \tag{2}$$

σ_{ij} represent constraints tensors respectively within the structure and air.
e_{ij} is the strain tensor in the frame.
θ^s and θ^f are structure and air expansions.
Φ is the medium porosity.

The elasticity coefficients P, N, Q and R are expressed by the Biot model and evaluated with three experiments (gedanken experiment) [3]. Since then the frame is composed of solid material that is slightly compressible four elasticity coefficients may be approximated according the following relations:

$$P=\frac{3}{4}N+K_b+\frac{(1-\Phi)^2}{\Phi}K_f$$

$$N=\frac{3K_b(1-2\nu)}{2(1+\nu)}$$

$$Q=(1-\Phi)K_f$$

$$R=\Phi K_f \tag{3}$$

K_b is the compressibility modulus of the solid structure in vacuum and can be evaluated by formula (4). Meanwhile K_f is the compressibility modulus of fluid inside the material pores and corresponds to the dynamic compressibility established for the equivalent fluid particularly for a rigid porous structure in the model of Allard [5].

$$K_b=\frac{E}{3(1-2\nu)} \tag{4}$$

$$K_f=K(\omega)=\frac{\gamma P_0}{\gamma-(\gamma-1)H(\omega)}$$

$$H(\omega)=\frac{1}{1+\frac{8\eta}{j\Lambda'^2\rho_0 P_r\omega}(1+\frac{j\Lambda'^2\rho_0 P_r\omega}{16\eta})^{1/2}} \tag{5}$$

E and v are respectively Young's modulus and Poisson's coefficient for the deformable solid matrix.

2.2. Acoustic signature V(z)

The micro characterization of a material is determined by its acoustic signature V(z) that is expressed by the following equation [6,7]:

$$V(z)=\int_0^{\theta_{max}}P^2(\theta)R(\theta)\exp(2jk_0\, z\, coc\theta)\sin\theta\cos\theta\, d\theta \tag{6}$$

Where: $j = \sqrt{-1}$.

θ_{max} is the maximal lens aperture.

$P^2(\theta)$ is the pupil function.

$R(\theta)$ is the function of the specimen reflection.

k_0 is the wave vector in the coupling liquid.

The acoustic microscope is used in a reflection mode and the piezoelectric transducer converts the output acoustic signal from the sample into a V(z) tension without losing the relative information for the studied material.

The measured amplitude of V(z) is a function of the distance z between the focal plan of the lens and the interface sample - liquid.

This model enables to calculate the speeds of the different modes from the period Δz as follows:

$$v_S = \frac{v_0}{\sqrt{1-\left(1-\frac{v_0}{2f\Delta z}\right)^2}} \quad (7)$$

3. Results and discussion

The digital processing is performed according to the model that has been achieved in the laboratory by means Matlab software package to determine the phase velocities V_{p1}, V_{p2} and V_s against the porosity after processing of the acoustic signature V(z) and the reflectance R (θ) where by the various mechanical parameters are calculated [8,9].

3.1. Coupling liquids influence on acoustic signatures V(z) for porous silicon

Fig.1 shows the acoustic signature V(z) of porous silicon (P = 20%) depending on three coupling liquids i.e. water, mercury and ethanol according to Schoch and Biot models as shown respectively in Fig.1a and Fig.1b.

For the modeling of acoustic signatures V(z) we use empirical formulas for the of liquids and materials density therefore, one is devoted for the material saturated with water while the other is for the material whose pores are empty.

$$\begin{aligned} \rho &= \rho_0(1-\varphi) \\ \rho &= \rho_0(1-\varphi)+\rho_{liq}\varphi \end{aligned} \quad (8)$$

We note that the effects of these three liquids are present. In this case we use ethanol and mercury because these two liquids have different properties notably viscosity and density.

These two plots also show that the period of oscillations ΔZ for ethanol is greater than that of mercury which implies that the speed of the surface wave (Rayleigh) for ethanol is greater than that obtained for mercury (relation between V_r and ΔZ) [10]. This difference between periods of oscillations for the former liquids appears in both models, on the other hand it is considerably marked in the model of Biot.

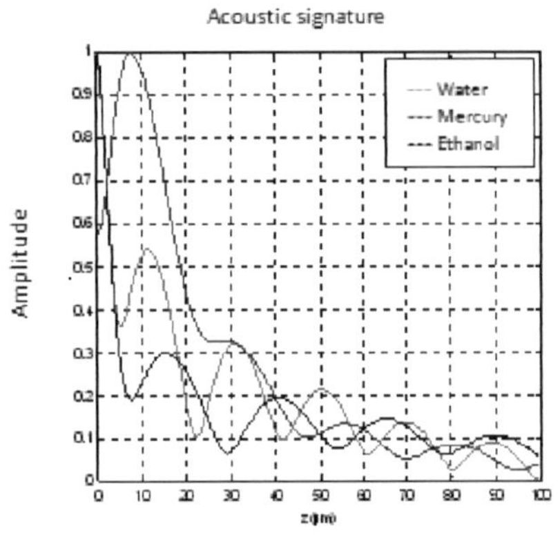

Figure 1 a. Acoustic signature of porous silicon (P=20%) saturated by different coupling liquids for Schoch's model

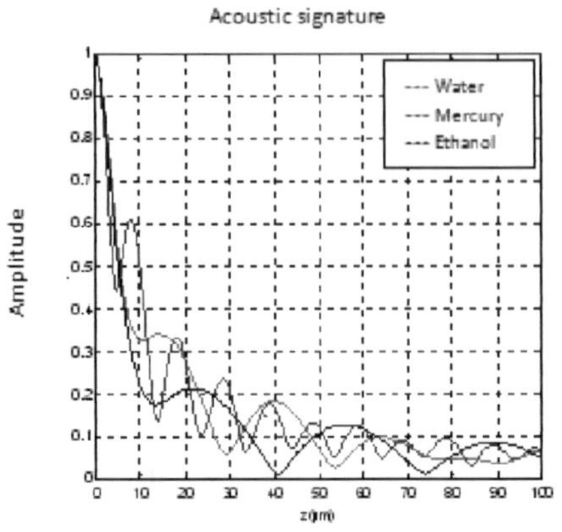

Figure 1 b. Acoustic signature of porous silicon (P=20%) saturated by different coupling liquids for Biot's model

3.2. Variation of reflection coefficient amplitude and phase against the incidence angle θ for different porosity ratios.

Fig.2 and Fig.3 reveal that the reflection coefficient values are different for the two models at various porosity ratios [11]. In Fig.3 the amplitude variation for the model of Biot reaches a maximum value for a porosity of 50% as the phase and amplitude variation are operated for an angle more than 50 ° that is defined as the acoustic lens aperture, moreover it is shown in Fig.4 that the phase variation obeys to the same rule as the amplitude change i.e. an increasing of critical angles as the porosity increases [12].

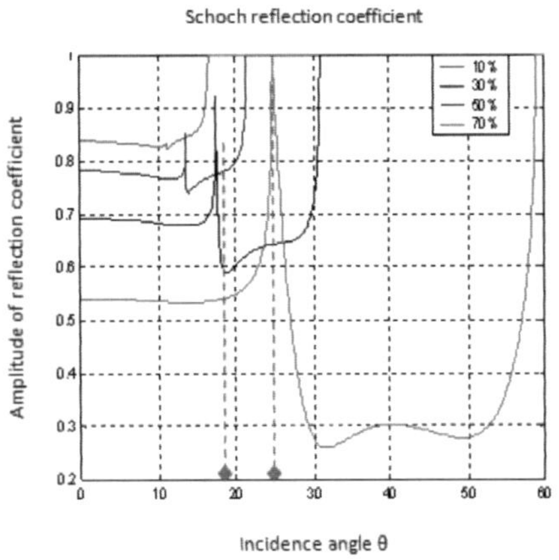

Figure 2. Amplitude variation for Schoch reflection coefficient vs. incidence angle θ

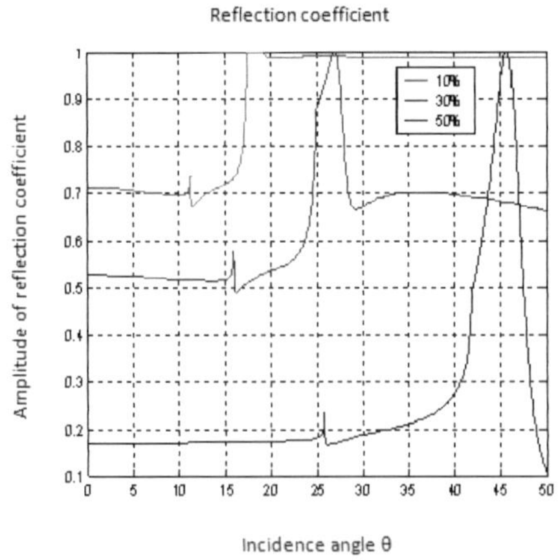

Figure 3. Amplitude variation for Biot reflection coefficient vs. incidence angle θ

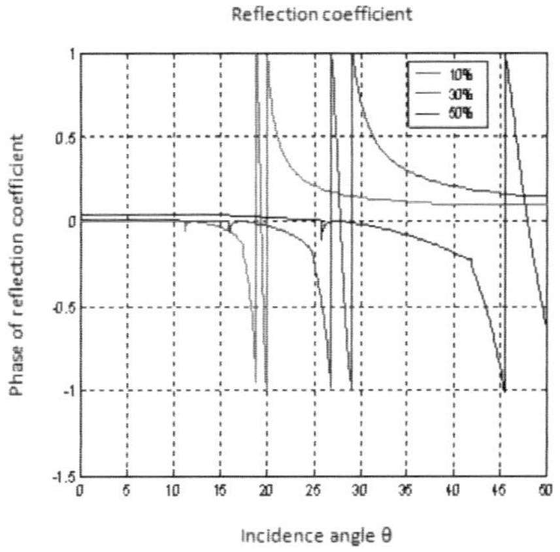

Figure 4. : Phase variation for reflection coefficient vs. incidence angle θ

3.3. Calculation of acoustic signature V(z) for different porosity ratios

The phase and amplitude variations for the coefficient R(θ) enable to calculate acoustic signatures for diverse porosities. The acoustic signatures V(z) are illustrated in figures (5,6) respectively for the models of Schoch and Biot.

Since the porosity increases in Fig.5 one notes an attenuation for acoustic signature while for the Biot's model in Fig.6 the greater is the porosity ratio, the broader is the signal attenuation and there is no signal. Obviously, the differences between the acoustic signatures for both models are larger, on the other hand several parameters such as the viscosity, the liquid and the nature of the pores are taken into account to for the model of Biot.

Therefore, the porosity ratio directly affects the acoustic signature V(z) for porous materials.

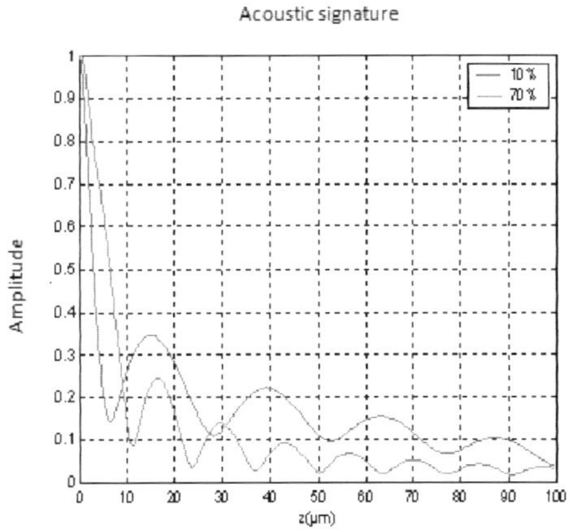

Figure 5. Acoustic signature variation V (z) of Schoch vs. the focal length z

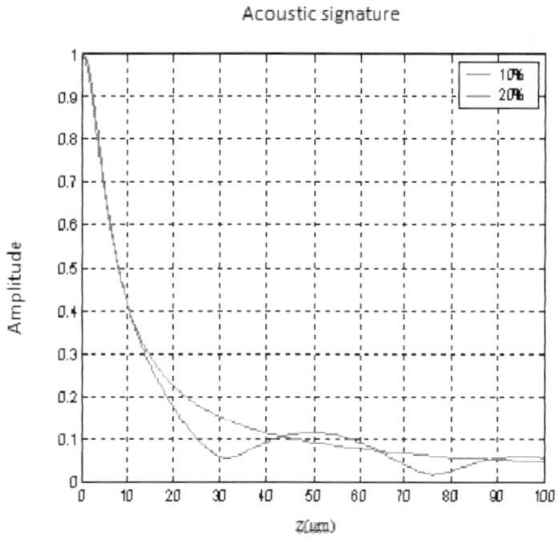

Figure 6. Acoustic signature variation V (z) of Biot vs. the focal length z

3.4. Determination of elastic parameters

The reflection coefficients at different porosity ratios lead to determine various elasticity parameters viz Young modulus, shear modulus and Poisson coefficient. The evolution of different parameters depending on the porosity according to the results in table 1 and table 2 for both models is shown in figures (7,8,9). It appears that the values obtained by the model of Schoch are larger in comparison to those obtained by the theory of Biot, this difference is higher when the porosity ratio increases and that finding is highlighted in Fig.9.

Table 1. Elasticity moduli estimation for various porosity ratio according to Schoch model

Porosity (%)	Young's modulus (E) GPa	Shear modulus (G) GPa	Poisson's coefficient (ν)
0	165.7	79.4	0.0428
10	130.5	60.7	0.0752
20	100	45.0	0.1104
30	74.1	32.2	0.1489
40	52.5	22.07	0.1911
50	35.1	14.2	0.2372
60	21.5	08.36	0.2879
70	11.5	04.3	0.3431

Table 2. Elasticity moduli estimation for various porosity ratio according to Biot model

Porosity (%)	Young's modulus (E) GPa	Shear modulus (G) GPa	Poisson's coefficient (ν)
0	210.6	76.7	0.371
10	128.38	55.7	0.151
20	84.9	37.9	0.118
30	55.1	24.8	0.110
40	34.06	15.3	0.112
50	19.5	08.7	0.123
60	10.1	04.4	0.149
70	04.45	01.82	0.217

It is noted in figures (7,8) a net decrease where the solid matrix trends to a porous matrix whose attenuation considerably increases, meanwhile fig.9 reveals that the coefficient of Poisson increases against the porosity due to the influence of the ratio that involves transverse and longitudinal velocities on the latter coefficient. On one hand the three graphs show differences between the parameters E, G and υ, on the other hand the curve evolutions behave in similar way for both models. Once again, less parameters affect the model of Schoch in comparison to those for model of Biot [12].

Therefore the porosity has great effects on the elastic properties of porous materials.

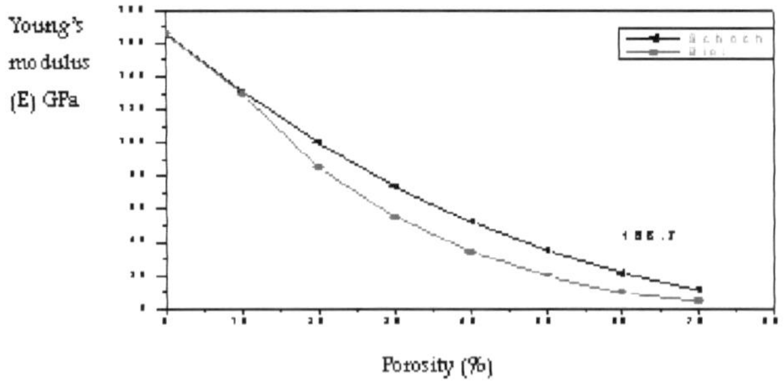

Figure 7. Evolution of Young's modulus vs. porosity

IOP Conf. Series: Materials Science and Engineering **28** (2012) 012009 doi:10.1088/1757-899X/28/1/012009

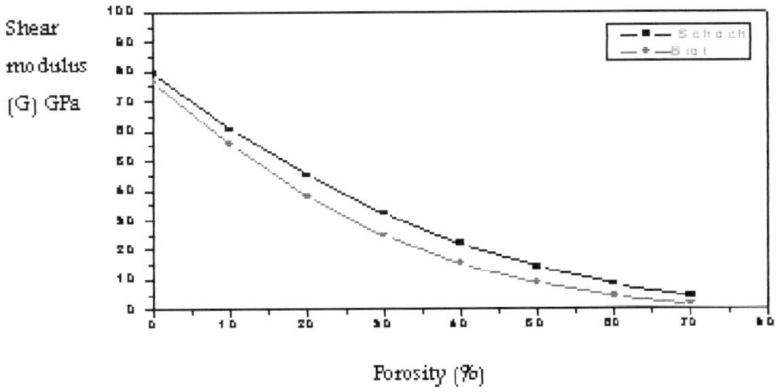

Figure 8. Evolution of shear modulus vs. porosity

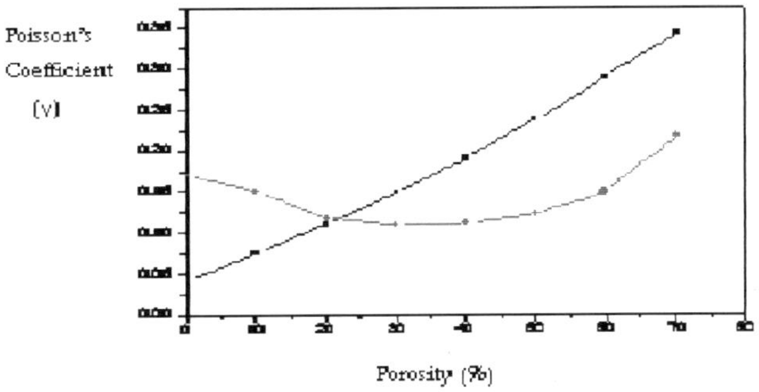

Figure 9. Evolution of Poisson's coefficient vs. porosity

4. Conclusion

The porosity is very important for some materials that are used in aerospace, microelectronics and acoustics. We obtained innovative results especially for the characterization of porous silicon, and we are able to make evident the relation of porosity to elastic properties and the speed of surface waves propagation, longitudinal waves and acoustic impedances.

The parameters used in the Biot theory can be taken into account for further investigations even though we highlighted Biot slow wave (second case).

References

[1] Da Fonseca R. J. M, Saurel J. M, Foucran A, Camassel J, Massone E, Taliercio T, Boumaiza Y 1995 *Journal of Materials Science* **30** 35-39.

[2] Quate C. F, Lemons R. A 1974 *Appl. Phy. Lett.* **24** 193.

[3] Biot M. A 1956 *J. Acoust. Soc. Am.* **28** 168-178.

[4] Biot M. A 1962 *J. Acoust. Soc. Am.* **34** 1254-1264.

[5] C. DEPOLLIER, J.F. ALLARD, W. LAURIKS 1988 *J. Acoust. Soc. Am.* **84** 2277-2279.

[6] Brekhovskikh L. M, O.A. Godin O. A 1990 Acoustic of layered media I: plane and quasi-plane waves, Springer-Verlag, Berlin.

[7] Cheeke J. D. N 2002 Fundamentals and Applications of Ultrasonic Waves, *CRC Series in Pure and Applied Physics*.

[8] Benbelghit A, Boutassouna D, Helifa B, Lefkaier I. K 2006 *NDT&E International* **39** 76-81.
[9] Briggs G. A. D, Kolosov O. V 2010 Acoustic Microscopy, Oxford University Press Inc..
[10] Da Fonseca R. J. M, Ferdj-Allah L, Despaux G, Boudour A, Robert L, Attal J 1993 *Advanced Materials* **5** (7/8) 508-519.
[11] Schoch A 1952 *Acoustic* **2** 1-17.
[12] Allard J. F, Depollier C, Nicolas J, Lauriks W and Cops A 1990 *J. Acoustique* **3** 29-38.

IOP Conf. Series: Materials Science and Engineering **28** (2012) 012010 doi:10.1088/1757-899X/28/1/012010

Dispersion of the carbon nanotubes (DWNTC) by the cationic surfactants imidazolinium type

A. Zafour[a], D. Abdelkader[a1], E. Flahaut[b], N. Rebizi[a] and A. Skender[a]
[a]Laboratory of Chemical Engineering, University Saad Dahlab of Blida, Road of Soumaa, P.B 270, Algeria
[b]CIRIMAT, University Paul Sabatier, Toulouse, France

E-mail: djalabk2003@yahoo.fr

Abstract. Carbon nanotubes (CNTs) discovered over the past fifteen years are very interesting materials because of their structural, mechanical, chemical and electronic properties. CNTs have arisen considerable attention owing to their attracting potentialities, and the great interest for their applications in wide and varied fields.
CNTs are in the form of soot sparse, brittle and isotropic, they form bundle like structures with very complex morphologies with a high Vander Waals interactions, causing a poor solubility in aqueous or organic medium. The first challenge to use them effectively is to separate their initial aggregated assemblies in homogeneous suspension. Therefore, significant research efforts have already been devoted now to optimize and develop new processes, but the simple route to achieve this task based in particular the use of surfactants.Moreover, to cross the technological barrier of dispersion, the different techniques are often associated (hybrid processes).
In this study, we propose new approach for the stabilization of dispersion based on the use of cationic surfactants (Imidazoliniumtype) synthesized from raw materials available in Algeria. Two surfactants used in this study were Imidazolinium Iodide and Imidazolinium Bromide. The results shows a homogeneous dispersions of CNTS with sonication with Imidazolinium Iodide (10 CMC) for two months.

1. Introduction

In recent years, the discovery of carbon Nanotubes (CNTs) [1], revealed a new class of materials with particular and extraordinary properties. For this purpose, they find wide applications in several sectors and attractive for various applications. After their synthesis, the nanotubes of carbon are presented in the form of not very dense and isotropic soot. Hence, the exact properties of CNTs are extremely sensitive to their degree of graphitization, diameter (or chirality), and whether they are in single wall or multi wall form. Furthermore, CNTs show great potential in biomedical, cosmetic and drug delivery. Although, they have a high Vander Waals interaction energy of tube–tube contact, which causes a strong bundling. This high interaction force makes the dispersion stability a challenging task [2]. There has been a large amount of literature published on the solubilization of carbon nanotubes through oxidation and/or surface modification. Thus, in view of the preceding, there has been an immense effort to establish the most suitable conditions to break down the agglomerates and obtain a

[1] To whom any correspondence should be addressed.

better dispersion. Obviously, a surfactant's property of accumulation at surfaces or interfaces has been widely utilized to promote stable dispersions of solids in water or organic medium. A wide variety of surfactants have been used to date for dispersion of CNTs, such as anionic and no ionic ones. However, to obtain stable CNTs dispersions, one usually needs high concentration of surfactants [3].

The pioneering works in order to estimate the effective use of CNTs is to have uniform dispersion more adapted to the suitable applications. This is why a number of different strategies have been reported in the literature, and many laboratories provide at present time main efforts to optimize and to develop a new processes in this field. However, the future projection of technology for many industrial applications based on the use of the CNTs, will depends on the knowledge of the unique structural, electrical, chemical and the mechanical properties of CNTs [4-7].

To cross the technological bolt of the dispersion of the CNTs at the present time, several techniques of dispersion were been used, among them, the process of the chemical functionalization of CNTs, by grafting a chemical compounds on their surfaces in the way to have an homogenous dispersion in booth organic or aqueous medium. [8-10]

In this present study, we proposed a new concept of dispersion of double walled carbon nanotubes (DWCNTs), based on the direct contact between a DWCNTs and a dispersant molecule using two cationic surfactants of the imidazolinium type, synthesized and characterized by spectroscopic and physicochemical methods. The surfactant is with a high degree of purity, and the fatty-acid used in this experiments, is from the Algerian raw material.

2. Experimental

2.1- Synthesis of the surfactant imidazoliniums type

As the availability of the fatty-acid locally and one of the cationic surfactant imidazolinium type are containing this matter, we were interested in the synthesis of this type of surfactant. The imidazolinium Iodide and the Bromide were prepared and compared to the commercial cationic surfactant CTAB (Cetyl Trimethyl Ammonium Bromide)

The structure of the surfactant is shown in Fig.1.

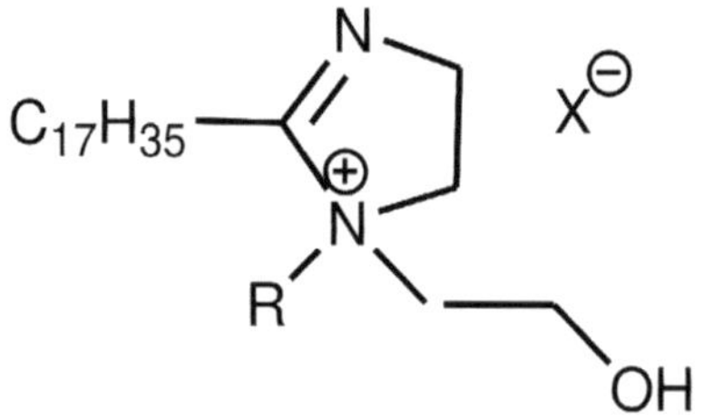

Figure 1. Chemical structure of the surfactant.

2.2. Characterization of the products of synthesis

The synthetized products were characterized by the determination of some physicochemical parameters by plotting the curves of the conductivity of the different cationic surfactants. Afterward, we deduced their critical micellar concentration (CMC).

The values of CMC were given as follows.

- CTAB: CMC =0, 84.10^{-3} mol/l
- Imidazolinium Iodide: CMC = 0, 8.10^{-3} mol/l
- Imidazolinium Bromide: CMC = 0, 78. 10^{-3} mol/l

With the FTIR spectroscopy, we could highlight the presence of functional groupings which characterize this type of products, herein on Figure. 2.

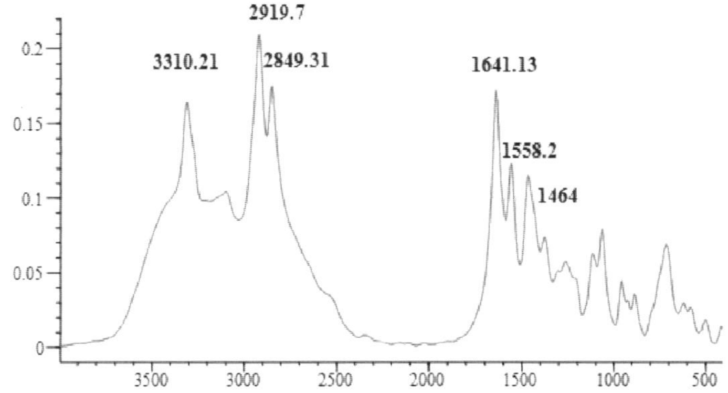

Figure 2. FTIR Spectrum of 1-methyl-1-(2-hydroxyethyl)-2-stearyl-imidazolinium

Imidazoline and imidazolinium were characterized by FTIR. Our interpretation was confirmed by UV. The instrumental analysis of imidazoline and imidazolinium confirmed the chemical structure proposed of imidazoline and cationic surfactant. The formation of imidazoline was confirmed by results FTIR indicating CH_3, CH - strong to 2919, 7 cm^{-1}; (CH_2) N skeletal to 714.27 cm^{-1}; N-H which extends to 3310, 21 cm^{-1}; A band at 1641 (cm^{-1}) corresponds to the N-C=N. Pour grouping the iodide imidazolinium one has the following results CH_3, CH - strong to 2918,73 cm^{-1}; (CH_2) N skeletal at 582.361cm^{-1}; N-H which extends to 3448,1 cm^{-1}; a band with 1637,59 (cm^{-1}) corresponds to grouping N-C=N and for the imidazolinium bromide there are the following results CH_3, CH - strong. to 2918, 73 cm-1; (CH_2) N skeletal to 583.89 cm^{-1}; N-H which extends to 3434, 6 cm^{-1}; a band with 1642, 09 (cm^{-1}) corresponds to grouping N-C=N

2.3. Dispersion of the carbon nanotubes

In our experiments, the suspensions of DWCNTs were prepared by adding 2mg of DWCNTs in 10ml of bidistillated containing a concentration of 3CMC of surfactants. Then the solutions were sonicated for 1 hour, which is the useful technic used for dispersion.

This mechanical process is the use of the vibration mechanism to break the tangles of CNTs. After the use under the following conditions: concentration of CNTs: 0.002g, and concentration of the surface-active one: ½ CMC, CMC, 2CMC, after agitation in the tank during 15min, a bad dispersion is obtained.

In this case, one increases the operating conditions, and the experimental protocol proceeds as follows: The concentration in cationic surfactant of imidazolinium used for dispersion is always largely higher than its critical micellar concentration, that is to say (3 CMC, 6 CMC, and 20 CMC) and 0.002g carbon nanotubes is taken. Dispersion is carried out by sonication in a bath with ultrasounds (tank ultrasound) of low power (60 W) and of frequency equal to 45 KHz during 1hour; and finally dispersions obtained are stable during several days

The experiments conditions are summarized as follows:

- 2mg DWCTNs + 10 ml of Imi-Iodide (3CMC)
- 2mg DWCTNs + 10 ml of Imi-Iodide (6CMC)
- 2mg DWCTNs +10 ml of Imi-Iodide (20CMC)
- 2mg DWCTNs + 10 ml of Imi-Bromide (3CMC)
- 2mg DWCTNs + 10 ml of Imi-Bromide (6 CMC)
- 2mg DWCTNs + 10 ml Imi-Bromide (20 CMC)

All the samples are dispersed and homogenized under the conditions described previously (1 hour with the tank of ultrasounds 60 W). Photos taken for T = 0; 0.5; 1; 2; 3; 2days; 3days; 10days; 20days. (Fig.2) and (Fig.3).

The dispersion by the Imidazolinium Iodide in this case is done by different concentration (20CMC, 6CMC, 3CMC) according to time and the characterization of the prepared solutions is done by visual analysis of the preparations according to the concentrations of surfactants and time. The results of dispersing samples are represented in photos herein:

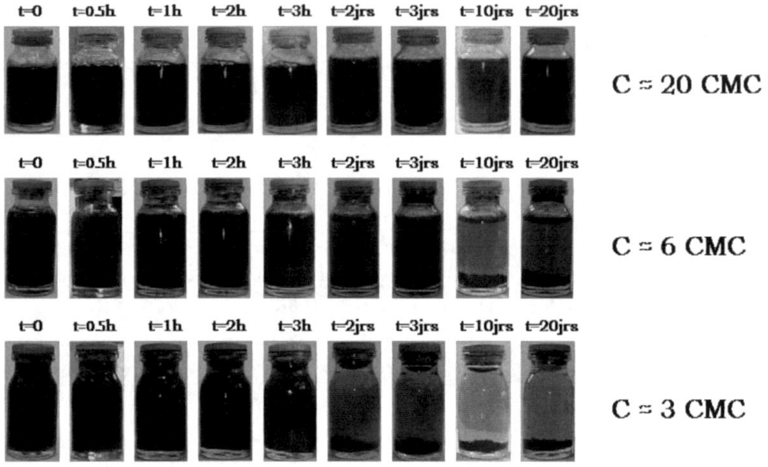

Figure 3. Illustrations of the sedimentation of the dispersion of DWCTNs in imidazolinium iodide according to time.

Same the proceeded operational one was followed with the second cationic surfactant Imidazolinium Bromide**:**

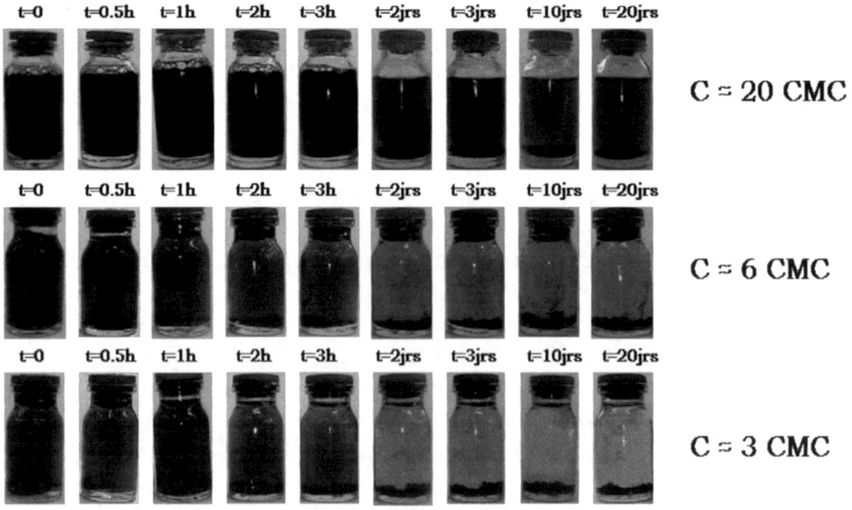

Figure 4. Illustrations of the sedimentation of the dispersion of DWCTNs in imidazolinium bromide according to time.

Immediately after homogenization, the six samples seem identical (dispersion black and homogeneous). However, sedimentation occurs with time. After three weeks with the abandonment, without any disturbance, the level of sedimentation is definitely different as attests the two figures of them.

The visual observation already makes it possible to highlight various things. We observe in particular that cationic surfactant of the imidazolinium type have different behaviors, and also the two surface-active ones compared do not have the same effectiveness.
Generally visually homogeneous and stable dispersions are obtained. One can say that in the case of the dispersion of CTNs by the cationic surfactant of type imidazolinium thus improves much the homogeneity and the stability of aqueous dispersion of CTNs.

3. Comparison with cetyl trimethyl ammonium bromide (CTAB):

We have to test the dispersion of the carbon nanotubes with surface-active the commercial CTAB to compare it with cationic surfactant to imidazolinium. After 30 days, one takes the samples, to 20 CMC, and 6CMC:

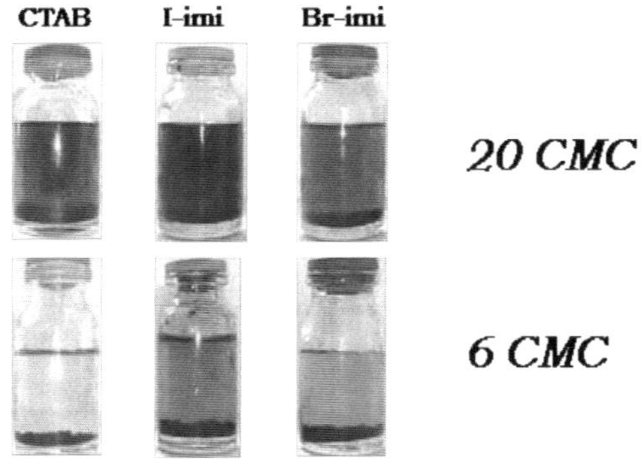

Figure 5. Comparison of the sedimentation of dispersions after 30 days in different surfactant (CTAB, imidazolinium iodide, imidazolinium bromide).

The visual observation shows that dispersion with surface-active imidazolinium and particularly the Iodide Imidazolinium gives a good and durable stability compared those with CTAB.

4. Conclusion

In summary, during this study we have demonstrated in these promising results an opening way for several prospects in the fields of CTNs for the improvement of the multipurpose properties for the carbon nanotubes.

Within the framework of this study, we were interested in the use of the Algerian raw material (fatty-acid) to produce cationic surfactant of the Imidazolinium type.

By the cumulative effect of adsorption of surfactants and a long sonication, the DWCNTs are dispersed in water. At the time of a comparative study, it was shown that the cationic surfactant of the Imidazolinium type insure to obtain good performances, i.e. stable aqueous suspensions of DWCNTs. The procedure of dispersion is also very important to manage to solubilize the nanotubes and to disaggregate them. It is necessary to subject the solutions of nanotubes and surfactants to a prolonged sonication enhance separations of the aggregates. However, it is preferable that the power of the ultrasounds is not too important because with the strong powers, the carbon nanotubes break and are shortened by the vibrations effect. On the other hand, for low powers, the duration of sonication must be higher than 1hour to obtain a stable dispersion.

References:

[1] Iijima S 1991 *Helical microtubules of graphitic carbon. Nature* **354** 6348 56 58
[2] Girifalco L. A, Hodak M, Lee R.S 2000 Carbon nanotubes, buckyballs, ropes, and a universal graphitic potential, *Phys. Rev B.* **62** 13104
[3] Madnia I, Chi-Yong Hwangb, Seong-Dae Parkc, Yong-Ho Choaa B, Hee-Taik Kima B 2010 *Mixed surfactant system for stable suspension of multiwall carbon nanotubes, Colloids and*

Surfaces A: Physicochemical. Eng. Aspects **358** 101 107

[4] Rosen J. Mr 1978 *Surfactant and interfacial phenomena. In Wiley interscience publication. New York*

[5] Bognolo G 1997 *Nonionic Surfactant. In Lipid Technologies and Applications*

[6] Monthioux Mr, Kuznetsov V. L 2006 *Who should be given the credit for the discovery off carbon nanotubes? Carbon* 44 9 1621 1623

[7] Heller D.A, Mayrhofer R.M, Baik S, Grinkova Y. V, Usrey Mr. L, Strano Mr. S 2004 *Concomitant length and diameter separation off individual-walled carbon nanotubes. Newspaper off the American Chemical Society* **126** (44) 14567-14573.

[8] Huang S, Cai X, Liu J 2003 *Growth off Millimeter-Length and Horizontally Individual-Walled Aligned Carbon Nanotubes one Flat Substrates. Newspaper off the American Chemical Society* **125** (19) 5636 5637.

Corrosion evolution of the aluminum alloys used in overhead transmission lines

E. Rhaiem[a 1], **T. Bouraoui**[a, b], **F. El Halouani**[a]
[a]Laboratoire des Systèmes ÉlectroMécaniques, LASEM–ENIS, Université de Sfax, B.P.W 3038, Sfax, Tunisie.
[b]Ecole Nationale d'Ingénieurs de Monastir, ENIM, Rte Ibn El Jazzar, 5019 Monastir, Tunisie

E-mail: recheldi@yahoo.fr

Abstract. The present paper is concerned with the corroded cables used few years ago. Microstructural evolution of the investigated cables is carried out by scanning electron and optical microscopies (SEM and OM). The results obtained from the comparison between the original (as-received) and used cables show the degradation of the surface and the presence of polluting elements. The evolution of mechanical properties is studied by the microhardness technique. The results reveal a deterioration of the mechanical properties of the used cables if compared with those obtained from the original cables. The main cause of the corrosion of the conductor was atmospheric corrosion elements in the case of high aggressive species such as Cl^- and SO_2.

1. Introduction

The high technological value and wide range of industrial applications, especially in aerospace and household industries, make of Aluminium and its alloys an important category of materials, whose use in light installations is widespread. Indeed, aluminium is very well-known by its low price, high electrical capacity and high energy density [1]. The most important feature of aluminium is its corrosion resistance owing to the presence of a thin adherent and protective surface oxide film. Aluminium and its alloys, however, are reactive materials and are prone to corrosion [2]. That is why their corrosion has been the subject of numerous studies due to their importance in the recent civilization. Many researchers have devoted their work to the study of the corrosion of aluminium in different aqueous solutions [2, 3], as well as to the research into their electrochemical behaviour and corrosion inhibition in a wide variety of conducting solutions [4].

Fortunately, they can develop an insulating film in the ambient atmosphere as a result of their natural tendency for direct oxidation with atmospheric oxygen under normal conditions [5]. The stability of the surface oxide formed on these materials and its protective nature represent a major field of research [6-9]. It was reported that the barrier film formed on Al and Al-alloys is of a dual Nature. It consists of an adherent, compact and stable inner layer of oxide film covered with a porous, less stable outer layer which is more susceptible to corrosion [7-12]. The surface oxide film on these materials is stable, in the pH range 4-9. Aggressive Anions, like Cl- lead to increased corrosion rates

[1] To whom any correspondence should be addressed.

and film breakdown [13-15]. The effect of Cl^- on the corrosion rate of Al and its alloys in the passive pH range has been investigated intensively [13-19]. In these solutions, pitting corrosion takes place [19, 20].

Aluminium and its alloys are widely-used metals in high power electrical transmission lines and distribution nets thanks to their low density, which brings about a real advantage compared with copper [21, 22]. As a consequence, aluminium requires twice less mass than copper for an equivalent linear electrical resistance and therefore becomes lighter and less expensive for a given current-carrying capability [21-24]. Several variations of aluminium conductors are available [25]. Among the mostly used, we can cite the All-Aluminium Alloy Conductor (AAAC) and the Aluminium Conductor Alloy Reinforced (ACAR). In the AAAC, the conductor is made from aluminium-magnesium-silicon alloy of high electrical conductivity with an optimum composition for optimum mechanical properties after treatment. These conductors are generally made out of aluminium alloy 6201. In ACAR type, aluminium strands are wound-about cores of stranded steel for a higher mechanical strength.

As far as corrosion is concerned, the AAAC conductor can be said to have better corrosion resistance compared with ACAR conductor, especially in coastal and chemically-polluted industrial areas [26, 27]. Some experimental work was carried out in order to evaluate the joint effect of marine and industrial pollutants on the atmospheric corrosion of aluminium alloys conductors. Vera et al. [28] have shown a good correlation between weight loss, attack depth and tensile strength to rupture, and time and pollutants contents of aluminium alloy used in high power electrical conductors.

It is in this context that the present work lies to present the corrosion detection of 6201 aluminium alloy used in overhead transmission lines, aiming at the identification of the causes of the observed corrosion and thus contributing to a better knowledge of the operational limits of aluminium alloys in power transmission lines. This research work undertakes the study of two different states of as-received cable and used cable. In order to successfully accomplish the diagnostic of the corrosion, microscopic and spectroscopic analyses as well as mechanical characterization were carried out on the used and original cables. It also proposes anti-corrosive solutions for an optimal use.

2. Experimental procedures

The studied material is an aluminium AA6201 alloy. It is moderate and especially developed for the manufacture of the electrical conductor. The characteristics of the cable are given in Table 1.

Table 1. Cable characteristics .

Cable section (mm^2)	Strands Number	Strand diametr (mm)	Externel diametr (mm)
148 ,01	19	3,15	15,75

Macroscopic and microscopic observations were carried out using a Philips XL30 scanning electron microscope (SEM) equipped with an electron dispersive spectrometry X-ray microanalysis system (EDX) for the elemental analysis of the selected areas. Philips XL30 SEM was also used to perform observations on fracture surfaces of tensile samples. The tensile and bending tests are carried out on a Lloyd machine with a displacement speed of 10 mm/mn. The tensile and bending tests are carried out on a Lloyd machine with a displacement of 10mm/mn. Vickers microhardness (Hv) was measured on samples using VH-6 hardness, at a load of 500g used for 15 Seconds. The given values of hardness are an average of 6 tests.

3. Results and discussion

3.1. Macrographic analysis

The macrographic analyses of the as-received and used cables are presented in Figure 1. It is worthy to note that the corroded microstructure of the cable is due to the corrosion phenomenon which could well be atmospheric [12]. Although the as-received cable with the delivery condition presents one clear and clean surface (Fig.1a), the used cable presents traces and cavities of corrosion (Fig.1b).

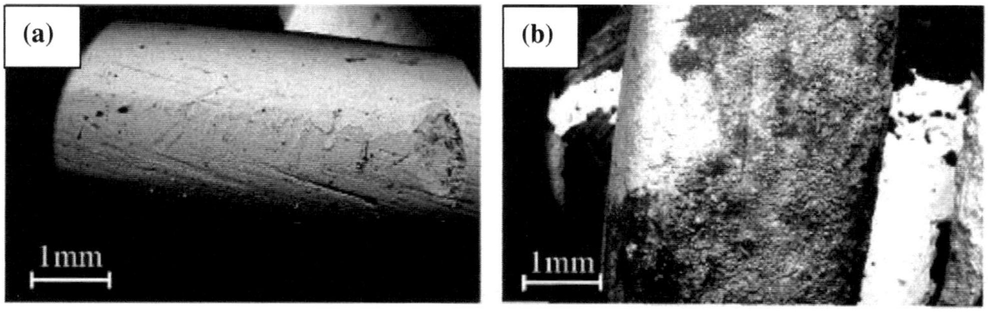

Figure 1. Macroscopic observations of cable strand: (a) as received cable and (b) used cable

The increase in the size of the cable is due to the accumulation of the elements pollutants to form a surface layer after the degradation of the natural protective coating of aluminium oxide. This surface layer formed after polluting elements (see analyses) can reach 40 μm in thickness, with the existence of swollen sectors in which there is one corrosion differential per crack.

3.2. Micrographic study with the SEM and analysis spectroscopic

The original and used cables were analyzed with the SEM and the obtained results are presented on Figures 2 and 3. The comparison of the observations of the cross sections clearly shows the evolution of the corroded zone (> 100 μm) until the appearance. The chemical analyses of the various zones from the investigated sample to the surface show the reduction in aluminium with the percentage mass increase of the corrosive elements (Cl^-, SO_2, K, O...).

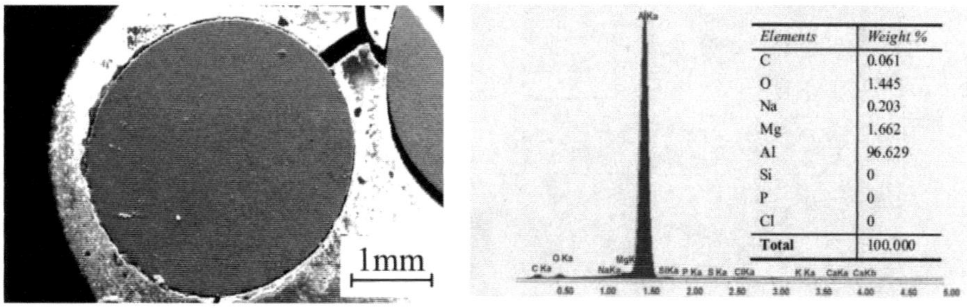

Elements	*Weight %*
C	0.061
O	1.445
Na	0.203
Mg	1.662
Al	96.629
Si	0
P	0
Cl	0
Total	100.000

Figure 2. SEM-micrograph and corresponds EDX analyses of as-received cable strand.

Moreover, this change of surface composition is strongly related to the time of exposure of the cables. Indeed, sulphur and chlorine do not exist in the surface of the original cable, whereas they increase with the time of use. The chloride ions, in great quantity, have a paramount role in the degradation of the surface in atmospheric exposure. The presence of these ions causes the rupture of the protection oxide film [29]. It is recognized that the chlorides in the atmosphere constitute a considerable factor leading to the corrosion of aluminium and especially on coastal areas [30].

Figure 4 shows micrographic observation of the cable in longitudinal and cross section of the original and the used cable. There is a variation of the grain size and the microstructure between the observations, in fact the longitudinal observation show elongation of grains (beads) with time of use due to the work hardening by stretching.

IOP Conf. Series: Materials Science and Engineering **28** (2012) 012011 doi:10.1088/1757-899X/28/1/012011

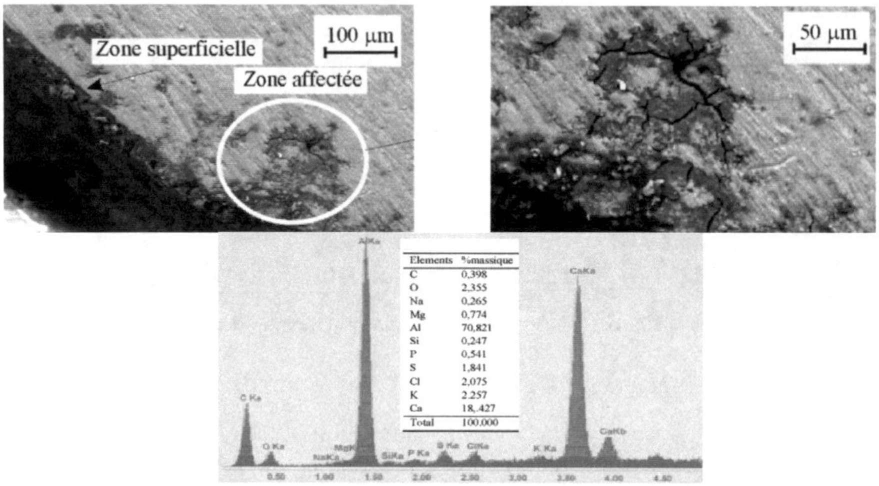

Figure 3. SEM images and EDX analyses of cross section of the affected zone of the cable strand after

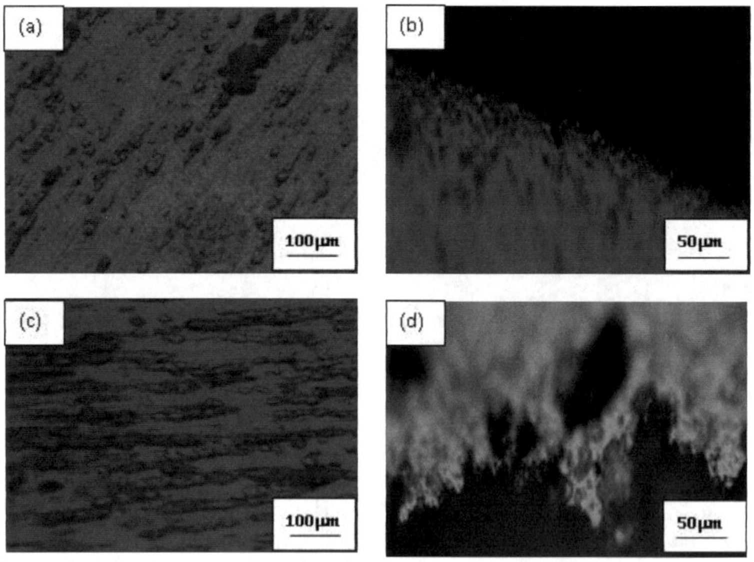

Figure 4. SEM images of (a) longitudinal section and (b) cross section of the original cable (c) Longitudinal section and (d) cross section of the used cable.

3.3. Mechanical proprieties

In order to characterize the mechanical behavior of the conductors, Vickers hardness tests whose results are summarized in Table 2, are carried out. The measured values show the increase in the hardness of the cable in surface. This increase of hardness is due to the retirement of the cable in the time of use.

Table 2. Micro hardness of as-received and used cables.

	As-received	used
Microhardness (HV)	82.4	86,4

On the other hand, the microhardness evolution of the two cables of the middle towards the outside is conducted and its results are presented in Fig. 5. What should be noted is that there is an increase in the hardness of both original and used cables, but this increase is more important for the used cable owing to the variation of microstructure and decrease of the plasticity. The results show the weakening evolution of the surface because of its corrosion.

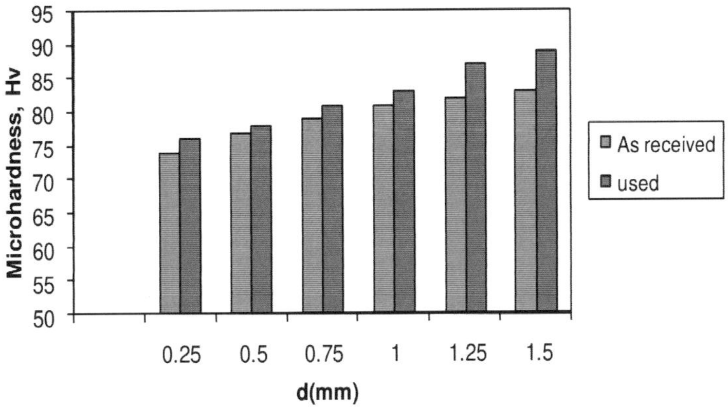

Figure 5. The measured values of Vickers hardness from the investigated sample to the surface.

Fig. 6 displays the stress-strain curves of the as-received and used cables. At these tempers, the obtained total elongations are approximately 9% and 6% for the as received and the used cables, respectively. The obtained yield strengths are 160 MPa for the as-received cable and 190 MPa for the used cable.

The SEM-micrographs of the fracture surfaces corresponding to the tested samples A, B and C are also given in Figure 6. The ductile fracture as characterized by the dimpled features can clearly be seen. It starts on the particles (precipitates, inclusions) present in the alloy. These results confirm those obtained for hardness measurements. So, the elongation at fracture decreases considerably after use and does not exceed 6 % per 9 % with the state of reception. In addition, for the used cable, we note a degradation of the plasticity and the ductility.

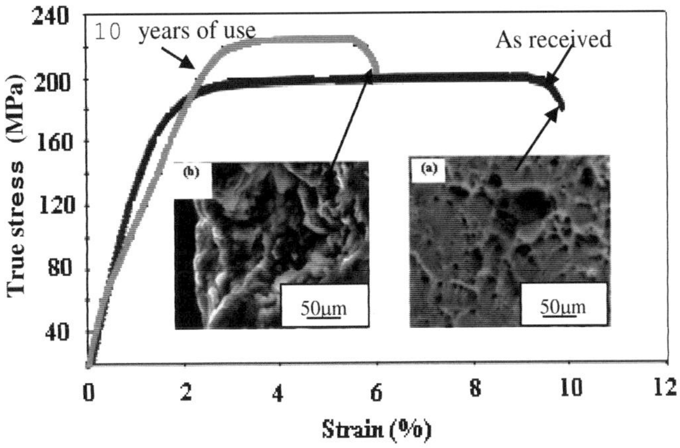

Figure 6. Stress-strains curves and SEM-images of a typical fractographies after the tensile tests of the sample: (a): As received, (b): used cable.

4. Conclusion

In the present work, the evolution of atmospheric corrosion on aluminium alloys used in electrical conductors, as-received and used cables, has been investigated by different methods. In fact, it has been proven that there are significant changes in macrostructure and mechanical behaviour after the use of the cable. Moreover, the microstructure of the used cables was characterized by change from ductility to fragility. Besides, the increase of corrosive elements which have a direct effect on mechanical properties has confirmed the correlation between the corrosion phenomenon and time of use. The main cause of the corrosion of the conductor was atmospheric corrosion elements in the case of high aggressive species such as Cl^- and SO_2. These results show a good correlation between microstructure and mechanical properties in the purpose of identifying the corrosion of the conductor. Actually, the microstructure properties give a significant idea of the decrease in tensile and resistance metal. So, the natural oxide film formed on aluminium does not resist aggressive factors in atmospheric corrosion.

References

[1] Addel-Gaber A. M, Abd-El-Nabey B. A, Sidahmed I. M, El-Zayady A. M, Saadawy 2006 *M. Mater. Chem. Phys.* **98** 291 297.
[2] Mountarlier V, Gigandet M. P, Normand B, Pagetti J 2005 *J. Corros. Sci.* **47** 937.
[3] Maayta A. K, Al-Rawashdeh N. A. F 2004 *J. Corros. Sci*. 2004 **46** 1129
[4] Abd El Rahim S. S, Hassan H. H, Amin M. A, *J. Mater. Chem. Phys.* 2002 **78** 337
[5] Young L 1961 *Anodic Oxide Films. Academic Press, New York* **4** 9
[6] Despic A, Parkhutik V 1989 *in: J.O'M. Bockris, R.E. White, B.E. Conway (Eds). Modern Aspects of Electrochemistry.* **20**, *Chap, 6, Plenum, NewYork.*
[7] Diggle J. W, Downie T. C, Goulding C 1970 Electrochim W, *Acta* **15** 1079.
[8] badawy W. A, El-Basiouny M. S., Ibrahim M. M 1986 *Indian J. Technol.* **24** 1.
[9] Al-Kharafi F. M, Badawy W. A 1996 *Ind. J. Chem. Technol.* **3** 212.
[10] Bret C. M. A, Gomes I. A. R, Martins J. P. S 1994 *Corros, Sci.* **36** 915.
[11] Badawy W. A, Al-Kharafi F. M 1990 *Bull. Electrochem.* **20** 1000
[12] Al-kharafi F. M, Badawy W. A 1995 *Electrochem. Acta* **40** 1811
[13] Mazhar A. A, Badawy W. A, Abou-Romia M. M 1986 *Surface and coatings Technol.* **29** 335.
[14] Cabot P. L, Centellas F. A, Garrido J A, Perez E, Vidal H 1991 *Electrochim, Acta* **36** 179.
[15] Carrol W. M, Breslin C. B 1991 *Br. Corros. J.* **26** 255.
[16] Brett C. M 1990 *J. Appl. Electrochem.* **20** 1000.
[17] Feres S. E, Stefenel M. M, Mayer C, Chierche T 1990 *J. Appl. Electrochem.* **20** 996.
[18] Srinivasan H. S, Mital C. K 1994 *Electrochim.acta* **39** 2633.
[19] Baumgaertner M, Kaesche H 1990 *Corros. Sci.* **31** 231.
[20] Wiersma B. J, Herbert K. R 1991 *J. Electrochem. Soc.* **138** 48.
[21] Davis J. R 1999 *(Ed.), Corrosion of aluminum and aluminum alloys, ASM International,* **25** 49.
[22] Hollingsworth E. H, Hun sicker H. Y 1987 Metals *handbook, 9th Edition, ASM International* **13** 583 609.
[23] Hatch J 1984 *(Ed.) Aluminum properties and physical metallurgy, Metals Park, ASM* **58** 319.
[24] Krishnamurthy S. R, Selvan P 1995 Use *of AAAC in a distribution network-a strategy for energy and cost reduction, Power Engineering Journal* **9** (3) 133 136.
[25] Short T.A 2003 *Electric power distribution handbook, CRC PRESS.*
[26] Elshawesh F, El Agaili M, Elwaer A 1997 *Atmospheric corrosion of aluminum conductor. Br. Corros. J.* **32** (1) 77 80.
[27] Elola A. S, Otero T. F, Porro A 1992 *Evolution of the pitting of aluminum exposed to the atmosphere, Corrosion;* **48**(1) 854 863.
[28] Vera R, Delgado D, Resales B 2007 *Effect of atmospheric pollutants on the corrosion of high power electrical conductors, Corrosion science* **49** 2329-2350.
[29] Foley R. T 1986 *Corrosion* **42** 277.
[30] Berukshtis G. K, Klark G. B 1996 Corrosion of metals and alloys, in N. D. Tomashov and E. N. Mirolyubev (Eds) 281-297.

IOP Conf. Series: Materials Science and Engineering **28** (2012) 012012 doi:10.1088/1757-899X/28/1/012012

Sound velocity variation as function of polarization state in Lead Zirconate Titanate (PZT) Ceramics

W. Essolaani[a] and N. Farhat[1 b]
[a]Unité de physique des solides, Faculté des Sciences de Monastir - Tunisia.
[b]Ecole supérieure des sciences et technologie de Hammam Sousse – Tunisia.

E-mail: najeh_farhat@yahoo.fr

Abstract. There are several ultrasonic techniques to measure the sound velocity, for example, the pulse-echo method. In such method, the size of transducer used to measure the sound velocity must be in the same order of the sample size. If not, the incompatibility of sizes becomes an error source of the sound velocity measurement. In this work, the Laser Induced Pressure Pulse (LIPP) method is used as ultrasonic method. This method has been very useful for studying the spatial distribution of charges and polarization in dielectrics. We take advantage of the fact that the method allows the sound velocity measurement, to study its variation as function of polarization state in (PZT) ceramics. In a sample with a known thickness e, the sound velocity v is deduced from the measurement of the transit time T.
The sound velocity depends on the elastic constants which in turn they depend on poling conditions. Thus, the variation of the sound velocity is related to the direction and the amplitude of the polarization.

1. Introduction

The knowledge of the materials behavior under poling and depolarizing conditions is essential for materials performance optimization. In this paper, we use the Laser Induced Pressure Pulse (LIPP) method to evaluate the sound velocity variation of a number of lead titanate zirconate (PZT) ferroelectrics submitted to a poling or depolarizing field. It is a nondestructive technique, where a short pulsed laser impact on a target generates a short pressure pulse. That pressure pulse is then transmitted to the material to be studied. Its propagation through the sample bulk moves locally and successively the charges and dipoles encountered. In turn that generates an electric signal between the electrodes which can be measured by convenient circuit.

This method has been very useful for studying the spatial distribution of charges and polarization in dielectrics [1, 2]. We take advantage of the fact that the method allows the transit time measurement, to study its variation. The transit time variation is related to the sound velocity modification. On the other hand, since the sound velocity depends on the elastic constants which in turn they depend on poling conditions, the variation, increase or decrease, of the sound velocity is related to the direction and the amplitude of the polarization.

[1] To whom any correspondence should be addressed.

The used ultrasonic technique shows an increase of the sound velocity when the piezoelectric material is under a poling process, i.e., when the ferroelectric domains have a preferred orientation. In contrary, the sound velocity decreases during a depolarizing phase.

In the next section, titled experimental set up, we describe the samples and we present the LIPP method as an ultrasonic technique. The section III contains the experimental results and their discussion and we finish by a conclusion in the section IV.

2. Experimental set up

The samples under investigation are polycrystalline PZT ceramics, supplied by CTS corporation [3]. The ceramic samples are 140 to 150 μm thick and 14 mm diameter disks, coated with Ni electrodes as supplied by the manufacturer.

The LIPP method has been described in numerous papers since its introduction in 1977 [4]. In this section, we I'll only make a briefly remind of its principle with the intention to give sufficient basis to understand the measured signals and we explain how it is used as an ultrasonic method. For more details, the reader can refer to references [5, 6].

A pressure pulse P is generated by the impact of a short duration laser pulse (Nd:YAG, 35 ps) on an aluminum target. The pressure pulse penetrates the sample by the front electrode and exit after a transit time T by the back electrode (Fig. 1).

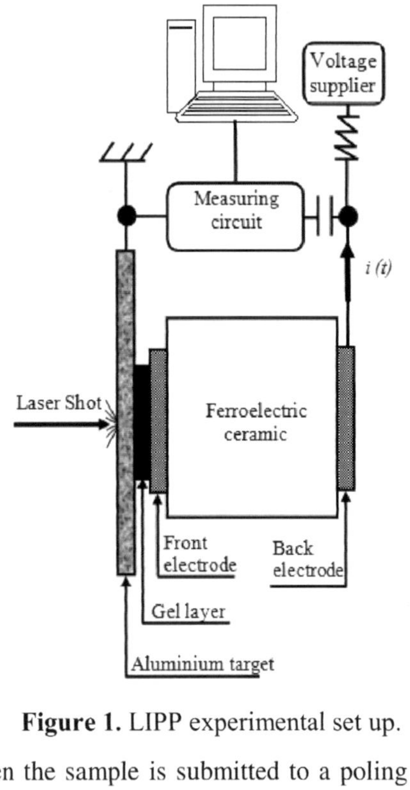

Figure 1. LIPP experimental set up.

It has been shown that when the sample is submitted to a poling field, the pressure propagation induces a current signal which can be measured by an appropriate external circuit [7, 8]. Assuming that the sample is free of real and polarization charges (Fig. 2.a), the measured current presents two peaks which are produced when the pressure pulse encounters the capacitance charge at both electrode insulator interfaces (Fig. 2.b). The first peak is proportional to the applied field and to the pressure pulse amplitude at the front interfacial region. It has the same form than the pressure pulse at the front and the same sign than the charges on the front electrode: $I_F(t) = \alpha E_a P_F(t)$, where E_a is the applied field and α is a constant. This peak indicates the pressure pulse entrance in the sample.

The second peak is proportional to the applied field and to the pressure pulse amplitude at the back interfacial region. It has also the same form than the pressure pulse but at the back region and the same sign than the charges on the back electrode: $I_B(t) = -\alpha\, E_a\, P_B(t)$. This peak marks the pressure pulse exit from the sample. The measured time, between these two peaks, is the transit time T of the pressure pulse through the sample. This time is also known as the time of flight in ultrasonic methods. It is related to the sample thickness e by the velocity of sound v in the material as: $v = \frac{e}{T}$. Thus, knowing the sample thickness, the transit time T measurement allows the sound velocity determination.

It is known that the sound velocity v depends on the poling state of the sample, then the transit T will depends also. The aim of this work is to study the variation of v as function of the polarization state of the sample.

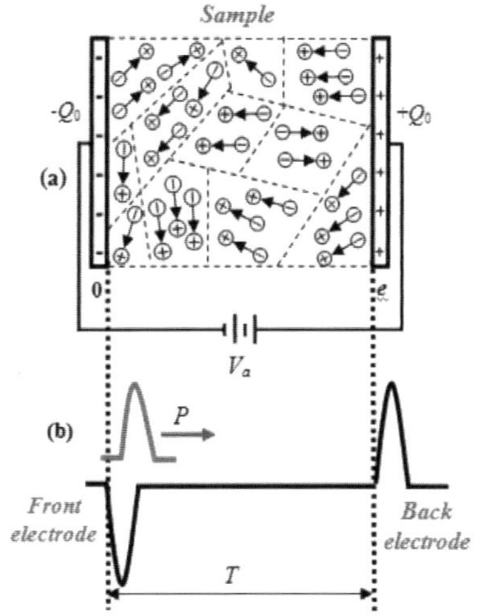

Figure 2. (a) Charges and dipoles configuration at the beginning of the stress cycle: free sample charges.
(b) Corresponding signal predicted by the LIPP method

3. Experimental result and discussion

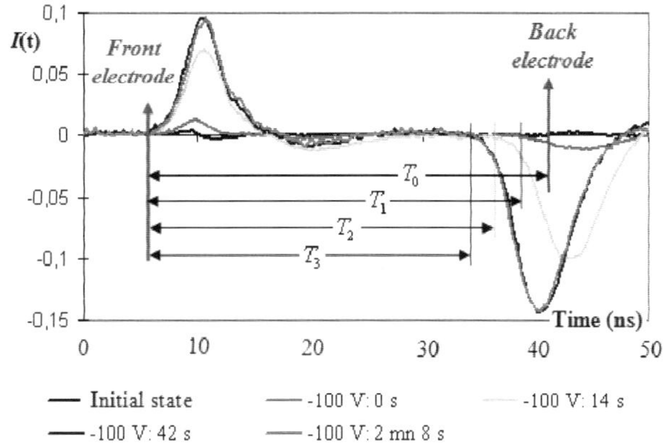

Figure 3. Transit time of the initial state and its variation during the poling process.

Figure 3 shows the evolution of measured transit time. At the initial state the sample is not poled ($T=T_0$). After that, we apply a negative potential to the back electrode whereas the front electrode is grounded. We follow the transit time variation during the poling process under the poling field. From this figure, we can see a clear decrease of the transit time as function of the poling duration.

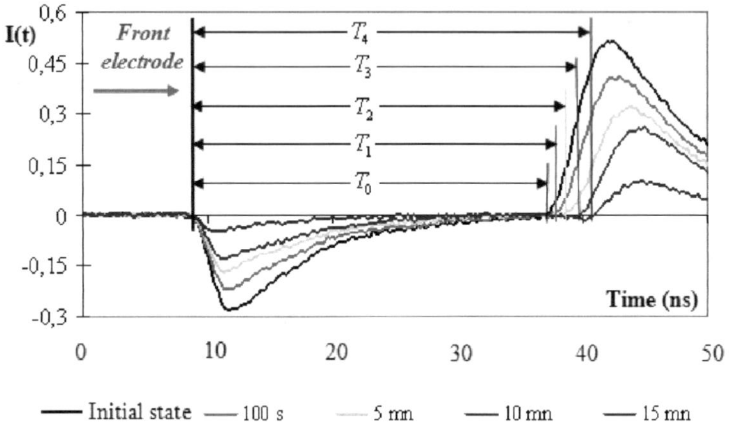

Figure 4. Transit time of the initial state and its variation during the depolarizing process.

Figure 4 shows the evolution of measured transit time of a sample already poled. Contrary to the precedent sample, the back electrode is submitted to a positive potential. Consequently the initial state present remnant polarization, the peak at the front electrode is negative and the peak at the back electrode is positive. To depolarize the sample, we submit the back electrode to a negative potential. In this case, we follow the transit time variation during the depolarizing process under the depolarizing field. From this figure, we can see a clear increase of the transit time as function of the depolarizing duration.

The detected transit time variation is in evidence related to the sound velocity variation. The sound velocity of a longitudinal wave propagating in the direction of the poling field, always chosen as axis 3, is correlated to the electromechanical characteristics of the material as [9, 10]: $v = \sqrt{\frac{c_{33}^D}{\rho}} = \sqrt{\frac{c_{33}^E}{\rho} + \frac{e_{33}^2}{\rho \varepsilon_{33}^S}}$, where c_{33}^E is the elastic constant for constant electric field, e_{33} is the piezoelectric constant, ε_{33}^S is the clamped dielectric permittivity and ρ is the density. Thus a variation of one among these constants leads to sound velocity variation. Indeed, it was mentioned [11] that the piezoelectric constant e_{33} and the dielectric constant ε_{33} will all change under different poling condition and produce a change of the sound velocity.

The transit time T and the sound velocity v are inversely proportional. Then, the decrease of the transit time observed in the figure 3 is due to an increase of the sound velocity. Whereas the increase of the transit time in the figure 4 is due to a decrease of the sound velocity. In fact, it has been reported [12, 13] that the electromechanical properties of poled ferroelectric ceramics depend strongly on the poling conditions, namely the poling time and the field strength. The poling time necessary to reach the saturation of the piezoelectric properties in piezoelectric ceramics is larger than 5 min.

V. Scherer [14] has reported that the sound velocity of a longitudinal wave propagating in the direction of the polarization vector is from 10 to 30 % much greater than a longitudinal wave velocity propagating in a same but no poled sample.

J. Yin [15] has reported that there are two reasons that might cause the velocity to decrease: one is the reduction of piezoelectric contribution, which softens the effective elastic constant along the poling direction, and the other is due to the rotation of domains, which reduces the velocity due to elastic anisotropy.

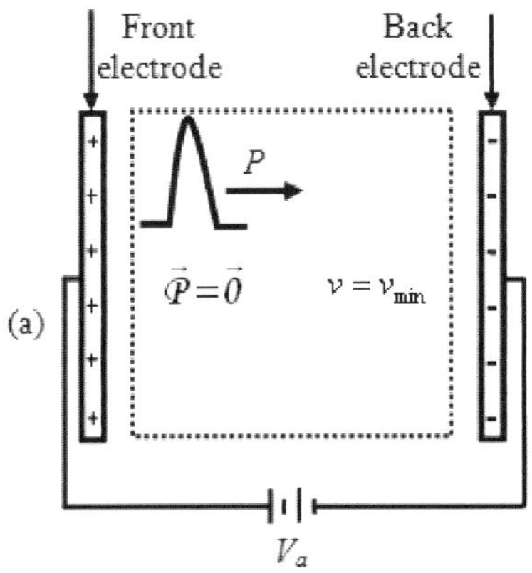

Figure 5. (a): Initial no poled state: Minimum velocity

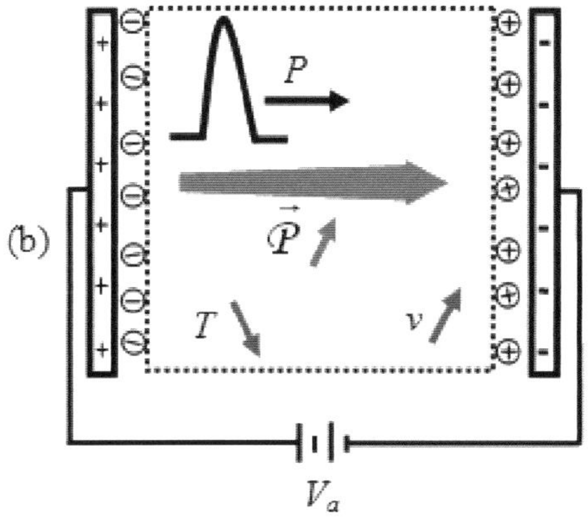

Figure 5. (b): Polarization phase: Velocity increase.

In our case, in all measurements the pressure wave (P) propagates from the front electrode to the back electrode. We have noticed that, when the polarization vector is in the direction of the pressure propagation and that the module of the polarization vector increases as function of time (Fig. 5.a, b), the velocity increases from a minimum value to a maximum value. On the other hand, when the polarization vector is in opposite direction of the pressure propagation and that the module of the polarization vector decreases as function of time (Fig. 6.a, b), the velocity decreases from a maximum value to a minimum value. The velocity decrease corresponds to the depolarization phase and the increase to the poling phase.

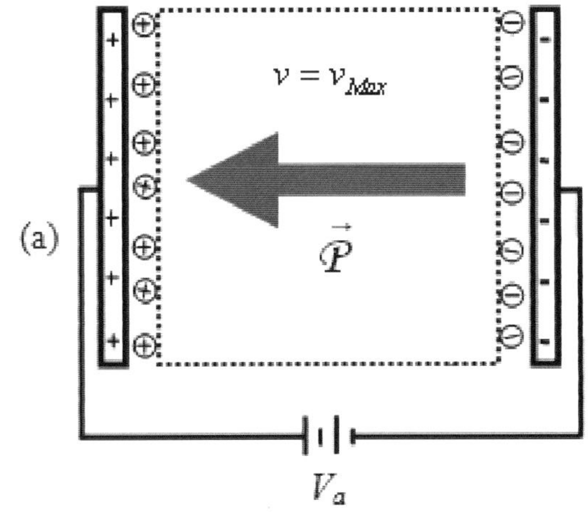

Figure 6. (a): Initial poled state: Maximum velocity

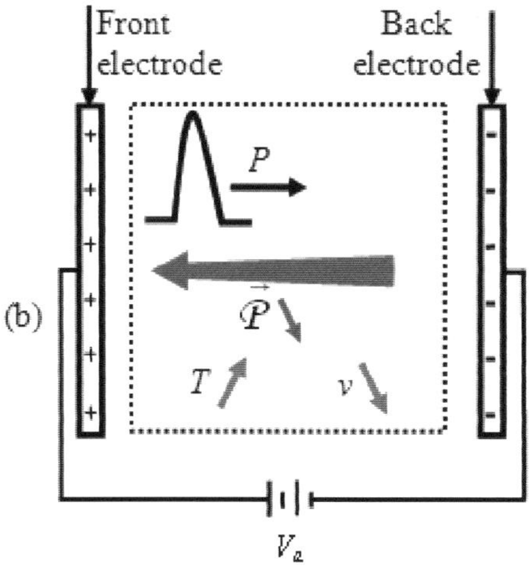

Figure 6. (b): Depolarization phase: Velocity decrease.

Figure 7 illustrates the sound velocity variation corresponding to measures indicates in figure 3. It shows how from a minimum value of the non poled state, the sound velocity begins to increase as function of poling time. Thus, the sound velocity value $4.46\ 10^3 m/s$, of the initially non poled state, is almost 14 % less than the value $5.08\ 10^3 m/s$ of the poled sample.

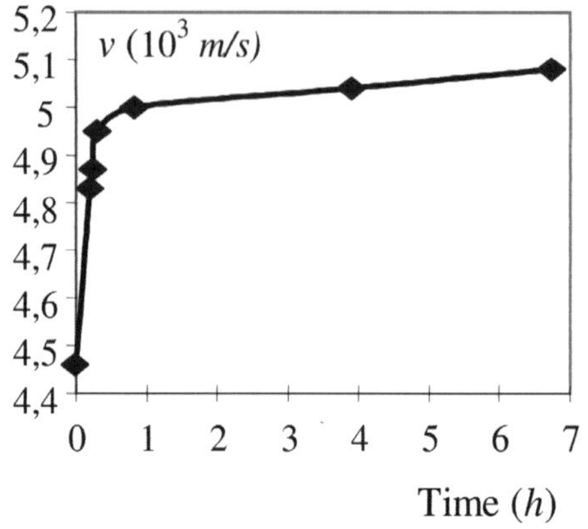

Figure 7. The sound velocity variation as function of time.

Finally, we specify that our discussion is based on the sound velocity variation related to the transit time change. But, since the transit time T links the sample thickness e and the sound velocity v of the pressure wave as: $T = \frac{e}{v}$, a variation of the transit time can be associated with variation of either e or v or both. In fact, it's known that piezoelectric materials change dimension when they are submitted to an electric field. The expansion of the thickness e under poling is a known phenomenon in the technology of actuators. Nevertheless such phenomenon induces very small displacements which they are out of our ability to detect them with our method in this thickness range.

4. Conclusion

In this paper, we have used LIPP as an ultrasonic method to study the variation of the sound velocity v through the variation of the transit time T. We found that the sound velocity depends on the polarization conditions. It decreases during the depolarization phase and increases during the poling one. The sound velocity of an initially non poled state, is almost 14 % less than of the poled sample.

References

[1] Farhat N, Ditchi T, Holé S and Lewiner J 2002 *J. Appl. Phys* **91** 5288.
[2] Farhat N, Holé S, Ditchi T, Lewiner J, Button T. W and Su B. 2006 *J. Appl. Phys.* **100** 054102.
[3] www.ctscorp.com
[4] Laurenceau P, Dreyfus G and Lewiner J, 1977 Phys. Rev. Lett. **38** 46.
[5] Holé S., Ditchi T., and J. Lewiner 2000 *Phys. Rev. B* **61** 13528.
[6] Holé S, Ditchi T and Lewiner J 2003 *IEEE Trans. Dielectr. Electr. Insul.* **10** 670.
[7] Alquié C and Lewiner J 1985 *Rev. Phys. Appl.* **20** 395
[8] Farhat N, Ditchi T, Holé S And Lewiner J 2002 *J. Appl. Phys.* **91** 5288.
[9] Jiang W and Cao W 2002 *Appl. Phys. Let.* **80** 2466.
[10] Friend J, Jamieson E, Pennell M and Huebner W 2001 *Appl. Phys. Let.* **79** 2794.
[11] Zhang R, Jiang B and Cao W 2002 *J. Appl. Phys.* **12** 10194.
[12] Ueda I and Ikegami S 1968 *Jpn. J. Appl. Phys.* **7** 236.
[13] Ikegami S, Ueda I and Nagata T 1971 *J. Acoust. Soc. Am.* **50** 1.
[14] Scherer V, Hirsekorn S, Rabe U and Arnold W 2000 *AIP Conference Proceedings* **509** 1493.
[15] Yin J and Cao W 2001 *Appl. Phys. Let.* **79** 4556.

IOP Conf. Series: Materials Science and Engineering **28** (2012) 012013 doi:10.1088/1757-899X/28/1/012013

Low polarization switching in Lead Zirconate Titanate (PZT) Ceramics

N. Farhat[a1] and W. Essolaani[b]
[a]Ecole Supérieure des Sciences et de Technologie – Hammam Sousse
[b]Unité de physique des solides- Faculté des sciences de Monastir

E-mail: najeh_farhat@yahoo.fr

Abstract. Usually to study the polarization reversal phenomenon, the applied field is varied gradually and the polarization is measured which give the hysteresis *P–E* loop. In many applications, such as in memory components, a fast switching (less than 100 ns) is required. For a fast switching the applied field must generally be of order $2E_C$, where E_C is the coercive field. In this case the polarization is balanced between saturation polarizations $+P_s$ and $-P_s$ very quickly. Thus it is difficult to study the polarization reversal evolution.
In this paper we have chosen to apply on poled PZT ceramic samples a depolarizing field slightly greater than the coercive field which permits a low polarization switching.

1. Introduction

It is thought that the silicon technology will soon reach its limit as a result of the miniaturization. In fact, the continuous reduction in size of the electronic components will lead to an important degradation of their performance. Ferroelectric ceramics based in lead zirconate titanate (PZT) are considered among the materials which have the capacities to be an alternative of some silicon dioxide applications. An important characteristic of these materials is the reversibility of the polarization which may be utilized to store data in binary form. So, ferroelectric ceramics can be used for both dynamic random access memory (DRAM) and non-volatile ferroelectric random access memory (FRAM).

For that reason they have intrigued an increasing interest last years. Various methods have been used for studying the polarization reversal such as the measurements of ferroelectric hysteresis loop (*P – E* loop) [1], transient reversal current [2, 3] ultrasound method [4]. Each method reveals a particular aspect of the polarization reversal process. The *P – E* method, considered as the common method in reversal polarization studies, measures the average polarization of poled state. However the spatial distribution of the polarization across an electroded sample during the polarization reversal has still not been investigated.

In the next section, we describe the samples, the LIPP method and the measurement procedure. The experimental results are presented in section III and the conclusion is given in the section IV.

---

[1] To whom any correspondence should be addressed.

2. Experimental set up

2.1. Samples

The samples under investigation are polycrystalline PZT ceramics, supplied by CTS corporation [5]. The ceramic samples are 140 to 150 μm thick and 14 mm diameter disks, coated with Ni electrodes as supplied by the manufacturer. The samples were supplied after poling. The poling was done in the thickness direction. For each disk, a V shape is graved on one of the electrodes. The side with the V shape was used as cathode during the poling and marks the negative side. The coercive field, for such samples, is approximately equal to 8 kV/cm.

2.2. LIPP method description

In this work, we use the Laser Induced Pressure Pulse (LIPP) method. This method has been very useful for studying the spatial distribution of charges and polarization in dielectrics [6, 7].
The LIPP method has been described in numerous papers since its introduction in 1977 [8]. In this section, we I'll only make a briefly remind of its principle with the intention to give sufficient basis to understand the measured signals. For more details, the reader can refer to references [9,10].

A short pulsed laser impact on a target generates a short pressure pulse. The pressure pulse p is generated by the impact of a short duration laser pulse (Nd:YAG, 35 ps) on an aluminum target. That pressure pulse is then transmitted to the material to be studied. The transmission, from the target where the pressure pulse is generated to the sample where it will propagates, is assured by an adequate coupling gel between the target and the front electrode (Fig. 1). During the pressure pulse propagation, the atomic structure of the sample is locally compressed resulting in a local displacement of bonded charges and dipoles. This creates a variation of image charges on the sample electrodes, which in turn produces a current in the external circuit connecting the electrodes. This current, which is the measured signal, depends directly on the distribution of space charges and dipoles [11].

One advantage of such method is that, in addition to be nondestructive, it gives directly the image of the dipole distributions within the bulk.

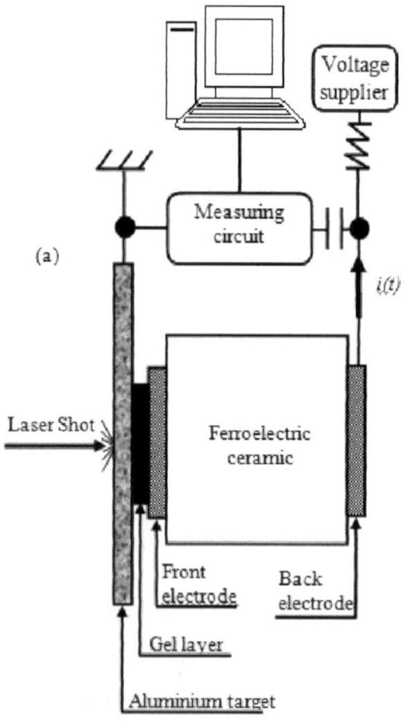

Figure 1. LIPP experimental set up.

2.3. Analyze of some charge configurations using the LIPP Method

2.3.1. Measurement at the beginning of the stress cycle

It has been shown [12,13], that few seconds after the poling voltage application and assuming that $\vec{P} = \vec{0}$, the sample is free of real and polarization charges (Fig. 2.a). The measured current presents two peaks which are produced when the pressure pulse encounters the capacitance charge at both electrode insulator interfaces (Fig. 2.b). The first peak is proportional to the applied field and to the pressure pulse amplitude at the front interfacial region. It has the same form than the pressure pulse at the front ($P_F(t)$) and the same sign than the charges on the front electrode: $I_F(t) = \alpha E_a P_F(t)$, where E_a is the applied field and α is a constant. This peak indicates the pressure pulse entrance in the sample. The second peak is proportional to the applied field and to the pressure pulse amplitude at the back interfacial region. It has also the same form than the pressure pulse but at the back region ($P_B(t)$), and the same sign than the charges on the back electrode: $I_B(t) = -\alpha E_a P_B(t)$. This peak marks the pressure pulse exit from the sample. Null amplitude of the signal between these two peaks confirms the real and polarization space charges absence.

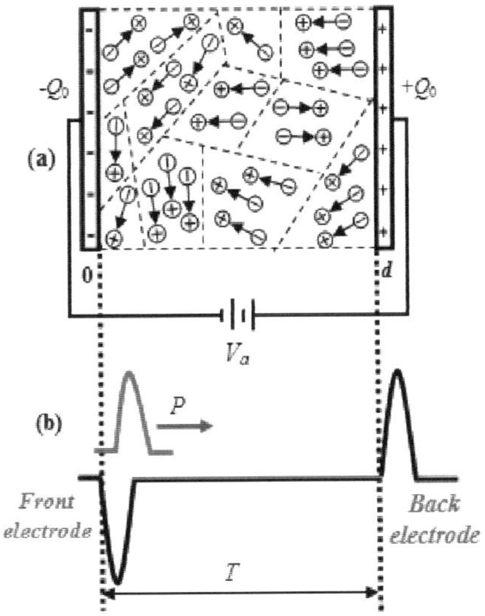

Figure 2. (a) Charges and dipoles configuration at the beginning of the stress cycle: free sample charges.
(b) Corresponding signal predicted by the LIPP method.

2.3.2. Measurement at the end of the stress cycle

Before poling, ceramics consist of many crystallites containing ferroelectric domains with random orientations. The dipoles moments averaged over the volume of the sample are zero and no global polarization, at macroscopic scale, is detected in the ferroelectric (Fig. 2.a). During the poling process, applying a sufficiently high dc electric field, domains and domain walls tend to align nearly parallel to the electric field. The alignment of the dipoles, in the case of a uniformly poled sample, is shown in idealized form after long voltage application. In such situation, the dielectric can be seen as a piling of thin polarized slabs (Fig. 3.a). Away from the edges of the dielectric, the surface charges of two adjacent slabs neutralize each other in such manner the interior of the dielectric remains electrically neutral. However a net negative charge $-Q_P$ accumulates at the dielectric surface adjacent to the anode capacitor electrode and a positive charge $+Q_P$ appears at the dielectric surface adjacent to the cathode resulting from the global dielectric polarization. $-Q_P$ and $+Q_P$ assure the whole electric neutrality of the dielectric and induce, by influence, additional charges $+Q_P$ on the anode and $-Q_P$ on the cathode. So, due to the polarization, the positive charge on the anode increases from $+Q_0$ to $+Q_0+Q_P$ and, in a similar manner, the negative charge on the cathode increases from $-Q_0$ to $-Q_0-Q_P$ (Fig. 3.b) [14].

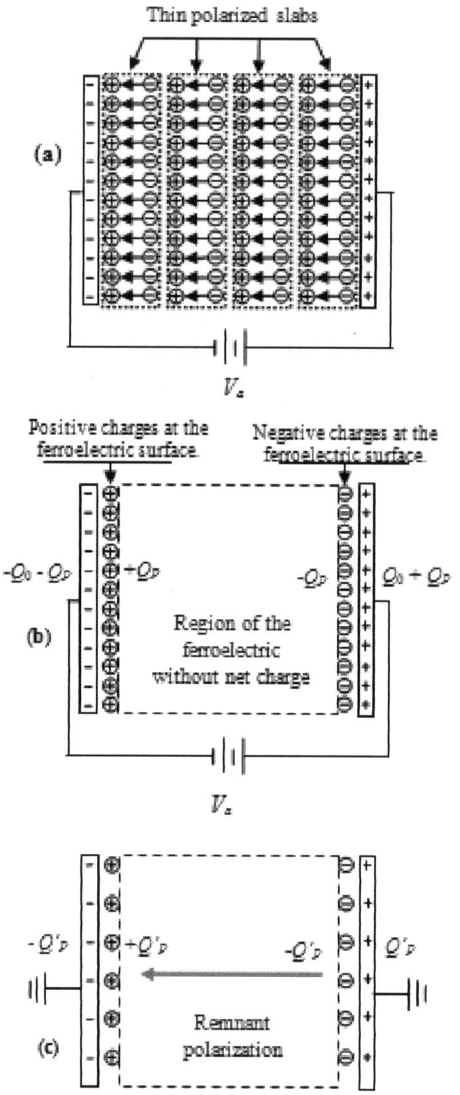

Figure 3. (a) Dipoles alignment after long voltage application.
(b) Charges configuration after long voltage application.
(c) Charges configuration after the voltage switching off.

Thus, the response to the pressure pulse propagation in the dielectric will be similar to what we have seen for the case at the beginning of the stress cycle where only charges on electrodes exist and where the response is non zero only at interfaces. Nevertheless the charges on electrodes are, this time, the sum of the capacitive component and the polarization component.

A signal with no significant deviation from the base line, in the bulk and far from the electrodes, proves that the sample is uniformly poled and without real space charges.

2.3.3. Short circuit measurement

When the applied field is reduced to zero, the ferroelectric displays a polarization called the remnant polarization $\vec{P}_r$. In fact, if the poling field was sufficiently intense, when it is removed the majority of the domains will remain in the direction of the previous field, requiring compensating charge to remain on the plates of the capacitor (Fig. 3.c). Here also, the response to the pressure pulse propagation in the dielectric will be similar to the case of the beginning of the stress cycle but with the contribution of the only polarization component.

3. Experimental results

As mentioned above, the samples were poled by the supplier. The marked face used as cathode during the poling performed by the supplier is chosen as the front electrode that is coupled to the target. In the experimental set up, this electrode is always grounded. The back electrode used as anode by the supplier can be subjected to a DC voltage, positive or negative with reference to the ground.
The charges configuration of the samples when they are received from the supplier is, in principle, the same to what is showed in figure 3.c.

3.1. Initial state measurement:

A first measurement is required in order to check the initial electrical state of the sample as delivered. This measurement is indicated as 0 V in the fig. 4. We can see from this measurement that the sample is uniformly poled since the signal presents only two peaks, a negative peak at the front electrode which has been used as cathode and a positive one at the back electrode which has been used as anode. Afterwards, to perform a reversal polarization, we apply an electric field opposing the polarization done by the supplier.

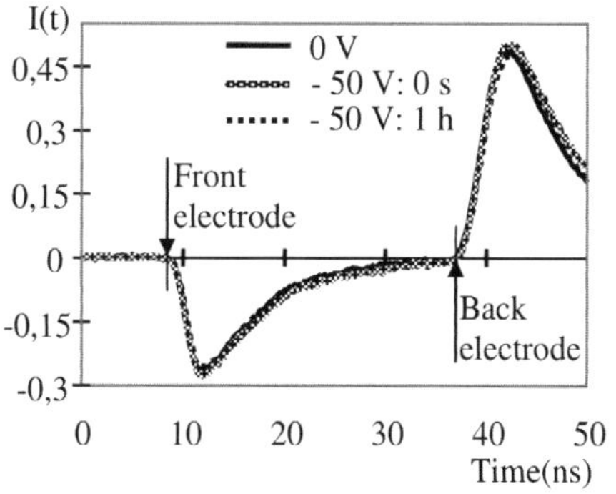

Figure 4. Initial state and low applied field signals.

3.2. First electric stress: Applied field more lower than the coercive field

We start by applying – 50 V to the back electrode of a 145 µm thick sample and we carry out as soon as possible a measurement shown by the trace 0 s in the fig. 4. A voltage of – 50 V applied to the back electrode means that this electrode is the cathode and the front electrode is the anode. So, our applied field polarity is the opposite of that applied by the supplier in the first poling. Thus if the applied field is sufficiently intense, the sample will undergoes a depolarizing process. We maintain the voltage applied and after one hour we performed again a measurement shown by the trace 1 h.

From the fig. 4, we can notice that the curves at 0 s and 1 h coincide with the curve of the initial state. Thus, as predicted, the voltage of – 50 V is a low voltage without any effect neither on the dipoles reorientation, because the signals amplitudes haven't exchanged, nor on the sound velocity since the transit time remains constant.

After that we switched off the voltage and we performed a control measurement in short circuit. This measurement, not shown here, is identical to the initial state measurement.

IOP Conf. Series: Materials Science and Engineering **28** (2012) 012013 doi:10.1088/1757-899X/28/1/012013

3.3. Second stress: Applied field slightly greater than the coercive field

After the first stress cycle, we increase the depolarizing field by applying a voltage of - 120 V. The applied field, 8.3 kV/cm, is slightly greater than the value, 8 kV/cm, of the coercive field. This time, the applied field is sufficient to depolarize the sample and reverse slowly the polarization. The Fig. 5 shows the measurements and the signals evolution as function of time during the first 15 mn.

From the Fig. 5, we can see that the applied voltage is sufficient to launch a depolarizing process. Indeed, we assist to a decrease of the amplitude in the 22 firsts minutes following the voltage application.

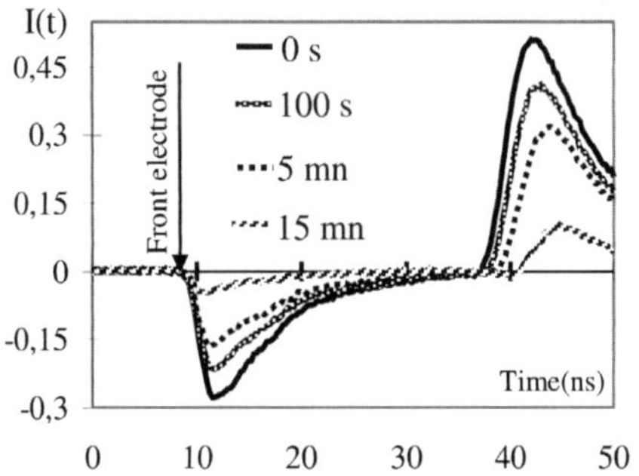

Figure 5. Signals evolution under applied field as function of time: depolarizing stage

After 15 minutes under applied field, we notice the beginning of a switching phenomenon. The switching is very clear in measurements between 18 and 26 minutes (Fig. 6). In fact, in the measurement carried out at 18 minutes, the signal presents a negative peak at the front electrode and a positive one at the back electrode whereas in the measurement done at 26 minutes the signal is inverted.

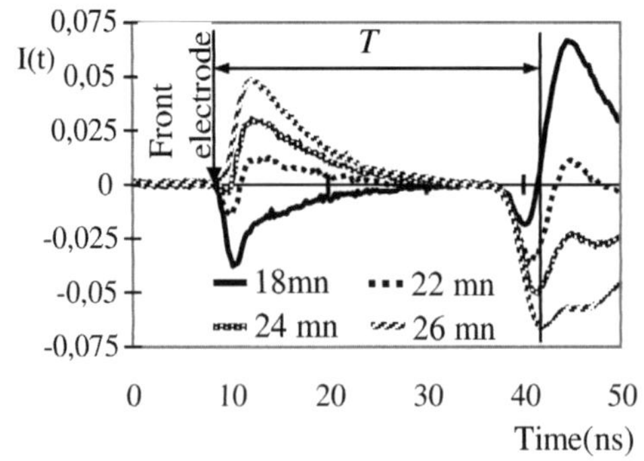

Figure 6. Signals evolution between 18 and 26 minutes: Switching polarization.

According to the amplitude evolution, we suggest that the sample is quasi depolarized after almost 18 minutes. In fact, the negative peak at the front electrode of the measurements done between 18 and 26 minutes is very small compared to the measurement at 0 second. For example the amplitude at the front electrode, obtained at 22 minutes, is reduces by a factor 20 relatively to the amplitude at 0s.

For durations more longer than 26 minutes under applied field, the switching is completed and the sample is poled again but in opposite direction which lead progressively to an increase of the amplitude (Fig. 7).

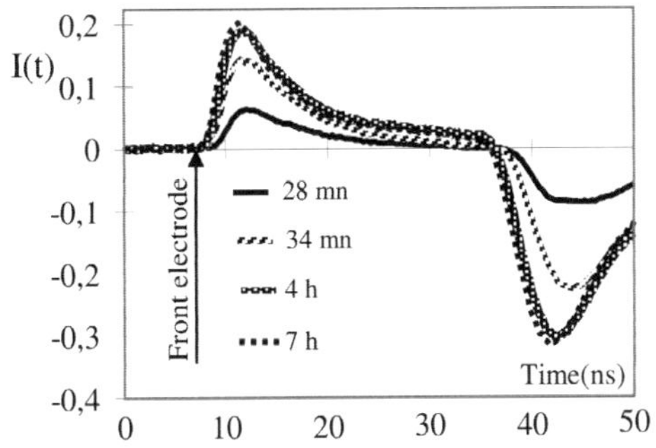

Figure 7. Signals evolution after the polarization switching: poling stage

The Fig. 8 summarizes the applied field effect for short and long times. We notice that the amplitude decreases until a very weak value, it exchanges the sign and after that it increases.

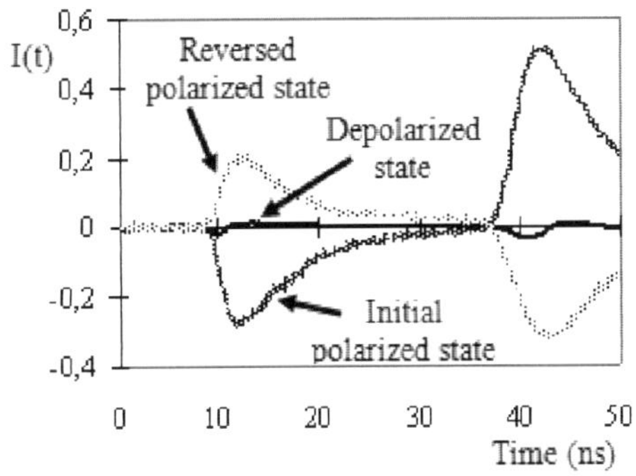

Figure 8. Summary of the applied field effect for short and long times.

4. Discussion

The results report the distribution of the poling quality within PZT ferroelectric samples during a slow polarization inversion. They show that the sample is uniformly poled and that the uniformity subsists during the switching.

As we have seen above, the ferroelectric poling process leads to the coalescence of all domains and the formation of a single domain. From this state (Fig. 9.a), where the sample presents a remnant polarization the inversion is performed by a depolarizing field slightly greater than the coercive field. The switching is characterized by an amplitude decrease of the measured signal until a very weak value, the inversion of the signal sign and after that the amplitude increase. This can be explained by the main stages of the polarization reversal process. In fact, it is known that when a switching field is applied to the single domain sample, new domains appear, mainly at the surface, and then grow through the sample (Fig. 9.b). It is generally believed that the domain growth parallel to the applied electric field (the forward growth (Fig. 9.b)) is much faster than that of the perpendicular one (the sideways growth (Fig. 9.c) [15,16]. Nevertheless sometimes the opposite situation, where the sideways growth is much faster than the forward growth, is observed [17].

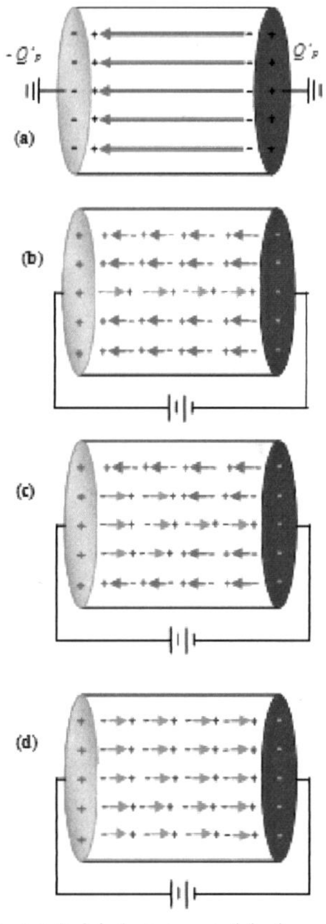

Figure 9. (a) Initial state, stable in zero field
(b) Polarization reversal in applied depolarizing field begins with nucleation and nucleus propagates to opposite electrode: Forward switching
(c) Lateral propagation of domain walls: sideways switching
(d) Switching process is completed.

Usually arising of new domains with antiparallel orientation continues during the whole switching process. Consequently, the global macroscopic polarization is reduced in the first stage (depolarizing stage), which is manifested by the signal amplitude decrease. The sample retrieves for a moment a state with an equal competition between parallel and antiparallel domains orientations which corresponds to the weakest amplitude. After that the phase reversal starts to occur and domains with antiparallel orientations, relatively to the initial state, prevail. In the second stage (poling stage), these domains begin to join together (to coalesce) in the opposite direction of the initial polarization and the amplitude increases until the switching process is completed.

The switching polarization is usually considered as the contribution of two mechanisms: domain nucleation (domain reorientation) and domain wall motion (domain wall rearrangement). It has been reported [18] that the domain reorientation occurs at short time (typically less than 30 ms) and that the domain wall rearrangement is predominant at longer times.

5. Conclusion

In this paper, the distribution of the poling quality within PZT ferroelectric samples during a slow polarization inversion has been studied by the LIPP method.

The switching is characterized by an amplitude decrease of the measured signal until a very weak value, the inversion of the signal sign and after that the amplitude increase. This can be explained by the main stages of the polarization reversal process, where new domains with antiparallel orientation appear at the surface and then grow through the sample. Consequently, the global macroscopic polarization is reduced in a first stage (depolarizing stage), which is manifested by the signal amplitude decrease. The sample retrieves for a moment a state with an equal competition between parallel and antiparallel domains orientations which corresponds to a signal with a weakest amplitude. In a second stage (poling stage), domains begin to join together (to coalesce) in the opposite direction of the initial polarization and the amplitude increases until the switching process is completed.

References

[1] Viehland D and Chen Y. H 2000 *J. Appl. Phys.* **88** 6696.
[2] Yu H, Gopalan V, Sindel J and Randall C. A 2001 *J. Appl. Phys.* **89** 561.
[3] Lente M. H and Eiras J. A 2001 *J. Appl. Phys.* **89** 5093.
[4] Yin J. and Cao W 2001 *Appl. Phys. Let.* **79** 4556.
[5] www.ctscorp.com
[6] Farhat N, Ditchi T, Holé S, And Lewiner 2002 *J. Appl. Phys.* **91** 5288.
[7] Farhat N, Holé S and Ditchi T, Lewiner J, Button T. W and Su B 2006 *J. Appl. Phys.* **100** 054102.
[8] Laurenceau P, Dreyfus G and Lewiner J 1977 *Phys. Rev. Lett.* **38** 46.
[9] Holé S, Ditchi T and Lewiner J 2000 *Phys. Rev. B* **61** 13528.
[10] Holé S, Ditchi T, and Lewiner J 2003 *IEEE Trans. Dielectr. Electr. Insul.* **10** 670.
[11] Alquié C, G. Dreyfus and J. Lewiner 1981 *Phys. Rev. Lett.* **47** 1483.
[12] Alquié C and Lewiner J 1985 *Rev. Phys. Appl.* **20** 395.
[13] Farhat N, Ditchi T, Holé S and Lewiner J 2002 *J. Appl. Phys.* **91** 5288.
[14] Callister W. D 2000 *Materials Science and Engineering*: An Introduction. 5th edition, John Wiley & Sons Ch.19.
[15] Shur V. Y. A. and Rumyantsev E. L 1997 *Ferroelectrics* **191** 319.
[16] Hong S, Colla E. L, Taylor D. V, ATagantsev K, Muralt P, No K and Setter N 1999 *J. Appl. Phys.* **86** 607.
[17] Kuroda A, Kurimura S and Uesu Y 1996 *Appl. Phys. Let.* **69** 1565.
[18] Lente M. H and Eiras J. A 2001 *J. Appl. Phys.* **89** 5093.

IOP Conf. Series: Materials Science and Engineering **28** (2012) 012014 doi:10.1088/1757-899X/28/1/012014

Structural and magnetic properties of Pr_2Co_7 powders

R. Fersi [a,b] [1], N. Mliki[a] and L. Bessais[b]
[a]Laboratoire Matériaux Organisation et Propriétés, Faculté des Sciences de Tunis, Université Tunis El Manar, 2092 Tunis, Tunisie.
[b]CMTR, ICMPE, UMR7182, CNRS – Université Paris 12, 2-8 rue Henri Dunant F-94320 Thiais, France.

E-mail: FERSI@glvt-cnrs.fr

Abstract. The Pr_2Co_7 intermetallic compounds are known to crystallize in two polymorphic forms. A hexagonal of the Ce_2Ni_7 type structure and a rhombohedral of the Gd_2Co_7 type structure were obtained by mechanical milling and subsequent annealing in high vacuum. The highest intrinsic coercivity value 18 kOe at room temperature and magnetization ratio (M_r/M_{max} = 0.64) were determined for alloys milled during 5 h and annealed at 800°C for 30 min. The enhanced remanence (M_r/M_{max} > 0.50) suggests strong exchange interaction between the adjacent crystallites. This is in good agreement with the small average crystallite size (D ≈ 50 nm). The high magnetic properties observed in these nanostructured Pr_2Co_7 intermetallic alloys have their origin in its relatively high uniaxial magnetocrystalline anisotropy field, and in the homogeneous nanostructure developed by mechanical milling process and subsequent annealing.

1. Introduction

The study of magnetic properties of rare earth (R)-3d transition metals (M) intermetallics has been a topic of great interest both from a fundamental point of view as applied one. This interest is explained by the exceptional properties of these alloys which arise from the presence, in the same compounds, of the outer well delocalized 3d electrons and well localized and 4f electronic shell [1]. This combination of R and M elements can give rise to materials exhibiting high Curie temperature, given by the strong exchange interaction between 3d electrons of M metals, and a strong anisotropy provided by the rare earth 4f electrons. Among the family of intermettallics, R_2Co_7 (R= rare earth element) have attracted interest due to their applications as permanent magnets. They exhibit excellent magnetic properties such as high Curie temperature T_c, large magnetocrystalline anisotropy H_A, and important saturation magnetization M_s [2, 3].

In the present work, we discuss the structural and magnetic properties of Pr_2Co_7 powders produced by high-energy milling which we have optimized the grain size in order to obtain a high coercivity.

2. Experiment

Pr_2Co_7 samples with praseodymium excess, were prepared by the technique of high energy ball milling and subsequent annealing from 600 °C to 1050 °C during 30 min. X-ray diffraction with Cu radiation has been carried out on a Brucker diffractometer with an internal Si standard to insure a unit

[1] To whom any correspondence should be addressed.

cell parameter accuracy of ± 1 ×10^{-3}. The intensities were measured from 2□ =20° to 110° with a step size of 0.04° and counting time of 22 s per scanning step. The structure refinement was performed by the Rietveld technique supported, in the assumption of Thompson-Cox- Hastings line profile, by the FULLPROF computer code. The detailed information was given in Ref. [4].
The magnetization was measured using a differential sample magnetometer (MANICS). When necessary, the samples were sealed in silica tubes to avoid oxidation upon heating. Thermomagnetic data were measured under an applied field of 1000 Oe with a heating rate of 10 K/min. The Curie temperature T_C was determined from the M-T curves by extrapolating the linear part of the M-T curve and finding the temperature values of the intersection with the extended baseline [5,6]. For magnetic hysteresis measurements at T = 293 K, we used a Physical Properties Measurement System (PPMS) Quantum Design equipment in fields up to 9 kOe.

3. Results and discussion

3.1. Structure analysis

Lets note that the Pr_2Co_7 compound was found to crystallize in two polymorphic forms: a high temperature phase modification of the hexagonal Ce_2Ni_7 structure and the low temperature phase modification of the rhombohedral Gd_2Co_7 type structure [7], can be obtained by stacking the hexagonal structural blocks for $PrCo_5$ (structure of the $CaCu_5$ type) and the cubic blocks $PrCo_2$ (structure of the $MgCu_2$) type along the common hexagonal (trigonal for $PrCo_2$) axis.

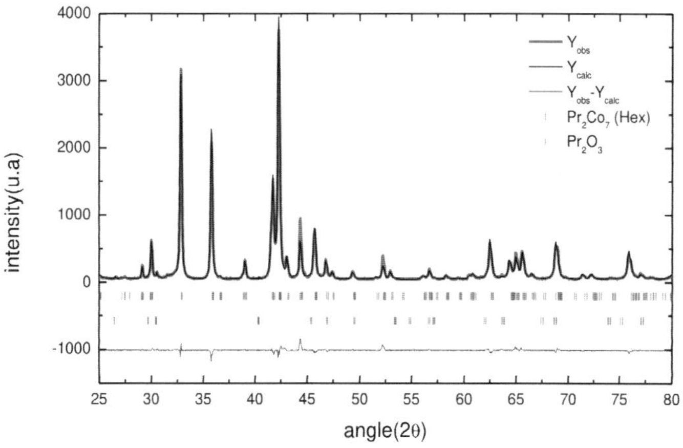

Figure 1. Figure Rietvield analysis for X-ray diffraction pattern of Pr_2Co_7 annealed at 750 °C for 30 min.

Fig.1 presents, as an example, the analysis result of XRD pattern of Pr_2Co_7 sample annealed at 750°C using Rietveld analysis. The result of the structure refinement performed for the Pr_2Co_7 alloys show the presence of main phase with the hexagonal structure. The lattice constants were a = 5.078 Å and c = 24.471 Å. This result suggests that the Pr_2Co_7 structure is isotopic with the crystal structure of Ce_2Ni_7 described by Cromer and Larson [8]. When the annealing temperature was increased from 700°C to 1050°C, we have seen emerging the Pr_2Co_7 rhombohedral phase at the expense of the hexagonal phase of Pr_2Co_7. However, we note minor quantities of the oxide phases appearing due to selective oxidation of Pr (PrO, PrO_2, Pr_2O_3). For the sample annealed at 1050 °C, the structure refinement show the presence of main phase with the rhombohedral Pr_2Co_7 alloys. The lattice constants were a = 5.070 Å and c =36.691Å. A comparison with X-ray reflexions of the Gd_2Co_7 compound showed that Pr_2Co_7 and Gd_2Co_7 are isotopic. The structure of Gd_2Co_7 has been described by Bertaut, Lemaire and Schweizer (1965) [9].

3.2. Intrinsic magnetic properties: Curie temperature and magnetic moments

The Curie temperature, also called the magnetic ordering temperature, is a direct measure of the exchange interaction, which is the origin of ferromagnetism. This interaction depends manifestly on the interatomic distance. Generally, the Curie temperature in rare-earth transition-metal intermetallic compounds is governed by three kinds of exchange interactions, namely, the 3d-3d exchange interaction (J_{Co-Co}) between the magnetic moments of the Co sublattice, 4f - 4f exchange interaction (J_{Pr-Pr}) between the magnetic moments within the Pr sublattice and the intersublattice 3d-4f exchange interaction (J_{Pr-Co}) [10]. The interactions between the rare earth spins 4f-4f are assumed to be weak and negligible in comparison with the other two types of interactions [11]. The thermal variation of magnetic susceptibility obtained at 293 K for the two samples, indicate that Pr_2Co_7 phase is ferromagnetic with a Curie temperature around 600 K. These results are in good agreement with literature [10,12]. However, the low temperature in Pr_2Co_7 compounds is due to the predominance of the Co-Co interactions.

3.3. Extrinsic magnetic properties: coercivity

The hysteresis loops of Pr_2Co_7 have been measured at T = 293 K (see Fig.2). One can see that for annealing temperature (T_a) equal to 800°C, the coercive field rich 18 kOe. It is well established that the coercivity decreases with increasing grain size i.e. with increasing annealing temperature, which is the reason of the relatively small H_C value for T_a = 1050°C.

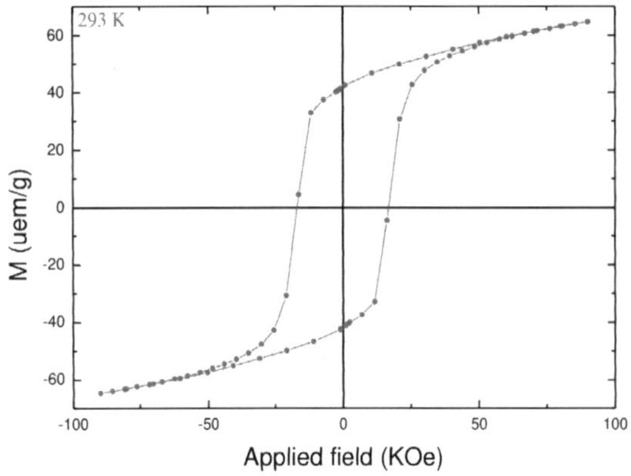

Figure 2. Magnetization vs $\mu_0 H$ of Pr_2Co_7 annealed at 800°C, measured at 293 K.

We have obtained for T_a = 293 K, H_c = 18 kOe, M_r = 41.5 uem/g and M_r/M_{max}= 0.64. Enhanced remanence ratio (M_r/M_{max} > 0.50) suggests magnetic exchange interactions between adjacent crystallites in these nanostructured powders [13,14].

The high magnetic properties observed in these nanostructured Pr_2Co_7 intermetallic alloys might have their origin in their relatively high uniaxial magnetocrystalline anisotropy field and in the homogeneous nanostructure developed by mechanical milling process and subsequent annealing.

IOP Conf. Series: Materials Science and Engineering **28** (2012) 012014 doi:10.1088/1757-899X/28/1/012014

4. Conclusion

Depending on the annealing temperature, we have shown that, by mechanical milling we have elaborated Pr_2Co_7 intermetallic compounds in two polymorphic forms a hexagonal of the Ce_2Ni_7 type structure and a rombohedral of the Gd_2Co_7 type structure. This compounds is a highly anisotropie uniaxial ferromagnet with the easy magnetization direction parallel to the c-axis. No saturation reached at room temperature for magnetic applied field of 90 kOe. High coercivity of about 18 kOe at 293K is obtained after 5h of mechanical-milling and annealed at 800°C for 30 min. Pr_2Co_7 alloys are good candidate for applications as permanent magnet.

References

[1] Deryagin A. V 1979 *J. Phys. C* **5** 165.
[2] Buschow K. H. J, Less-Common J 1973 *Met.* **33** 311.
[3] Kirchmayr H and Polldy C 1978 *J. Magn. Magn. Mater.* **8** 1.
[4] Rietveld H 1967 *Acta Crystallogr.* **22** 151.
[5] Bessais L, Djega-Mariadassou C, and Beaunier P 2006 *J. Appl. Phys.* **99** 093906.
[6] Khazzan S, Mliki N and Bessais L 2009 *J. Appl. Phys.* **105** 103904.
[7] Khan Y 1974 *Acta Crystallogr.* **30** 1533.
[8] Cromer and D. T, Larson A. C 1959 *Acta Crystallogr.* **12** 855.
[9] Bertaut E. F, Lemaire R, Schweizer J 1965 *Bull. Soc. Fr. Mineral. Cristallogr.* **88** 580–585.
[10] Buschow K. H. J 1977 *Rep. Prog. Phy.* **40** 1179.
[11] Givord D and Lemaire R 1974 *IEEE Trans. Magn.* **10** 109.
[12] Zhaoy X, Zhang Z, Yao Q, Liu J, Ren W, Lui W and Geng D 2006 *J. Mater . Sci . Technol.* **22** 843.
[13] Fischer R and Kronmüller H 1996 *Phys. Rev. B* **54** 7284.
[14] Fischer R and Kronmüller H 1999 *J. Magn. Magn. Mater.* **191** 225.

IOP Conf. Series: Materials Science and Engineering **28** (2012) 012015 doi:10.1088/1757-899X/28/1/012015

Material Choice for spindle of machine tools

S. Gouasmi[a1], B. Merzoug[a], G. Abba[b], L. Kherredine[a]
[a]Laboratoire de recherche en mécanique des matériaux et maintenance industrielle, Département Génie Mécanique, Faculté des Sciences de l'Ingénieur, Université Badji Mokhtar, B.P 12, Annaba 23000 Algérie
[b]Laboratoire de Génie Industriel et de Production mécanique, ENIM, Metz, France

E-mail: sassigouasmi@yahoo.com

Abstract. The requirements of contemporary industry and the flashing development of modern sciences impose restrictions on the majority of the elements of machines; the resulting financial constraints can be satisfied by a better output of the production equipment. As for those concerning the design, the resistance and the correct operation of the product, these require the development of increasingly precise parts, therefore the use of increasingly powerful tools [5]. The precision of machining and the output of the machine tools are generally determined by the precision of rotation of the spindle, indeed, more this one is large more the dimensions to obtain are in the zone of tolerance and the defects of shape are minimized. During the development of the machine tool, the spindle which by definition is a rotating shaft receiving and transmitting to the work piece or the cutting tool the rotational movement, must be designed according to certain optimal parameters to be able to ensure the precision required. This study will be devoted to the choice of the material of the spindle fulfilling the imposed requirements of precision.

1. Introduction

Usually, the material choice is made according to economical, technological, functional, and availability considerations, etc. Working stress being known, one then dimensions the part in such way that this one resists to the imposed loads.

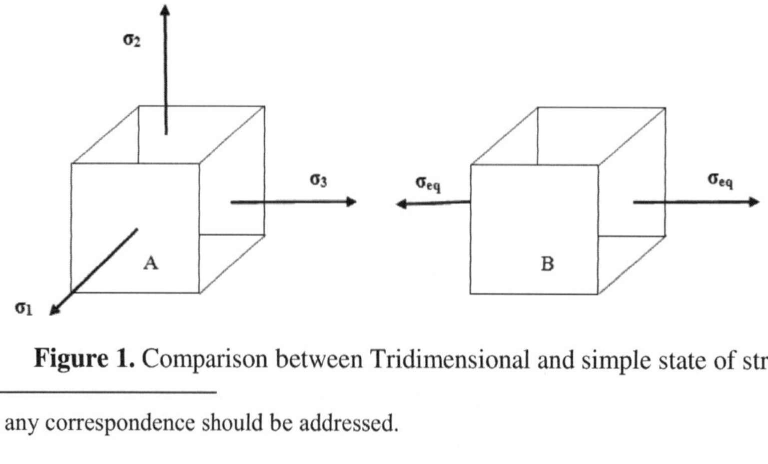

Figure 1. Comparison between Tridimensional and simple state of stress

[1] To whom any correspondence should be addressed.

In our case the problem arises differently: dimensions are imposed by the calculation of precision and required rigidity but without taking into account the matter implemented. The loads to which the spindle is subjected not being able to modify dimensions, it is nevertheless necessary to guarantee resistance as well volumetric as superficial of the spindle. For that it is necessary to determine the maximum equivalent stress the spindle undergoes as shown in Fig. 1. Our goal will be reached thanks be to the choice of a factor named criterion of resistance [2], [1] Once this one calculated, it will not remain any more that to choose a material whose resistance is higher than that evaluated previously

2. Determination of the stresses

We are in the general case of loading, since in the cross-section under consideration (bearing C) Fig. 2, there is a simultaneous action of simple bending, torsion and compression in addition to the radial and tangential stresses due to rotation.

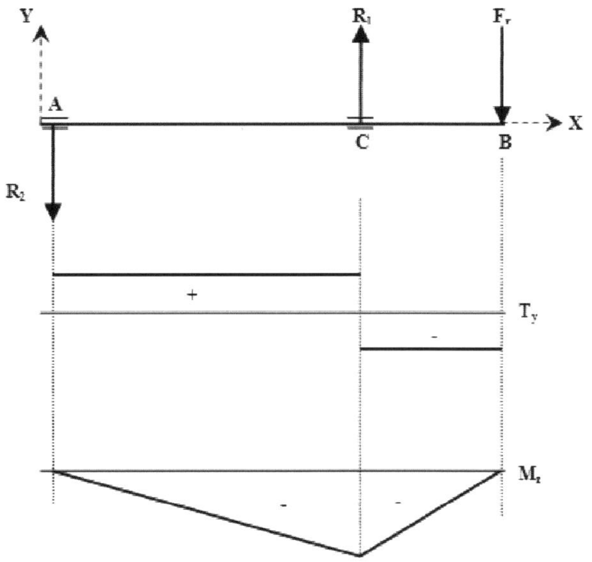

Figure 2. Shear and moment diagrams respectively.

2.1. Bending and compression

$$\sigma_f = \pm\frac{M_z \cdot y}{I_z} - \frac{4F_a}{\pi \cdot (D^2 - d^2)}$$

σ_f = stress at a critical point in tension or compression due to a combination of bending and axial load, N/mm^2
M_z = bending moment, Nmm
y = distance from neutral axis to outer surface, mm
I_z = axial moment of inertia of cross section, mm^4
F_a = axial load, N
D, d = outer and iner diameter of cross section respectively, mm

$$\tau_f = \frac{T_y \cdot Q_z}{I_z \cdot b_y}$$

τ_f = transverse shear, N/mm^2
T_y = transverse shear load on the cross section, N
b_y = with of the section containing the critical point, mm
Q_z = moment of the cross-sectional area under consideration, mm^3

Torsion:

$$\tau_t = \frac{M_t \cdot \rho}{I_p}$$

M_t = torsional moment, N-mm
ρ = distance from neutral axis to point of interest, mm
I_p = polar moment of inertia of cross section, mm^4

2.2. Stresses due to interference and rotation

We determine the value of these stresses at the location of bearing C by considering the fitted parts as thick-walled cylinders [4], as shown in Fig. 5 and superposing [1], [3] the stresses due to the pressure with those due to rotation.

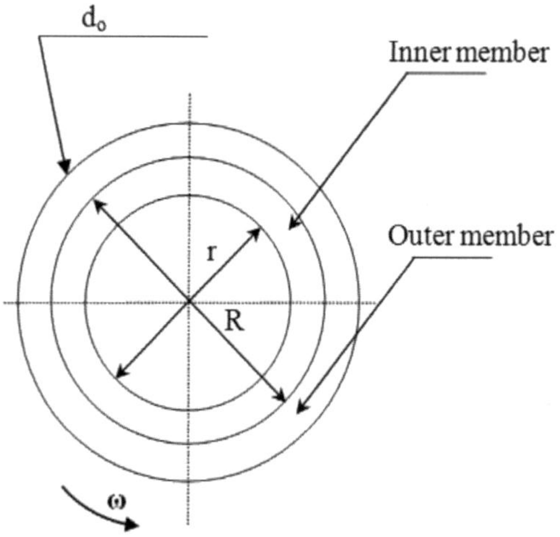

Figure 3. Interference fits.

2.2.1. Stresses due to pressure

$$(\sigma_r)_p = -\frac{p \cdot R^2}{R^2 - r^2}(1 - \frac{r^2}{r_1^2})$$

σ_r = radial stress, N/mm^2
R = outer radius of the spindle at C, mm
r = inner radius of the spindle at C, mm
r_1 = distance from neutral axis to point of interest, mm

$$(\sigma_\theta)_p = -\frac{p \cdot R^2}{R^2 - r^2}(1 + \frac{r^2}{r_1^2})$$

σ_θ = tangential stresses, N/mm^2
p = pressure at the contact surface, N/mm^2
The pressure at the contact surface may be calculated by the following equation[4]:

$$p = \frac{\delta}{d_c[\frac{d_c^2 + d_i^2}{E_i(d_c^2 - d_i^2)} + \frac{d_c^2 + d_i^2}{E_o(d_o^2 - d_i^2)} - \frac{\nu_i}{E_i} - \frac{\nu_o}{E_o}]}$$

δ = the total interference, mm
d_c = outer diameter of of the spindle, mm
d_i = inner diameter of of the spindle, mm

d_o = outside diameter of outer member, mm
E_i = modulus of elasticity of inner member, N/mm^2
E_o = modulus of elasticity of outer member, N/mm^2
v_i = Poisson's ratio for inner member
v_o = Poisson's ratio for outer member

2.2.2. stresses due to rotation

$$(\sigma_r)_R = \frac{3+\nu}{8}\mu\omega^2 R^2[(\frac{r}{R})^2 + 1 - (\frac{r_1}{R})^2 - (\frac{r}{r_1})^2]$$

ν = Poisson's ratio for inner member
μ = masse density, Kg/mm^3
ω = speed of rotation for spindle, rad/s

$$(\sigma_\theta)_R = \frac{3+\nu}{8}\mu\omega^2 R^2[(\frac{r}{R})^2 + 1 - (\frac{1+3\nu}{3+\nu})(\frac{r_1}{R})^2 + (\frac{r}{r_1})^2]$$

σ_θ = tangential stress

3. Discussion

According to the Maximum principal stress theory [7] (Von Misès criterion), the material fails by yielding when the maximum principal stress exceeds the tensile yield strength, or the minimum principal stress exceeds the compressive yield strength. Although it used in many softwares, this theory has some limitations. Furthermore, failure in ductile materials is fundamentally a shearing phenomenon, and one would therefore assume the failure criteria for such materials to rely upon shearing rather than tensile stresses. The maximum shear stress theory [6] (Tresca or Guest criterion in recognition of their contribution to its application), is an outgrowth of the experimental observation that a ductile material yields as a result of slip or shear along crystalline planes. This theory predicts that yielding will start when the maximum shear stress in the material equals the maximum shear stress at yielding in a simple tension test (for ductile material). If the point M (σ_1, σ_2, σ_3) is out of the envelope of the threshold of Tresca shown in Fig. 4, yielding will occur. The stresses σ_1, σ_2 and σ_3 are called principal stresses and are shown in Fig.1. Good agreement with experiment has been realized for ductile materials.
Numerical example:
Lets take σ_1= 208 N/mm^2, σ_2= -108 N/mm^2, σ_3= 200 N/mm^2 and y_p= 270 N/mm^2.
The criterion of tresca predicts that this state of stress will produce the flow in the material.

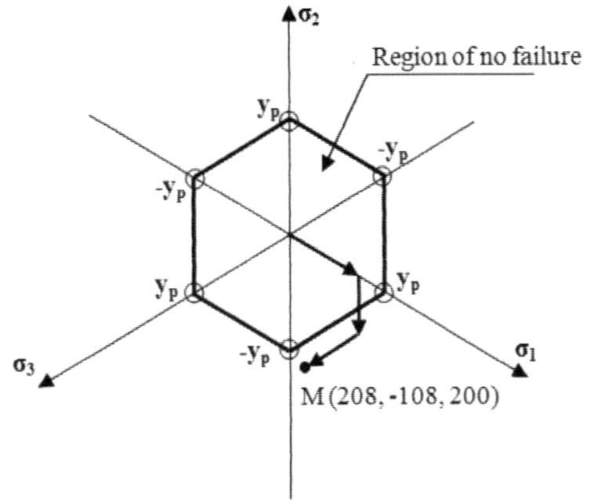

Figure 4. graphical representation of Tresca criterion.

IOP Conf. Series: Materials Science and Engineering **28** (2012) 012015 doi:10.1088/1757-899X/28/1/012015

4. Conclusion:

From the viewpoint of mechanical design, it is imperative that some practical guides be available to predict yielding under the conditions of stress as they are likely to exist in service. To meet this need and to understand the basis of material failure, the maximum principal theory (Von Mises criterion) and the maximum shear stress theory (Tresca criterion) have been used. The latter seems to be more advantageous.

References:

[1] Bazergui A, Bui-Quoc T, Biron A, Mcintyre G, Laberge C 1985 *Resistance des matériaux, Edition de l'Ecole polytechnique de Montreal.*

[2] Anissimov A 1993 *Calcul des barres dans le cas de sollicitation composée (Partie 1-de RDM2), U.S.T.H.B.*

[3] Fanchon J. L 2001 *Guide de Mécanique Sciences et technologies industrielles, Nathan.*

[4] Hall A. S, Holowenko A. R, Laughlin H. G 1961 *Machine Design, McGraw-Hill.*

[5] Selivanov S 1984 *Calcul et élaboration des ensembles broches-paliers, Université de Annaba.*

[6] Ugural A. C, Fenster S. K 1975 *Advanced Strength and Applied Elasticity, Teaneck, New Jersey.*

[7] Barthod Ch 2010-2011 *Module SPM2A, Résistance des Matériaux, IUT Annecy.*

IOP Conf. Series: Materials Science and Engineering **28** (2012) 012016 doi:10.1088/1757-899X/28/1/012016

Poly [(pyrrole-2, 5-diyl) co(4-Nitro benzylidene)]: A new synthetic approach and electronic properties

H. Gherras[a], A. Hachemaoui[a1], A. Yahiaoui[a], M. Belbachir[b] and A. Belfedal[a]
[a]Laboratoire de chimie Organique, Macromoléculaire et des Matériaux (LCOMM), Université de Mascara, Faculté des Sciences et de la Technologie, BP 763 Mascara 29000, Algérie, Fax: +213 45 81 39 98
[b]Laboratoire de Chimie des Polymères, Faculté des sciences, Université d'Oran, BP 1524 El- Menouer 31000 Oran, Algérie

E-mail: hachemaoui2002@yahoo.fr

Abstract. A new poly (heteroarylene methylene) containing one to four α-pyrrole rings alternating with an sp3-carbon atom in the main chain were synthesized and characterized. The synthesized poly [(pyrrole-2, 5-diyl)-co-(4-nitrobenzylidene)] was found to be highly dehydrogenated, thus resulting in black materials with conjugated segments larger than a single pyrrole ring. The conjugated polymers containing alternating aromatic and quinoid heteroarylene moieties in the main chain were prepared by polycondensation of 4-nitrobenzaldehyde and pyrrole catalysed by Maghnite-H+. The new polymer is useful precursors to small-bandgap polymers based on poly (heteroarylene methines) and is also useful model system for the investigation of the effects of π-conjugation length on the electronic and optical properties of conjugated polymers.

1. Introduction

Among conducting electroactive polymers, polypyrrole and its derivatives have received a great deal of attention due to its high electrical conductivity, good environmental stability, relative ease of synthesis and good redox reversibility. These properties are favourable to its applications in the areas of secondary batteries, electrocatalysis, electrochromic display devices, light-emitting devices, chemical sensors and biosensors [1-5]. However, polypyrrole and other conducting electroactive polymers are limited in practical use due to their very fragile structure and insolubility, which give rise to processing difficulties [6, 7]. Several approaches have been taken to improve processability of conducting polymers [8-11]. One effective method is to introduce insulating polymeric matrices into them due to the excellent processability of classical insulating polymers [12-18]. This can be achieved by blending, composite formation or copolymerization. Copolymerization could be a desirable way because the chemical linkage between the insulating matrix and the conjugated polymer can improve the chemical stability of the polymer [17, 18]. The resulting copolymers with new functional groups showed different properties from polypyrrole homopolymer, which probably widen the application of conducting polymer.

Several kinds of copolymers containing pyrrole and other insulating units, such as styrene, tetrahydrofuran, methyl methacrylate, ε-caprolactone, acryloyl chloride, etc, have been prepared and studied [18-23]. All results showed the success in improving the mechanical and physical properties of

[1] To whom any correspondence should be addressed

polypyrrole. However, these synthetic methods contained many steps and strict condition, leading to limited application of these copolymers. A simple and alternative way is direct electrochemical copolymerization of monomers present in the mixture in proper solution [24-26].

Recently, an Algerian proton exchanged montmorillonite clay called Maghnite-H+ (Mag-H+), a new non toxic cationic initiator, was used as a catalyst for cationic polymerization of a number of vinylic and heterocyclic monomers [27-30].

In the present work, we present a new approach to design poly[(pyrrole-2,5-diyl)-co-(4-nitrobenzylidene)] (PPNB) in one shot, namely by the condensation of pyrrole and 4-nitrobenzaldehyde catalyzed by Mag-H+ (scheme 1).

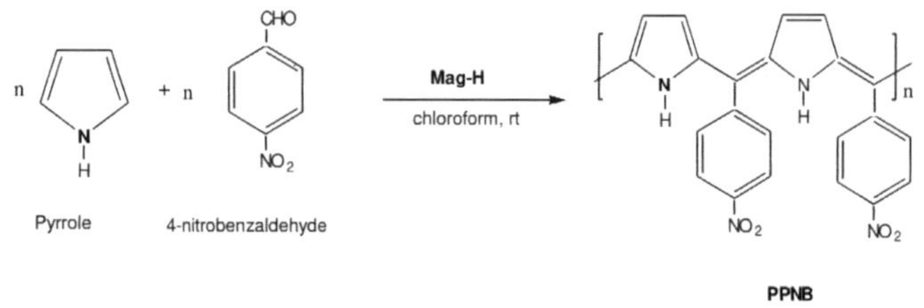

Schema. Synthetic route for PPNB polymer

The conjugated polymer was characterized by means of 1H-NMR, UV, FT-IR spectroscopy. The X-ray data showed the presence of a backbone form of the poly [(pyrrole-2,5-diyl)-co-(4-Nitrobenzylidene)] formed.

2. Experimental

2.1. Materials

Pyrrole was purchased from Aldrich Chemical Co. and distilled under reduced pressure. Dichloromethane and 4-nitrobenzaldehyde were used as received.

The double bond content was obtained by UV spectroscopy (UV/ visible OPTIZEN 2120 PC apparatus) in CH_2Cl_2. The molecular structure of the polymer was characterized by FT-IR spectroscopy (Perkin-Elmer System). 1H-nuclear magnetic resonance (NMR) measurements were carried out on a 300 MHz Bruker NMR Spectrometer equipped with a probe BB05 mm, in CDCl3. Tetramethylsilane (TMS) was used as the internal standard in these cases.

2.2. Polymer preparation

The condensation of 4-nitrobenzaldehyde with pyrrole in the presence of Mag-H+ as a catalyst was carried out by condensation in solution under nitrogen for 6 h.

Each mixture was prepared with a weighted quantity of Mag-H+ dried just before use for 1 h in a drying oven at 100°C. 4-nitrobenzaldehyde (8 mmol) and pyrrole (8 mmol) were dissolved in 10 ml of 1,2-dichloroethane, and 10 % by weight of Mag-H+ were added. The reaction was carried out at 25°C for 6 h. At the end of the reaction, the resulting mixture was filtered to remove the clay and then slowly added to methanol with stirring and then the polymer was dried under vacuum at room temperature for 24h. The yield was 93.86 %

Anal. found: C, 63.61; N, 13.27; O, 16.94; H, 4.48. IR (film on KBr, cm^{-1}): 3419, 1650, 1596, 1517, 1429, 1347, 1109, 1017,862, 772, 732, 466. 1H NMR ($CDCl_3$, δ: ppm, TMS reference): 7.13 (2H, pyrrole ring), 7.60 (2H, phenylene ring), 8.20 (2H,phenylene ring).

3. Results and discussion

The synthesis of soluble poly(heteroarylene methylenes) by acid-catalyzed polymerization of thiophene or pyrrole with aldehydes has previously been demonstrated.[30-34] This article reports the synthesis of pyrrole with aldehydes possessing electron acceptor side groups. In contrast to most of the other conductive polymers, PPNB is a soluble polymer in common organics solvents. The catalyst can be easily separated from the polymer product and regenerated by heating at a temperature above 0 °C [27]. The effect of the amount of catalyst and of time on the polymerization yield and on the viscosity of the polymers was studied. These polymers can be dissolved in high polar solvents such as DMSO, DMF.

Most of the PPNB were found to be soluble in organic solvents such as CH_2Cl_2, N, N-dimethylformamide (DMF), and sulfolane. Although polymers have highly conjugated chains due to the high degree of dehydrogenation, they were very soluble in organic solvents such as THF, giving grey solutions of high concentrations. The very good solubility of polymers in spite of their high degree of π-conjugation is due largely to the bulky side groups (ϕ) at the methane carbon =C (ϕ) link and also to the low molecular weight to some extent.

3.1. Polymer synthesis and structure

3.1.1. FT-IR spectrum

Figure 1 shows the FT-IR absorption spectrum of PPNB. A distinct peak at 767 cm^{-1} of (PPNB) is due to the Cβ–H the out-of-plan vibration, characteristic of α-linkage in pyrrole ring. The peak at 1423 cm^{-1} of Fig. 1 is assigned to the C=C symmetric stretching vibration bands of the pyrrole rings. This assignment is based on the FT-IR spectrum of pyrrole. The broad band at 3742 cm-1 is N–H stretching vibration of pyrrole ring. Compounds containing the NO_2 group have strong absorption bands from symmetric and asymmetric deformations (1339 cm^{-1} and 1511 cm^{-1}, respectively) [35]. The exact position of these bands depends on the substitutions and unsaturations within the NO_2 group. Collman et al [36] assigned the 1542-1526 cm^{-1} and 1356-1340 cm^{-1} absorption ranges to the asymmetrical and symmetrical vibrations, respectively, for the NO_2 groups in 5,10,15,20-tetrakis-(o-nitrophenyl)porphyrin), Goldberg et al [37] reported these absorptions at 1500 and 1330 cm^{-1}, respectively, for the same porphyrin. Similar FT-IR bands at 2840–2960 cm^{-1} region are observed in nitrobenzene. An important feature of the FT-IR spectrum of PPNB is the 1648 cm^{-1} absorption band, which can be assigned to the C=C stretching vibration of quinoid ring and methine conjugated with quinoid ring.

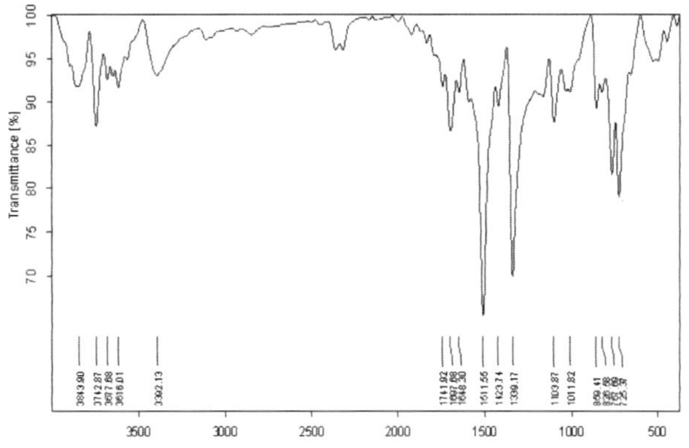

Figure 1. FT-IR spectrum of intrinsic PPNB..

3.1.2. Thermal properties

Figures 2 shows the TGA curves of PPNB obtained in a nitrogen atmosphere at a heating rate of 10°C/min. These polymers show good thermal stability. For example, the intrinsic PPNB has an onset of thermal decomposition of 200 °C.

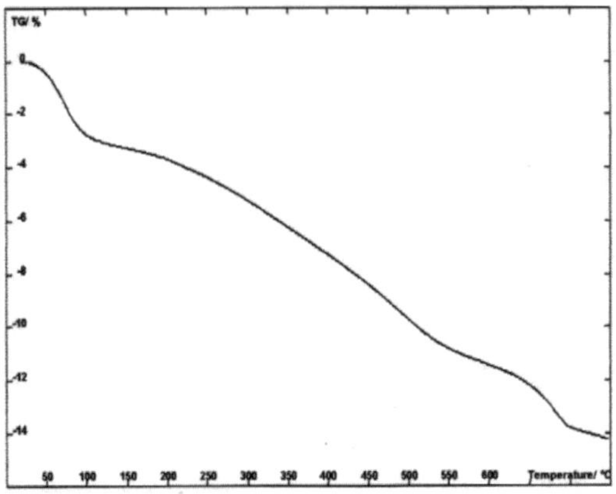

Figure 2. TGA spectrum of intrinsic PPNB.

3.1.3. Optical results

When light passes through a medium, its behavior will depend on the proprieties of the medium and of the light (reflection, absorption, refraction and scattering). The refractive index in its real and imaginary parts or dielectric function is the response to interaction between the material and light. To well study these parameters it is necessary to make the optical measurements (transmission and/or reflection) in the UV wavelength range. We present in Fig. 3 the transmittance T and absorptance A spectra normalized with regard to the basic line in the range wavelength 190 to 1100 nm [38]. We distinguish clearly tow regions in these spectra:

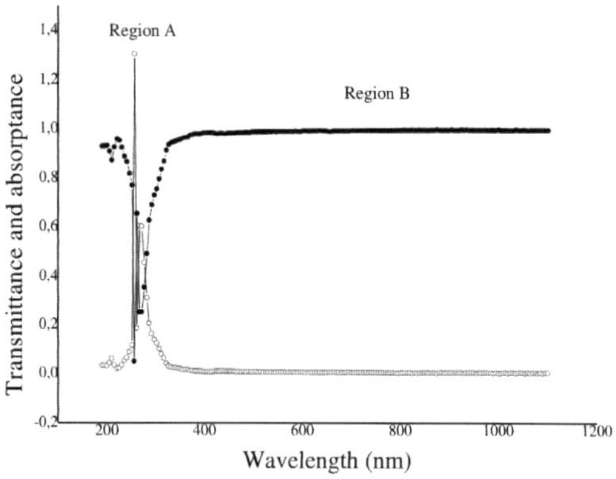

Figure 3. Transmittance and absorptance variation in UV-range

i- Region A or absorption region: in this range the transmittance and absorptance spectra show a strong variation with diminution the transmittance spectrum and increase the level of absorptance. In this range we can deduce from the transmittance spectra the absorption coefficient $\alpha(\omega)$ at frequency ω, by the following relation [39]:

$$\alpha(\omega) = -\frac{1}{d}\ln[T] \qquad (1)$$

where d is the thickness of the film and T its transmittance normalized by background.
For the material no dispersive, the reflectance spectrum R is calculated by:

$$R = 1 - A - T \tag{2}$$

ii- Region B or transparency region: in this range of wavelength the optical transmittance remains practically constant at 0.97 (97%) and the absorptance is typically equal to zero. In this region we can determined the refractive index n. Where R is related to refractive index n for the normally incidence by relation:

$$R = \left(\frac{n-1}{n+1}\right)^2 \tag{3}$$

From this expression we can deduce the refractive index in transparency region:

$$n = \frac{1+\sqrt{R}}{1-\sqrt{R}} \tag{4}$$

The values found for the refractive index at room temperature for unpolarized light are represented in the Fig. 4.

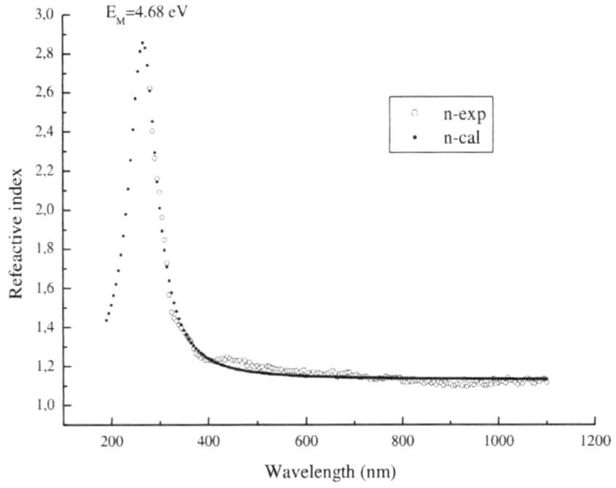

Figure 4. Refractive index variation in UV-range and simulated curve by Lorentz model

From the results shown in this figure it clearly appears that the refractive index decrease rapidly (1.5 at 1.1) in the wavelength range 190 to 500 nm and remains constant in the region between 500 and 1100 nm. This behavior is a good agreement with results found for semi-conductors such as silicon (Si) and germanium (Ge) [40, 41]. For determined the refractive index in all wavelength range studied we simulated this parameter by Lorentz model [42]. We notice that both spectra are perfectly superposed in the high wavelengths range from 400 to 1100 nm. We can by this model generalize the refractive index variation on the low wavelengths range. The Lorentz model works surprisingly well, provided we remember that real materials correspond to a collection of Lorentz oscillators with different frequencies. The outer, or valence, electrons predominantly determine the characteristics of the optical properties a material. Semiconductors are covalently bonded materials where the electrons are evenly shared between neighboring atoms. (Some insulators are covalently bonded, too.). This means that the electrons are smeared out into broader bands and that their resonance frequencies are lower than for ionically bonded materials. Usually these materials can be described by a single energy gap and single broad absorption band above the energy gap (example Silicon see work [43]). From these simulations by Lorentz model, we can determined the average gap EM centered at 260 nm equivalent to energy EM=4.68 eV. The average gap usually considered as the energy separation between the centers of both the conduction and valence bands. This gap such as defined by this method is relatively high than the optical gap which separate localized and delocalized states defined by Tauc [44]. In order to understand this phenomenon well, the values of the static refractive index n_0 [refractive index at zero energy: i.e $n_0 = n(\hbar\omega = 0)$] is determined from the analysis of the Wemple and Didominico one oscillator model [45]. According to this model, the dispersion relation of the refractive index is given by:

$$n^2(\hbar\omega) = 1 + \frac{E_D.E_m}{Em^2 - (\hbar\omega)^2} \tag{5}$$

Where Em denotes the average gap, E_D denotes the dispersion energy related to the coordination number of the atoms. The static refractive index n_0, average gap Em and dispersion energy E_D are deduced by considering the variation of $[n^2(\hbar\omega)-1]^{-1}$ as function of $(\hbar\omega)^2$ (figure 5). The values obtained for these parameters by this analysis are: Em=4.16 eV, $E_D \approx 1$ eV and $n_0 = 1.24$

The difference obtained in values for average gap Em by these models due to principally to prediction of each. Nevertheless, the values found for Em show clearly that our material is a good semi-conductor and can be compared to silicon or germanium. Although the value of E_D is relatively lower.

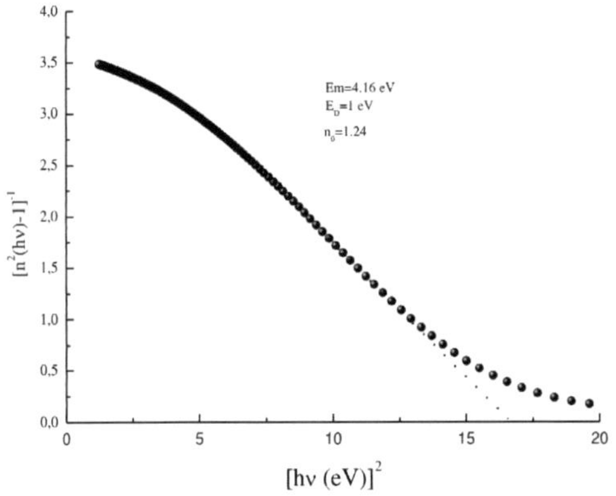

Figure 5. Show the Wemple Didominico model for determined the average gap Em, dispersion energy E_D and static refraction index n_0.

In the figure 6, we represent the variation of the imaginary part, ε_2, of dielectric function deduced from refractive index n and excitation coefficient k by:

$$\varepsilon_2 = 2.n.k \tag{6}$$

Where k related to absorption coefficient α by relation:

$$\alpha = \frac{4\pi.k}{\lambda} \tag{7}$$

λ is the wavelength studied.

Experimentally the dielectric function also can be obtained by spectroscopic ellipsometry (SE) when employed to study thin film growth or modification of the surface in various liquid environnement though in-situ or real-time measurement [46]. It can be seen clearly that this spectra present the look of a semiconductor a microstructure. In fact, in the high energy range, two peaks are observable. One mean peak around 4.6 eV and one intense shoulder around 4.9 eV. They correspond to a convolution of direct electronic transitions in UV region [47]. These also features are an obvious signature of the presence of crystallites structure in this film.

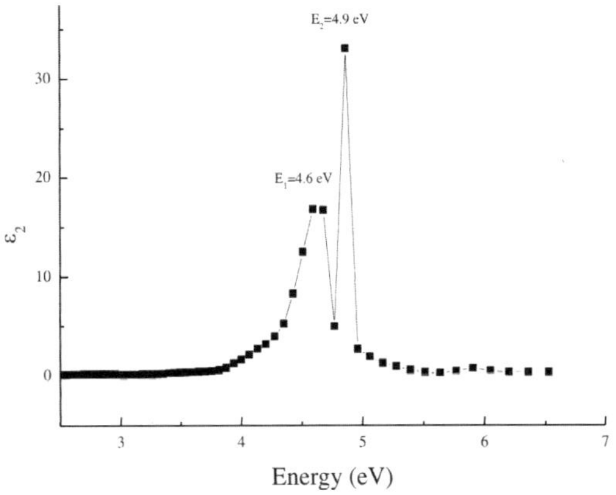

Figure 6. Imaginary part of dielectric function ε_2

Our material studied here is a mixture of poly [(pyrrole-2, 5-diyl)-co-(4-nitrobenzylidene)] and dichloromethane, then two or more liquids are mixed together physically, we may apply the Bruggeman effective medium theory (EMT) to deduce the optical function of the mixture [48] by the formula:

$$f_A \frac{\varepsilon_A - \varepsilon}{\varepsilon_A + 2\varepsilon} + (1 - f_A)\frac{\varepsilon_B - \varepsilon}{\varepsilon_B + 2\varepsilon} = 0 \quad (8)$$

Here, ε_A and ε_B are the dielectric function of liquid A and liquid B respectively. f_A is the volume fraction of liquid A and ε is the resultant dielectric function of mixture. We are going to return on this study in detail in our next works when we employed the spectroscopy ellipsometry for these materials.

4. Conclusion

In conclusion, a novel polymer, poly [(pyrrole-2, 5-diyl)-co-(4-nitrobenzylidene)], which has a π-conjugated chain was synthesized by polycondensation of 4-nitrobenzaldehyde and pyrrole catalysed by Maghnite-H+. The resultant polymer showed good solubility in common organic solvents and good film formability. Such results may serve primarily to illustrate a new strategy to increase the solubility of low band gap polymers through the arrangement of different aromatic heterocycles in conjugated polymer backbones.

The results obtained with optical measurements and analyzed by different models, show clearly that our material (PPNB) is a good semi-conductor, when the average gap around 4eV. In fact, the analysis the imaginary part of dielectric function, ε_2, show also the structure of the semiconductor present a crystallites by presence the two peaks of Van Haove.

References

[1] Mermilliod N, Tanguy J, and Petiot F 1986 *J. Electrochem. Soc.* **133** 6 1073-1079.
[2] Mac Diarmid A. G and Epstein A. J 1997 *Acs. Symp. Ser.* **672** 395-407.
[3] Mao H. Y and Pickup P. G 1989 *J. Electroanal. Chem.* **265** 1-2 127-142.
[4] Ramanavicius A, Ramanaviciene A and Malinauskas A 2006 *Electrochim Acta* **51** 27 6025-6037.
[5] Xue H. G 1995 and Mu S. L *J Electroanal Chem* **397** 1-2 241-247.
[6] Lindenberger H, Schafersiebert D, Roth S and Hanack M 1987 *Synth. Met.* **18** 1-3, 37-41.
[7] Qian R. Y, Qiu J. J and Shen D. Y 1987 *Synth. Met.* **18** 13-18.

[8] De Paoli M. A, and Gazotti W. A 2002 *Macromol. Symp.* **189** 83-103.
[9] Li Y. F and Ouyang J. Y 2000 *Synth. Met.* **113** 1-2, 23-28.
[10] Corres M. A, Mulgica A, Carrasco P. M, and Cortazar M. M *Polymer* 2006 **47** 19 6759-6764.
[11] Balci N, Toppare L, Akbulut U, Stanke D and Hallensleben M. L 1998 *J. Macromol. Sci.-Pure Appl. Chem.* **A35** 10 1727-1739.
[12] Lee Y, Shin D, Cho J, Park Y. H, Son Y, and Baik D. H 1998 *J. Appl. Polym. Sci.* **69** 2641-2648.
[13] Levine K. L, and Iroh J. O 2004 *J. Porous. Mater.* **11** 87-95.
[14] Yin W. S, Yan T. J, Gan L. M, Chew C. H, Liu H. W and Gan L. H 1998 *Eur. Polym. J.* **34** 1763-1766.
[15] Mano V, Felisberti M. I, Matencio T and DePaoli M. A 1996 *Polymer* **37** 5165-5170.
[16] Brahim S, and Guiseppi-Elie A 2005 *Electroanal* **17** 556-570.
[17] Park Y. H, Shin H. C, Lee Y, Son Y, and Baik D. H 1999 *Mol. Cryst. Liq. Cryst.* **327** 221-224.
[18] Stanke D, Hallensleben Ml and Toppare L 1995 *Synth. Met.* **73** 261-266.
[19] Kizilyar N, Toppare L, Onen A and Yagci Y 1998 *Polym. Bull.* **40** 639-645.
[20] Park Y. H, Kim K. W, and Jo W. H 2002 *Polym. Adv. Techn.* **13** 670-677.
[21] Tarkuc S, Sahin E, Toppare L, Colak D, Cianga I, and Yagci Y 2006 *Polymer* **47** 2001-2009.
[22] Kizilyar N, Toppare L, Onen A and Yagci Y 1999 *J Appl. Polym. Sci.* **71** 713-720.
[23] Jerome C, Martinot L, Louette P and Jerome R 2000 *Macromol. Symp.* **153** 305-319.
[24] Borole D. D, Kapadi U. R, Mahulikar P. P and Hundiwale D.G 2006 *Mater. Lett.* **60** 2447-2452.
[25] Mu S. L 2006 *Electrochim Acta* **51** 3434-3440.
[26] Dhanalakshmi K and Saraswathi R 2001 *J Mater Sci* **36** 4107-4115.
[27] Belbachir M and Bensaoula A 2001 *US patent (Houston)* **6** 274 527.
[28] Yahiaoui A and Belbachir M 2006 *J Appl Polym Sci*, **100** 1681-1687.
[29] Hachemaoui A, Yahiaoui A and Belbachir M. 2006 *J Appl Polym Sci* **102** 3741-3750.
[30] Meghabar R, Megherbi A and Belbachir M 2003 *Polymer* **44** 4097-4100.
[31] Jenkhe S. A 1986 *Nature* **322** 345-347.
[32] Chen W. C and Jenkhe S. A 1995 *Macromolecules* **28** 454-464.
[33] Chen W. C and Jenkhe S. A 1995 *Macromolecules* **28** 465-480.
[34] Yi W. H, Wang L. L, Yan W and Wan M. X 2000 *Acta. Polym. Sinica.* **1** 36-40.
[35] Silverstain R. M, Bassier G. C and Morril T. C 1974 *Spectrometric Identification of Organic Compounds, John Wiley and Sons, Inc.*, 3a ed. 256.
[36] Collman J. P, Brauman J. I, Doxee K. M, Halbert T. R, Bunnenberg E, Linder R. E, Lanar G. N, Gaudio J. D, and Spartalian K 1980 *J. Am. Chem. Soc.* **102** 4182.
[37] Kumar R. K, Balasubramanian S and Goldberg I 1998 *Inorg.Chem.* **37** 541-552.
[38] Monfredotti C, Fizzoti F, Borco M, Pastorino P, Polessello P and Vittone E 1994 *Phys. Rev.* **50** 24 18046 -18053.
[39] Brodsky M. H, Cardona M and Cuomo J 1977 *Phys. Rev.* **16** 8 3556-3571.
[40] Bouizem Y, Belfedal A, Sib J. D, Kebab A and Chahed L 2007 *J. Phys. Condens. Matter.* **19** 35 356215.
[41] Bouizem Y, Belfedal A, Sib J. D, Kebab A and Chahed L 2005 *J. Phys. Condens. Matter.* **17** 33 5149-5158.
[42] Marquardt D. W 1963 *J. Soc. Indus. Appl. Math.* **I1** 431-441.
[43] Belfedal A, Bouizem Y, Benlakehal D, Baghdad R, Sib J. D, Kebab A, Chahed L and Zellama K 2010 *Phys. Status. Solidi. C7* **3-4** 565-570.
[44] Tauc J and Meuth A 1972 *J. Non-Cryst. Solids* **8-10** 569-585.
[45] Wemple S. H and Didomenico M 1971 *Phys. Rev. B* **3** 1338-1351.
[46] Arwin H 1998 *Thin Solid Films* **313-314** 764-774.
[47] Van Hove 1953 L *Phys. Rev.* **89** 1189-1193.
[48] Brugemann A. G 1936 *Ann. Phys. (Leiptzig)* **24** 636-665.

Synthesis and crystal structure of $Sr_2As_2O_7$ from single-crystal data

F. Edhokkar[a1], A. Hadrich[a], M. Graia[b], T. Mhiri[a]
[a] Laboratoire de l'Etat Solide, Faculté de Sciences, Université de Sfax, BP 1171-3000 Sfax, Tunisie.
[b] Laboratoire de Matériaux et Cristallochimie, Faculté des Sciences, Université de Tunis Elmanar, BP W 1060, Tunis, Tunisie.

E-mail: fadhila_edk@yahoo.fr; mohseng2002@yahoo.fr

Abstract. The title compound, strontium diarseniate $Sr_2As_2O_7$, was prepared by solid-state method. This compound crystallizes in the tetragonal system, space group $P4_1$ with a= 7.142 (5) Å; c= 25.693 (5) Å and Z= 8. Its structure is composed of $SrO_{7\text{-}9}$ polyhedra and $[As_2O_7]$ bitetrahedral units. These groups are linked together by corner and edge-sharing forming a three dimensional framework. The crystal structure of the title compound has been the subject of a previous investigation based on X-ray powder data [M. Weil, T. dordevic, C. L. Lengauer, U. Kolitsch. Solid State Sciences 11 (2009), 2111–2117]. The present investigation confirms the previous study, but with higher precision and with all displacement parameters refined anisotropically. $Sr_2As_2O_7$ is isotypic with the high-temperature polymorphs of $Ca_2P_2O_7$ and $Sr_2V_2O_7$.

1. Introduction

In the past years an increasing interest in the synthesis and characterization of alkaline earth metals arsenate crystals has been observed due to their interest in various sectors such as the chemistry, the geology, the environment and the technology [1-4].

The crystal structure of strontium diarsenate $Sr_2As_2O_7$ compound has been the subject of a previous investigation based on X-ray powder data [1]. Strontium diarsenate(V) is isotypic with the high-temperature polymorphs (b-forms) of $Ca_2P_2O_7$ [5] and $Sr_2V_2O_7$ [6] and with the ambient condition A-type polymorphs of the $Ln_2Si_2O_7$ disilicates (Ln = La, Ce, Pr, Nd, Sm) [7].

Here, we report the synthesis by solid-state methods and the crystal structure of $Sr_2As_2O_7$ determined from single-crystal x-ray diffraction data.

2. Experimental

2.1. Preparation:

Single crystals of the title compound were obtained from a mixture of As_2O_5, and $SrCO_3$ with a molar ratio of 7/3. The calcinations of the mixture are performed through three steps: first at 550°C for 12 h, second heated to 900°C for 48 h, the molten mixture is then cooled to 650°C/ 12 h and finally slowly cooled to room temperature. The obtained crystals were separated from the excess flux by washing the product in boiling water.

[1] To whom any correspondence should be addressed.

IOP Conf. Series: Materials Science and Engineering **28** (2012) 012017 doi:10.1088/1757-899X/28/1/012017

2.2. Structure determination:

A suitable single crystal with dimensions 0.16 x 0.11 x 0.03 mm^3 was chosen for the structure determination and refinement. It was selected under a polarizing microscope and was mounted on a glass fibre. The crystal structure is determined from single-crystal X-ray diffraction data collected at room temperature using an Enraf-Nonius CAD-4 diffractometer with monochromated graphite MoK_α radiation [8]. The reflections were corrected for Lorentz and polarization effects; absorption correction was obtained via a psi-scan [9] and secondary extinction correction was applied too [10]. The crystal structure was solved and refined against F^2 in the space-group $P4_1$ using the SHELX-97 computer programs [10, 11] included in the WingX software package [12].

Table 1. Crystallographic data, recording conditions, and refinement results for $Sr_2As_2O_7$.

Crystal data	
Chemical formula	$Sr_2As_2O_7$
Formula Weight	437.08
Crystal system	Tetragonal
Space group	$P4_1$
Cell dimensions	a= 7.142 (5) Å; c= 25.693 (5) Å
Cell volume	1311.6 (1) $Å^3$
Z	8
data collection	
Diffractometer	Enraf-Nonius CAD-4
Temperature	296 K (2)
Shape	platelet
Radiation,graphite monochromator:	Mo Kα (λ=0.71069 Å)
scan mode	w – 2θ
Theta range for data collection	2.9 – 25.0
Range of h, k, l	$-8 \leq h \leq 3$, $-1 \leq k \leq 8$, $-1 \leq l \leq 30$
Measured reflections	2356
Independent reflections	1235 [R_{int}= 0.11]
refinement	
Refinement method	Full-matrix least-squares on F^2
parameters	200
$S(F^{2)}$	1.03
Final R indices [I > 2σ(I)]	R_1= 0.047; wR= 0.127
$(\Delta\rho)_{max}$ / $(\Delta\rho)_{min}$	1.9/-2.1 (e.$Å^{-3}$)

A least-square refinement of the unit cell dimensions, using the four circle diffractometer angular data, leads to the following quadratic parameters: a = 7.142 (5) Å; c= 25.693 (5) Å. These values illustrate the isotypy of the obtained phase with $Ca_2P_2O_7$ and $Sr_2V_2O_7$ [5, 6].

We note that the attempt to solve the structure in each of the two space groups $P4_1$ and $P4_3$ gives similar results. So we opted for the first space group ($P4_1$) according to the previous study [1].

Experimental conditions used for the intensity data collection and the refinement results are listed in Table 1. All atoms were refined with anisotropic thermal parameters. Refinement converged with R = 0.047 for 1235 observed reflections. The atomic coordinates and the displacement parameters are reported in Tables 2 and 3. The significant bond lengths and angles are listed in Table 4.

Table 2. Atomic coordinates and U_{eq} temperature factors for $Sr_2As_2O_7$.U_{eq} = 4/3 Σi Σj $a_i a_j$ U_{ij}

Atome	x	y	z	U_{eq}(Å^2)
Sr(1)	0.2434 (3)	0.1256 (3)	0.76338 (10)	0.0110 (6)
Sr(2)	0.2736 (3)	0.2364 (3)	1.12523 (9)	0.0102 (5)
Sr(3)	0.6547 (3)	0.1039 (3)	0.87748 (11)	0.0141 (6)
Sr(4)	-0.1521 (3)	0.5262 (3)	0.77205 (9)	0.0086 (5)
As(1)	0.1454 (3)	0.2512 (3)	0.88804 (10)	0.0074 (6)
As(2)	0.7211 (3)	0.0164 (3)	0.74683 (10)	0.0076 (5)
As(3)	0.3954 (3)	0.3133 (3)	0.98926 (10)	0.0071 (5)
As(4)	0.6459 (3)	0.2651 (3)	0.64473 (10)	0.0093 (6)
O(1)	0.234 (2)	0.200 (2)	1.0232 (7)	0.010 (3)
O(2)	0.112 (2)	0.406 (2)	0.8391 (8)	0.013 (4)
O(3)	0.283 (2)	0.083 (2)	0.8653 (9)	0.019 (4)
O(4)	0.555 (2)	-0.020 (2)	0.7901 (7)	0.018 (4)
O(5)	-0.115 (2)	0.167 (2)	0.7694 (7)	0.010 (4)
O(6)	0.575 (2)	0.130 (2)	0.6977 (7)	0.010 (3)
O(7)	0.504 (2)	0.493 (2)	1.0128 (8)	0.019 (4)
O(8)	-0.040 (2)	0.171 (2)	0.9175 (7)	0.012 (4)
O(9)	0.554 (2)	0.158 (2)	0.9677 (7)	0.009 (3)
O(10)	0.803 (2)	-0.175 (2)	0.7163 (7)	0.015 (4)
O(11)	0.448 (2)	0.329 (3)	0.6164 (9)	0.027 (5)
O(12)	0.271 (2)	0.389 (2)	0.9313 (7)	0.009 (3)
O(13)	0.775 (2)	0.123 (2)	0.6066 (8)	0.015 (4)
O(14)	0.783 (2)	0.4430 (19)	0.6642 (8)	0.017 (5)

Table 3. Anisotropic thermal parameters U_{ij} for $Sr_2As_2O_7$.

Atome	U_{11}	U_{22}	U_{33}	U_{23}	U_{13}	U_{12}
Sr(1)	0.0134 (10)	0.0129 (10)	0.0068 (14)	0.0015 (9)	-0.0010 (9)	0.0029 (8)
Sr(2)	0.0117 (10)	0.0130 (10)	0.0060 (13)	-0.0024 (9)	0.0006 (10)	0.0021 (8)
Sr(3)	0.0144 (10)	0.0149 (10)	0.0130 (15)	-0.0012 (10)	-0.0017 (11)	-0.0016 (8)
Sr(4)	0.0112 (10)	0.0098 (10)	0.0049 (14)	0.0015 (9)	0.0002 (9)	0.0011 (8)
As(1)	0.0066 (10)	0.0076 (10)	0.0080 (15)	0.0006 (9)	-0.0006 (10)	0.0012 (8)
As(2)	0.0074 (10)	0.0093 (10)	0.0061 (14)	0.0010 (10)	-0.0007 (10)	0.0004 (8)
As(3)	0.0084 (10)	0.0072 (10)	0.0059 (14)	0.0001 (9)	-0.0006 (10)	0.0013 (8)
As(4)	0.0089 (10)	0.0087 (10)	0.0101 (15)	0.0008 (10)	-0.0008 (10)	-0.0018 (8)
O(1)	0.014 (8)	0.016 (8)	0.000 (10)	0.002 (7)	0.000 (7)	-0.001 (6)
O(2)	0.011 (7)	0.014 (8)	0.013 (10)	0.006 (8)	-0.005 (8)	-0.002 (6)
O(3)	0.010 (7)	0.014 (7)	0.033 (13)	-0.003 (9)	-0.007 (8)	-0.005 (7)
O(4)	0.021 (9)	0.021 (9)	0.010 (11)	0.010 (8)	0.005 (8)	0.006 (7)
O(5)	0.007 (7)	0.018 (8)	0.003 (10)	-0.005 (7)	-0.002 (7)	-0.003 (6)
O(6)	0.006 (7)	0.013 (7)	0.010 (10)	-0.001 (7)	-0.001 (6)	0.000 (6)
O(7)	0.022 (9)	0.012 (8)	0.024 (13)	-0.009 (8)	-0.008 (8)	-0.008 (8)
O(8)	0.007 (7)	0.019 (8)	0.011 (10)	-0.008 (8)	0.007 (7)	-0.003 (6)
O(9)	0.014 (7)	0.012 (7)	0.000 (9)	0.001 (7)	0.004 (7)	0.007 (6)
O(10)	0.022 (9)	0.019 (8)	0.005 (11)	-0.001 (7)	0.007 (8)	-0.001 (7)
O(11)	0.002 (7)	0.035 (10)	0.042 (15)	-0.008 (10)	-0.009 (9)	-0.008 (7)
O(12)	0.014 (7)	0.012 (7)	0.000 (9)	0.000 (7)	-0.013 (7)	-0.001 (6)
O(13)	0.013 (7)	0.014 (7)	0.019 (12)	0.002 (8)	0.003 (8)	0.010 (6)
O(14)	0.010 (7)	0.001 (6)	0.038 (14)	-0.010 (8)	0.007 (8)	0.002 (6)

Table 4. Main interatomic distances (Å) and bond angles (°) for $Sr_2As_2O_7$.

Sr(1) -	O(4)	2.548 (17)	Sr(2) -	O(10)^{iii}	2.442 (19)
	O(5)	2.584 (15)		O(14)^{iv}	2.532 (16)
	O(8)^{i}	2.589 (18)		O(3)^{v}	2.584 (16)
	O(1)^{i}	2.603 (16)		O(13)^{iii}	2.631 (17)
	O(9)^{ii}	2.644 (17)		O(2)^{vi}	2.624 (17)
	O(3)	2.65 (2)		O(1)	2.648 (18)
	O(6)	2.911 (17)		O(12)^{vi}	2.820 (16)
	O(2)	2.945 (18)			
	O(7)^{ii}	3.193 (18)			
Sr(3) -	O(9)	2.458 (17)	Sr(4) -	O(7)^{viii}	2.555 (17)
	O(8)^{vii}	2.454 (16)		O(5)	2.576 (16)
	O(11)^{vi}	2.480 (15)		O(10)^{iv}	2.591 (18)
	O(4)	2.52 (2)		O(2)	2.697 (16)
	O(14)^{vii}	2.577 (16)		O(9)^{ii}	2.682 (16)
	O(3)	2.674 (15)		O(11)^{v}	2.79 (2)
	O(13)^{vii}	2.885 (17)		O(13)^{v}	2.82 (2)
	Sr(3) —O(5)^{vii}	3.257 (18)		O(14)^{ix}	2.87 (2)
				O(1)^{ii}	3.042 (16)
As(1)-	O(8)	1.632 (15)		O(8)—As(1)—O(3)	113.2 (8)
	O(3)	1.658 (17)		O(8)—As(1)—O(2)	117.3 (8)
	O(2)	1.690 (18)		O(3)—As(1)—O(2)	107.1 (10)
	O(12)	1.731 (16)		O(8)—As(1)—O(12)	108.8 (8)
				O(3)—As(1)—O(12)	109.2 (8)
				O(2)—As(1)—O(12)	100.3 (8)
As(2)-	O(4)	1.647 (18)		O(4)—As(2)—O(10)	116.1 (9)
	O(10)	1.682 (17)		O(4)—As(2)—O(5)^{vii}	111.5 (9)
	O(5)^{vii}	1.692 (14)		O(10)—As(2)—O(5)^{vii}	115.9 (8)
	O(6)	1.825 (17)		O(4)—As(2)—O(6)	97.2 (8)
				O(10)—As(2)—O(6)	103.6 (8)
				O(5)^{vii}—As(2)—O(6)	110.4 (7)
As(3)-	O(7)	1.616 (16)		O(7)—As(3)—O(1)	121.5 (10)
	O(1)	1.654 (16)		O(7)—As(3)—O(9)	109.0 (9)
	O(9)	1.677 (15)		O(1)—As(3)—O(9)	108.7 (8)
	O(12)	1.817 (17)		O(7)—As(3)—O(12)	107.9 (8)
				O(1)—As(3)—O(12)	103.7 (8)
				O(9)—As(3)—O(12)	104.8 (8)
As(4)-	O(11)	1.653 (17)		O(11)—As(4)—O(14)	114.7 (9)
	O(14)	1.678 (15)		O(11)—As(4)—O(13)	112.3 (10)
	O(13)	1.689 (17)		O(14)—As(4)—O(13)	108.1 (8)
	O(6)	1.744 (18)		O(11)—As(4)—O(6)	104.5 (9)
				O(14)—As(4)—O(6)	110.8 (9)
				O(13)—As(4)—O(6)	106.0 (8)

Symmetry codes: **(i) y, -x, z-1/4 ; (ii) y, -x+1, z-1/4; (iii) -x+1, -y, z+1/2; (iv) -x+1, -y+1, z+1/2; (v) -y, x, z+1/4; (vi) -y+1, x, z+1/4; (vii) x+1, y, z; (viii) y-1, -x+1, z-1/4; (ix) x-1, y+1, z**

3. Experimental Description of the structure and discussion

A perspective representation of the three-dimensional $Sr_2As_2O_7$ structure is given in Fig. 1. The atomic arrangement can be described as a succession of layers extending along the [001] direction. AsO_4

tetrahedra, (forming As_2O_7 groups via corner-sharing) form layers. The strontium atoms are almost embedded in the layers and connect them to form a three-dimensional framework. Two adjacent layers are linked together by corner and edge-sharing between SrO_n (n = 7-9) polyhedra and $[AsO_4]$ tetrahedra (Fig 1).

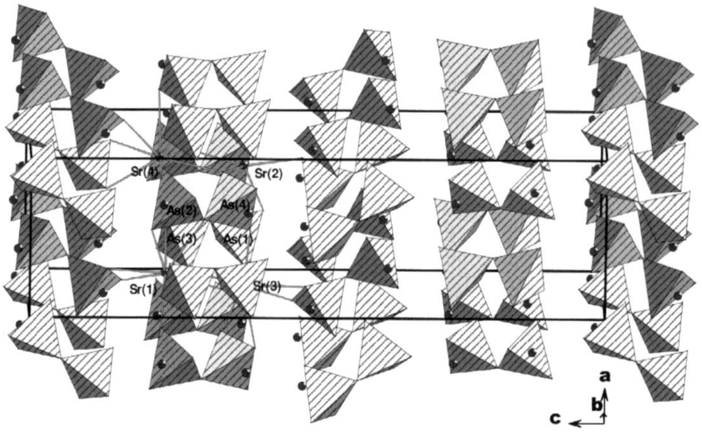

Figure 1. A perspective view of the atomic arrangement of $Sr_2As_2O_7$.

$Sr_2As_2O_7$ is isostructural with the diphosphates calcium $Ca_2P_2O_7$, as it is often the case when As atoms are only tetrahedrally coordinated. In other compounds (when As atoms exhibit an octahedral coordination) no isotypy between arsenates and phosphates can be observed [4, 13].

The average distances for AsO_4 tetrahedra are $\langle$As(1)—O$\rangle$ = 1.68 (2) Å, $\langle$As(2)—O$\rangle$ = 1.71 (2) Å, $\langle$As(3)—O$\rangle$ = 1.69 (2) Å and $\langle$As(4)—O$\rangle$ = 1.69 (2) Å. The average values for the Sr—O distances are 2.74 (2) Å, 2.61 (2) Å, 2.66 (2) Å and 2.74 (2) Å, respectively for $Sr(1)O_9$, $Sr(2)O_7$, $Sr(3)O_8$ and $Sr(4)O_9$ polyhedra. These geometrical characteristics are very similar to those observed in the related oxide compounds so far known [4, 14-17].

4. Conclusion

The present investigation confirms the previous study, but with higher precision and with all displacement parameters refined anisotropically. $Sr_2As_2O_7$ is isotypic with the high-temperature polymorphs of $Ca_2P_2O_7$ and $Sr_2V_2O_7$ [5, 6], but studied compound dont show any polymorphism.

References

[1] Weil M, Dordevic T, Lengauer C. L, Kolitsch U 2009 *Solid State Sciences* **11** 2111–2117.
[2] Harvey C. F, Ashfaque K. N, Ye W, Badruzzaman A. B. M, Ali A. M, Oates P. M, Michael H. A, Beckie R, Islam S, Ahmed M. F 2006 *Chem. Geol.* **228** 112.
[3] Itali T, Masuda H, Seddique A. A, Mitamura M, Maruoka T, Li X, Kusakabe M, Dipak B. K, Farooqi A, Yamanaka T, Nakaya S, Matsuda J, Ahmed K. M 2008 *App. Geochem.* **23** 3155.
[4] Graia M, 2002 Thèse de doctorat, Faculté des Sciences de Tunis.
[5] Boudin S, Grandin A, Borel M. M, Leclaire A, Raveau B 1993 *Acta Crystallogr.* **49** 2062.
[6] Baglio J. A, Dann J. N 1972 *J. Solid State Chem.* **4** 87.
[7] Deng B, Ibers J. A 2005 *Acta Crystallogr **E61*** i76.
[8] CAD-4 Express Software 1994 Enraf-Nonius, Delft, The Netherlands.
[9] North A.C.T, Phillips D.C, Mathews F. S 1968 *Acta Crystallogr.* A **24** 351–359.
[10] Sheldrick G.M 1997 SHELXS-97 A Program for Crystal Structure Determination, University of Göttingen Germany.

[11] Sheldrick G. M 1997 SHELXL-97. A Program for the Refinement of Crystal Structures, University of Göttingen, Germany.
[12] Farrugia L. J, 1999 *J. Appl. Crystallogr.* **32** 837–838.
[13] Driss A, Jouini T 1989 *J. Solid State Chem.* **78** 126.
[14] M. Graia, A. Driss, T. Jouini 1999 *Acta Crystallogr.* **C55** 1395-1398.
[15] Graia M, Driss A, Jouini T 2002 *Journal de la Société Chimique de Tunisie* **4** N°12 1657-1663.
[16] Graia M, Driss A, Jouini T 1999 *Z. Kristallogr.* NCS **214** 1-2.
[17] Karlsruhe F 2008 Inorganic Crystal Structure Database (ICSD). Release **02** Germany.

IOP Conf. Series: Materials Science and Engineering **27** (2012) 012018 doi:10.1088/1757-899X/27/1/012018

Study of the valorization of phosphogypsum in the region of Gafsa as filler in paper

S. Hassen[a1], Z. Anna[a], E. Elaloui E[a], M. N. Belgacem[b], E. Mauret[b]
[a]Research Unit of materials, environment and energy "(06/UR/12-01), Faculty of Sciences of Gafsa, Gafsa University Tunisia
[b]Laboratory of Process Paper Engineering, Grenoble, France (LGP2)

E-mail: hassen.sana@yahoo.fr

Abstract. Four different fractions of phosphogypsum were characterized by several analytical techniques. The size and shape of particles were determined by scanning electron microscopy. Analyses of infrared spectroscopy and X-ray diffraction have shown that these particles are mainly composed of calcium sulfate more or less hydrated. Calcination tests were carried out showed that treatment of these powders at 600°C for one hour increase their initial britghtness to values high enough to be used as a filler in paper.

1. Introduction

Fertilizer manufacturers are required to produce phosphoric acid by reacting sulfuric acid with natural ores containing phosphates. This process is set up to produce more concentrated phosphate fertilizers. This production is not without consequences on the immediate environment of the production sites. Indeed, the activity of processing phosphate generates gaseous emissions, liquid discharges and leads to the appearance of enormous amounts of solid residue: phosphogypsum.

Indeed, cleaning of industrial phosphogypsum is an environmental problem that affects many countries, given the large amount of waste and its chemical component. Numerous studies have focused on the use of phosphogypsum in several areas as a substitute for natural gypsum, particularly for producing gypsum [1] or add cement, or by including in the manufacture of bricks [2]. The industrial process used in Tunisia produces about four tones of apatite ore and 2.5 tons of sulfuric acid, approximately one ton of phosphoric acid and more than five tons of phosphogypsum. So large quantities of phosphogypsum [3] (10 million tons per year for Tunisia) are produced each year. Currently, the primary mode of management of this byproduct is the storage heap, mostly near the plant and the coastline [4].

Pollution problems caused by this by-product were cited by several authors worldwide. The waste causes various environmental problems when discharged directly to the environment [5]. Rouis indeed showed that the phosphogypsum is a by-product which causes harm to environment if not stored properly [6]. The "Tunisian Chemical Group" and the Tunisian Ministry of Environment, in recent years, focused in the problems of storage and the use of phosphogypsum [7].

There are three options for managing phosphogypsum: (i) disposal or dumping, (ii) staking, (iii) use-in, for example, agriculture, construction, or landfill [8].

[1] To whom any correspondence should be addressed.

The phosphogypsum is a by-product of the manufacture of phosphoric acid. The main raw material industry phosphoric acid and phosphate ore natural calcium, which varies from fluorapatite $Ca_5(PO_4)_{3-x}(F,Cl,OH)_x$ ($x \leq 1$) to the apatite fluoro-carbonate $Ca_{10}F_2(PO_4)_6$ $CaCO_3$[9].
This process leads to the production of large quantities of phosphogypsum: 1, 7 tons of phosphogypsum per ton of phosphate rock are used, the equivalent of 5 phosphogypsum per ton of P_2O_5 produced [10].

2. Experimental

The characterization of phosphogypsum is made by different techniques. Indeed, the measurement of pH and conductivity is made using a pH-conductivity meter type portable Thermo Scientific (Orion 4 STAR), morphological analysis and mapping are determined by environmental scanning electron microscopy of type "Quanta 200 -FEI "with an accelerating voltage of 15KV. This SEM is equipped with an X-ray source for mapping atomic samples tested. The spectrophotometer used for the infra-red Fourier Transform is a FTIR Perkin Elmer (Paragon 1000). To make the measurements of the brightness of the different samples of phosphogypsum we used a spectro colorimeter (Color Touch Technidyne). Finally, the powder X-ray diffraction experiment were performed on a philips X'pert X-ray diffractometer(Cu K_α radiation; $\lambda = 1.54$ Å).
Four samples are taken at different depths of the heap, as summarized in Table 1:

Table 1. The different heights of the samples of phosphogypsum

	Deep in the heap (m)
E1	0
E2	20
E3	25
E4	30

3. Results and discussion

3.1. pH and conductivity measure

All samples gives acidic aqueous suspension (Table 2). These results confirm the acidic nature of phosphogypsum. This acidity is more pronounced for samples of recent productions E1. Samples from older productions have lost their acidity, probably due to leaching to lower layers.

Table 2. Measurement of pH and conductivity of the samples of phosphogypsum

	pH	conductivity (μS/cm)
E1	3,76	3,24
E2	2,94	6,06
E3	3,00	5,90
E4	3,10	4,63

A study by Felfoul [11] shows the need for twelve successive washes with deionized water to achieve the neutralization of fresh phosphogypsum.

The conductivity values vary from one sample to another. We note that the sample E2 has the highest value. This can be explained by the presence of other ions in this sample.

3.2. Rate of dry matter and ash

The ash content in the four samples was established, as summarized in Table 3, moisture content or humidity is naturally high for the new production of phosphogypsum (E1). This concentration decreases with time to reach an equilibrium value of about 33% under the great heap. The measured values correspond to samples taken at the top of the heap are variable and depend on weather conditions.

Table 3. Percentage of ash and dry matter in different samples

Echantillon	Ash content (%)	Rate of dry matter (%)
E1	6,61	59,14
E2	3,55	82,77
E3	3,16	67,29
E4	3,47	67,31

We note that the existing quantity of ash in all samples is low. This is due to the extent that the organic matter present in the phosphogypsum is unimportant.

3.3. Morphological analysis SEM

Examination with scanning electron microscopy showed that the studied phosphogypsum samples have a tubular shapes (platelets) or hexagonal well known (Figure1-4 [8]), with over a certain district of form that is clearer for the samples more great depths.

Figure 1. SEM photograph of E_1.

Figure 2. SEM photograph of E_2.

Figure 3. SEM photograph of E_3.

Figure 4. SEMphotograph of E_4.

This trend resulted from erosion dissolution causes a narrowing of particle size with time. Indeed, the sample of fresh phosphogypsum E1 has a rectangular platelet morphology such that the particles have lengths of between 0.2 and 180µm in length and about 50 microns median.

The photograph of the second sample E2, taken at a depth of 20 m from the slag heap of phosphogypsum, highlights several types of particles. This explains the presence of other ions, other than SO_4^{2-} and Ca^{2+}, leading to changes in facies samples. You can find the form plate similar to the case of the first sample, there are also forms needles, beads, sponges.

3.4. X – ray analysis

Following the morphological analysis (SEM) of phosphogypsum samples, we studied the X-ray analysis using the apparatus of SEM coupled with EDS. The main elements present in all samples are calcium, oxygen, sulfur. Added to this is the fluorine for the third sample E3.

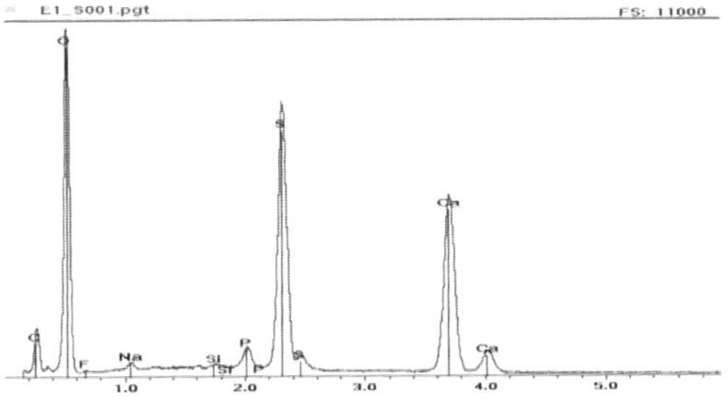

Figure 5. EDS spectrum of sample E1-E2

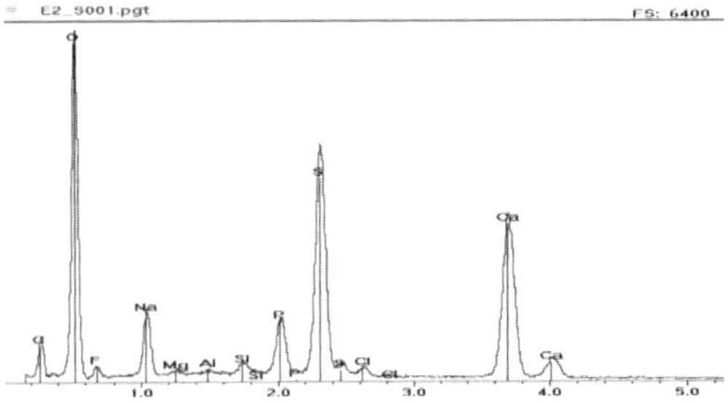

Figure 6. EDS spectrum of sample E_2

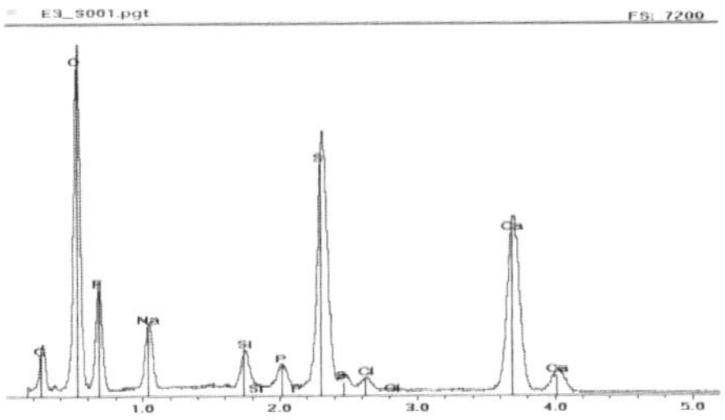

Figure 7. EDS spectrum of sample E_3- E_4

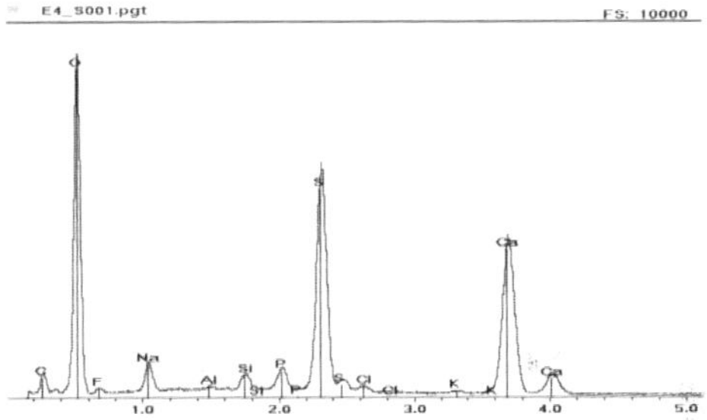

Figure 8. EDS spectrum of sample E_4.

From these spectra, we can deduce the key chemical elements in different samples of phosphogypsum. Indeed, the elements that are present in all samples are carbon, oxygen, calcium, sulfur, silica and phosphorus.

3.5. Sample's atomic cartography

Scanning Electron Microscopy coupled with EDS also determines the mapping of the atomic sample to be studied is to show, the distribution of each element present in the sample throughout its surface.

The main elements present in all samples are calcium, oxygen, sulphur and fluorine. In sample (E3), the fluorine presents a poor dispersion as schown in figure 15.

Only E1 and E3 are taken as exemple because E2 and E4 are similar than E1.

Figure 9. Distribution of oxygen in E1

Figure 10. Distribution of sulfur in E1

Figure 11. Distribution of calcium in E1.

Figure 12. Distribution of oxygen in E3.

Figure 13. Distribution of sulfur in E3.

Figure 14. Distribution of calcium in E3.

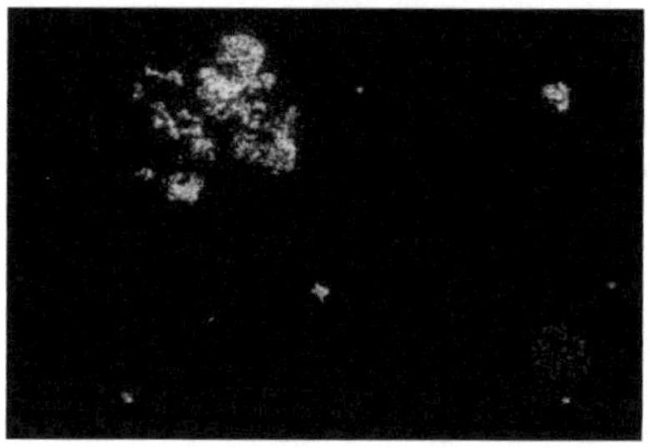

Figure 15. Distribution of fluoride in E3.

According to this distribution of elements on the surface of the sample, we note the absence of impurities and greater uniformity in the case of E1, when the sample E3 contains fluorine with poor dispersion (regional concentration) (cf. Figure 15).

3.6. FTIR spectrometry analysis

The positions and the allocation of IR absorption bands of phosphogypsum are available in the literature and mentioned by several authors such as Dumitras [12].

The IR spectra of our samples are almost identical and each has a main peak around 1099 cm^{-1} corresponds to the stretching vibration of the sulfate ion (SO_4^{2-}) in the network of gypsum and another less intense peak around 594 cm^{-1} corresponding to the deformation of SO_4^{2-}. The presence of water is marked by two peaks around 1620 cm-1 and 3590 cm^{-1}.

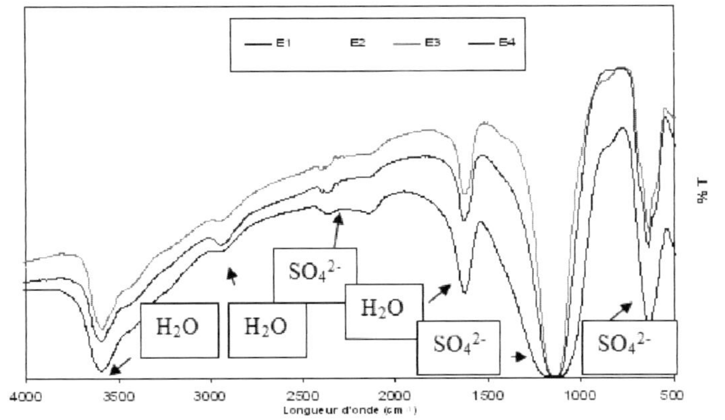

Figure 16. FTIR spectra of E1, E2, E3 and E4.

3.7. X-rays diffraction

The samples of phosphogypsum dried at 105° C for 48 hours before performed tests of X-ray diffraction.The X-ray diffraction patterns of the four samples studied are shown in the following figures (Figures17-20):

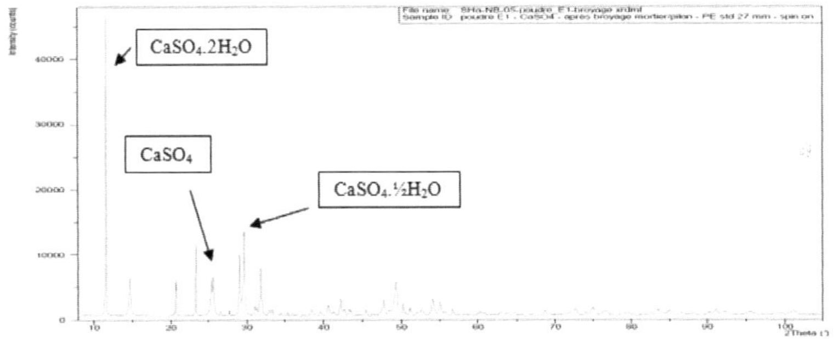

Figure 17. Diagram of X-ray diffraction of sample E1.

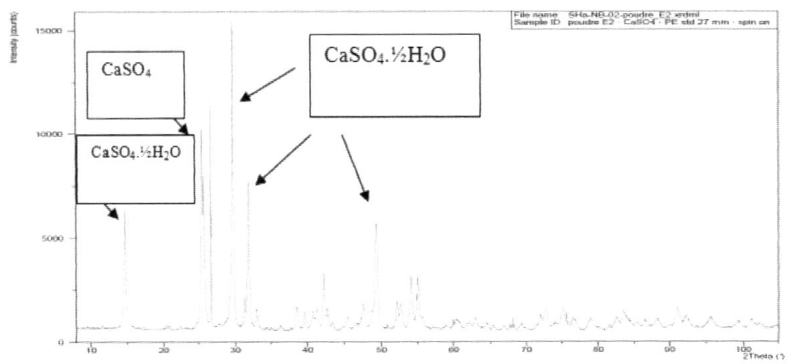

Figure 18. Diagram of X-ray diffraction of sample E2.

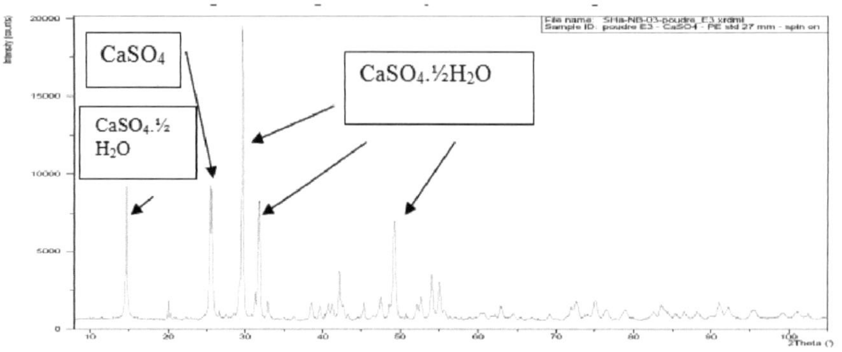

Figure 19. Diagram of X-ray diffraction of sample E3.

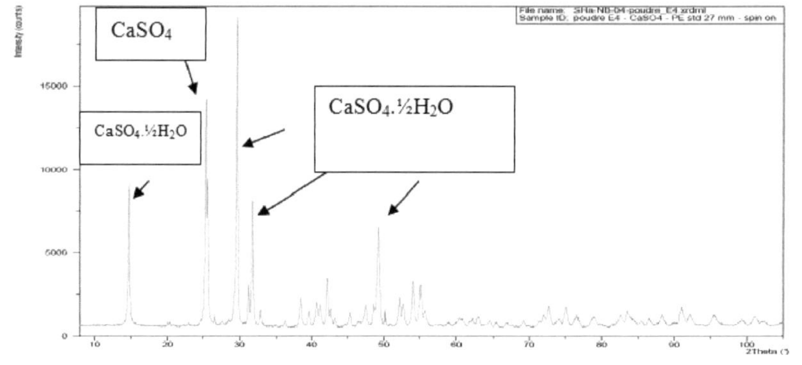

Figure 20. Diagram of X-ray diffraction of sample E4.

The difractograms obtained show the presence of hydrated calcium sulfate $CaSO_4.1/2\ H_2O$ and anhydrite $CaSO_4$, SiO_2 quartz and gypsum $CaSO_4$, $2H_2O$, with different proportion. Indeed, anhydrite has the highest percentage for sample E2 26.5%. This sample also contains the largest amount of quartz 15.5% (see Table 4):

Table 4. Percentage of different composition in the four samples

	E_1	E_2	E_3	E_4
$CaSO_4$	6, 2 %	26,5 %	22, 2 %	40 %
$CaSO_4.0.5\ H_2O$	22, 7 %	59,2 %	50,5 %	41, 1 %
$CaSO_4.2H_2O$	9,2 %	2,1 %	0,9 %	0, 5 %
SiO_2 (quartz)	1,6 %	15,5 %	3,1 %	3, 1 %

3.8. Degree of whiteness

The brightness is an important feature loads of paper that is why we conducted this test for our 4 samples, because we are interested in the recovery of phosphogypsum in the field of papermaking. Table 5 shows the whiteness of the four samples, as shown and after calcination at different temperatures.

Table 5. Degree of whiteness of the various samples calcined for 6 hours

	E1	E2	E3	E4
Dried sample	42,7	53,4	49,8	57,5
Calcined at 200°C	50,9	53,6	50,7	55,8
Calcined at 400°C	52,8	50,4	58,9	62,6
Calcined at 600°C	83,1	83,7	87,4	89,9

These results shown that the variation in the value of brightness depends on the calcination temperature. Indeed, for each sample of phosphogypsum, increasing the calcination temperature increases the brightness gradually until it reaches a maximum at 600°C.

By varying the calcination duration at 600 ° C, degrees of whiteness of the various samples also vary (table 6).

Table 6. Degree of whiteness of phosphogypsum samples calcined at 600 ° C

	E1	E2	E3	E4
Dried sample	42,7	53,4	49,8	57,5
30 min	71,4	76,6	88,4	84,5
1 h	80,1	83,9	89,9	85,3
4 h	82,8	83,9	89,7	85,3
6 h	83,1	83,7	87,4	89,9

It's clear that the brightness increases gradually for all the samples to reach values near 80 and more for the calcination of 60 minutes.

So for samples tested just calcine at 600° C for 60 minutes to reach a good value white enough to integrate the powders produced in the paper-making.

4. Conclusion

In this work characterization by different analytical methods (SEM, FTIR, XRD, Data Color ...) of phosphogypsum in the region of Mdhilla-Gafsa in order to study its possible recovery in paper (as an expense). The whiteness test of the samples shows that after calcination at 600 ° C for 1 hour, a high value of whiteness is obteined. This value is sufficient for the integration of these powders in the manufacture of paper. Given the local context, these results are promising and have potential to more rational management of industrial waste: the phosphogypsum.

References

[1] J-Lvignes, Centre de Ressources pédagogique de chimie (Economie –Industrie-EXC-1ENS de Cachan -94235 Cachan Cedex), CEM-CNRS-94400 Vitry-sur-Seine, 1997, ESSADDAM H IPEST-La Marsa-Tunisie et DALIGAND D: Syndicat National des Industries du plâtre.

[2] Kurandt H. F 1968 Emploi de gypse de l'acide phosphorique dans l'industrie du bâtiment; Congrès technique de l'IFA 10-13.

[3] Choura M, Evaluation environnementale de l'activité de transformation de phosphate côtière sud du grand Sfax, 2008. Projet SMAP III-TUNISIE (2006-2008), stratégie de gestion intégrée de la zone côtière sud du grand sfax.

[4] Rutherford P. M, Dudas M. I, Samek R. A 1994 *Sci. Total environ* **149** 1-38.

[5] Taher M. A 2007 *Ressources conservation and recycling*, **52** 28-38.

[6] Rouis M. J, Bensalah A 1990 phosphogypsum in Tunisia: environmental and required solution, in proceeding of the third international symposium on phodphogypsum 87-105.

[7] Ajam L, Ben Ouezdou M, Sfar-Falfoul H, Mensi R. El 2009 Construction and Building Materials **23** 3240-3247.

[8] Kuryatnyk T, Angulski Da Luz C, Ambroise J, Pera J 2008 *Journal of Hazardous materials* 681-687.

[9] Sfar-Felfoul H 2002 Propriétés et perspectives de valorisation du phosphogypse l'exemple de la Tunisie; Procceding of International Symposium on Environemental pollution control and waste Management, Tunis(EPCOW M'2002), 7-10 January 2002, pp.510-520.

[10] Charfi F 1999 Substitution du gypse naturel par le phosphogypse dans la fabrication du ciment, Thèse de Doctorat, Faculté des Sciences de Tunis.

[11] Sfar-Felfoul H, Ben Ouezdou M, Claster P 2004 Etude du phosphogypse de Sfax en vue d'une valorisation en technique routière, Laboratoire de Génie Civil, ENIT, Tunis, Laboratoire "Matériaux et Durabilité Des Constructions". LMDC-INSA-UPS, Toulouse, France.

[12] Gorlé D, le phosphogypse comme matériau de remblai routier, CR28/85, Centre de Recherches Routières, Bruxelles.

The porous nature of ZnO thin films deposited by sol–gel Spin-Coating technique

M. Karyaoui [a1], A. Ben Jaballah[a], R. Mechiak[b], and R. Chtourou[a]
[a]Laboratory of Photovoltaic Research and Technology Centre of Energy, Borj Cedria Technopark, BP.95, Hammam-Lif 2050, Tunisia.
[b] Department of Medicine, Faculty of Medicine, Hadj Lakhdar University, Batna, Algeria.

E-mail: karyaoui@yahoo.fr

Abstract. Zinc oxide (ZnO) thin films were deposited on silicon and quartz substrates, by sol-gel method, using zinc acetate dehydrate [$Zn(CH_3COO)_2.2H_2O$] dissolved in isopropanol and glycerol. The structural, morphologic and optical properties of ZnO thin films subsequently annealed at 700°C in air for 30 min have leads to a porous nature of these films. To calculate, the refraction index and the extinction coefficient values, Cauchy formalism is used to evaluate the Spectroscopic Ellipsometry results. Two distinct configurations were proposed for each sample: in the first, the film is considered as mixture of randomly distributed voids and ZnO crystallites while in the second, the effect of porosity gradient is highlighted. This optical analysis gives a better agreement between experiment and theory for a wide range of wavelengths regarding the second configuration.

1. Introduction

During the last two decades, wide band gap semiconductors such as zinc oxide (ZnO) have attracted a great interest due to their potential applications in optoelectronic devices. In particular, it is largely used as: gas-sensors [1], surface acoustic wave devices [2], solar cell windows [3, 4], UV light emitting devices [5, 6] and so on. In previous works, ZnO thin films have been grown by means of various deposition techniques like sputtering [7], Molecular Beam Epitaxy (MBE) [8], Pulse Laser Deposition (PLD) [9] and Metal-Organic Chemical Vapor Deposition (MOCVD) [10]. These methods, so expensiveè, require no simple experimental conditions. The developments of chipper-growth technologies for photovoltaic applications are prerequisite, hence simple, low cost and methods such as spray pyrolysis [11, 12], dip-coating and spin-coating are in demand.

In the case of Spray Pyrolysis, the poor homogeneity of the layers and the large amount of material loss are the main drawbacks during the preparation of ZnO thin films. On the other hand, the layer developed by dip-coating, are shown to have a nonuniform thickness along the substrate surface. To overcome these disadvantages, the spin-coating technique has been chosen to prepare ZnO thin films. This technique has many advantages such as the possibility of preparing at low cost a small as well as a large-area thin films with high purity, for various and suitable applications [13, 14].

---

[1] To whom any correspondence should be addressed.

Finally, the structural and morphological analyses of ZnO prepared films are performed by means of by X-ray diffraction and atomic force microscopy techniques, respectively. The combination of the transmission and ellipsometry investigations let us to reach the optical properties of such oxide films. To date, these two proposal models are not yet been tested to reach the porosity nature of ZnO thin films obtained by spin coating as a simple route.

2. Experimental protocols

2.1. ZnO Films preparation

The used solution was prepared using 4 g of zinc acetate dehydrate [Zn $(CH_3COO)_2.2H_2O$] dissolved in 50 ml isopropanol corresponding to x10^{-2}M as a molarity of zinc ion (Zn^{+2}). A room temperature agitation during 20 min was performed before adding 4 ml of Glycerol. After 10 min, the obtained mixture has been agitated under reflux at 100°C for 40 min, to obtain a homogeneous and transparent solution. For high quality ZnO thin films, prior to coating, all substrates were cleaned ultrasonically, first in acetone, and subsequently in ethanol for 10 min each run and then rinsed in bi-distilled water. The rotation speed of the coating unit was 2600 rpm and the duration of the single coating was 30 s. Under these experimental conditions, ZnO films were then deposited using the spin-coating technique on p-type silicon or quartz substrates.

2.2. Investigations set up

First, the lattice parameter variations and the crystalline structure, a Siemens D5005 automated powder diffractometer, was used. The X–rays diffraction patterns were recorded with a copper anticathode (I=20 mA and V=40 kV), using step-scanning, between 10° and 70° at 0.1°/s. Second, the morphological state was performed using Atomic Force Microscpy (AFM) in contact mode by a Topometrix TMX 2000 Explorer AFM. Optical properties of the films were examined by a UV–VIS spectrophotometer (UV3101PC). Finally, the spectroscopic Ellipsometry (SE) measurements were performed at room temperature using a GES5 automatic ellipsometer from SOPRA. The system is equipped by: a xenon lamp (75W), a rotating polarizer, an autotracking analyser, a double monochromator and a photomultiplier tube and a GaInAs photodiode as detectors. Data were recorded at an incidence angle of $\theta = 75°$ under 0.25–1.5 μm wavelength domain with 0.005 μm as step.

3. Results and discussions

3.1. Structural and morphological properties

Figure 1 depicts the diffraction pattern of ZnO thin film deposited on Si (100) substrate by spin-coating process and annealed at 700°C under atmospheric conditions for 30 min. The peaks in the pattern are identified as (100), (002), (101), (102) and (110) reflections of wurtzite structure. XRD results point out that all the films are polycrystalline in nature and the grains are randomly oriented. Undoubt, under the preparation conditions previously indicated in the experimental section, it is found that ZnO obtained films show the same crystalline aspects as those reported by S. Mridha, D. Raoufi and J. Lee [15, 16, 17] using other precursors . However, with the same precursors, J. Petersen and al. [18] have found that the main orientation growth of the crystallites is (002). This may be explained by the effect of the substrate orientation nature.

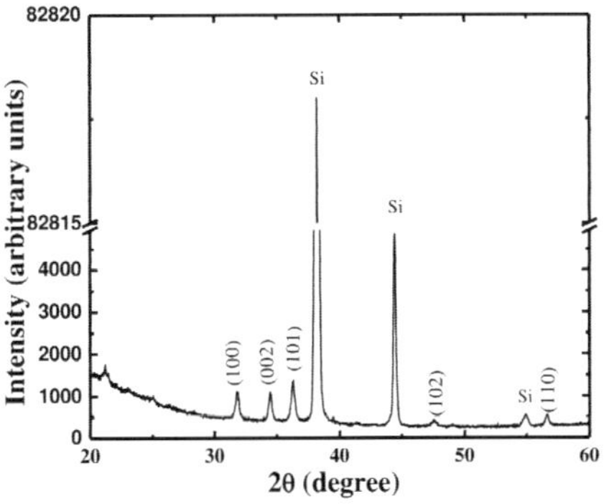

Figure 1. X-ray diffraction (XRD) patterns of ZnO thin film annealed at 700°C in air for 30 min.

Moreover, figure 2 shows 2D(a) and 3D (b) AFM images of 3x3μm^2 area scan on the same ZnO film surface. They indicate a porous state and fine structure with considerable micro- and nano-pores related to the grain boundaries observed as the film is much thicker. This usually seems to be the main characteristic of sol–gel ZnO films [19]. The surface roughness and average grains size of ZnO thin film are 3.07 nm and 34.5 nm respectively.

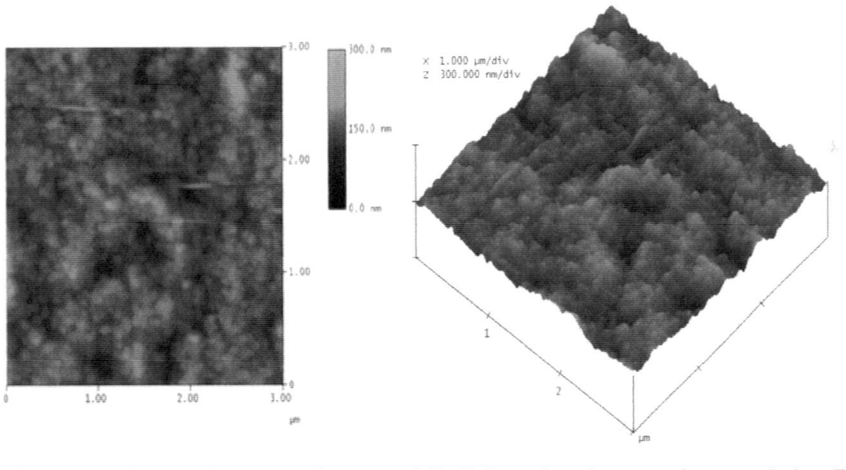

Figure 2. Atomic force microscopy images of ZnO deposit subsequently annealed at 700°C under atmospheric conditions for 30 min: a) top view and b) three dimension view probing the porous aspect of the film.

3.2. Optical properties

The optical properties of ZnO thin films and especially the high transmission factor in the visible region of such films is an important factor for some physical applications. Endeed, figure 3 shows the UV–VIS–IR transmission spectra of ZnO films having different thicknesses, in the spectral range of 200 nm to 1000 nm. The optical transmission in the visible region was high (>93%) but it decreased substantially at short wavelengths near the ultraviolet region whatever is the film thickness. However, with increasing film thickness one should observe a drastic decrease of the transmission factor for short wavelengths. As indicated above, the morphological aspects of the elaborated films consist of a porous material with increasing porosity as the thickness of grown films decreases (the porosity gradient increase from the top to the bottom). This feature is understood from the fact that the increase of film thickness can lead to an increase in scattering, reflection and optical absorption on the film surface.

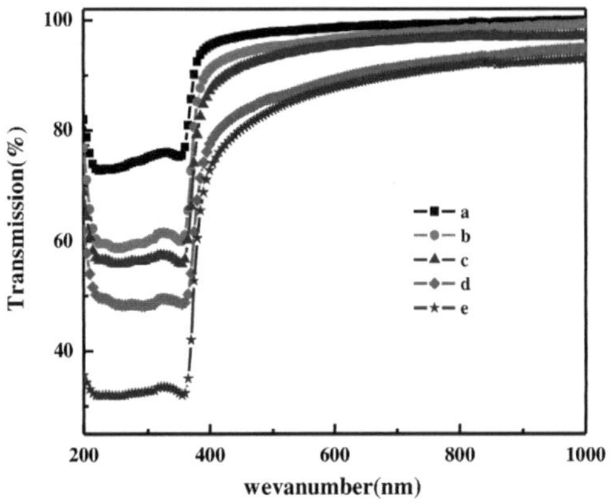

Figure 3. Optical Transmittance of ZnO thin films versus wavelength. The effect of the film thickness: a) simple run, b) double run, c) third run, d) fourth run, e) fifth run, of spin-coating. It clearly shows that the films are more transparent for short wave lengths as the film thickness increase due to the increase of light scattering on the surface of the porous films.

3.3. Ellipsometry investigations

Spectroscopic ellipsometry measurements were performed at 300 K for wavelengths ranging from 0.6 eV to 5.5 eV, by means of a photometric ellipsometer with rotating polarizer. Various theoretical models have been used to determine the optical properties of ZnO thin films [20, 21]. In our case, the experimental results are adjusted to the theoretical ones using the Cauchy formalism. The following formulas were used to define the refractive index (n) and the extinction coefficient (k):

$$n(\lambda) = A + \frac{B}{\lambda^2} + \frac{C}{\lambda^4} \quad (1)$$

$$k(\lambda) = \frac{D}{\lambda} + \frac{E}{\lambda^3} + \frac{F}{\lambda^5} \quad (2)$$

The six fitting parameters in the model A, B, C, D, E and F are all summarized in table 1. The Cauchy model is used for two sets of sample configurations (Figs. 4-a and b). In first configuration, the thin monolayer is treated as a full arrangement of ZnO inclusions and voids having $v_{(ZnO)}$ and $v_{(void)}$ as volume fractions respectively. When the film becomes thicker, based on the morphological and optical observations, it is worth to study the optical properties taking into consideration the effect of the gradient porosity which vanishes in the bottom of the film. Then, a second configuration can be considered with a porous top layer and pure ZnO thin film in the bottom.

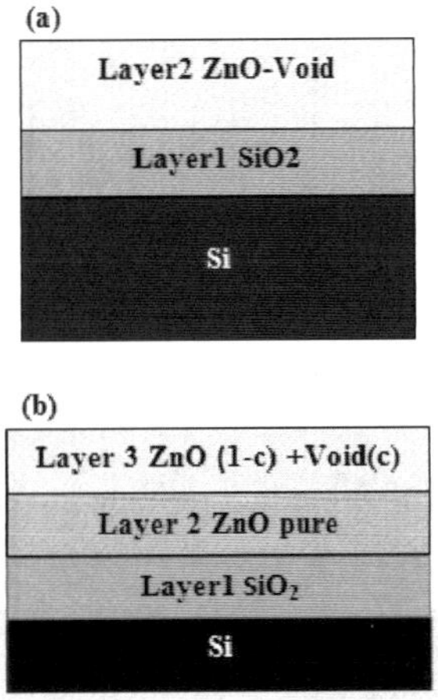

Figure 4. The effectiveness of the Cauchy formalism to determine the refractive index and the extinction coefficient: Comparison between experiment and calculations.

Table 1. Accurate fitting parameters of the Cauchy model to adjust the theoretical parameters to determine the refraction index and extinction coefficient of ZnO thin films

	Structure 1		Structure 2	
Layer 1[SiO_2] (μm)	0.018132±0.0020276		0.018248±0.005967	
			Layer2(μm)	0.119301±0.004623
Thickness of ZnO / Void	**Layer2**(μm)	0.163756±0.003414	**Layer3(μm)**	0.056713±0.005218
Void volume fraction	13 %		4 %	
Sigma	0.001359		0.001010	
R^2	0.989899		0.995273	
A	2.4457±0.5928		2.192069±0.02104	
B	0.629266±0.050734			
C			-0.003172±0.000079	
D	-9.5631±1.7200			
E	1.440500±0.216569		0.021682±0.001615	
F	-0.047446±0.008286			

Figure 5 shows the experimental values of tan (*Ψ*) and cos (*Δ*) which were adjusted by the calculated curves for both configurations. The model consists of a c-Si substrate with a thickness of about 525 μm and two-layers assigned to layer1 and layer 2 with thicknesses e_1 and e_2 respectively. Layer 1 is a silicon oxide layer SiO_2 caused by an oxidation of the substrate while layer 2 is best modeled by the Bruggeman effective medium approximation composed of ZnO and void using the Cauchy law.
It can be seen in figure 5 a good agreement between experimental and calculated curves of tan (*Ψ*) and *cos(Δ)* along the spectral range varying from 0.6 eV to 5.5 eV. This confirms the choice of the Cauchy formalism as a theoretical model for the study of optical and electrical properties of ZnO thin films by ellipsometry [22, 23, 24].

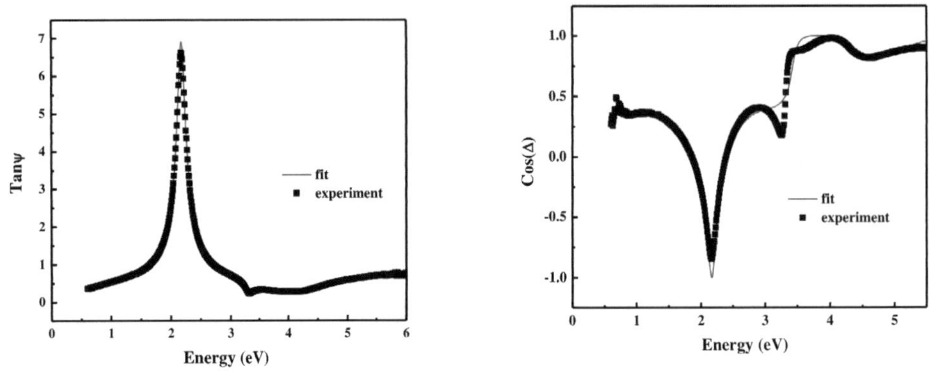

Figure 5. The multilayer model of the ZnO thin film on Si Substrate prepared by sol gel method.

Figures 6 a and b show the changes of the refractive index (n) and the extinction coefficient (k) of ZnO thin films in the wavelength range 260–2000 nm, calculated, from the model fitting of the experimental spectroscopic ellipsometric data. A good fit is therefore found between the model fit and experimental data in the entire wavelength range.

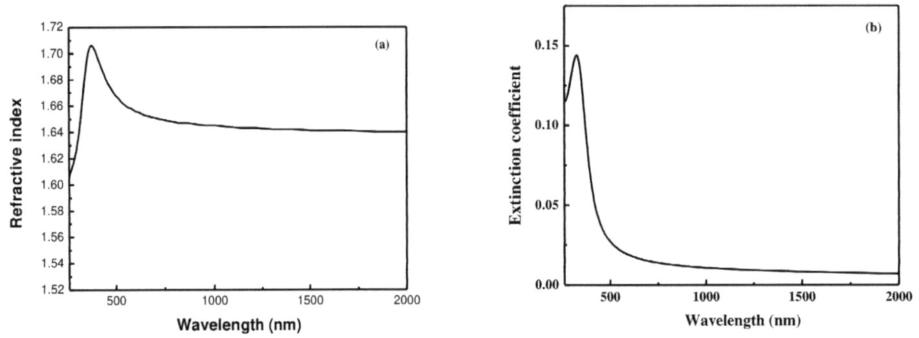

Figure 6. Refractive index (a) and extinction coefficient (b) as a function of wavelength for ZnO thin film annealed at 700°C in air for 30. It clearly shows a full agreement between experiment and theoretical calculations based on

Moreover, the thickness and porosity of ZnO thin layers are about 163 nm and 13% respectively achieved by adjusting the theoretical values to the experimental results. As can be seen, with increasing of the wavelength, the refractive index increases until λ reaches 780 nm, and becomes nearly constant with decreasing photon energy. Also, the maximum reduction of light appears for λ equal to 377 nm including the wavelength corresponds to the energy gap.

In addition, the extinction coefficient reduces to zero when the wavelength is larger than 400 nm. There is a little absorption in the wavelength range of 400–2000 nm, showing that ZnO films are transparent in this wavelength region, which is consistent with the results of transmission spectra shown in figure 3.

It is worth noting that the refractive index determined in this study lies below the expected value in comparison with the work reported in the literature [24, 25], which is due to the porosity effect [26]. To determine the optical properties of ZnO thin films with a good accuracy, the influence of porosity is considered, the Cauchy model is tested over another structure (figure 6-b) where the volume fraction of void is taken into account.

Figure 7 shows the experimental and calculated spectra of tan (*Ψ*) and *cos(Δ)* in the spectra range varying from 0.6 eV to 5.5 eV with the Cauchy modified model. The final values of the fitting parameters, i.e. the thickness and porosity (void in percentage) of each sublayer are also presented in figure 6.b. The thickness and the porosity were evaluated to be 119 nm and 4%, respectively. It is found that the thickness of the oxide layer (SiO_2) remains almost the same in both structures (18 nm).

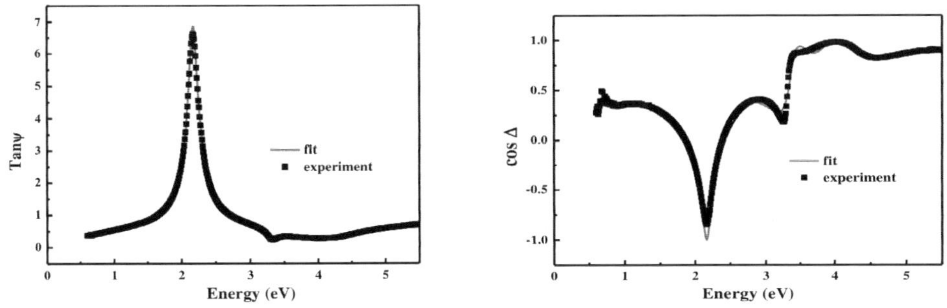

Figure 7. Comparison between experiment and calculations.

Figure 8 shows the variation of the refractive index in both cases. We find that in both configurations the refraction index shows the same behavior in the wavelength range between 260 and 2000 nm, except that the determined values increase when the porosity deceases.

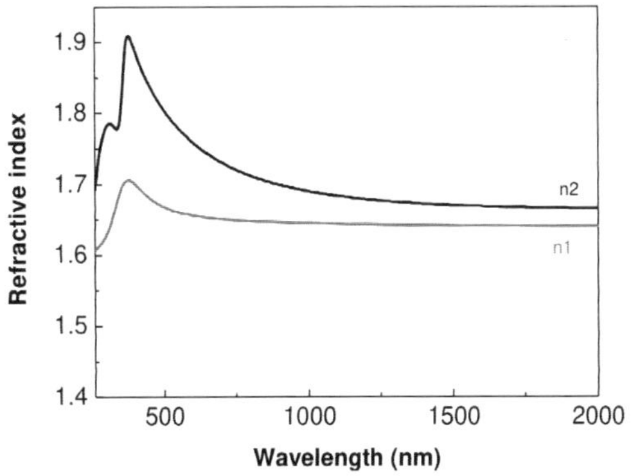

Figure 8. Refractive index as a function of wavelength of ZnO thin film.

4. Conclusion

Wurtzite phase ZnO thin films have been synthesized from zinc acetate with isopropanol and glycerol using the sol–gel procedure. Both microstructures analysis based on atomic force microscopy and optical characterizations have shown that the obtained films were porous in nature. Moreover, thicker films behave as an efficient window for light absorbing for photovoltaic components as well as for UV light emitting devices. This feature is due to the fact that the increase of film thickness can yield an increase in scattering, reflection and optical absorption on the film surface. Spectroscopic ellipsometry investigation processed under the Cauchy model revealed that the changes of the refractive index and extinction coefficient of the films are mainly due to the existence of a high volume fraction of voids for thinner layers. The full agreement between experiment and theory when varying the porosity let us think that ZnO films with appropriate porosities are convenient for chemical and gas sensing devices. Ellipsometry investigations have been carried out to perform the optical properties of pourous ZnO thin films deposited on various substrates.

IOP Conf. Series: Materials Science and Engineering **28** (2012) 012019 doi:10.1088/1757-899X/28/1/012019

References

[1] Pizzini S, Butta N, Narducci D, Palladino M 1989 J. *Electrochem. Soc.* **136** 1945.
[2] Kino G, Wagner R 1973 *J. Appl. Phys.* **44** 1480.
[3] Aranovich J, Golmayo D, Fahrenbruch A, Bube R *J. Appl. Phys.* **51** 4260.
[4] Jin Z, Hamberg I, Granqvist C, Sernelius B, Berggren K 1988 *Thin Solid Films* **164** 381.
[5] Thareja R. K, Mitra A 2000 *Appl. Phys. B. Lasers Opt.* **71** 181.
[6] Makino T, Chia C, Nguen T, Segawa Y 2000 *Appl. Phys.Lett.* **77** 1632.
[7] Fu Z, Lin B, Liao G, Wu Z 1998 *J. Cryst. Growth* **193** 316.
[8] Esaki L, Chang C. L 1976 *Thin Solid Films* **36** 285.
[9] Singh A. V, Mehra R. M, Yoshida A, Wakhara A 2001 *J. Appl. Phys* **90** 5661.
[10] Gruber Th, Kirchner C, Thonke K, Sauer R, Waag A 2002 *Phys. Stat. Sol. A* **192** 166.
[11] Mondragon S, Maldonado A, Reyes A 2002 *Appl. Surf. Sci.* **193** 52.
[12] Lee J, Park B 2004 *Mater. Sci. Eng. B* **106** 242.
[13] Alaya A, Nouiri M, Ben Ayadi Z, Djessas K, Khirouni K, El Mir L 2009 *Mater. Sci. Eng. B* **2** 159-160.
[14] Reisfeld R 2001 *Opt. Mater.* **16** 1.
[15] Mridha S and Basak D 2007 *Materials Research Bulletin* **42** 875.
[16] Raoufi D and Raoufi T 2009 *Appl. Surf. Sci* **255** 5812.
[17] Lee J, Easteal A, Pal U, Bhattacharyya D 2009 *Current Applied Physics* **9** 792.
[18] Petersen J, Brimont C, Gallart M, Cregut O, Schmerber G, Gilliot P, Honerlage B, Ulhaq-Bouillet C, Rehspringer J, Leuvrey C, Colis S, Slaoui A, Dinia A 2009 *Microelectronic Journal* **40** 239.
[19] Brinker C, Scherer G 1975 Sol–Gel Science: The Physics and Chemistry of Sol Gel Processing, *Academic Press*, New York **87**.
[20] Liu Y, Asieh J, Tung S 2006 *Thin Solid Films* **510** 32.
[21] Dumont E, Dugnoille B, Bierfait S 1999 *Thin Solid Films* **353** 93.
[22] Tompkins H, McGahan W, Spectroscopic Ellipsometry and Reflectometry, John Wiley & Sons, Inc., New York 93.
[23] Swanepoel R 1983 *J. Phys. E: Sci. Instrum.* **16** 1214.
[24] Moustaghfir A, Tomasella E, Ben Amor S, Jacquet M, Cellier J, Sauvage T 2003 *Surface and Coatings Technology* **193** 174–175.
[25] Ashour A, Kaid M, El-Sayed N, Ibrahim A 2006 *Appl. Surf. Sci.* **252** 7844.
[26] Diaz-Parralejo A, Caruso R, Ortiz A, Guiberteau F 2004 *Thin Solid Films* **458** 92.

IOP Conf. Series: Materials Science and Engineering **28** (2012) 012020 doi:10.1088/1757-899X/28/1/012020

Structural and magnetic investigation on tetragonal R-Fe alloy with 1:12 stoechiometry

S. Khazzan[1 a, b], N. Mliki[a], L. Bessais[b]
[a]Laboratoire Matériaux Organisation et Propriétés, Faculté des Sciences de Tunis, Campus Universitaire, BP 2092, Tunis, Tunisie
[b]CMTR, ICMPE, UMR7182, CNRS, Université Paris 12, 2-8 rue Henry Dunant BP 94320 Thiais, France.

E-mail: khazzan@glvt-cnrs.fr

Abstract. Structure and magnetic proprieties of ball-milled $Sm(Fe,Mo)_{12}$ compounds were carried out under an Ar atmosphere. Milled powders were subsequently annealed at a temperature ranging from 700°C up to1200°C. The effects of heat treatment, on structure and magnetic property changes, have been investigated by means of x-ray diffraction using the Rietveld method, and differential sample magnetometer. At low temperature, Rietveld analysis has revealed the presence of new phase with the 1:10 stoechiometry. Insertion of carbon atoms was carried out via a solid - solid reaction either. Upon carbonation, the Curie temperature of samarium based alloy was enhanced. Besides, systematic analyses of the structural aspects have been undertaken upon insertion.

1. Introduction

Rare earth transition metal intermetallics R-M are attractive compounds for their magnetic proprieties and technological applications. The R- M alloys with the 1/12 stoechiometry with tetragonal $ThMn_{12}$ type structure has been subject of investigations since the higher 3d transition metal content which provides higher magnetization and Curie temperature. However, binary RFe_{12} does not exist for any of the rare earths, nevertheless the $ThMn_{12}$ structure can be stabilized in pseudobinaries $RFe_{12-x}M_x$ with M= Si, V, W, Ti [1, 2]

It is well known that the magnetic proprieties of iron- rich rare earth intermetallics materials can be drastically altered when light elements such as nitrogen, hydrogen or carbon are inserted in interstitial sites. To get a better insight into the structural and magnetic behavior of $ThMn_{12}$ type structure substitutional Mo series are prepared. In this paper, we will investigate the effect of heat treatment on the structure transformation, the modification versus Mo concentration of magnetic proprieties and the effect of interstitial atoms on structural and intrinsic proprieties.

2. Experiment

Various alloys with nominal composition of $SmFe_{12-x}Mo_x$ (x=1, 1.5 and 2) have been prepared by mechanical alloying technique. High purity Sm and Mo powders were added to Sm_2Fe_{17} prealloy. In order to maintain an overpressure of samarium on the samples an appropriate of this material is added. Next, the mixture was ball milled in a high energy planetary. The as-milled powder were wrapped in

[1] To whom any correspondence should be addressed.

tantalum foils and subsequently annealed under vacuum for 30 minutes at a temperature ranging from 700°C up to1200°C. The samples structure was checked by means of a Brucker x-ray diffractometer (XRD) (CuKα radiation λ= 1.54178 Å). The Rietveld method was used to analyze the crystal structure and calculate the lattice parameters and the autocoherent-diffraction domain size. Thermomagnetic data were measured under an applied field of 1000 Oe using a differential sample magnetometer (DSM) MANICS.

3. Results

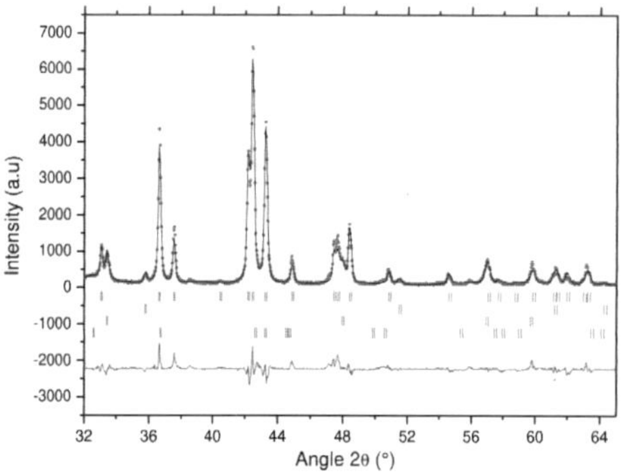

Figure 1. Rietveld fitting of XRD pattern of $SmFe_{10}Mo_2$

Fig. 1 illustrates, as an example, the XRD diagram of $SmFe_{12-x}Mo_x$ relative to a molybdenum content equal to x=2 annealed at temperature 1200°C. Owing to Rietveld refinement [3], all the samples crystallize in the tetragonal $ThMn_{12}$ (I4/mmm) structure type. The additional weak lines are assigned to a small amount of samarium oxides (Sm-ON, Sm_2O_3) which are logically present owing to the samarium excess. The later, reacts either with small traces of oxygen or air during handling. It was demonstrated that the Mo preferential occupation site is the *8i* this result is in agreement with [4]. It appears that the unit cell parameter of the tetragonal phase increase with increasing x.

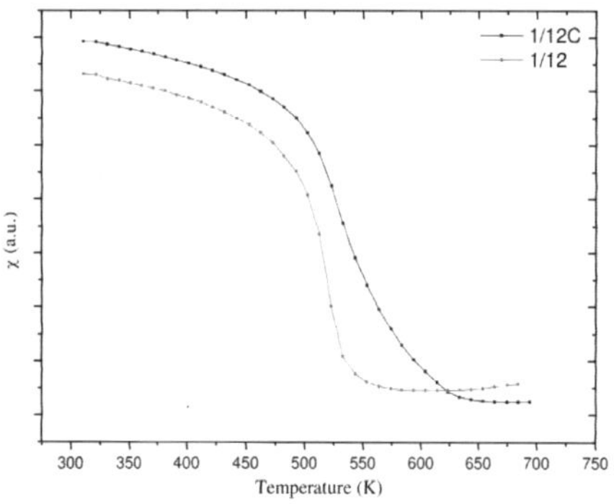

Figure 2. Curie temperature versus x Mo content.

At lower annealing temperature a new phase appears. This phase is identified as hexagonal (P6/mmm structure) with a stoechiometry equal to 1/10. The structure has been refined on the basis of the vacancy model proposed for overstoichiometric Sm –Co based compounds [5]. The unit cell parameter for this out of equilibrium phase are *a*= 4.888, *c*=4.253Å. The Mo occupation is identified to be the *6l* site.

The X- ray carbide patterns reveal a unique carbureted phase, meaning that the carbon diffuses homogenously within the material. The introduction of interstitial atoms does not change the crystal structure but an increase of the lattice parameters *a* and *c* with a consequent increase in the unit cell volume is observed upon carbonation. This result agrees well with [6] Neutron diffraction studies performed on nitrogenated $Y(Fe,Mo)_{12}$ and $Nd(Fe,Mo)_{12}$[7] which have shown that the nitrogen atoms occupy the *2b* (0,0,1/2) octahedron site. It is reasonable then to assume that carbon atoms occupy the same site.

Table 1. Unit cell parameters of the carbureted $Sm(Fe, Mo)_{12}$ phase with the corresponding host alloy.

Phase	1/12	1/12C
a (Å)	8.548	8.568
c (Å)	4.772	4.798
$\chi 2$	1.89	22.16
R_B	4.40	9.604
X{8i}	0.359	0.351

The crystallographic parameters are gathered and compared in table.1 for the carbide with the corresponding host alloy.

The Curie temperature increases with increasing Mo content at first, going through a maximum at $x \approx 1$ then decreases monotonously as seen in figure 2. This behavior was also observed in 2/17 phases[8].

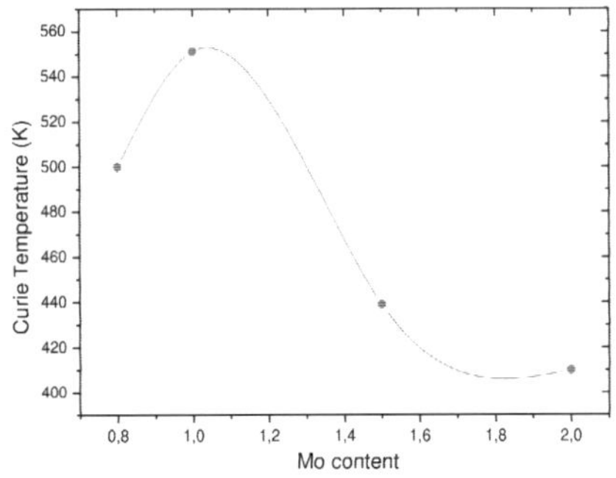

Figure 3. Curie temperature versus x Mo content.

From fig 3 it can be seen a Curie temperature enhancement of 60 K compared to that of the host alloy. The carbides T_c evolution versus the Mo content pursues the same tendency than the non carbonated starting alloys. It is well known that the Curie temperature is mainly governed by Fe-Fe exchange interaction and this interaction depends markedly on the interatomic distances. The interatomic distances between the near neighbors of the atoms at the different sites are different. It is obvious that the Fe-Fe interaction is strongly enhanced by interstitial carbon atoms, but this interaction is weakened by Mo substitution in both original and carbureted compounds. Nevertheless, interstitial carbon atoms were found to decrease the coercive field. This result is in agreement with that found in the case of nitrogenated and carbureted Nd based alloy [9].

4. Conclusion

Homogenous $Sm(Fe,Mo)_{12}$ were successfully elaborated by mechanical milling. Conclusive results as regarding the crystallographic data were obtained by mean of the Rietveld method. The unit cell parameters increase with Mo increment, in agreement with a steric effect. The tetragonal $ThMn_{12}$ phase is obtained when annealing at high temperatures. Below 800°C a structure transformation is observed. The new phase is identified to adopt the hexagonal P6/mmm space group. It was observed that carbon uptake results in Curie temperature enhancement and volume expansion.

References

[1] Verhoef R, De Boer F.R, Zhang Zhi-dong, Buschow K.H.J, *J. Magn. Magn. Mater.* 1988 **75** 319.
[2] Buschow K. H. J, *J. Magn. Magn. Mater.* 1991 **100** 79.
[3] Rodríguez-Carvajal, *J. Physica B* 1993 **192** 55.
[4] Li Z. W, Zhou X. Z and Morrish A. H, *J. Appl. Phys.* 1991 **69** 5602.
[5] Djéga-Mariadassou C, Bessais L, *J. Magn. Magn Mater.* 2000 **210** 81.
[6] Fruchart D, Gignoux D, Inacio C, Vert R, Skolozodra R. V, *J. Magn. Magn. Mater.* 1999 **196** 731.
[7] Tomey E, Isnard O, Fagan A, Desmoulins C, Miraglia S, Soubeyroux J. L, Fruchart D, *J. Alloys. Comp.* 1993 **191** 233.
[8] Kamimori T, Koyama K, Mori Y, Asano M, Konishita K, Mochimaru J, Konishi K, Tange H, *J. Magn. Magn. Mater.* 1998 **177** 1119.
[9] Wei Gong, Hadjipanayis G. C, *J. App. Phy.* 1993 **73** 6245.

IOP Conf. Series: Materials Science and Engineering 28 (2012) 012021 doi:10.1088/1757-899X/28/1/012021

Evaluation and measurement of the damping properties of laminated CFRP composite plates

L. Kherredine[1], S. Gouasmi, R. Laissaoui, N. E. Zeghib
Laboratoire de recherche en mécanique des matériaux et maintenance industrielle (LR3MI), Département Génie Mécanique, Faculté des Sciences de l'Ingénieur, Université Badji Mokhtar, B.P 12, Annaba 23000 Algérie.

E-mail: Kherredine_lamri@yahoo.fr

Abstract. The main objective of the present investigation was to examine how the damping properties of laminated composite plates vary with changes in the plate dimensions. Also, it was very important for plates with different ply-orientations to see how the dynamic characteristics vary from one plate to another or when the fibre-orientation was changed. The theoretical prediction of the natural frequency, mode shape and specific damping capacity for each mode obtained using the finite element method was to be compared with experimental results.

1. Introduction

The use of modern advanced composite materials has gained wide acceptance over the last few decades because of their high strength-to-weight and stiffness-to-weight ratios. Compared to metallic structures, composite laminates offer some unique engineering properties while presenting interesting but challenging problems for analysts and designers. A considerable amount of research work has been done on the damping analysis of fiber-reinforced composites [1-2]. In 1969, Ashton et al [3] used so-called beam characteristic functions to investigate the natural frequencies and mode shapes of anisotropic plates with various boundary conditions. The plates considered were laminated from plies of orthotropic boron-epoxy material. Good results were obtained with this analytical method for certain boundary conditions. However, the error between the theoretical and experimental results was very large for the case of free edges. It is even now difficult to use this method with all edges being free. With increasing application of the finite element method to structural mechanics, many authors have used this technique for analysing the vibration of plates. Nevertheless, most of them did not make any comparison of their results with experimental values. Cawley and Adams [4] used a finite element program which included the effect of transverse shear deformation to predict the natural modes of free-free carbon fibre reinforced plastic plates. This method produced improved accuracy for theoretical predictions and the predicted results gave generally good agreement with the measured natural frequencies. Reddy [6] has used the same technique as [4-5], and also included the effect of shear deformation and rotary inertia. The work was concentrated on antisymmetric angle-ply laminates. Some numerical results were given for a simply-supported rectangular plate, but no experimental values. Ju et al [7] suggested a new high-order transverse shear deformation model to modify the theory. This theory was compared with the Kirchhoff, Mindlin and Cawley-Adams models

[1] To whom any correspondence should be addressed.

to observe the effect of shear deformation and it was suggested that it gave a better approximation to the behaviour of real plates. Investigators in the damping of anisotropic plates, such as Hashin [8], and Bert [9] and have used the Kimball-Lovell (the so called complex modulus) model. This method has had only a few applications with composite plates. Adams and Bacon [10] suggested the damped element model and used an energy method to predict the damping of narrow, angle-ply, laminated composite beams. This method was developed further by Ni and Adams [11] who showed good agreement between their predictions and measurements of damping and stiffness of carbon fibre-reinforced plastic beams cut at various angles from a plate. Grootenhuis [12] also used a complex modulus model for treating isotropic sandwich plates. Lin et al [13] predicted the damping of carbon fibre and glass fibre reinforced plastics plates using a combination of the finite element and the damped element model.

In this work, investigations for predicting the natural modes and damping of anisotropic free-free laminated plates using the finite element technique [4] and the damped element model [11] are presented.

The main feature is to observe the effects of the plate dimensions and ply-orientations on the damping properties of laminated composite plates. Experimental results are then compared to the theoretical prediction of the natural frequency, mode shape and specific damping capacity for each mode obtained using the finite element method.

2. Experimental procedures

1.1. Preparation of specimens

The preparation of the carbon fibre plates was is follows:

- Cut the pre-preg to the size of the mould [240 mm x 240 mm).
- Lay up the pre-preg to the required thickness, one sheet on top of another according to the angle of orientation. Trim the edges to 0.5 mm less than the mould size.
- Pre-heat the metal mould to 170° C in the hot press and, when it is at the required temperature, put the laid up pre-preg (1.6 mm thick) into the mould, put the lid on, and transfer to the hot press. Press the mould very lightly and wait for the excess resin (10%) to flow out of the top of the mould and down the sides. When the resin gels (approximately 3 minutes), press the mould to the stops using no more than 5 KN pressure.
- Leave in hot press at 170 °C for 2 hours.
- After 2 hours, take the mould out of the hot press and leave to cool to room temperature.
- Remove carbon or glass fibre plate from mould.

1.2. Material properties

Table 1. Plate data.

Plate number	Plate N° 1		Plate N° 2	
*	sq f	s t	sq f	s t
Material	CFRP		CFRP	
Number of layers	12		8	
Density (Kg.m^{-3})	1636,4		1941,4	
Thickness (mm)	2,02		1,62	
Length (mm)	215,6		215,0	
Width (mm)	215,0 (sq) 167,9 (s)	195,4 (f) 107,5 (t)	215,0 (sq) 167,9 (s)	195,4 (f) 107,5 (t)
V_f	0,61		0,494	
Ply orientation	$(0, -60, 60, 0, -60, 60)_S$		$(0, 90, 45, -45)_S$	

The material used in this investigation was HM-S carbon fibre in DX-210 epoxy resin. The plates were made by hot-pressing 8 and 12 layers of pre-impregnated fibre, such that different laminate orientations could be obtained; details of the plates used are given in Table 1.

The material properties used in the theoretical prediction are given in Table 2. All the values in this table are established using beam specimens cut from a unidirectional plate. The longitudinal and transverse Young's moduli (E_1, E_2) and damping (ψ_1, ψ_2) were established using 0° and 90° beam specimens respectively in the flexural apparatus [14].

Table 2. Moduli and damping values for material used in the plate.

Material	HMS/DX-210
E_1 (GPa)	172,7
E_2 (GPa)	7,20
G_{12} (GPa)	3,76
ψ_1 (%)	0,45
ψ_2 (%)	4,22
ψ_{23} (%)	7,05
υ_1, υ_2	0,3
V_f	0,5

The modulus and damping in shear (G_{12}, ψ_{12}) were determined using a 0° fibre orientation beam in the torsional apparatus figure 1. However, the value of torsional damping (ψ_{23}) of the 90° fibre orientation beam was not important in the prediction, since, when this value was changed from 6% to 15%, there was no discernible difference in theoretical results [10].In the prediction, ψ_{23} taken as the same value as ψ_{12} as both values involved essentially shear in the matrix.

For HM-S carbon fibre in DX-210, the fibre volume fraction of the plates varied, so the material properties in the theoretical predictions were then corrected from a standard set given for a 50% fibre volume fraction using the method presented in [15].

Figure 1. General view of the torsional pendulum with a spare coil assembly

2.2. Specimen specification

The plates made from HM-S carbon fibre in DX-210 were first tested as squares. They were then cut along edge to form a series of rectangular plates with a plan aspect ratio of 1.1. In addition to these two series, the same plates were cut a second and third time from the same side to give two new series of rectangular plates with plan aspect ratios of 1.28 and 2. The dimensions of this series of four plates are given in table 1.

2.3. Test equipments

The plate was supported on soft sponge blocks to provide the minimum constraint and damping so as to approximate closely to the free-free condition. Excitation was via an electrodynamic shaker. The positions of the sponge supports were adjusted to coincide with the nodes, thus maximising the amplitude of vibration by minimising the damping at the supports. A condenser microphone was positioned over an antinode and the frequency of the signal generator was tuned until resonance was obtained. Dry sand was spread lightly over the surface and this migrated to the nodal lines. Pictures were then taken of the mode shape of the nodal lines. This procedure was carried out for the first and second modes of vibration for each plate.

The damping measurement was achieved by suspending the plate at two nodal points by means of a nylon thread figure 2. Detection of the response was by means of a small accelerometer which was attached to the plate at an anti-node. The specimen was lightly tapped at an anti-node, and the transient response of the plate to the single pulse excitation was processed by the zoom FFT analyser and the frequency spectrum was obtained for each natural frequency.

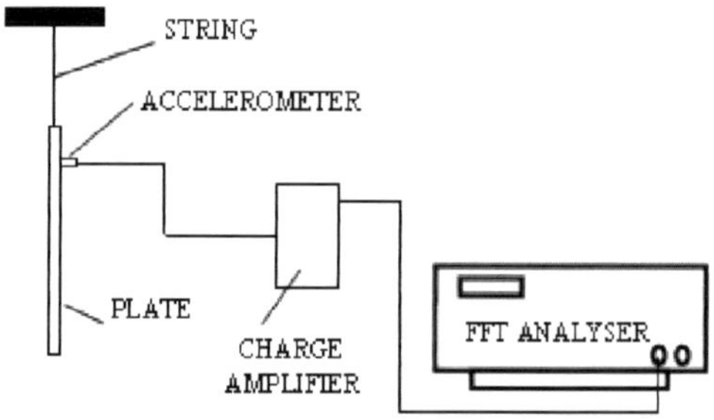

Figure 2. Apparatus for measuring specific damping capacities and natural frequencies of the plate.

3. Vibration of Composite Plates

For a laminated plate with midplane symmetry, there is no extensional–bending coupling, and the in-plane response can be assumed to be negligible compared with the flexure-induced displacement. For a symmetric laminate plate subjected to bending only, the laminate moment–curvature relationships are given by

$$\begin{Bmatrix} M_x \\ M_y \\ M_{xy} \end{Bmatrix} = \begin{bmatrix} D_{11} & D_{12} & D_{16} \\ D_{12} & D_{22} & D_{26} \\ D_{16} & D_{26} & D_{66} \end{bmatrix} \begin{Bmatrix} k_x \\ k_y \\ k_{xy} \end{Bmatrix}$$

where (M_x, M_y, M_{xy}) and (k_x, k_y, k_{xy}) are the moment and the curvature, respectively. $D_{ij}(i, j = 1,2,6)$ is the flexural stiffness.

The partial differential equation governing the transverse motion of a symmetrically laminated thin plate is

$$D_{11}\frac{\partial^4 w}{\partial x^4} + 4D_{16}\frac{\partial^4 w}{\partial x^3 \partial y} + 2(D_{12} + D_{66})\frac{\partial^4 w}{\partial x^2 \partial y^2} + 4D_{26}\frac{\partial^4 w}{\partial x \partial y^3} + D_{22}\frac{\partial^4 w}{\partial y^4} = -\rho h\frac{\partial^2 w}{\partial t^2}$$

For the orthotropic plate, we have $D_{16} = D_{26} = 0$. For free harmonic vibrations at frequencyω, we can assume that

$$\omega(x, y, t) = W(x, y) \sin \omega t$$

where $W(x, y)$ is the maximum deflection and can be represented as a linear series of assumed functions:

$$W(x, y) = \sum_{m=1}^{p}\sum_{n=1}^{q} A_{mn} X_m(x) Y_n(y)$$

The assumed functions $X_m(x)$and $Y_n(y)$ must be admissible, such that they satisfy the essential boundary conditions of the plate.

The characteristic equations of vibrating beams are used as the assumed functions X_m and Y_n. The characteristic equations of vibrating beams are chosen so that the boundary conditions of the beam matches those of the plate, guaranteeing satisfaction of the essential boundary conditions. The method of determining the set of characteristic function sand a summary of the properties of these functions was given by Young [16]. The characteristic functions $X_m(x)$ for the free–free beam used in this study is given as follows:

$$X_1(x) = 1, \qquad X_2(x) = \sqrt{1}[1 - (2x/l)]$$

$$X_m(x) = cosh(c_m x/l) + cos(c_m x/l) - \alpha_m[sinh(c_m x/l) + sin(c_m x/l)] \qquad m = 3,4,5,...$$

where l is the length of the beam. The numerical values of c_m and α_m are given by Young [16]. The functions $X_1(x)$and $X_2(x)$ represent a rigid-body translation and rotation, respectively, and are included to obtain a complete orthogonal set. Similar expressions also are used for $Y_n(y)$.With the stationary potential energy theory; the Rayleigh–Ritz technique will provide a discrete number of stationary values ω_6, which are the lowest upper bounds of the actual resonant frequencies [17].

4. Discussions

4.1. Mode shapes and natural frequencies

Table 3. Natural frequency and damping capacity of various modes of a 12 layer (0, -60, 60, 0, -60, 60)s HMS/DX210 CFRP (Plate N° 1)

(a)

Modes	(1-1)	(0-2)	(2-0)	(1-2)	(2-1)	(0-3)
Dimensions $a = 215{,}60$ (mm) $b = 215{,}60$ (mm)		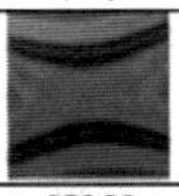	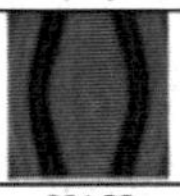		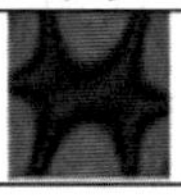	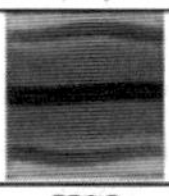
Frequencies(Hz)	163,42 (161,0)	275,25 (274,2)	384,38 (374,7)	426,11 (412,6)	507,3 (490,5)	776,9 (785,16)
SDC %	1,41 (1,49)	0,90 (0,84)	0,61 (0,73)	1,21 (1,25)	0,94 (1,00)	0,89 (-)

(b)

Modes	(1-1)	(0-2)	(2-0)	(1-2)	(2-1)	(0-3)
Dimensions $a = 215{,}60$ (mm) $b = 196{,}0$ (mm)	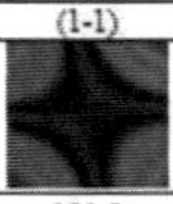		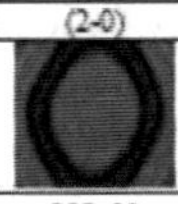			
Frequencies (Hz)	180,3 (175,6)	324,35 (326,2)	393,66 (388,1)	481,06 (461,7)	539,4 (523,6)	936,54 (949,1)
SDC %	1,41 (1,64)	0,88 (0,96)	0,63 (0,81)	1,19 (1,33)	0,97 (1,25)	0,90 (-)

(c)

Modes	(1-1)	(2-0)	(0-2)	(1-2)	(2-1)	(3-0)
Dimensions $a = 215{,}60$ (mm) $b = 168{,}40$ (mm)	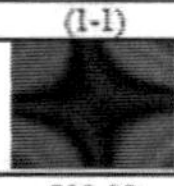	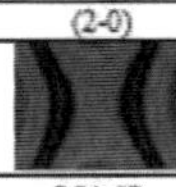	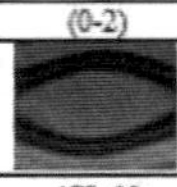	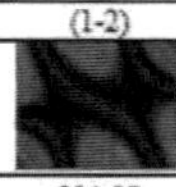	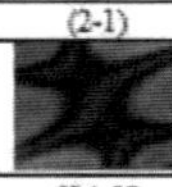	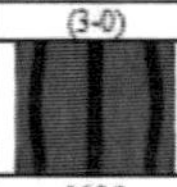
Frequencies (Hz)	210,22 (206)	364,67 (366,3)	472,65 (458,6)	554,99 (546,1)	634,59 (619,2)	1036 (1029,7)
SDC %	1,40 (1,65)	0,71 (0,78)	0,82 (0,99)	1,12 (1,21)	1,06 (1,25)	0,62 (-)

(d)

Modes	(1-1)	(2-0)	(2-1)	(3-0)	(0-2)	(1-2)
Dimensions $a = 215{,}60$ (mm) $b = 107{,}80$ (mm)	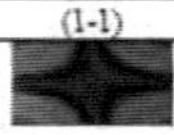	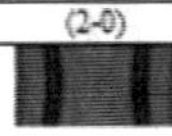		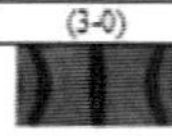		
Frequencies (Hz)	325,63 (326,7)	372,12 (362,1)	759,17 (761,3)	1017,7 (1021)	1120,6 (1142,1)	1315,8 (1343,3)
SDC %	1,39 (1,52)	0,65 (0,87)	0,96 (-)	0,69 (-)	0,91 (-)	1,21 (-)

Fibres direction ⟶

The results presented in tables 3a, 3b, 3c and 3d show that the theoretical prediction obtained by the finite element program and the experimental results for the first six modes of CFRP plate N° 1 with a variety of plan aspect ratios. The modes shown on these tables were obtained by computer, and then reduced to a smaller size of the tables. The materials from which the plates were made are listed in Table 1 for various fibre orientations.

On the whole, there is good agreement between the predicted and measured values of frequency. The highest error was 4,33 % in mode 2 Table 3d. The average error in the results obtained from the plates presented in these tables is 2,07 %. The difference between the predicted and experimental frequencies is believed to be mainly due to slight variations in the thickness and the nominal angle of the fibres. Hence, some modes have higher predicted frequencies and some lower than those measured experimentally.

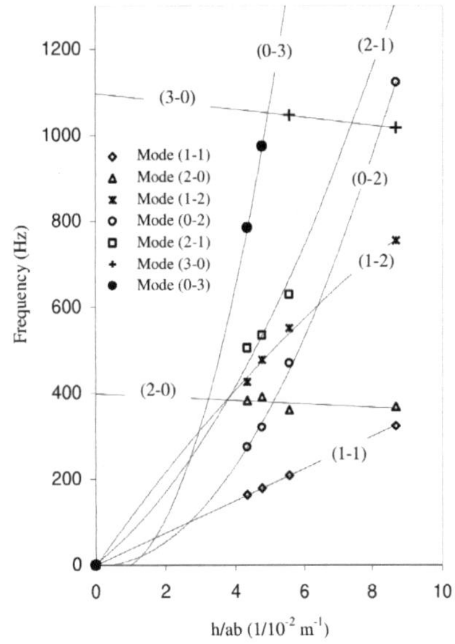

Figure 3. Variation of natural frequencies with the ratio of thickness to the area (Plate N° 1)

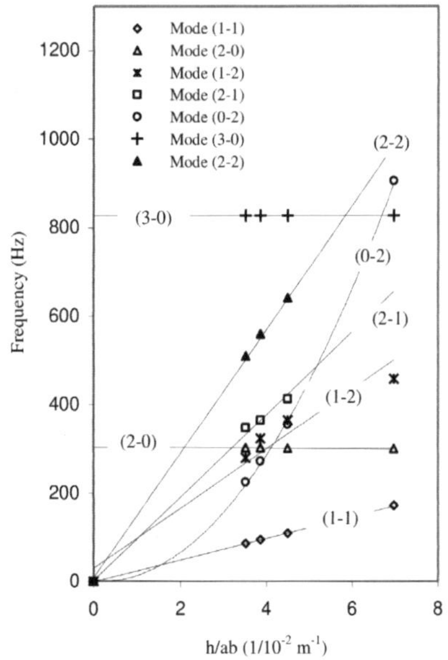

Figure 4. Variation of natural frequencies with the ratio of thickness to the area (Plate N° 2)

It should be noted from Fig. 3 and Fig. 4 that when the plate is reduced from a square shape towards a beam, the frequencies of the different modes change differently with the parameter *ab* (a and *b* are the side lengths). Whereas the (1-1) mode is the lowest for a square plate, the (2-0) mode is the lowest for a beam. Anisotropy will change the position (*ab* product) at which the lowest frequency of a rectangular plate will be the (1-1) or the (2-0) mode.

Table 4. Natural frequency and damping capacity of various modes of a 12 layer (0, -60, 60, 0, -60, 60)s HMS/DX210 CFRP (Plate N° 1)

Modes	(1-1)	(0-2)	(2-0)	(1-2)	(2-1)	(0-3)
Dimensions a = 215.60 (mm) b = 215.60 (mm)	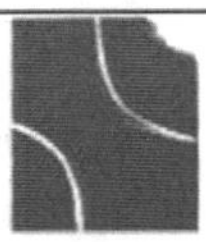	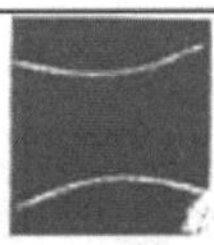		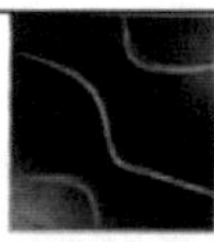	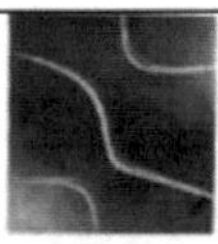	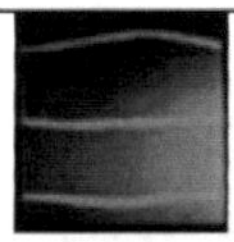
Frequencies (*Hz*)	163,42 (161,0)	275,25 (274,2)	384,38 (374,7)	426,11 (412,6)	507,3 (490,5)	776,9 (785,16)
SDC %	1,41 (1,49)	0,90 (0,84)	0,61 (0,73)	1,21 (1,25)	0,94 (1,00)	0,89 (-)

The nodal line patterns observed using sand were in good agreement with those predicted using the finite element program, except when there are two close natural frequencies with symmetrical modes such as modes 4 and 5 in Table 4. In such cases, the patterns observed depend on the position of excitation. Another way to obtain the predicted mode shape is by clamping the plate at a point which lies on a nodal line of one of the possible modes but at which there is no motion in the other mode. Vibration of the alternative mode was then prevented. Figure 5 shows the experimental nodal lines; these nodal patterns were obtained by taking pictures when sand was spread over the surface of the plate when resonance occurred.

The theoretical nodal patterns were drawn from the graphical output plots resulted from the finite element program. It can be said that the F.E program does not only give the right succession of modes, but also the predicted and experimental nodal lines are very close.

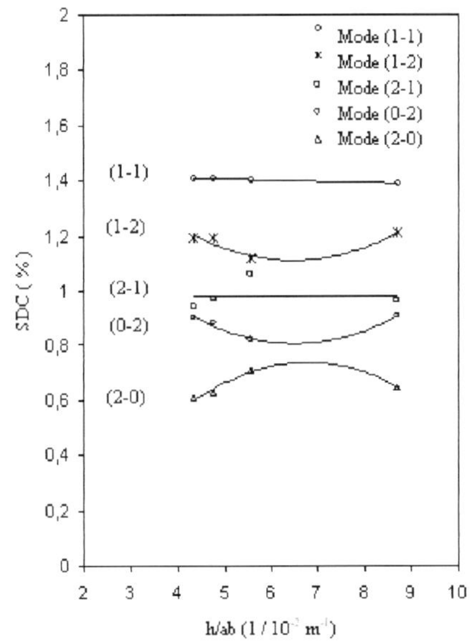

Figure 5. Variation of the specific damping capacity with the ratio of the thickness to the area. (Plate N^0 1).

For any mode of vibration, the nodal patterns are defined by *m* and *n*, the number of nodal lines in the *x* and *y* directions respectively. Figures 3 and 4 show the variation of the natural frequencies with respect to the plan aspect ratio (b/a) for any particular mode. It can be seen from these figures that, for the domain where the plates were tested, having an aspect ratios b/a equal to 0.50, 0.78, 0.90 and 1 for CF reinforced plastic plates, the variation of the natural frequency with respect to the plan aspect ratio (b/a) is linear. Furthermore, the rate of change of frequency with this ratio is different for different modes, such that two different modes can occur at the same frequency, depending on the aspect ratio and the laminate lay-up. However, for a very low values of (b/a), the theory shows that the relationship between the frequency and the plan aspect ratio is non linear.

When the vibration results essentially in nodal lines parallel to one side of the plate, the frequencies may be expected to bear some relation to those of a beam. As can be seen in figures 3, the variation of the natural frequencies for mode (2-0) and (3-0) of the CF reinforced plastic plate with respect to the plan aspect-ratio (b/a) is roughly constant. This is in contrast to the other modes, where the frequency increases with increase in the ratio (b/a).

It can be seen from figures 3 and 4, corresponding to the square series plates N° 1 and N° 2 made from CF reinforced plastic that the dynamic characteristics of composite laminated plates vary considerably with the ply orientation. The plates had almost the same size length being 215,5 and 215,5 mm and were between 1,62 and 2,02 mm thickness.

The lowest frequency of the (0, -60, 60, 0, -60, 60)s plate N° 1 is over 52 % higher than the (0°,90°,45°, -45)s plate N° 2.

However, if more plates were taken with the same number of layers, same thickness and same side lengths, a good relationship could be found and a prediction of the lowest frequencies would be easy to obtain from plates having different ply-orientations. Nevertheless, this could not be investigated in this work; unfortunately the plates supplied were of different thickness, which is a very important factor in varying the dynamic characteristics.

Similarly, as with ply-orientation, the natural frequencies vary considerably with the plan aspect-ratio (b/a). Therefore, the designer can exercise considerable control over the dynamic response of laminated composite structure by changing the ply-orientation or the ratio (b/a). By using these graphs, therefore, it is now a simple matter with any aspect-ratio (b/a) in the range tested to estimate the natural frequencies and mode shapes of the first six modes of plates which are made with the same material and have the same fibre orientations.

4.2. Damping

It can be seen from tables 3a, 3b, 3c and 3d that the measured specific damping capacity values are very close to and mostly higher than those predicted. This is due to extraneous energy dissipation caused by air damping and energy transmission to the accelerometer cable.

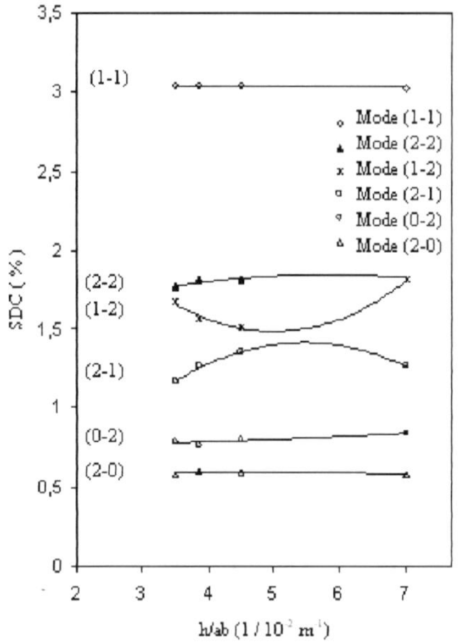

Figure 6. Variation of the specific damping capacity with the ratio of the thickness to the area. (Plate N^0 2).

It can be seen from figures 5 and 6 that, for a similar material, the SDC is roughly constant for the same mode with different aspect ratios. The SDC of mode (1-1) of plate N° 1 is almost constant and equal to 1.41, 1.41, 1.40 and 1.39 respectively with respect to the plan aspect ratios of 0.5, 0.78, 0.90, 1,0 respectively. In addition, the SDC of mode (2-0) are 0.61, 0.63, 0.71 and 0.65 respectively, and 0.90, 0.88, 0.82 and 0.91 respectively for mode (0-2).

It can be seen that although the frequencies for a given mode shape (say 2-0) change with the (b/a) ratio, there is a little if any change in the damping. Although the SDC is independent of the dimensions of the plates, it is very dependent on the mode shape for plates which have the same fibre orientation.

Similar to the frequencies, using figure 5 and 6, the SDC could be estimated as far as the variation against the plan aspect ratio is fairly constant. Therefore, if the SDC of a particular mode of a square plate is known, the change in dimension of one side does not change the value of the SDC of that particular mode significantly.

5. Conclusions

The results show good agreement between the theoretical predictions and the experimental values of the natural frequencies and damping, the worst error in frequency being 8.07 per cent. Therefore, the accuracy obtained using the analysis given in Chapter 3 appears to be as good as is warranted by the uniformity of the materials.

These results have been obtained using a fairly coarse mesh comprising 36 elements. The elements are readily general sable to shells and will accommodate irregularly shaped boundaries. Crawly & Adams (3) comparing a 6x6 mesh with a 5x5 mesh showed a variation in natural frequencies of less than 2%.

The natural frequencies of laminated plates are affected by the fibre volume fraction, the mode shape, the number of layers, the ply-orientation and the plate dimensions. As with the natural frequencies, the damping values are also affected by these factors but not significantly with the plan aspect-ratio (b/a).

If an increase of the natural frequencies of some modes is required, the fibre-orientation of each layer of the laminates or one of the size parameters of the plates should be tailored.

It should be noted that, when varying the plate plan-aspect ratio (b/a), the order of the modes may change.

Finally, graphs for simplifying the prediction of natural frequencies and nodal lines of several modes of vibration of rectangular plates have been presented.

References

[1] Lin D. X, Ni R. G, Adams R. D 1984 *J. Composite Materials,* **18** (2) 132 152.
[2] Shokrieh M. M, Najafi A 2006 *Composite Structures* **75** 1 4 472 478.
[3] Ashton J. E, Waddoups M. E 1969 *J. Composite Materials* **3** 148.
[4] Cawley P, Adams R. D 1978 *J. Composite Materials* **12** 336 347.
[5] Malekzadeh K, Khalili M. R and Mittal R. K 2007 *International Journal of Impact Engineering* **34** 3 522 543.
[6] Reddy J. N 1979 *J. Sound and Vibration* **66**(4) 565.
[7] Ju F, Lin D. X, Xiao C. Z and Ji Y. Z 1987 *3rd National conf. on vibration theory and application*, China.
[8] Hashin Z 1970 *International Journal of Solids and Structures* **6** 6 797 807.
[9] Bert C. W 1980 *Int. Conf. University of Southampton*, 7 11 July.
[10] Adams R. D, Bacon D. G. C 1973 *J. Composite Materials* 7 402.
[11] Ni R. G, Adams R. D 1984 *J. Composite Materials* **18** 104.
[12] Lonnides E, Grootenhuis P 1979 *J. Sound and Vibration* **67**(2) 203.
[13] Ni R. G, Lin D. X and Adams R. D 1984 *J. Composite Materials* **15** 297.
[14] Adams R. D, Fox M.A.O, Flood R. J. L, Friend R. J and Hewitt R. L 1969 J. *Composite Materials* **3** 594.
[15] Fox M. A. O 1972 *Ph.D. Thesis, Dept. Mec. Eng., Bristol University.*
[16] Young D 1950 *Journal of Applied Mechanics* **17** 4 448 453.
[17] Leissa A. W 1973 *Journal of Sound and Vibration* **31** 3 257 293.

IOP Conf. Series: Materials Science and Engineering **28** (2012) 012022 doi:10.1088/1757-899X/28/1/012022

Study by AES, EELS Spectroscopy of electron Irradiation on InP and $InPO_4$/InP in comparison with Monte Carlo simulation

Z. Lounis[a1], M. Bouslama[a], K. Hamaida [a], C. Jardin[b], A. Abdellaoui[a], A. Ouerdane[a], M. Ghaffour[a], N. Berrouachedi[a]

[a]Laboratoire Matériaux (LABMAT), Ecole Normale Supérieure d'Enseignements Technologiques (ENSET) d'Oran, Oran Mnaouer, B.P. 1523, Oran, Algeria

[b]IUT B Rue de France Villeurbanne, Université Claude Bernard Lyon 1, 43 Bd du 11 Novembre 1918, Villeurbanne 69 622, France

E-mail: lou_zakia@yahoo.fr

Abstract. We give the great interest to characterise the InP and $InPO_4$/InP submitted to electron beam irradiation owing to the Auger Electron Spectroscopy (AES) associated to both methods Electron Energy Loss Spectroscopy (EELS). The incident electron produces breaking of (In-P) chemical bonds. The electron beam even acts to stimulate oxidation of InP surface involving on the top layers. Other, the oxide $InPO_4$ developed on InP does appear very sensitive to the irradiation due to electron beam shown by the monitoring of EELS spectra recorded versus the irradiated times of the surface. There appears a new oxide thought to be In_2O_3. We give the simulation methods Casino (Carlo simulation of electron trajectory in solids) for determination with accuracy the loss energy of backscattered electrons and compared with reports results have been obtained with EELS Spectroscopy. These techniques of spectroscopy alone do not be able to verify the affected depth during interaction process. So, using this simulation method, we determine the interaction of electrons in the matter.

1. Introduction

There are a lot of methods applied to characterize the material but it is necessary to choose the one which would be very adapted to material surfaces [1,3].Owing to electron energy loss spectroscopy (EELS) and Auger electron spectroscopy (AES) we display the morphology and structural change of InP and $InPO_4$/InP submitted to irradiation electron beam. The indium phosphide has attracted attention due to his potential device application, but the nature and quality of surface preparation in semiconductor technology is of the utmost importance during device fabrication and has an influence on the performance and reliability of these devices [4, 5]. The most important results so found are the instability of InP against all physical treatment such as cleaning by argon ion bombardment or heating. So that appears an excess of element III on the surface. In order to restore the surface stoichiometry many suggestions have been done namely evaporation of element V such as phosphorus or antimony on InP previously cleaned in UHV [6]. These methods have given good results for the technological applications from indium phosphide. In this paper we give some results about a thin film $InPO_4$ grown

[1] To whom any correspondence should be addressed.

on InP. Regarding to monitoring of InP and $InPO_4$/InP irradiated by the electron beam of 1 keV energy of the surface in UHV.
So we use a numerical calculation carried out following the programme related to casino (Carlo Simulation of electron trajectory In Solids), in order to determinate a backscattering coefficient of electron, determination of energy loss spectra and depth affected by the beam of incident electrons during the interaction process in InP and $InPO_4$/InP.

In this study we give some results about the behaviour of InP and $InPO_4$ oxide was grown on InP by using powerful spectroscopy methods such as EELS, AES. We show the effect of the electron irradiation on such system compared with results obtained by Monte Carlo simulation.

2. Experiments

2.1. Effect of electron beam on InP

Although, the InP (100) surface was washed ex-situ by H_2SO_4 solution, acetone and deionised water but this chemical treatment reveals insufficient because the AES spectra display the presence of carbon and oxygen [7]. The cleaning of the InP (100) surface is necessary to eliminate these impurities. The argon ion bombardment is achieved according to the experimental conditions as follow: Normal incidence; current 1 μA; pressure 10^{-5} Torr.

The energy choice of Ar^+ is based on the fact that we must minimise the disturbance of the sample in volume. Auger electron spectroscopy is performed to monitor the InP surface evolution while it is irradiated by an electron beam of 4 KeV energy and 10^{-3} $A.cm^{-2}$ current density during long time. Owing of the good spectrometer analysis there appears a structural modification of the Auger signals shapes $InM_{45}N_{45}N_{45}$, P-LMM and O-KLL spectra (see figure 1). There gives rise to an oxide including In resulting from (In-P) chemical bonds breaking. The structural modification of PLMM occurs in conjunction with the OKLL one. The intensity of the P-LMM Auger signal of phosphorus bounded to indium (P–In) is reduced in intensity displaying the phosphorus depletion from the surface. We also note the appearance of O-KLL Auger signal. The AES analysis does not reveal any presence of carbon as impurity on the surface [8].

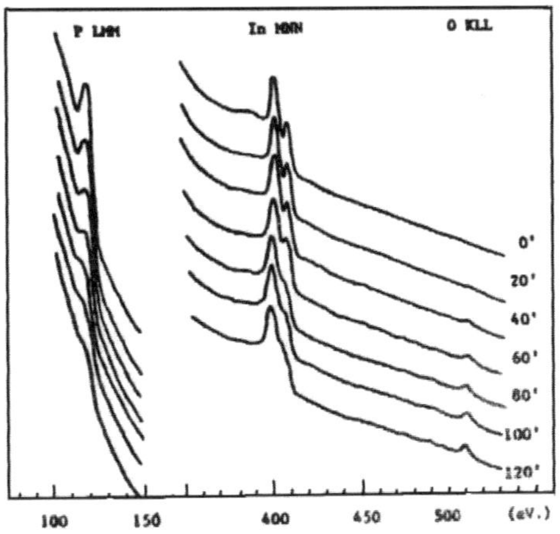

Figure 1. Evolution of Auger spectra P-LMM In-MNN and O-KLL showing the oxidation on InP surface actived by electronic irradiation.

In figure 2, we give AES spectra of the peak attributed to indium oxide located at low energy about 2.2 eV of peak characteristic $InM_5N_{45}N_{45}$ of InP. This shift energy resulted from modification in bond energy and relaxation energy, this energy is sensitive to chemical environment.

In figure 3, the EELS spectra of InP gives rise to attenuation of the peaks in spectrum (b) compared to spectrum (a) recorded after cleaning in 1000eV, there is the structural located at 5.5 eV attributed to interband transition from the level O_{2p} of oxygen and the bulk Plasmon in 17 eV [8].

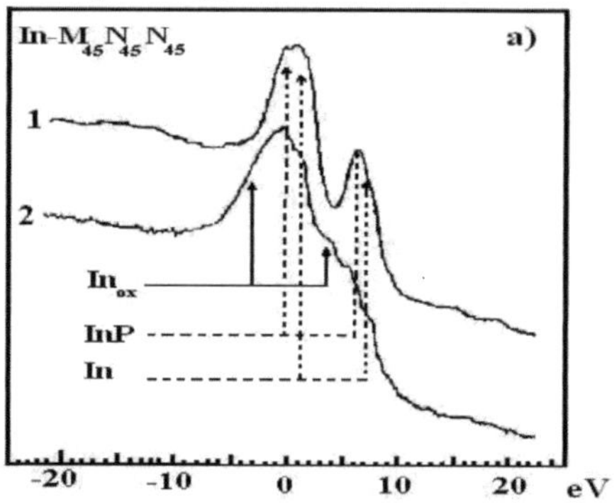

Figure 2. Structure fine comparison of In-$M_{45}N_{45}N_{45}$ 1): before 2): after stimulated oxidation on InP surface.

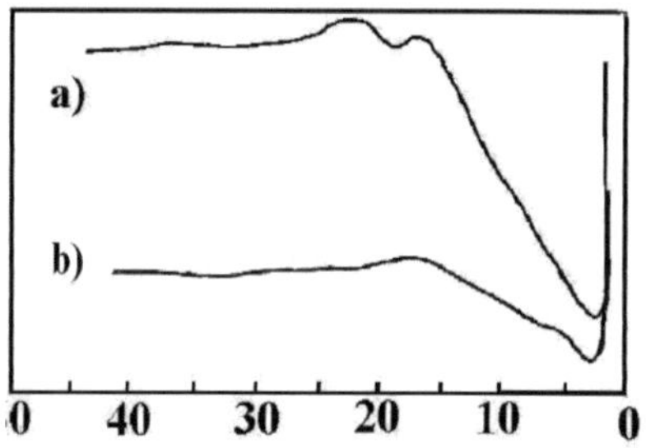

Figure 3. Comparison of InP loss energy spectra a): after cleaning by ion bombardment b) stimulated oxidation at 4keV energy.

2.2. Effect of electron beam on InPO4/InP

In Fig. 4, we show the EELS spectra of clean $InPO_4$/InP recorded at the primary energy of 1 keV for different incidence angles θ varying from 20° (spectrum 1) up to 14° (spectrum 4). We see the structural modification of the EELS spectra as a function of the incidence angle. We note that the structure of these spectra is different from those described by Pellerin [9] and Bouslama [10] related to the contaminated InP surface by carbon and oxygen. The energy loss peaks appeared at 26 and 6 eV. The intensity of the energy loss peak located at 15.4 eV decreases as a function of the incidence angle θ. The energy loss peak intensity at 15.4 eV related to the bulk plasmons of InP is attenuated when the incidence angle decreases. The energy loss peaks located at 5.8 and 7.8 eV are related to the interband transitions of $InPO_4$ [6]. Their intensities remain independent on the incidence angle. The energy loss peak located at 22.4 eV is related to plasmons energy of $InPO_4$. Its intensity practically remains unchanged because of the presence of that oxide on the outmost layers on the surface.

In Fig.5, we give the EELS spectra recorded at the same primary energy Ep = 1000 eV. So, the effect of the electron irradiation on the system $InPO_4$/InP leads to modify the EELS spectra. There is displacement of the energy loss peak located at 22.4 eV to the higher energy at 23.4 eV. Such a displacement is due to the modification of the chemical composition of the surface. The irradiation of the system $InPO_4$/InP by the electron beam also leads to the attenuation of energy loss peak intensity related to InP bulk plasmons (located at 15.4 eV). Such a modification is produced in conjunction with the increase in the intensity due to the Auger transition O-KLL. The electron beam stimulates the surface oxidation with breaking of chemical bonds between the chemical species of $InPO_4$ and even between indium and phosphorus (In–P). This leads to the formation of another oxide whose composition is not clearly known for the moment. However, we presume it to be In_2O_3.

So, after the electron irradiation of $InPO_4$/InP, the chemical composition of the surface has changed. The modification is confirmed by the variation of the elastic peak intensity. We see that it decreases because of such a change in the chemical composition of the surface on which a new atomic arrangement occurs. The interfacial defects are suspected to be present in the $InPO_4$/InP system to stimulate this oxidation process by the electron irradiation advantaged by the partial pressure of oxygen into the spectrometer chamber [11].

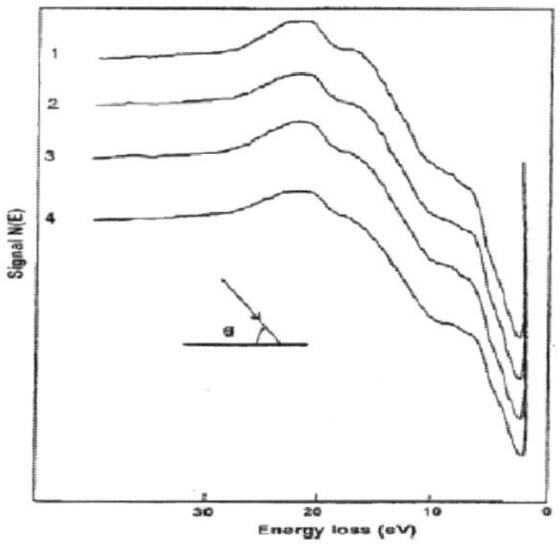

Figure 4. EELS spectra of InPO4/InP recorded for different incidence angle θ at the primary energy Ep=1000eV, the angle θ decreases from the spectrum number1 to spectrum number 4.

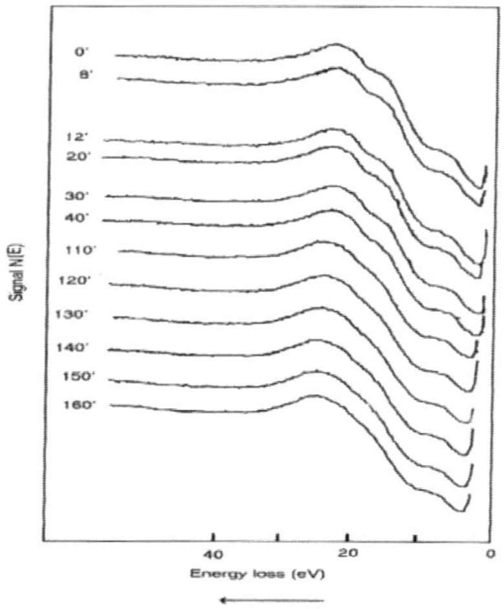

Figure 5. EELS spectra of $InPO_4$/InP recorded while the surface is irradiating by the electron beam at Ep = 1000eV, the irradiation time is given in minutes (mn).

3. Monte Carlo Simulation CASINO

Dominique Drouin et al. [12] have developed recently a good simulation method CASINO. This method is based on the interaction process between the incident electrons of different energies and the matter. The volume of electron interaction may be modelled from first principles using a Monte Carlo method, where the paths of a series of incident electrons are modelled probabilistically with equations for elastic and inelastic scattering determining the scattering angles, mean free-paths, and the rate of energy-loss. Each electron trajectory is simulated in stepwise fashion. Monte Carlo results may be used to model not just the electron trajectories, but the distributions of the resulting of electrons.

3.1. Study of InPO4/InP by Carlo simulation method

Measurements were been displayed in figure 6 of the effect of the electron beam of 1000eV energy in the change of incidence angular. From amplified spectra in the range of the loss energy ≤ 50 eV, we illustrate the peaks of the loss energies. The peaks values located at ~8.5, 14 eV can be probably attributed to surface and bulk plasmon of InP. The valor 18 eV is assigned to transition band of indium metal, but for both values 23.5 and 40 eV is estimated respectively to plasmon and transition band of $InPO_4$. These values are en agreement with determinate by EELS spectroscopy.

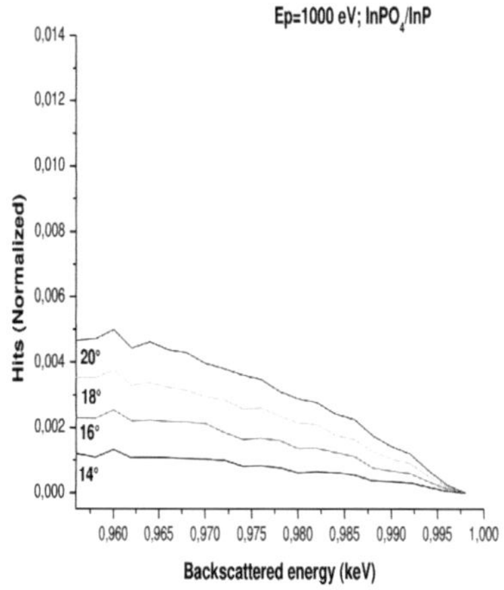

Figure 6. Carlo simulation of loss energy spectra recoded for InPO4/InP at different incidence angle for Ep= 1000eV.

For lower incidence angle θ=14° there is attenuation of the peaks attributed to plasmons of InP which confirm the presence of InPO4 on the surface of the system. In figure 7 the same calculates of loss energy are carried out with 4 keV of primary energy. We illustrate the almost attenuation of the peaks given in figure 6 but the loss peak in 24 eV persisted even for higher time of irradiation. We attributed the Loss peak in the beginning of spectra to transition band of oxygen. That approved the oxidation stimulated by electron beam on the surface. Where is the formation of the oxide of In2O3.Which is in agreement with the result determinated precedely [10].

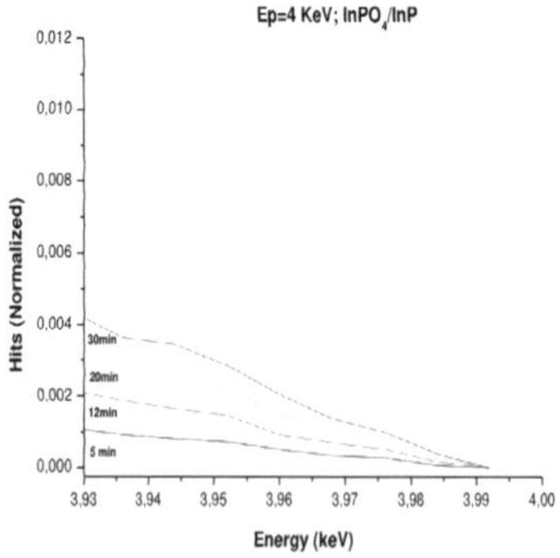

Figure 7. Carlo simulation of loss energy spectra recoded for InPO4/InP at different times of irradiation recorded for Ep= 4000eV

We give the number of backscattered electrons per times and incidence angle has been calculated for $InPO_4/InP$ in the primary energy 4k eV have been showing in the fig.8.

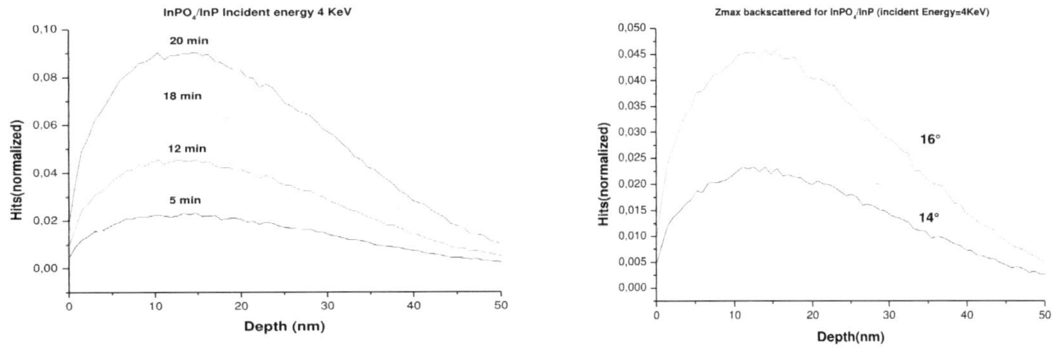

Figure 8. Carlo simulation spectra of $InPO_4/InP$ recorded for different (I): times of irradiation and (II): incidence angle; showing the depth of interaction process at incident energy 4KeV.

Where the data is sufficient to estimating the magnitude of interaction depth, in the surface and the bulk target. However the powerful EELS and AES spectroscopes can not approve this information experimentally.

3.2. Study of InP by Carlo simulation method

We use a numerical calculation carried out following the programme casino for InP. The number of backscattered electron per 1000 eV and 4 keV primary beam electron has been determinate for many thick of In on InP. The choose of this thickness 3, 4 and 5 A° of indium is just to simulate the presence of indium metallic on the upper layers of InP when it was submitted to electronic irradiation for few times. From this hypothesis we presented on figure 9 the backscattered electron coefficient increase at higher thickness of In on InP. This approved the dependence of this coefficient with the atomic number. The backscattered electrons are sensitive to chemical composition of the matter [13]. Also there gives direct information on the inclination of the specimen surface relative to the primary beam at each probing point. Besides this, there is additional calculation presented in figure 10. concerning the loss energy of electrons during interaction process with the matter.

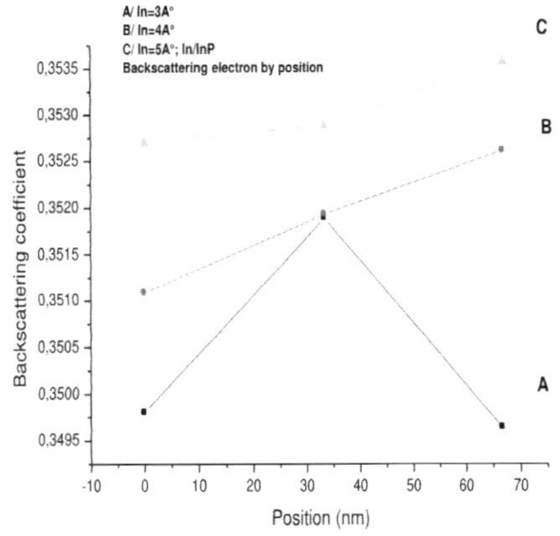

Figure 9. Carlo simulation spectra of In/InP recorded for Backscattering coefficient in different layers of Indium (3,4, 5A°).

IOP Conf. Series: Materials Science and Engineering **28** (2012) 012022 doi:10.1088/1757-899X/28/1/012022

We gives the peaks located at~ 7, 12, 16 eV from, the amplified spectra which we can be corresponding respectively to surface, bulk plasmons of metallic indium and bulk plasmon of InP. Which have been determinate by EELS spectrometer and reviewed prescendently.

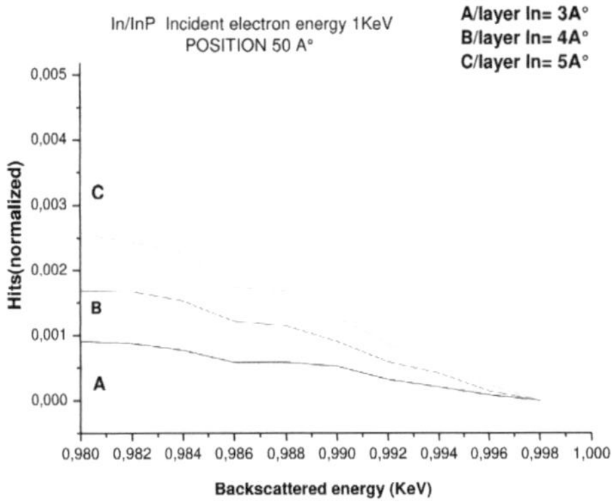

Figure 10. Carlo simulation of loss energy spectra for In/InP recorded for different layers

4. Conclusion

By using a two analysis methods EELS, AES spectroscopy, we are able to determinate the structural and composition of InP and $InPO_4$/InP submitted to physical treatment such as electronic irradiation. Because of the great sensitivity of the analysis method AES, we give in this paper the information of InP induced by irradiation electronic with 4 keV energy. Same results are given about irradiation of a thin layer of oxide $InPO_4$ developed on InP, the oxide can not protect the InP surface because there is the participation of InP by indium oxided resulted from the breaking bonds between indium and phosphorus. Besides these experimental results we give several data carried out by using Carlo simulation. However we give the data concerning the loss energy of inelastic backscattered electrons have been showing from the amplified zone for the low loss energy backscattered to both InP and $InPO_4$/InP targets. The backscattering coefficients for InP determinate for different layers of indium. We reported the bulk and magnitude of the depth of interaction of beam electron in $InPO_4$/InP target realised for different times of irradiation and angle of incidence.

References:

[1] Jardin C, Bouslama M 1988 *Microsc .Spectrosco. Electron.* **13** 395.

[2] Bouslama M 1988 doctorate these, University Claude Bernard Lyon1 France.

[3] Wager J. F, Ellsworth D. L, Goodnick S. M, Wilsen C. W 1981 *J.Vac. Sc. Technol.* **19** 513.

[4] Feng M. N, Holonyak B. C, Walter G and Chan R 2004 *Appl. Phys. Lett.* **84** 4792.

[5] Xu X. G, Hu J, Watkins S. P, Matine N, Dvorak M. W and Bolognesi C. R 1999 *Appl. Phys. Lett.* **74** 976.

[6] Bouslama M, Ghamnia M, Gruzza B, Miloua F and Jardin C 1994 *J. Elect. Spect and Related Phenomena* **68** 377.

[7] Lounis Z, Bouslama M, Berrouachedi N, Jardin C, Auvray L, Abdellaoui A, Ouerdane A, Ghaffour M 2008 *Vacuum* **82** 529.

[8] Bouslama M, Monteil Y, Younes G, Jardin C, Ghamnia M 1996 *Vacuum* **47** 27.

[9] Pellerin F 1981 *Es-science doctorate these, Orsay, Paris-sud.*

[10] Bouslama M, Lounis Z, Ghaffour M, Ghamnia M, Jardin C 2002 *Vacuum* **65** 185.

[11] Ouerdane A, Bouslama M, Ghaffour M, Abdellaoui A, Hamaida K, Lounis Z, Monteil Y, Berrouachedi N, Ouhaibi A 2008 *Appl. Surf. Science* **254** 7394.

[12] Drouin D, Alexander R. C, Gauvin B, Hovington P, Horny P, Demers H, *Program Casino Version 2.42 University of Sherbrooke Quebec Canada.*
[13] Myklebust R. L, Newbury D. E. 1991 The R Factor: The x-ray loss due to electron backscatter. In: Electron Probe Quantitation Heinrich KFJ, Newbury DE (eds). New York, Plenum Press, pp 177-190.

IOP Conf. Series: Materials Science and Engineering **28** (2012) 012023 doi:10.1088/1757-899X/28/1/012023

Synthesis of Plaster for moulers's elaboration of imprint denture

M. Hamiane[a1], **N. Rabahi**[a], **M. Saidi**[a], **M. Salhi**[b]
[a]University M'Hamed Bougara of Boumerdes. Algeria.
[b]Center for Studies and Materials Technology Service of Boumerdes. Algeria.

E-mail: hamiane@umbb.dz / mess_h@yahoo.fr

Abstract. Our goal was the synthesis of plaster from local raw materials to be used in denture mouler's. The plaster type α and β was synthesis by hydrothermal and dry method from gypsum (CaSO4. 2H2O) of the west Algerian. After crushing and mineralogical analysis, gypsum has undergone through cooking in an oven at a temperature (T = 200 ° C), for a time t = 4 hours. The synthesis and characterization of the product has involved several ways investigated as diffraction RX, X-ray fluorescence, time taken, electron microscope (MEB), hardness, SSB, compressive and bending strength and Ph. A digester horizontal type Toni - technical laboratory was used for the synthesis of plasterα .
The results are conformable with the standard and plaster synthesized can be a basic material in the manufacture of moulers imprint denture.

1. Introduction

The plaster is among the oldest building materials produced by humans, as it is known since the beginning of the 6th millennium before Jesus Christ.
It is still used today in its traditional form of wasted powder with water to make coatings. Nevertheless, new uses in the form of prefabricated grow ever be able meet the needs of the construction. The plaster is obtained by dehydration of natural gypsum rock or by using a by-product of certain industries like a hydrated calcium sulfate.

Research is being actively pursued in many areas, either to improve techniques and traditional employers, or to find other forms of applications. However, special manufacturing processes and implementation will improve the quality and expand the use of plaster, particularly in the medical and imprint denture. In this objective, we formulated our work (1), which is to synthesize plaster by two methods:
- Hydrothermal Method
- Dry Method
- Characterization of the finished product and compare it with other plasters production local and foreign, for denture imprint.

2. Experimentation

The raw material is gypsum from Oran (Fleurus). The methods used to obtain the plaster alpha and beta; are respectively the hydrothermal method using an autoclave type Toni-technical ", and the dry

[1] To whom any correspondence should be addressed.

IOP Conf. Series: Materials Science and Engineering **28** (2012) 012023 doi:10.1088/1757-899X/28/1/012023

method using an oven drying. The denture plaster using for imprint is designed in the CETIM laboratory with local raw materials and was compared with other dental plaster used in Algeria namely:Italian Denture Plaster, Algerian Denture Plaster "used in the market", Fleurus Plaster (Oran).

The analysis of these materials included the following characteristic: Chemical and mineralogical analysis, the grain size distribution, and Analysis of physical, and mechanical properties.

The analysis and results have been achieved and compared according to ISO standard.(2)

The experimental realisation has involved several means of investigation as: X-ray diffraction, X-ray fluorescence, scanning electron microscope (MEB), autoclave, pH meter; oven drying, compressive and bending strength equipment, permiabilimeter of blain, vicat device, a set of screens and other devices and systems for the degree of whiteness, the time taken, the fluidity, true density, etc.

3. Results

3.1. Synthesis of "β and α » plaster from gypsum "Fleurus"

The different stapes of our process of synthesis of plaster in the CETIM laboratory are describe in the figure 1.

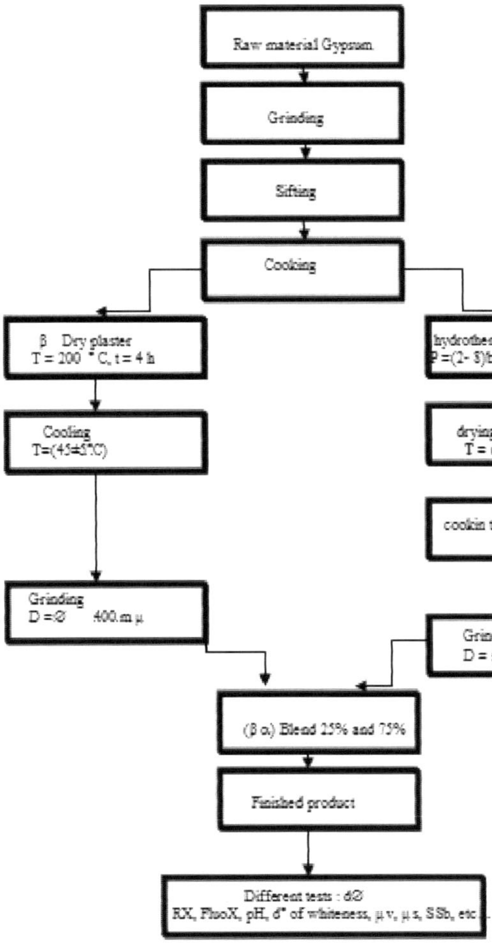

Figure 1. Methodology of synthesis plaster (β and α).

3.2. Mineralogical and chemical analysis

These analyses are made by X-ray diffractometry and X-ray fluorescence spectrometry of the center of study and technological building materials service (CETIM), of Boumerdes.

The results are given in Table 1 and 2, figures 2 to 6.The value of mineral composition are calculate from the chemical composition results.

Table 1. Mineralogical Analysis.

Raw Material	Mineral composition	Value (%)
Gypse	Gypse	89.0
	Calcite	7.5
	Quartz	3.5
Laboratory Plaster	Palastre	94.0
	Calcite	3.0
	Quartz	3.0
Italien Plaster	Plaster	93.0
	Calcite	6.0
	Quartz	1.0
Fleurus Plaster	Plaster	96.0
	Dolomie	1.5
	Quartz	2.5

Table 2. Chemical analysis.

Chemistry Composition	Gypsum oxyde (%)	Plaster labortory oxyde (%)	Plaster Italien oxyde (%)	Plaster Fleurus oxyde (%)
SiO_2	3.32	2.84	1.17	1.60
Al_2O_3	0.39	0.39	0.17	0.31
Fe_2O_3	0.15	0.15	0.11	0.08
CaO	32.13	37.12	38.03	37.41
MgO	0.24	0.23	0.13	0.54
SO_3	41.25	49.66	51.11	51.55
K_2O	0.05	0.07	0.04	0.06
Na_2O	0.08	0.06	0.05	0.05
PF	21.86	8.82	9.10	8.18
H_2O Cristallisation	18.59	5.81	6.67	5.98

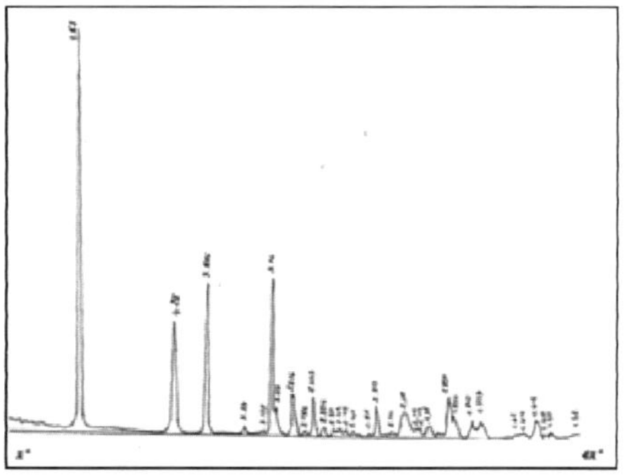

Figure 2. DRX Fleurus Gypsum

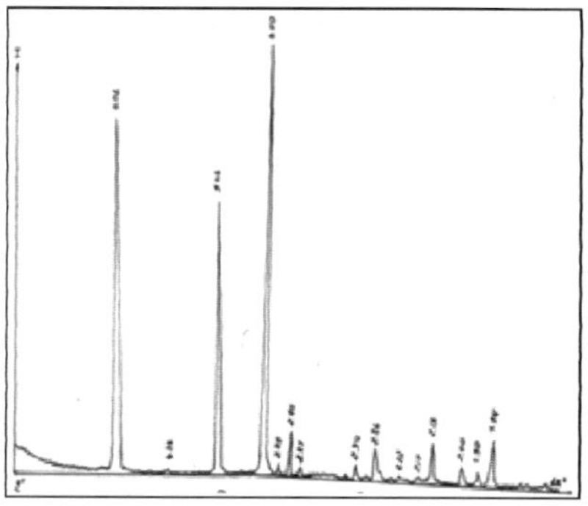

Figure 3. DRX Laboratory Plaster (PL)

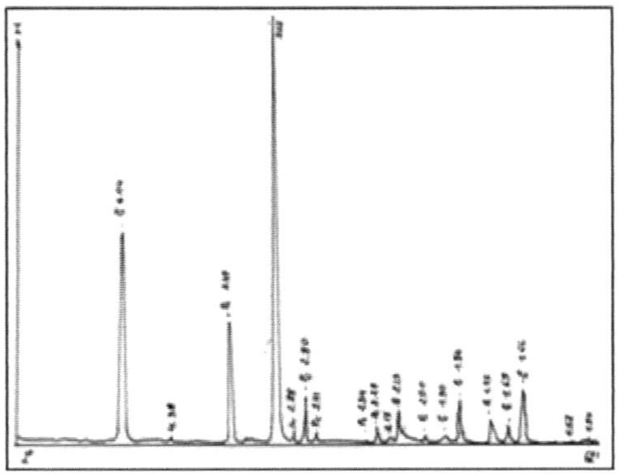

Figure 4. DRX Italien Plaster (PI)

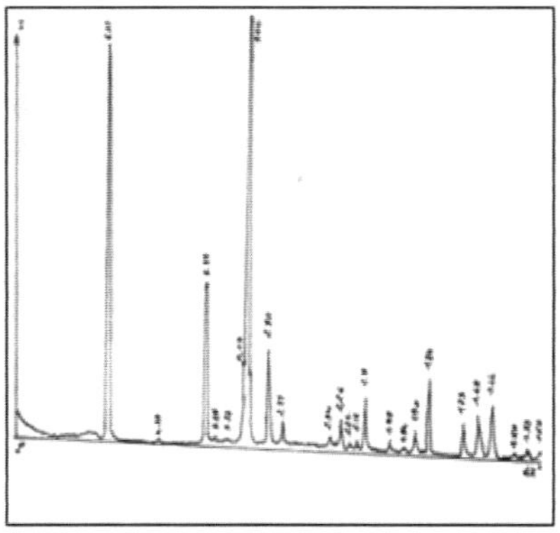

Figure 5. DRX Fleurus Plaster(PF)

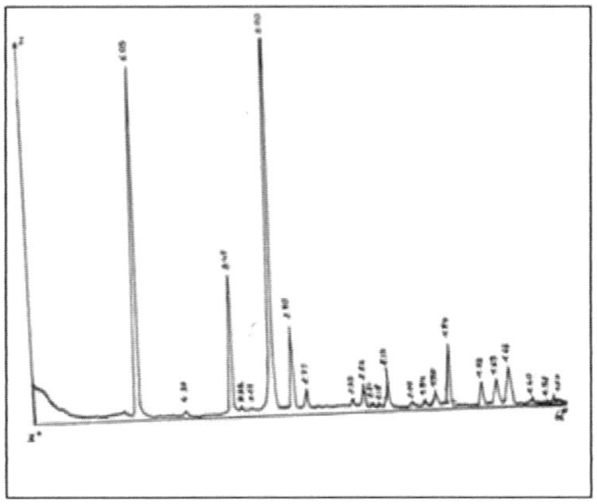

Figure 6. DRX Algerian Plaster (PA).

3.3. Results of grain size distribution

Physical testing and pH meter of different type of plaster are given in the table 3, 4 and 5.

Table 3. Grain Size distribution (Ø)

Plaster / Tamis	α (%)	β (%)	P.L (%)	P.F (%)	P.I (%)
800µm	0.2	0.2	0.4	0.2	0.0
400 µm	0.6	0.6	0.4	0.0	0.0
200 µm	2.4	2.2	3.6	0.4	0.0
100µm	6.6	5.6	8.6	2.0	1.0
75 µm	40.2	47.4	33.8	40.8	59
63 µm	17.2	16.4	17.8	30.6	12
< 63 µm	32.5	29.8	35.4	26.4	28

α = Plaster α PL= Plaster realized in laboratory
PI= Italien Plaster
β = Plaster β PF = Plaster of Fleurus

Table 4. Physical characteristics of different types of plaster.

Plaster / Properties	P.L	P.F	P.I	P.A	Gypse
True density (g/cm^3).µs	2.6415	2.7093	2.6895	2.7100	2.3000
SSB (cm^2/g)	5534	4093	3450	4089	---
Density tamped (g/l). µv	1146.03	1328.35	---	---	---
Density non tamped (g/l)	715.50	883.00	---	---	---
Rapport plaster/water	2.00	1.40	---	---	---
whitness (D.B)	76.6°	74.5°	rose	76.4	---
pH	8.62	8.11	9.39	8.9	7.82

PA = Algerian Plaster PL= Laboratory Plaster PI= Italien Plaster
SSB = specific surface PF = Plaster of Fleurus

Table 5. Mechanical Properties of laboratory and Fleurus plaster for different ages

Plaster	Ages	Strength to bending (Kgf/cm2)	Strength in compression (kgf/cm2)
	Iere Age	63.60	257.81
	IIème Age	66.94	270.62
(a) P. F	IIIème	68.25	276.56
	Iere Age	37.39	128.52
	IIème Age	47.50	128.68
(b) P. L	IIIème Age	49.02	129.37

4. Discussion

4.1-Mineralogical and chemical analysis

The chemical composition remains comparable to the different types of plaster, with the exception of silica, which remains slightly marked by the case of plaster laboratory.

The mineralogical analysis of gypsum from Fleurus has enabled us to assess the mineralogical composition of gypsum as a raw material able of producing plaster.

The percentage of 89% of gypsum; calcite 7.5% and 3.5% quartz leaves assume that the plaster is not very high purity and may still influence the quality of plaster synthesized.

The mineralogical composition of different plasters remains generally comparable except the percentage 6% of calcite in the Italian plaster, which remains significant.

4.2-Synthesis of plaster and comparison with other plasters

Samples of various plasters analyses X-ray have helped distinguish the different existing minerals and particularly the composition of important semi-hydrate (Ca SO4 1/2H2O). The data show that distances inter-reticular plaster synthesized in the laboratory under a pressure of saturation (8 bars) are very close to that of Italian plaster and that of Fleurus. The results obtained by the X-ray diffraction, show that the pressure appropriate for the synthesis of denture plaster is equal to 8 bars ,175 ° C , and a cooking time of 5 hours on a drying duration of 64 hours. The mineralogical composition of different plaster is quit comparable.

There is clearly a size distribution that the plaster developed in the laboratory contains more grain diameter less than 63 µ and therefore a greater responsiveness (reactivity) in relation to other plaster.

These results are confirmed by the specific surface of such different plaster. The absence of a level of great size grain for the Italian plaster and a lower content of Fleurus plaster is prove that the balance in the distribution size of laboratory plasters is better than the others. The results of physical and mechanical properties of the laboratory plaster are conformable with the standard.

4.3. The scanning electron microscope Analyses.
The textural was observed by electron microscope .The natural gypsum compact form, does not feature a crystalline apparent. However plasters alpha and beta crystalline form presents a well defined. The crystals are shaped more or less needles lying with a slight but significant difference in the swelling of crystals and the diameter of pores between the two plasters (α, β).

4.4. Testing Laboratory Plaster
Our product (laboratory plaster (75% (α) & 25 %(β)), was tested in the cabinet of stomatology. The dentist proved that the performance characteristic of our product for using in the mouler denture imprint is confirmed.

5. Conclusion

The synthesis method by hydrothermal plaster alpha at different pressures saturates (2, 5, 6, 7, 8 bars) and with temperature (120 to 175 ° C) for a period of 5 hours is fully confirmed by X.R diffractometrey, X-ray fluorescence and MEB, which enabled us to conclude that the synthesis of plaster by hydrothermal is better for the following parameters (P = 8 bars, T = 175° C, t = 5 hours).

The plaster prepared in the laboratory for imprint denture has good resistance, good size distribution and a better time taken.

The chemical, physical and mechanical proprieties show broadly that are conformity with the ISO standard.

Finally, it should be pointed out the need to add certain chemicals plaster alpha synthesized, such as a hardener, a self-timer or a color for commercial applications.

References

[1] Rebahi N, Hamiane M 2001 Development and characterization of denture plaster MFE Boumerdes Algeria.
[2] Standard ISO 6873 (1998) - Standard NFB 12401 (1987).

AES and EELS tools associated to TRIM simulation methods to study nanostructures on III-V semiconductor surfaces

A. Ouerdane [a1], M. Bouslama[a], M. Ghaffour[a], A. Abdellaoui[a], A. Nouri[a], K. Hamaida[a], Y. Monteuil[b]
[a]Laboratoire Matériaux ENSET d'Oran Mnaouer Oran 31000.
[b]Laboratoire des Multi-Matériaux et Interfaces, Université Claude Bernard Lyon 1, 43 Bd du 11 Novembre 1918, Villeurbanne 69622, France

E-mail: ouerdanea@yahoo.fr

Abstract. At low energy (300 eV), the Ar+ ions bombardment lead to the formation of small nanodots on the InP and the InSb surface compounds. We used the Auger electron spectroscopy (AES) and electron energy loss spectroscopy (EELS) to detect the presence of these features. However, these techniques alone do not allow us to determine with accuracy their disturbed dimension related to the height and periodicity. For this reason, we combine these spectroscopy methods with the TRIM (transport and range of ions in matter), SRIM (Stopping and Range of Ion in Matter) and Sigmund simulation methods to show the mechanism of interaction between the argon ions and the III-V compounds cited above and determine the dimension of disturbed areas as a function of Ar^+ energy during 30 min.

1. Introduction

The advent of nanostructures using ion bombardment on III-V semiconductor surfaces has opened a very interesting field of research. By such techniques, it will be possible, under special circumstances, to replace the more expensive technology using lithography and growth process for the development of nanodots and nanowires [1-6]. These nanostructures increase the surface area and form excellent black-body absorbers which are of great importance for solar cell. Furthermore quantum dots (QDs) of III-V narrow gap semiconductor, are applied as a potential platform for optoelectronic devices operating in the mid-infrared wavelength region [7-9]. InP and InSb are the III-V purpose semiconductors with direct band gap. It is interesting in understanding the effect of the argon ion bombardment on the surface topography at normal incidence [10-12]. Our paper deals first with the careful cleaning of the two surfaces achieved at low energy (300 eV). Afterwards, we study the ionic irradiation of the clean and stoichiometric InP and InSb surfaces. The effect of Ar^+ on these compounds will be shown by using sensitive methods such as AES and EELS to characterize the surfaces. However, it is not easy for us to determine the size of the nanostructures due to the etching process of InP and InSb surfaces by Ar+ ions. We associate these analysis techniques with the simulation methods, SRIM (stopping and range of ions in matter) and Sigmund methods. These methods allow us to deduce the roughness parameter and the wavelength λ or the spacing between bombardment-induced ripples [13].

[1] To whom any correspondence should be addressed.

2. Experimental

The Auger electron spectra (AES) and electron energy loss (EELS) spectra are recorded by using a hemispherical spectrometer. For the best compromise between the transmission and the resolution of the apparatus, we use pass energy of 80 eV between the deflectors of the analyzer operating in direct mode N(E) [13-15] The samples were chemically cleaned ex situ by diluted solution of H_2SO_4, then rinsed in methanol and finally washed in de-ionized water. They were then characterized by AES (electron beam of 3 keV energy, with a current density of 10^{-3} A/cm^2). The incident electrons were focused on an area of 1mm of diameter. These experimental parameters were chosen because they are appropriate for reducing the effect of the electron beam on the surface [13].The vacuum in the spectrometer chamber was about 10^{-9} Torr. The first recorded AES spectra C-KLL and O-KLL reveal that the InSb and InP surfaccs involved a contamination layer composed of carbon and oxygen because of its exposure to air. Subsequently, the surfaces were subjected in situ to argon ions bombardment. The Ar^+ ions of 2 mA current were accelerated at energy of 300 eV and targeted towards the surface for a 10 min exposure period.

3. Results and discussion

3.1. AES spectra

The irradiated area of the III-V compounds InSb and InP by the Ar^+ ions was of about 1 cm^2.We deduce a current density of about 2 mA/ cm^2. After such an ionic bombardment, the Auger signals C-KLL and O-KLL vanished completely. We show in Fig. 1, the Auger transitions In-$M_{45}N_{45}N_{45}$ of indium in InP and InSb, Sb-$M_{45}N_{45}N_{45}$ and antimony of InSb. The structure of these AES spectra is similar to those of the stoichiometric compounds. In another hand the EELS spectra of the two compounds cited above confirm the result given by AES analysis.

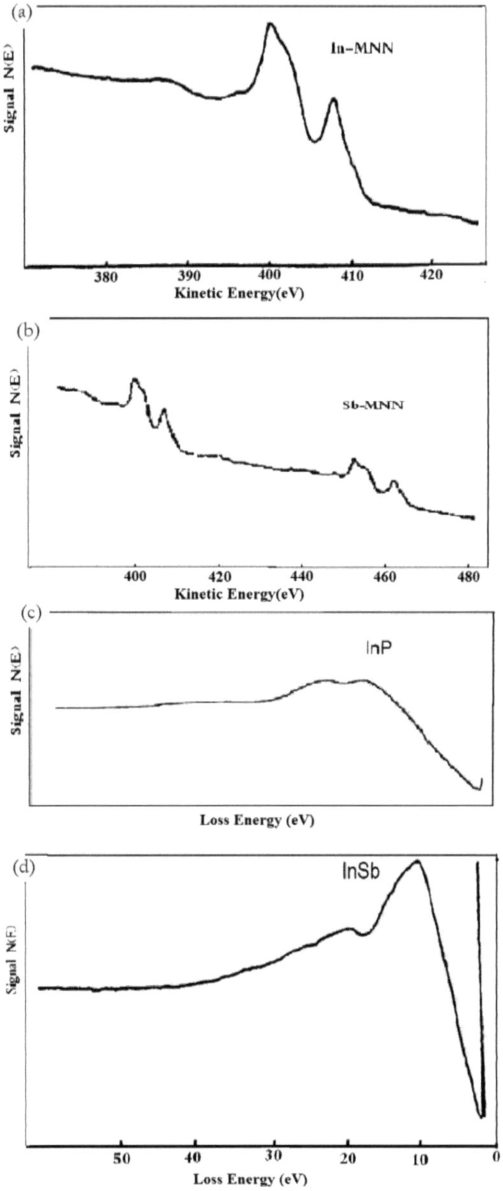

Figure 1. The Auger transitions In-$M_{45}N_{45}N_{45}$ and Sb-$M_{45}N_{45}N_{45}$ spectrum of indium and antimony and EELS spectrum of cleaned InP and InSb.

The AES Auger spectra for nanodots of InSb grown by MOCVD on the GaAs substrate were recorded by S. Shusterman et al. [1], as shown on fig .2.

The InSb dot signal is mentioned by solid line. Whereas the background signal is mentioned by dash line. The Sb peak is clearly seen in the dots while it is missing in the background. The In dot is shown by the high peak related to Auger transition of indium of InSb.

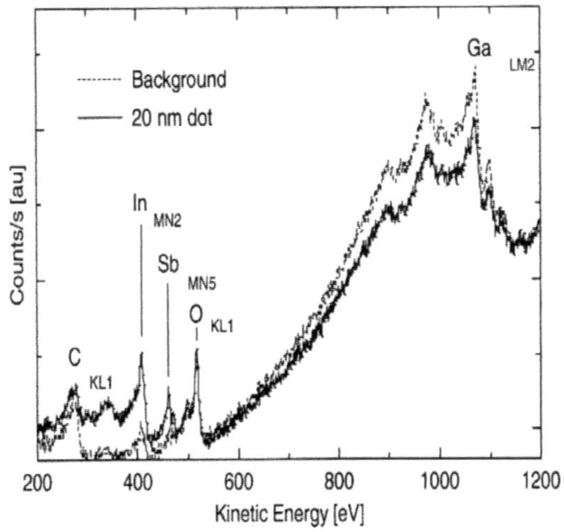

Figure 2. AES Auger spectra of InSb nanodots spectra.

3.2. EELS analysis of InSb

Our work is based on the study of the nano-spectra using electron loss spectroscopy (EELS). With increasing the energy and the time of bombardment (for 15 mn 30 mn and 45 mn) we recorded the spectra fig.3. On the recorded EELS spectra at low energy (250 eV) of clean InSb, we can see two distinct peaks located at 9 eV and 12.6 eV. The intensity of the peak located at 12.6 eV increases when the primary energy of the electron beam increases. However, the increase of the electron beam energy induces the decrease of the peak intensity located at 9 eV. On this basis in agreement with results reported by other authors, we interpret these two peaks as related to surface plasmons (9 eV) and bulk plasmons (12.6 eV) of InSb [14, 15]. We used the EELS at low and high energy in order to achieve the qualitative analysis of the outmost layers of the surface [17, 18]. In the second and the third step respectively, we irradiated the InSb surfaces by the argon ions at the same energy as in the previous case but the irradiation time was increased to 30 min and 45 mn. We also recorded the EELS spectra of InSb surfaces at the same primary energy (Ep = 250 eV). We see as shown in Fig. 3 change in the structure of EELS spectra. The effect of Ar^+ ions on InSb led chiefly to the appearance of other peaks related to the presence of In metal due to surface and bulk plasmons, which are located at 8.6 eV and 11.6 eV, respectively in agreement with results obtained by other authors [18-20]. The In metal is distributed on the InSb surface to form the physical system composed of In and InSb (In; InSb). Furthermore, the peak related to InSb surface plasmons at 9 eV diminishes in intensity. This led us to infer that the topography of the surface had changed.

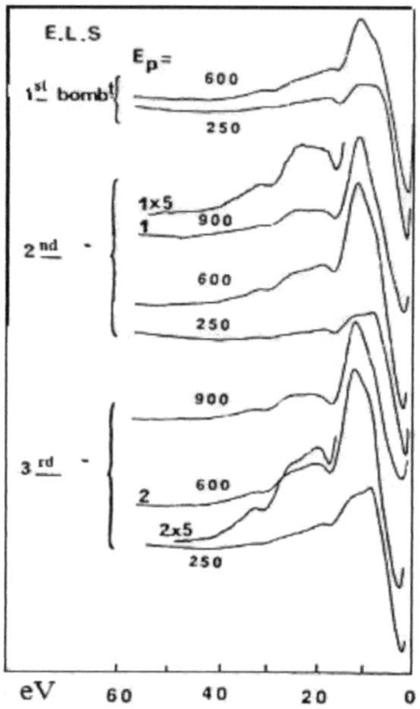

Figure 3. EELS spectra of InSb (100) submitted to ions beam bombardments. The analysis is led for different energies and for different times (15 mn; 30mn and 45 mn) respectively.

In order to investigate further this change of the surface, we varied the incidence angle Θ, defined between the direction of the incident electron beam and the surface. Nevertheless, we maintained the primary energy E_p equal to 250 eV. Fig.4 shows the change of the structure of the EELS spectra when Θ varies from the value 10–28°. We notice for the first spectrum (Θ = 10°) that the intensity of the peak located at 9 eV is higher than the other spectra. Besides, there are other interesting features related to bulk plasmons, interband transitions of indium and antimony located at 12.6 eV, 18 eV and 34 eV, respectively. Since the InSb compound has a small band gap, its valence electrons might be regarded as nearly free. The plasmons energy is large compared to its gap. The peaks due to indium interband transition at 18 eV and 9 eV increase when the incidence angle Θ decreases. The peak of antimony interband transition, in contrast, decreases when the incidence angle increases. The antimony contribution in the energy losses comes from the bulk, where Sb is bonded to indium in the InSb form Furthermore, for low values of Θ, we can observe in the recorded spectra that the peak at 9 eV is in good agreement with the energy loss corresponding to surface plasmons of the pure indium (In) taken for comparison.

It is known that the surface plasmons are more likely to occur at grazing incidence. The indium contribution is more important on the surface than in bulk because of the depletion of antimony from the surface. Furthermore, we did not observe any growth of metallic antimony on the surface, since the pure antimony spectrum (Sb) taken for comparison, does not correspond to any other spectra recorded for different values of angle Θ. We conclude that the Ar^+ ions bombardment leads to the development of an important superficial roughness on InSb. Indeed this result confirms the existence of InSb nanostructures formed on the surface. The roughness importance was displayed by the variation of the primary energy of the electron beam or the variation of the incidence angle. The In metal is distributed on the InSb surface to form the (In; InSb) system[21].The HREELS experiments performed on n-type and p-type InSb by Jones et al. [22] have shown that the monitoring of the post-bombardment annealing temperature displays the excitation of the conduction band plasmons. The peak position characterizing the plasmons energy is dependent on the annealing temperature. Furthermore, it has been shown, in same study, that increasing the incident electron energy E_p leads to the displacement of the plasmons energy (to free carriers) to the elastic peak. This plasmons energy practically disappears within the tail of the elastic peak when E_p reaches a value of about 120 eV.

Our own study, the use of EELS at the primary energy Ep = 250 eV, indicates that the loss features due to free-carriers are not resolved from the tail of the elastic peak. This result is in good agreement with the one reported by Jones et al. [22]. This EELS analysis of irradiated InSb leads us in one hand to conclude that the surface stoichiometry shifted to the In-rich side, probably due to the preferential sputtering of Sb, and, in another hand, to affirm the formation of small irregular surface structures occurring on the surface of irradiated sample. The In-rich regions or In islands formed by surface diffusion may act as the mask against the following surface erosion and are responsible for the generation of the ordered dot array [23].

It must be noted; however, that argon ion bombardment on InSb has been performed at room temperature. At the 350 K irradiated surface nanodots had an ordered distribution compared to that of the room temperature irradiated sample which was dispersed randomly, suggesting the thermally stimulated surface process is essential for dot generation.

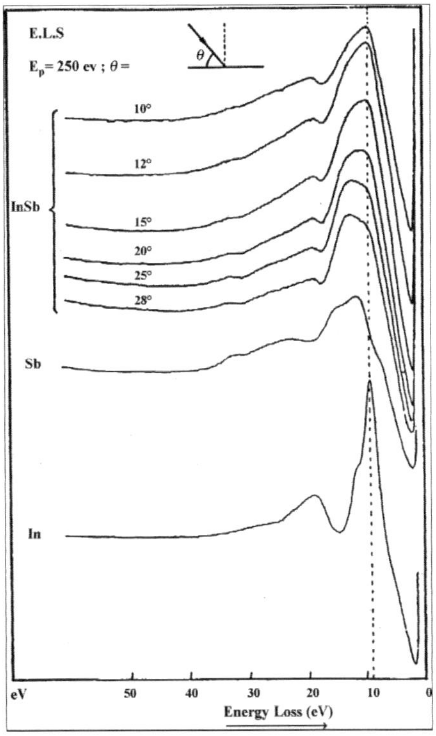

Figure 4. EELS spectra of InSb for different incidence angle after Ar+ ion erosion. during 30 mn.

Our results related to the formation of InSb dots separated by areas due to In metal, are in agreement with those reported by F.Krok et al and F.Frost et al. We give in figure 5 their results obtained by atomic force microscopy (AFM) and scanning electron micrography (SEM) [7, 8].

3.3. EELS analysis of InP

We know by the EELS analysis that argon ion bombardment of InP surface shift its stoichiometry to the In-rich side [27].

In Fig.6 we record the evolution of the EELS spectra of InP with the surface exposure time to the Ar^+ ions. The In-InP system is formed after about 45 mn of beam erosion (spectrum c) when compared to the In spectrum surface and bulk plasmons (e). The intensity of bulk plasmon increases while the surface plasmon remains stable which explain that InP nanostructures are present on the In-rich surface.

IOP Conf. Series: Materials Science and Engineering **28** (2012) 012024 doi:10.1088/1757-899X/28/1/012024

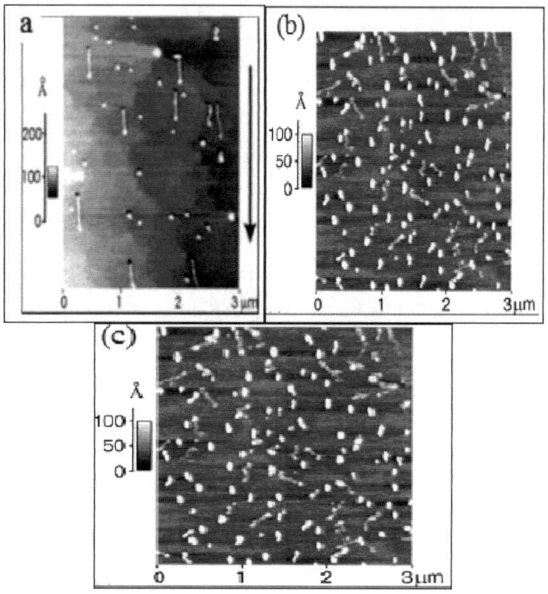

Figure 5. (a). Clean InSb before Ar+ erosion, (b). AFM image of nanostructured InSb(001) surface etched by Ar+ at 300eV energy The arrow indicates the normal direction of the ion beam on the surface, (c) InP(100) surface after the ion beam erosion

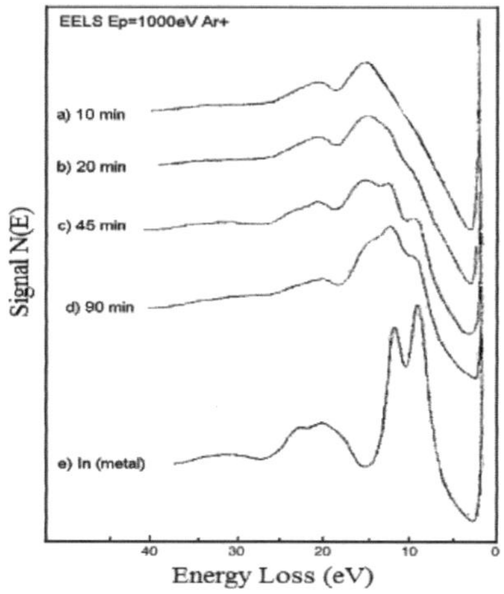

Figure 6. Evolution of InP surface roughness with the time exposure to Ar^+ ion bombardment.

The argon ions erosion of InP surface led to a development of superficial defects shown on fig 7. The EELS spectra in the same figure displays the decrease of both the elastic peak intensity and the collected signal with an increase of the relative intensity $I(\hbar\ \omega_s)/I(e.p)$, where $I(\hbar\ \omega_s)$, $I(\hbar,\omega_p)$ and I(e.p) are the intensities of the surface plasmon, bulk plasmon and elastic peak respectively. This variation of these intensities is due to the change of the InP surface. As shown before, we interpreted this variation as due to the presence of indium metal on the surface to form the diphase (In; InP) system.

Fig.7 shows also the Scanning Electron Micrography which represents the defects of the surface of InP after having been eroded by Ar^+ ions beam at 300eV energy giving rise to a nanostructures on InP surface. [15]

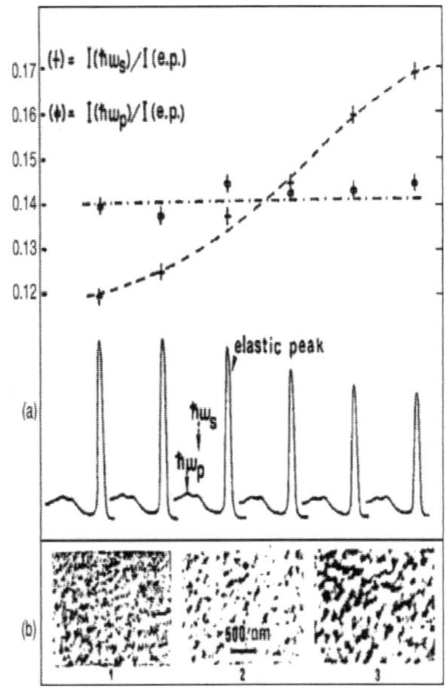

Figure 7. SEM image of nanostructures of InP (0 0 1) surface etched by Ar+ at 300eV energy followed by the EELS spectra showing the evolution of the surface roughness of InP.

4. Simulation methods

4.1. Simulation methods SRIM and TRIM

In this study, despite the interesting results concerning the effect of Ar^+ on InSb and InP compounds, allowing us to affirm the formation of nanostructures; we could not determine the size (height and periodicity) of these features. For this purpose, we used the simulation method TRIM established by Zeigler et al [24]

Ziegler has developed good simulation methods SRIM (Stopping and Range of Ion in Mater) and TRIM (Transport and Range of Ion in Mater). The Ar^+ ions bombardment is achieved according to the normal incidence with the surface of atoms of the target with the possibility to form vacancies on irradiated area.

Fig.8 gives the TRIM spectra :(the number of vacancy as function of the depth) or the ions Ar+ at the energy 300, 500,800 and 1000eV which enables us to deduce a range projectile of about R_p=22 **Å** for InP compound and R_p=19Å for InSb compound at energy set at 300 eV.

The Sigmund simulation method depends on the sputtering yield Y (Θ, E) (the number of atoms removed per incident ion), where Θ is the incident angle, formed by the direction of the incident ion beam and the surface of the target, and E the ion incident energy [25]. Also it depends on the nuclear stopping cross section computed by the SRIM simulation method which is based on the interaction process between the Ar^+ ions of different energies and the matter.

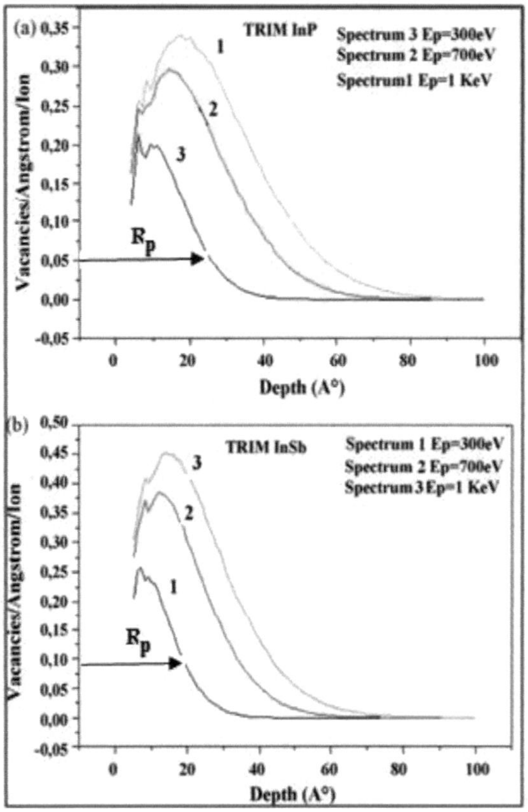

Figure 8. TRIM spectra of InP and InSb submitted to Ar^+ for different energies. The arrows indicate the values of Rp (Range projectile).

4.2. Sigmund simulation methods

The Sigmund theory is a successful theory describing the sputter yield Y(Θ,E) in the linear cascade regime[26].The energy dependence of the sputter yield is complex because it defines the three regimes of sputtering, i.e. the direct knock-on, linear cascade and spike sputter regimes [27,28].

The ion energy used in this study is usually the one associated with the linear cascade regime. The bombardment-induced topography can be quantitatively described in terms of a noisy Kuramoto–Sivashinsky partial-differential equation [29] which is extremely complicated. The analytical solution has been obtained for simplified form [30], giving the height h(x, y, t) of the roughness surface of the target as function of the bombarded ion energy E, the fluence ϕ and the sputtering yield
Y (0, E) [31]. The Y (0, E) at normal incidence (Θ=0°) was found to be proportional to the nuclear stopping range S_n given by SRIM program. Such a process is susceptible to lead to a new arrangement of atoms occurring in the material matrix. As mentioned previously, we interest to characterize the material after such a treatment by using the experimental means related to the AES and EELS to understand the interaction mechanism between Ar^+ InP and InSb targets.

Using SRIM 2008, we plot successively the spectra corresponding to the variation of the stopping range S_n and the sputter yield Y (0, E). However, our aim is to know the nanodots height h(x,t) formed by Ar^+ and the spatial periodicity λ for the two studied III-V compounds

According to our result in Fig.9., the roughness amplitude h(x, t) of the nanostructures due to Ar^+ ion bombardment is about 20Å and 30 Å for InSb and InP respectively at 300 eV incident ions energy[32].The distance λ between the bombardment-induced ripples depend on the ion flux J (number of ion by cm^2), the sputter yield of the bombarding ions Y(0,E) ,the density of the III-V compounds InP and InSb of energy deposition E, the surface diffusion parameter [33-37].; the range projectile R_p of energy deposition E computed in.fig.8 for the two compounds respectively. Using all these parameters, we compute the values of λ_{InSb}= 80 nm and λ_{InP} = 120 nm for Ar^+ energy set at 300 eV at normal incidence. These values are in concordance with experimental results [11, 30, 38 and 39].

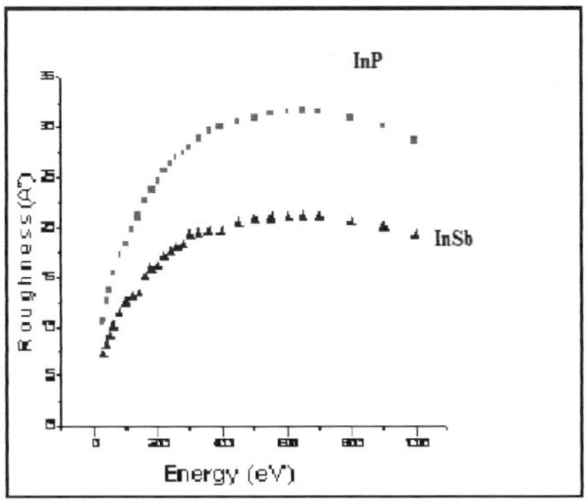

Figure 9. Roughness h(x,t) as function of Ar+ energy for the two III-V compounds InP and InSb

5. Conclusion

With respect to their behaviour during Ar^+ bombardment, two III-V compounds InSb and InSb were studied using AES and EELS experimental methods and TRIM, SRIM and Sigmund simulation methods.

The InP surface was very sensitive to argon bombardment than the InSb. The argon ions at low energy led to the removal of contamination layer composed of carbon and oxygen during well controlled time of the exposure of the surface to Ar^+. However when this exposure time increases, the In metal appeared on both surfaces of the two targets giving rise to the two diphases (In; InSb) and (In; InP) at room temperature. We deduce a formation of superficial nanostructures detected by AES and EELS experimental methods. The TRIM; SRIM and the Sigmund methods were useful to show the interaction process of Ar^+ with both InSb and InP. They also allowed us to determine the dimensions (height and periodicity) of the nanodots formed on both InSb and InP surfaces.

References

[1] Shustermana S, Paltiela S. Y, Shera A, Ezerskyb V, Rosenwaks Y 2006 *Journal of Crystal Growth* **291** 363.
[2] Wang Y, Xiu F, Wang Y, Zou J, Beyermann W. P, Zhou Y, Wang K. L 2011 *Nanoscale Res. Lett.* **6** 134.
[3] Grossklaus K. A, Millunchick J. M 2011 *Journal of Appl. Phys.* **109** 014319.
[4] Chin H. C 2009 *Optical Review* **16** 367.
[4] Thirumaleshwara N. B, Basanta R, Mohana K. R, Mahesh K, Krupanidhi S. B, Neeraj S. 2010 *Appl. Phys. Lett.* **97** 202107.
[6] Iwata N, Mori G, Arai N, Murakami Y, Takahashi A 2010 *IOP Science Nanotechnology* **21** 365301.
[7] Krok F, Kolodziej J. J, Such B, Piatkowski P, Szymonski M 2003 *Nucl Instrum Methods* **B 212** 264.
[8] Frost F, Ziberi B, Hoche T, Rauschenbach B 2004 *Nucl Instrum Methods* **B 216** 9-19.
[9] Malherbe J. B 1994 *Critical Rev. Solid State Mater. Sci.* **19** 129.
[10] Malherbe J. B Chakra Borty (Ed.) 2003 *Nova Science Publ., New York* 357.

[11] Malherbe J. B, Odendaal R. Q 1999 *Appl. Surf. Sci.* **144** 192.
[12] Eckstein W 2008 *Vacuum* **82** 930.
[13] Bradley R. M, Harper J. M. E 1988 *J. Vac. Sci. Technol.* **A6** 2390.
[14] Ouerdane A, Bouslama M, Ghaffour M, Abdellaoui A, Hamaida K, Lounis Z, Monteil Y, Berrouachedi N, Ouhaibi A 2008 *Appl. Surf. Sci.* **254** 7394.
[15] Berrouachedi N, Ouerdane A et al 2009 *Appl. Surf. Sci.* **256** 21.
[16] Nouri A, Lounis Z, Ouerdane A, Ghaffour M, Bouadi M, Dumont H, Auvray L and Bouslama M 2007 *Vacuum* **81** 979.
[[17] Abdellaoui A, Ghaffour M, Ouerdane A, Hamaida K, Monteil Y, Berrouachedi N, Lounis Z, Bouslama M 2008 *Appl. Surf. Sci.* **254** 4024.
[18] Bouslama M, Jardin C, Ghamnia M 1995 *Vacuum* **46** 143.
[19] Jardin C, Bouslama M 1988 *J. Microsc. Spectrosc. Electron.* **13** 395.
[20] Bouslama M, Ghamnia M, Gruzza B, Miloua F, Jardin C 1994 *J. Relat Phenom.* **68** 377.
[21] Ibach (Ed.) 1988 Electron Spectroscopy for Surface Analysis, Springer, Berlin, 1977
[22] Jones T. S, Ding M. Q, Richardson N. V, McConville C. F 1991 *Appl. Surf. Sci.***45** 85.Solids, Pergamon Press, New York, 1985.
[23] Yoshiko Y, Sumiaki H, Kinji G 2003 *Nucl Instrum Methods* **B 206** 648.
[24] Zeigler J. F, Beirsack J. P, Littman U 1985 the Stopping and Range of Ions in Solids, Pergamon Press, New York.
[25] Hundur Y, Güvenç Ziya B, Hippler R 2010 *Communication in Nonlinear Science and Numerical Simulation* **15** 1103.
[26] Haoyu H. C 2005 Thesis Harvard University.
[27] YiPing Ch 2002 *Journal of Nuclear Materials* **303** 99.
[28] Sanchez G, Castro M and Gago R 2007 *Cond-mat mtrl sci arXiv*: **2625v1** 0706.
[29] Tan S. K, Liu R, Sow C. H, Wee A.T.S., 2006 Nucl Instrum Methods **B248** 83
[30] Malherbe J. B. et al. 2005 *Nucl Instrum Methods* **B230** 533.
[31] Mammeri S, Ouichaoui S, Ammi H and Zemih R 2010 *Nucl Instrum Methods* **B 268** 140.
[32] Tolstogouzov A, Daolio S, Pagura C, Greenwood C. L, Karpuzov D. S 2003 *Nucl Instrum Methods* **B203** 198.
[33] Mutzke A, Eckstein W 2007 *Nucl Instrum Methods* **B266** 872.
[34] Malherbe J. B et al. 2006 *Solar Energy Material & solar Cells* **90** 1504.
[35] Sekhar B. R, Rundhe M, Besson D, Treilleux M and Hoareau H 2004 *Surface Science* **549** 165.
[36] Borg B. M, Messing M. E, Caroff P, Dick K. A, Deppert K, Wernersson L. E 2009 *Indium Phosphide and related Materials* **9** 249.
[37] Callegari V, Nellen P. M, Yang T, Hauert R, Muller U, Hernandez-Ramırez F, Sennhauser Urs 2007 *Applied Surface Science* **253** 8969.
[38] Malherbe J. B, Van Der Berg N. G, Claudel F, Osman S. O. S, Odendaal R. Q, Krok F, Szymonski M 2004 *Nucl Instrum Methods* **B 216** 9.
[39] Krok F, Kolodziej J, Such B, Piatkowski P, Szymonski M 2003 *Appl. Surf. Sci.***210** 112.
[40] Zhao Q, Zhai G. J, Kwock R. W. M 2006 *Appl. Surf. Sci.***253** 1356.
[41] Ben Khalifa S, Gruzza B, Robert-Goumet C, Biddeux L, Monier G, Saidi F, M'Ghaieth R, Hjiri M, Hamila R, Hassen F, Maaref H, Bremond G, Beji L 2007 *Surf. Scie* **601** 453.
[42] Hofer C, Abermann S, Teichert C, Bobek T, Kurz H, Lyutovich K, Kasper E 2004 *Nucl Instrum Methods* **B 216** 178.
[43] Abhilash V, Sumesh M, Nabhiraj A, Nathan S, Mohan S 2004 Computational Materials Science **30** 523.

IOP Conf. Series: Materials Science and Engineering **28** (2012) 012025 doi:10.1088/1757-899X/28/1/012025

A study on Fluoroanhydrite; a solid waste of the chemical industry: Characterization and valorization attempts

F. Rais[a] [1], A. Kamoun[a], A. Jelidi[b] and M. Chaabouni[a]
[a]Laboratoire de Chimie Industrielle, Ecole Nationale d'Ingénieurs de Sfax, BP W 3038 Sfax, Tunisia
[b]Département de Génie Civil, Institut Supérieur des Etudes Technologiques, Nabeul, Tunisia

E-mail: faicelrais@yahoo.fr

Abstract. In the present work the characterization of a fluoroanhydrite which is a by-product of a Tunisian plant for fluorhydric acid preparation and a formulation of a self-leveling composition for floor toppings were reported. The fluoroanhydrite was characterized by x-ray diffraction (XRD), Infra-red (IR) and differential scanning calorimetry (DSC) techniques. These analyses confirmed that the by-product is the anhydrite II. The formulation was mainly, based on a fluoroanhydrite. It also contains aggregates such as a neutralizing agent, a hydration accelerator and a plasticizer. The optimization of four responses (plasticity, compressive strength, initial and final setting times) has been carried out using preliminary experiments. This retained composition aimed at obtaining a product with the required characteristics requested by the users of this kind of material. The optimized formula has the following composition: 39.85% binder (ICF-Fluoroanhydrite; particle size 1200 μm), 39.85% aggregates (silica sand), 4.40% neutralizing agent ($CaCO_3$), 0.52% hydration accelerator (K_2SO_4), 0.23% plasticizer (sodium lingosulfonate) and 15.15% water.

1. Introduction

The accumulation of industrial by-products is considered one of the current major environmental pollution problems. These by-products often contain mineral and/or organic compounds that can serve as raw material in chemical processes to obtain products with high added value, contributing thus in efforts for pollution reduction.

Anhydrite or calcium sulphate exists naturally or as a synthetic product under different phases depending on physico-chemical characteristics. It can be obtained by heating gypsum at different temperature

$$CaSO_4,2H_2O \xrightarrow{150C^\circ} CaSO_4,\frac{1}{2}H_2O \xrightarrow{200C^\circ} AnhydriteIII \xrightarrow{400C^\circ} AnhydriteII \xrightarrow{1200C^\circ} AnhydriteI$$

or from certain chemical reactions such as industrial solid waste product [1].

One can distinguish two essential differences between the two shapes of the anhydrite:

- The anhydrite II density (2.93) is larger than the anhydrite III one (2.58).
- The stitch of the anhydrite II is orthorhombic. With the anhydrite III one is hexagonal.

---

1 To whom any correspondence should be addressed.

The anhydrite II crystallizes in orthorhombic system. This structure puts in evidence the existence of appreciably oblong section channels of which measurements (2A°) are not as water can penetrate there explaining thus the very restricted possibilities of rehydratation of the natural or synthetic anhydrite II, whereas anhydrite III or soluble anhydrite is unsteady, it itself rehydrates very quickly in semi - hydrate in contact with water in phase steam.

Usually, the production of hydrofluoric acid by reacting sulfuric acid with calcium fluoride results in a by-product which consists mainly of anhydrous calcium sulfate (anhydrite II). In addition to minor quantities of some impurities such as iron oxide, aluminum oxide and their sulfates and sometimes barium sulfate, the by-product contains also between 1% and 3% by weight of un-reacted calcium fluoride. These reaction products depend on the origin and degree of purity of the calcium fluoride used in the process. Also, there are varying quantities of sulfuric acid and hydrofluoric acid, depending on the method of production and operation conditions. It would be desirable from both the economic, the ecological standpoint to convert this by-product into a commercially usable form. In particular, an effective use for the product eliminates many problems which result from its disposal. It is well known that such by-products are very widely employed in the field of construction. In particular, the calcium sulfate anhydrite II can be used as a binder in a self-leveling floor [2–5]: a fluid mortar which is generally placed in position by pumping to give without any mechanical treatment (vibration, leveling with a straightedge, etc.) a perfectly flat floor free from any irregularities. Many other hydraulic binders such as Portland cements [6], Portland blast furnace slag cements [4] and alumina cements [7] are also used in the floor leveling compounds. Their compositions also include many kinds of additives such as dispersing agents to facilitate obtaining the desired plasticity required for processing, accelerators or retarders to adjust the compounds to the desired working time, setting time, fillers and water [4,8,9].

In this paper, we used the FTIR, DSC, DRX, etc techniques to characterize a by-product from the "Industries Chimiques du Fluor" (ICF) (Gabès, Tunisia) with the aim to confirm that the by-product is anhydrite II. Moreover, we report results relating to its application to the formulation of a fluoroanhydrite-based self-leveling floor having the required properties requested by the users of this kind of materials.

2. Materials and methods

The material used in the experiments was Fluoroanhydrite: A by-product of "the Industries Chimiques du Fluor" (ICF). The analyzed compositions for the investigated material are given in table 1. This by-product has been ground into a very fine particles in the range of 200–1200 μm.

Table 1. Analyzed compositions for the Fluoroanhydrite.

Element	$CaSO_4$	CaF_2	Na_2SO_4	$MgSO_4$	$FeSO_4$	$PbSO_4$
Wt(%)	94	1.8 to 3	1.63	0.120	0.108	0.024

The structure of the studied material was carried out with a wide angle X-Ray diffractometer using Cu-Kα radiation (λ=0.15406 nm). Scans were collected over a 2θ range of 20–55° with a step of 0.02°. For the phase analysis, X'Pert High Score plus program was used. Thermal behaviour was performed under Air atmosphere using differential scanning calorimetry (DSC) at a temperature ranging from 25 to 600°C and at a constant rate of 10 °C/min. The FTIR analysis was performed by using a Perkin–Elmer Paragon 1000 FTIR spectrometer. For this purpose well KBr pellets containing a mixture of fluoroanhydrite are prepared, and the FTIR spectra were recorded.

Mortars are prepared by mixing the different components as follows: the fluoroanhydrite is mixed with a neutralizing agent and a part of water for 5 min. The plasticizer, the hydration accelerator and a remaining quantity of water are then added. The whole is mixed for 2 min. Finally, the sand is incorporated and the mixing is continued for two more minutes. The properties such as plasticity, setting time and compressive strength of the self-leveling mortar are measured.

3. Results and discussion

3.1. Fluoroanhydrite characterization.

Figure 1 represents the typical X-ray pattern of the fluoroanhydrite sample. The revealed peaks, as compared to the ASTM fiche, correspond to the $CaSO_4$ phase with orthorhombic structure. This structure reveals the existence of channels of substantially rectangular section with dimensions (2 Å) where the water cannot penetrate. This fact can explain the very limited possibilities for rehydration of anhydrite strictly natural or synthetic anhydrous non soluble [3].

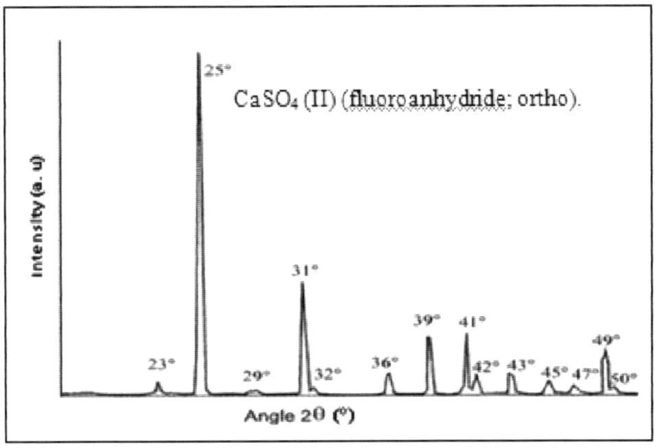

Figure 1. X-ray diffraction pattern of the investigated fluoroanhydrite sample.

In order to fully characterize the ICF-fluoroanhydrite, we compared the studied sample with anhydride III (Gypsum heated at 350 °C) and anhydrite II (Gypsum heated at 450 °C). FTIR spectra have been utilized as a powerful tool to provide supplementary information on the nature of studied material. Examination of the infrared spectrum of ICF-fluoroanhydrite can draw the following conclusions were drawn:

- The broad band located between 3600 and 3400 cm^{-1} is characteristic of water adsorbed on both the KBr and the analyzed product.
- The strong band located between 1160 and 1130 cm^{-1} is characteristic of the antisymmetric stretching vibration of the sulfate ion.
- The band of low intensity located at 1005 cm^{-1} can be attributed to the symmetric stretching vibration of the sulfate ion which is normally inactive in the infrared.

Moreover, the spectrum of anhydrite II (Fig. 2b) and ICF-fluoroanhydrite (fig. 2c) appear similar. This fact confirms that the ICF-fluoroanhydrite is anhydrite II.

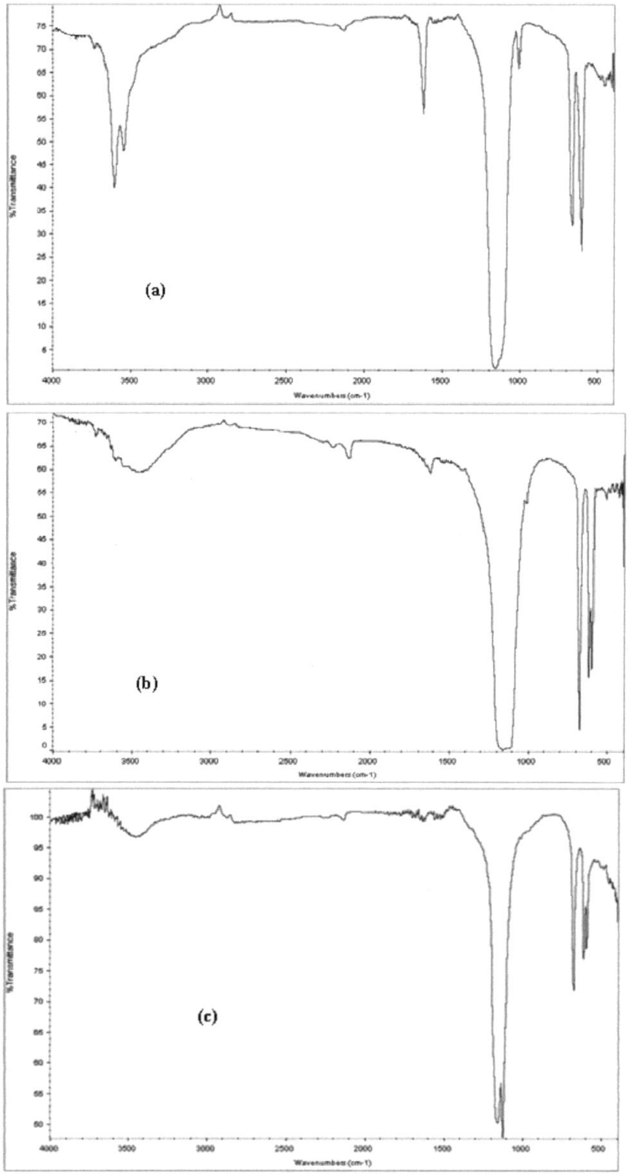

Figure 2. The IR spectra of the (a) Anhydrite III (b), Anhydrite II and (c) Fluoroanhydrite samples

Figure 3 shows the DSC traces of anhydrite II, anhydrite III and ICF-Fluoroanhydrite samples. These curves exhibit an endothermic peak occurring over quite a large temperature interval and may be originated from the dehydration of $CaSO_4$ samples (Figs. 3a-c). Moreover, one can notice an exothermic peak at 430°C (Fig. 3a), which can be attributed to the phase transition from the anhydrite III to the anhydrite II. This process seems to have a very slow kinetics which would explain the broad shape of this exothermic peak. Furthermore, one can suggest that the transformation of anhydrite III to anhydrite II begins only at 320 °C and does not end completely until 470 °C (Fig. 3a). On the other hand, the curve 4c appears similar to the curve 4b, which confirms also that the fluoroanhydrite is anhydrite II.

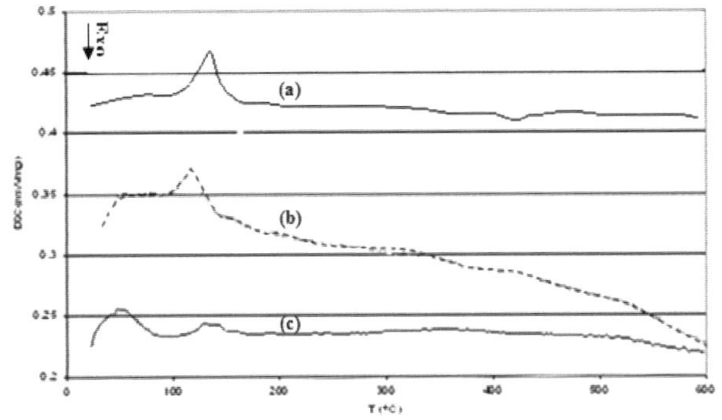

Figure 3. The DSC curves of the investigated samples: (a) anhydrite III, (b) anhydrite II and (c) fluoroanhydrite samples.

3.2. Fluoroanhydrite valorization

Preliminary experiments are carried out to look for a coarse composition of a fluoroanhydrite-based self-leveling floor, which is just about acceptable. To this end, we measure the characteristics of several mixtures obtained by changing the percentage of the components. In particular, we check the effect of the different components described in Table 2 on the mortar properties. These tests lead to conclude that the sodium lignosulfonate has a benefic effect on the plasticity of the self-leveling fluid mortar without a harmful effect on the setting times of the mortars. Moreover, the compressive strength is very sensitive to changes in the proportions of the neutralizing agent, the plasticizer and the water. The former exhibits a strong positive effect while the two other components exhibit a strong harmful effect when increasing their proportions. The setting times are very sensitive to changes in the proportion of the hydration accelerator or water. Indeed, a small increase in the amount of the accelerator can produce a sharp decrease in the setting times and, as expected, an increase of water can greatly delay the setting of mortars.

Table 2. Preliminary experiments of fluoroanhydrite-based self-leveling floor: composition and characteristics.

Fluoroanhydrite-based self-leveling floor composition					Fluoroanhydrite-based self-leveling floor characteristics			
N°Exp	Neutrali. agent	Hydratat. accelerator	plasticizer	water	Plasticity (cm)	Strength (MPa)	Initial Setting time (mn)	Final setting time (mn)
1	5.77	0.35	0.20	18.66	24.00	17.10	545.00	685.00
2	5.19	0.93	0.20	18.66	21.00	19.30	360.00	420.00
3	5.60	0.35	0.38	18.66	26.00	15.30	695.00	740.00
4	5.02	0.92	0.38	18.66	26.00	16.10	265.00	340.00
5	6.96	0.39	0.30	17.35	23.50	21.12	510.00	660.00
6	6.38	0.97	0.30	17.35	24.00	19.40	195.00	345.00
7	4.41	0.31	0.29	19.98	27.00	17.70	800.00	845.00
8	3.84	0.89	0.29	19.98	25.50	17.40	345.00	420.00
9	6.75	0.68	0.21	17.35	22.50	20.90	215.00	335.00
10	6.59	0.68	0.38	17.35	24.00	17.20	320.00	425.00
11	4.21	0.60	0.20	19.98	26.00	17.90	280.00	355.00
12	4.04	0.60	0.37	19.98	28.00	16.10	440.00	590.00
13	5.40	0.64	0.30	18.66	25.00	18.80	395.00	440.00
14	5.89	0.48	0.27	18.35	25.50	18.90	355.00	505.00
15	5.56	0.82	0.26	18.35	26.00	18.17	370.00	475.00
16	5.64	0.65	0.35	18.35	23.50	19.20	390.00	510.00
17	4.50	0.61	0.29	19.59	24.00	18.80	225.00	275.00

Results of this preliminary study allow us to retain the following composition which shows acceptable rheological properties and setting time of the fresh mortar and compressive strength of the hardened mortar. The optimal composition consists of the following: 39.85% binder (ICF-Fluoroanhydrite; particle size 1200 μm), 39.85% aggregates (silica sand), 4.40% neutralizing agent ($CaCO_3$), 0.52% hydration accelerator (K_2SO_4), 0.23% plasticizer (sodium lingosulfonate) and 15.15% water. To this composition corresponds the measured characteristics of the mortar given in table 3 as well as a comparison with the characteristics of a self-leveling floor reported in the literature.

Table 3. Characteristics of the optimal composition of the self-leveling fluid mortar.

Response	Value	Literature [5]
Plasticity (cm)	23.57	24
Strength (MPa)	21.22	20
Initial setting time (mn)	280.71	280
Final setting time (mn)	444.84	510

4. Conclusion

The present study provides a solution to the valorization of fluoroanhydrite, a by-product generated by the fluorhydric acid industry in Tunisia. The ICF-fluoroanhydrite was analyzed by XRD, IR and DSC techniques. These analyses showed that the by-product is composed almost exclusively of anhydrite II. More specifically, a fluoroanhydrite-based self-leveling floor composition is formulated. Preliminary tests have found a composition of a rough screed based fluoroanhydrite, with acceptable mechanical and rheological properties.

References

[1] Hinsche F and Schaupp K 1977 *U.S. Patent* **4**,016,240.
[2] Andree G and Juergen B 2000 *U.S. Patent* **6**,106,608.
[3] Conroy M 1995 U.S. Patent **5** 378,279 .
[4] Frouin L and Sari M 1999 U.S. Patent **5** 910,215 .
[5] Daligand D 1996 *Cent. Sci. Tech. Batim. Mag.-France* **94** 15–19.
[6] Guerinet J. P, Mosquet M, Bosc F and Chappuis J 2001, *U.S. Patent*, **6**,309,457.
[7] Gaines R. W, Fiala D. F and Babcock H. N 1979 *U.S. Patent*, **4**,157,263.
[8] Feldman J. G 1984 *U.S. Patent*, **4**,444,925.
[9] Jorgenson C. R, DeBoer H. M and Jorgenson R. M 1978 *U.S. Patent*, **4**,075,374.

IOP Conf. Series: Materials Science and Engineering **28** (2012) 012026 doi:10.1088/1757-899X/28/1/012026

TEM and stress relaxation analysis of 99.1% aluminum processed by Equal Channel Angular Extrusion

A. Rebhi[a1], T. Makhlouf[a], Y. Champion[b], J. P. Couzinié[b], N. Njah[a]
[a]Laboratoire de Métallurgie Appliquée, Université de Sfax, Faculté des Sciences de Sfax, BP.1171 – 3000, Sfax, Tunisie
[b]Institut de Chimie et des Matériaux Paris-Est (ICMPE), CNRS 2-8 rue Henri Dunant 94320 Thiais, France

E-mail: rebhi_atef@yahoo.fr

Abstract. Relaxation experiments were performed on 99.1% aluminum in addition to transmission electron microscopy and mechanical tests to follow the evolution of the microstructure and the mechanical properties changes after equal channel angular extrusion (ECAE). After deformation, weak beam investigations were conducted on selected specimens. The microstructure of deformed material shows the presence of almost parallel and elongated grains. Moreover, newly created subgrains separated by dislocation walls are observed. The tensile and Vickers hardness experiments for deformed samples showed an improvement of mechanical properties compared with the unextruded one. The activation volume, obtained from relaxation tests decreases with increasing the number of passes. This is due to change in mechanism from dislocation-dislocation interaction in coarse grains material (N=0) to dislocation-grain boundary or grain deformation in ultra fine grains material (N≥1).

1. Introduction

Severe plastic deformation (SPD) is an effective tool for producing bulk ultrafine grained (submicron grain sized or nanostructured) metals [1]. One of the most common SPD methods is equal channel angular extrusion (ECAE)—a technique that results in a homogeneous submicron grain structure of the workpiece [2]. During the ECAE process, a metal billet is pressed through a die consisting of two channels with equal cross-section and intersecting at an angle Φ. The billet undergoes essentially severe plastic deformation mainly shearing but retains the same cross-sectional geometry, so that it is possible to repeat the extrusion several times. By multiple pressings, a very large effective shear deformation can be developed in bulk products [3, 4].

In the last two decades, ECAE experiments have mostly focused on the effect of the number of pressings, extrusion route and die shape on the obvious microstructure and the resulting mechanical or physical properties [5–12]. However, dislocation history of metals subjected to ECAE is not sufficiently studied. For example, it's possible to control the dislocation interactions by the mean of the evolution of their activation volume. This parameter can be described as the number of atoms that have to be coherently thermally activated to move the dislocations over localized "obstacles" in their

[1] To whom any correspondence should be addressed.

slip planes. The most used methods to determine the activation volume are the stress relaxation and the rate-jump tests [13, 14].

In the present work, TEM technique was used to study the microstructural evolution of the material after ECAE. Vickers hardness and tensile tests were used to control the effect of ECAE on 99.1% Aluminum. Stress relaxations tests were performed to evaluate the changes in activation volume.

2. Experimental details

A 99.1% aluminum was used in the present study. The chemical composition is given in table 1. Prior to ECAE deformation, the material was annealed for 24 h at 500 °C in order to obtain a fully recrystallized homogeneous microstructure. The resulting grain size is about 85 µm. Specimens with (10x10) mm^2 cross-section and 70 mm length were machined from the annealed ingots. ECAE deformation was carried out at room temperature up to three passes with a 90° die angle corresponding to a total equivalent strain of about 1 for one passage. Route B in which the sample is rotated by 90° around its longitudinal axis after each passage was chosen.

Table 1. A Chemical composition of the investigated material (ppm weight)

Cr	Cu	Fe	Mg	Mn	Si	Zn	Al
345	409	3600	1700	385	3500	1900	Balance

Hardness changes during ECAE were followed by Vickers hardness testing with a load of 1Kg applied for 5sec. Tensile tests were performed using a HOUNSFIELD 50KS tensile machine in air at room temperature at a strain rate of $7.\ 10^{-5}\ s^{-1}$.

Samples of (5 x 3.5 x 3.5) mm^3 were prepared from the ECA extruded billets and compression tested at room temperature at a nominal strain rate of about $10^{-4}\ s^{-1}$.

The stress relaxation consists in interrupting the sample loading during a mechanical test (in compression in the present study). As a result, the stress decreases for increasing time.

Among the three known dependencies of the shear stress τ against relaxation time, i.e. logarithmic, semi-logarithmic and hyperbolic laws, the logarithmic one is the most used since it is found to fit well experimental data mainly in the early stages of the relaxation process. In this case, the shear stress variation $\Delta\tau = \tau - \tau_0$, as a function of the time (with τ_0 the stress at the onset of the relaxation) is given by [15]:

$$\Delta\tau = \frac{kT}{\Omega V^*} \ln\left(1 + \frac{t}{c}\right) \tag{1}$$

c is a time factor and kT has its useful meaning, ΩV^* is the apparent activation volume, where Ω is the correction factor depending on the non thermal strain hardening rate K which is a function of the relaxation conditions, an immobilization factor β related to the mobile dislocation density and the combined modulus of the sample and the testing machine G_{sm}. It's given by [15]:

$$\Omega = (1 + \frac{K}{G_{sm}}) \times (1 + \beta) \tag{2}$$

The microstructures of materials before and after ECAE processing were examined by transmission electron microscopy (TEM) using JEOL 2000EX microscope. For this purpose, samples of 3mm diameter were obtained by punching and finally thinned by twin-jet electropolishing in a 33% nitric acid and 67% methanol solution at –30 °C under 12V. The plane X, perpendicular to the longitudinal axis of the pressed sample [11], was chosen for study.

3. Results and discussions

Fig. 1 shows TEM micrographs, obtained using weak beam technique, of the microstructure of the material before deformation (N= 0) and after ECAE (N=1, 3 and 4). The aluminum of the present investigation contains 0.9 % of impurities, mainly iron, silicon, magnesium and zinc. Iron and silicon are known to form, in aluminum, several ternary compounds unequally rich in silicon. From XRD data and X-ray emission spectrometry coupled with SEM, these precipitates were found to be Al_8Fe_2Si of the hexagonal structure; they are located essentially at the grain boundaries [16]. Fig. 1 (a) shows that the undeformed sample contains a low density of dislocations lines.

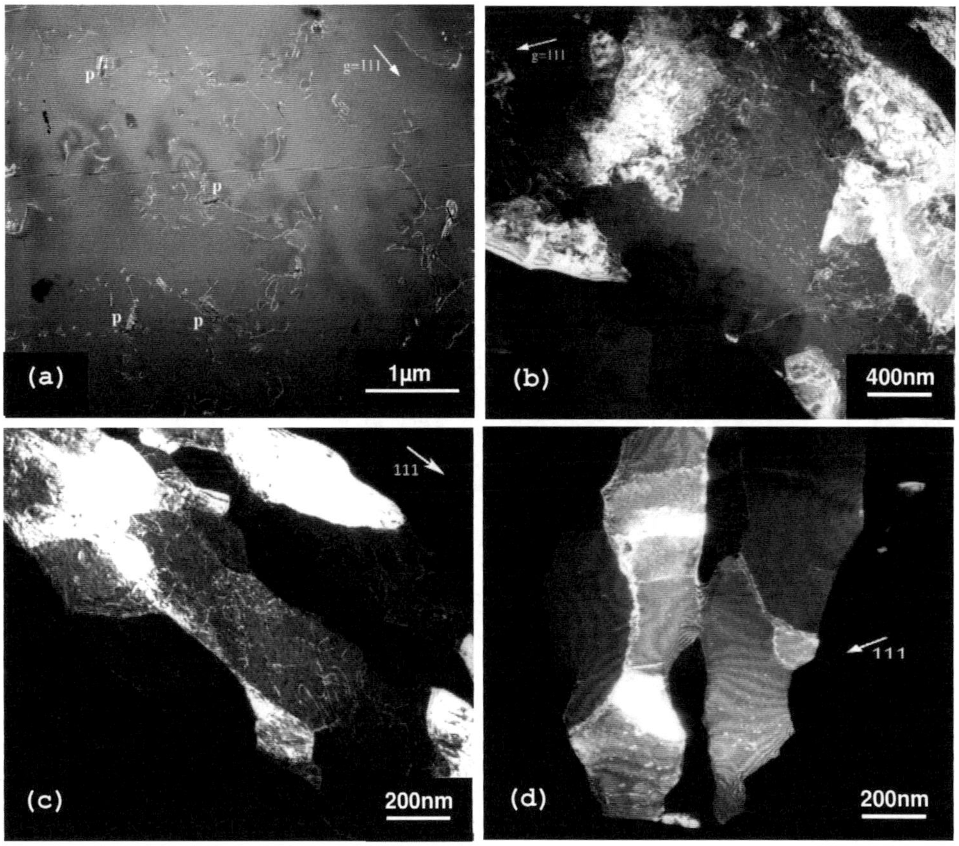

Figure 1. TEM images of the material: (a) before deformation N=0 and after deformation: (b) N=1, (c) N=3 and (d) N=4.

After ECAE, the microstructure of deformed material (N=1) shows the presence of new deformed grains with high dislocation density (Fig. 1 (b)). In addition, newly created subgrains separated by dislocation walls are observed. Moreover, many dislocations are observed at the boundaries as well as inside the "grains". Sub-grains are of about 400 nanometers in size with many dislocation walls. In addition, dislocation loops were observed within the deformed grains. After the third pass (Fig. 1 (c)), the contrast difference between the inner part and the outer part of the grains shows a bending of the latter indicating a non equilibrium state. The tendency of grains for bending can be related to size and orientation effects. High density of dislocations is retained after this stage of deformation.

Details of microstructural parameters of the material after severe plastic deformation have been presented in a previous work using XRD measurements [17]. It has been shown that the dislocation density reached 0.5 10^{16} m^{-2} after the first pass, the value of coherency length D achieved was 85 nm and the rms-strain ε was equal to 0.1 %.

Dislocation loops with a diameter of ~15nm are retained for N=4 (Fig. 1 (d)). The intersection of dislocations on different slip planes often creates vacancies, thus we have the most probable source. Indeed, the stress in the ECAE experiment applied to background of existing dislocations both within the grains and at grain boundaries will continually nucleate and cause dislocations to glide forming the

IOP Conf. Series: Materials Science and Engineering **28** (2012) 012026 doi:10.1088/1757-899X/28/1/012026

tangles. Intersections between dislocations in these high dislocation density (~10^{16} m^{-2}) tangles will subsequently provide localized concentrations of vacancies sufficient to form the loops.

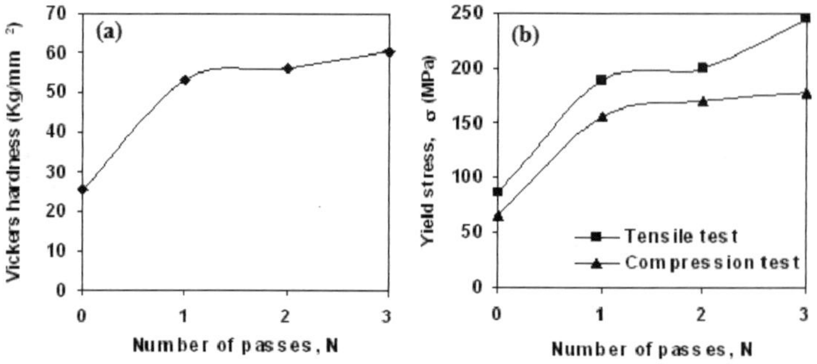

Figure 2. (a) Changes in Vickers hardness as a function of number of passes and (b)Yield stress variation in compression and tensile tests of the unextruded sample and the samples ECAE processed until N=3.

The Vickers hardness values after each ECAE pass are plotted as a function of the number of passes in Fig.2 (a). The figure shows that hardness increases remarkably after the first pass. Indeed, it becomes twice larger for N=1 than for the undeformed sample (N=0). This improvement can be allotted to the strain hardening generated by the creation of dislocations. Furthermore, the increase in hardness after three passes can be attributed to the considerable structure refinement that occurs after intensive plastic deformation. Yield stress variation in compression and in tension of the unextruded sample and the samples ECAE processed up to N=3 are shown in Fig. 2 (b). Significant strengthening was achieved after a single pressing. The superior strengthening in the extruded samples is believed to be primarily related to high dislocation accumulation rate generated by strain hardening during ECAE. It was also noticed that the yield strength saturates, this points toward that the dislocation density saturates with reaching the second pass and that it is merely a rearrangement of the already existing dislocations with the subsequent ECAE passes. The high value of hardness and yield stress of the material after ECAE can be explained by the strengthening model proposed by Lian et al. [18]. In fact, plastic flow in a nanocrystalline material is considered to be controlled by the stress required to attain dislocation loops (from Frank-Read sources) in a set of larger grains with the critical semicircle configuration. This model can still be applied to discuss the plastic flow of the ultrafine grains material.

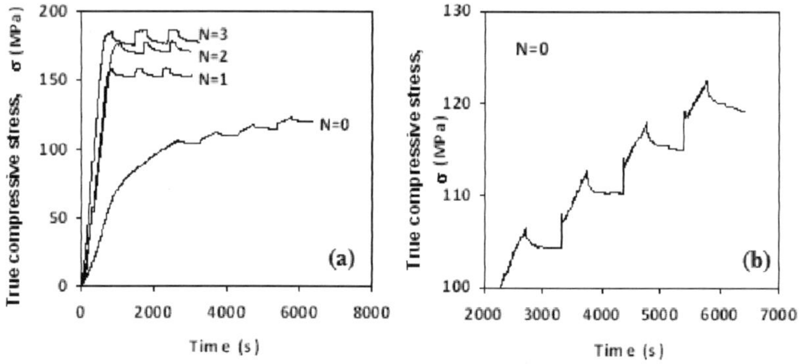

Figure 3. (a) Repeated relaxations for undeformed material (N=0) and at the constant stress regime (plateau) for extruded material (N=1, 2 and 3), (b) Repeated relaxations in the strain hardening domain for undeformed material (N=0).

Fig. 3 (a) shows the stress-time curves recorded in compression. The undeformed material (N=0) with coarse grains shows a steep quasi-linear strain hardening (Fig. 3 (b)), which is typical for stages I and II strengthening of f.c.c. metals. The strain hardening disappears for specimens subjected to ECAE. This is due to the fact that after having reached a steady state defect density by ECAE, the ability to work hardening ceases. This is due probably to a competition between the creation and the dynamic annihilation of dislocations.

The Shear stress variation as a function of time for the first relaxation is plotted in Fig. 4(a). All stress relaxation curves are fitted by the logarithmic relation. Fitting leads to a simultaneous determination of the kinetic parameters of equation 1, i.e. the activation volume V^* and the time factor *c*. The evolution of activation volume (V_a) with number of passes, reported in Fig. 4 (b), shows an abrupt decrease after the first pass from 500 b^3 to 140 b^3 (with b the Burgers vector). For the subsequent passes, the variation is less intense. Such a behavior was observed in the case of b.c.c. metals where the activation volume rapidly varies from 1000 b^3 to 20 b^3 at the microplastic-macroplastic transition [19].

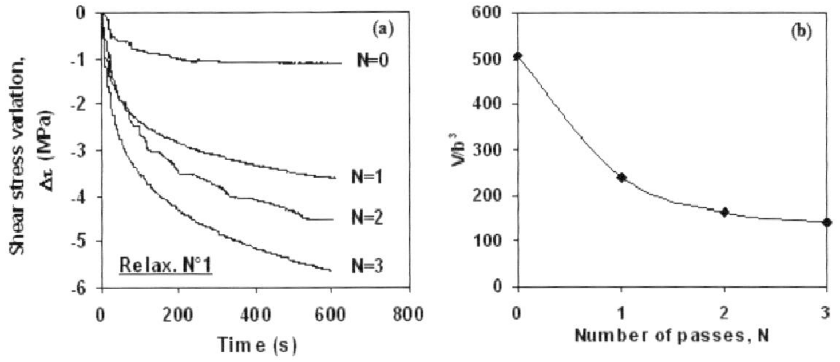

Figure 4. (a) Shear stress variation as a function of time for the first relaxation fitted by logarithmic relation Eq. 1), (b) Activation volume variation as a function of number of passes.

For microcrystalline metals, subjected to work-hardening, activation volume decreases as the stress increases. This is in connection to the interactions of the mobile dislocations with the forest dislocations which interspacing is getting narrow with stress. At a saturation volume of dislocation, the activation volume is of the order of 100 b^3 for the main f.c.c. metals (Al, Ni, Cu ...), which is practically the case of the present study.

When the grain size is sufficiently small and dislocation-dislocation interactions and then work-hardening is improbable, dislocations most likely interact with grain boundaries. In this domain of size, which ranges from few hundreds to few tens of nanometers, the deformation is believed to occur via dislocations activities and grain boundary sliding.

Usually a decreasing V_a with increasing starting stress (σ) at the onset of the relaxation process, is referred to as 'normal' behavior and is due to thermal activation of dislocations over short-range barriers. A decreasing V_a with decreasing σ is referred to as 'inverse' behavior. Conrad [20] found a similar 'inverse' behavior for ultrafine grains (UFG)/nano-grain-sized Cu and attributed it to grain boundary shearing promoted by dislocation pile-up within grains. In comparison, coarse grained (CG) Cu showed the 'normal' trend of decreasing V_a with increasing σ usually attributed to the dislocation-dislocation interaction process. However, in another study on UFG Cu obtained by ECAE [21], V_a, estimated from strain rate change tests, showed the 'normal' behavior of decrease in V_a with increase in stress for both CG and UFG Cu. In both studies [20, 21], the magnitude of V_a was significantly lower in UFG (100 b^3) as compared with CG (1000 b^3) Cu. In the present study, since σ increases when the number of passes (N) increases (Fig. 4(a)), the decrease of V_a (Fig. 4(b)) agrees with the results obtained in UFG alloys.

4. Conclusion

A 99.1% aluminum was severely deformed by ECAE for four passes using route B. The microstructure of ECAE deformed material show the presence of newly deformed grains. At this stage, ECAE leads to the formation of sub-grains of about 400 nanometers in size with many dislocation walls. In addition, dislocation loops were observed within the deformed grains. The tensile and Vickers hardness experiments of deformed samples showed an improvement in mechanical properties compared with the unextruded one. This is due to grain refinement and strain hardening. The activation volume obtained from relaxation experiments, decreases from 500 b^3 for the undeformed material to 140 b^3 after ECAE. This leads to change in deformation mechanism from dislocation-dislocation interaction in coarse grains material (N=0) to dislocation-grain boundary or grain deformation in ultra fine grains material (N≥1).

References

[1] Baik S. C, Estrin Y, Kim H. S, Jeong H. T and Hellmig R. J 2002 *Mater. Sci. Forum* **697** 408.
[2] Stolyarov V. V, Zhu Y. T, Alexandrov IV, Lowe T. C and Valiev R. Z 2003 *Mater. Sci. Eng. A* **343** 43.
[3] Segal V. M 1995 *Mater. Sci. Eng. A* **197** 157.
[4] Segal V. M 2004 *Mater. Sci. Eng. A* **386** 269.
[5] Iwahashi Y, Horita Z, Nemoto M and Langdon T. G 1997 *Acta Mater.* **45** 4733.
[6] Semiatin S. L, Berbon P. B and Langdon T. G 2001 *Scripta Mater.* **44** 135.
[7] Iwahashi Y, Horita Z, Nemoto M and Langdon T. G 1998 *Acta Mater.* **46** (9) 3317.
[8] Nakashima K, Horita Z and Nemoto M 2000 *Mater. Sci. Eng. A* **281** 82.
[9] Yamashita A and Yamaguchi D 2000 *Mater. Sci. Eng. A* **287** 100.
[10] Wang Z. C and Prangnell P. B 2002 *Mater. Sci. Eng. A* **328** 87
[11] Valiev R. Z and Langdon T. G 2006 *Progress in Materials Science* **51** 881
[12] Oh-ishi K, Hashi Y, Sadakata A, Kaneko K, Horita Z and Langdon T. G 2002 *Mater. Sc. Forum* **396** 333.
[13] Champion Y and Nowak S 2008 *Mater. Sci. Forum.* **399** 584.
[14] Spätig P, Bonneville J and Martin J. L 1993 *Mater. Sci. Eng. A* **167** 73.
[15] Baur and Lehr P 1975 *Mém. Sci. Rev. de Métallurgie* 551 **7–8** .
[16] Korchef A, Njah N, Champion Y, Guérin S, Leroux C, Masmoudi J and Kolsi A 2004 *Adv. Eng. Mater.* **6** 222.
[17] Rebhi A, Makhlouf T, Njah N, Champion Y and Couzinié J. P 2009 *Mater. Charact.* **60** 1389.
[18] Lian J, Baudelet B and Nazarov A. A 1993 *Mater. Sci. Eng. A* **172** 23.
[19] Meakin J. D 1967 *Can. J. Phys.* **45** 1121.
[20] Conrad H 2003 *Mater. Sc. Eng. A* **341** 216.
[21] Dalla Torre F, Pereloma E and Davies C 2004 *Scr. Mater.* **51** 367.

IOP Conf. Series: Materials Science and Engineering **28** (2012) 012027 doi:10.1088/1757-899X/28/1/012027

Granulometric distribution of natural and flux calcined chert from Ypresian phosphatic series of Gafsa-Metlaoui basin compared to diatomite filter aid

R. Saidi[a] [1], A. Tlili[b], A. Fourati[c], N. Ammar[c], A. Ounis[d], F. Jamoussi[a]
[a]Laboratoire de Valorisation des Matériaux Utiles CNRSM 02 Borj Cedria, BP 273, 8020 Soliman. Tunisie.
[b]Faculté des Sciences de Sfax, Département des Sciences de la Terre
[c] Groupe Chimique Tunisien de Sfax, Laboratoire de Recherche
[d]Faculté des Sciences de Tunis Département des Sciences de la Terre, El Manar, Tunis.

E-mail: Saidi_raja@yahoo.fr, alitlili@yahoo.fr, fjamoussi@yahoo.com

Abstract. The cherty rocks of the Chouabine Formation of the Gafsa-Metlaoui basin (south-western Tunisia), that is composed by biogenic silica, are treated using thermal treatment at 1000°C with flux calcination method in order to prepare a specific filter aids of melting sulfur filter used for the production of sulfuric acid. This work presents the effect of heating on the granulometry of chert. The mineralogical composition of natural starting chert is composed by opal CT (cristobalite/tridymite) and by the mineral mixture of quartz, smectite clay minerals, palygorskite-sepiolite fibrous clay minerals, calcite and hematite. After thermal treatment, at 1000°C, the crystallinity of chert increases significantly and the opal-CT, transforms mostly to opal-C. The comparison of infrared spectra of untreated and treated cherts shows systematic variation of the wavenumber and the intensity of the absorption bands, due to the apparition of 619, 795, 1094 and 1202 cm^{-1} absorption bands of cristobalite in the infrared spectrum of chert treated sample, which is similar to diatomite. Granulometric analysis show that the natural untreated chert sample displays unimodal distribution, whereas the treated chert sample display bimodal distribution, as same as diatomite. Thus, a new mode appears systematically, between 0.1 μm and 1 μm, for all thermal treated samples with alkaline flux as observed for diatomite.

1. Introduction

Biogenic silica from cherty rocks of the Gafsa-Metlaoui basin (south-western Tunisia) characterizes the Ypresian phosphatic series of Chouabine formation [1, 2, 3, 4, 5, 6, 7]. In general, the cherty levels are located at the base of the main phosphatic series between CVI and CVII phosphatic layers. Their thickness increases significantly from the East to the West of the Gafsa-Metlaoui basin [1, 2, 4]. The cherty layers are composed principally of opal CT (cristobalite/tridymite) [1, 4, 5, 8].

At this time, the cherty rocks constitute a by product of the phosphate industry. This work represent a contribution to the valorization of cherty rocks of Gafsa-Metlaoui basin in order to prepare a molten sulfur filter using thermal treatment at 1000°C, with adding some amount of alkaline flux. Thus, this

[1] To whom any correspondence should be addressed.

work is focused to study the comparison between the granulometry of treated and untreated chert with diatomite.

2. Experiments

2. 1. Materials

2. 1. 1. Cherty sample
The studied natural sample comes from the TK2 Mides trench of "Compagnie de phosphates de Gafsa" (CPG), taken from the intercalation III of chert which occurs between IV and V phosphatic layers.

2. 1. 2. Industrial diatomite filter
The industrial diatomite filter is imported by Tunisian Chemical Group from Ceca-French. DRX, IR and Granulometry Lazer analyses of this diatomaceous filtration aid are from the present work. Only silica content and major elements of the diatomite sample were taken from industrial sheet of CECA-France.

2. 2. Thermal treatment
The thermal treatment of cherty samples at 1000°C with 5% of alkaline flux (Na_2CO_3) was performed in a refractory brick container, using an electric furnace. Such that activated calcination and simple calcination denote the thermal treatment with and without alkaline flux respectively.

2. 3. Chemical analysis
Chemical analysis of the raw chert (Table 1) has been achieved in Laboratory of Research of the Tunisian Chemical Group (GCT). The silica content was determined with gravimetric dosage. Major elements were determined using atomic broadcast technique ICP by JOBIN YVON JY2000. Silica content and major elements of diatomaceous filtration aid came from the industrial sheet of CECA-France.

2.4. X-ray Diffraction
X-Ray Diffraction data for untreated chert, flux calcined chert and diatomite were measured at the Department of Earth Sciences (El Manar Faculty of Sciences, Tunis). Philips PW type diffractometer PANalytical X Pert Pro MPD (θ- 2θ) system equipped of a copper anticathode. The samples were checked by X-ray diffraction (Siemens D500, CuKα radiation, 45 kV, 40 mA diffractometer) to confirm the mineral identity and the crystallinity of the mineral phases. Conditions of the analyses are maintained constant for all samples in the interval 3-70° in 2θ, for a step of 0.017°/s (Fig. 2).

2.5. Infrared spectroscopy
Fourier Transformation Infrared (FTIR) absorption spectra of the powdered chert samples were obtained with Jasco FT-IR-420 spectrophotometer (at the Chemistry Department, Sciences Faculty, Sfax, Tunisia) using the KBr pellet method. The pellets were prepared by mixing approximately 1 mg of the sample with 200 mg of KBr powder. FTIR were recorded in the range 400 - 4000 cm^{-1}. Spectra were interpreted with comparison to reference spectra [9].

2.6. Laser Granulometry
Granulometric analysis for untreated chert, flux calcined chert and diatomite sample were determined with Granulometer Lazer Masteriser S at the Tunisian Chemical Group of Gabes, using wet method. The samples have been analyzed in water suspension. In order to obtain good particle dispersions, a suitable ultrasonic sound time has been applied (120s). In order to compare the granulometric parameters from different samples, medium diameter (MMD) has been calculated. D (v; 0.5)

corresponds to the particle calibre for that 50% of the sample having a lower size and 50% of sample having a higher size.

2.7. *Scanning Electron Microscope (SEM)*

A Philips ® 30 Analytical Scanning Electron Microscope (SEM) equipped with an EDX system for the characterization of the elemental composition of crystals was used to examine raw and treated samples of chert. SEM observations aimed to obtain additional information on size and shape of the granulometric distribution of raw and treated cherty samples.

3. Results and discussion

3.1. *Chemical analysis*

Chemical composition of untreated chert present a high SiO_2 content as same as diatomite Table 1. Chert and diatomite present also roughly the same percentage of the other elements, excepting CaO which is higher for the cherty sample.

Table 1. Chemical analysis of untreated chert and diatomite industrial filter aid

Oxydes	Raw chert (%)	Diatomite (%)
SiO_2	76,8	86
Al_2O_3	3.66	3.68
Fe_2O_3	1.49	2.6
CaO	6.61	1.11
MgO	1.42	0.78
K_2O	0.26	0.32
Na_2O	1.45	3.88

3.2. *X-ray diffraction*

The XRD patterns of natural cherty samples are given in Figure 1. The X-ray diffraction patterns of the powdered natural cherts show: silicate minerals as opal CT, quartz, smectite, palygorskite, sepiolite and carbonate as calcite, as founded in previous studies [1, 5, 8]. Chert samples contain Opal-CT as a major silica mineral, which is characterized by two broad reflections at the vicinity of 4.32-4.05 Å and 2.50 Å respectively [10, 11, 12, 13]. Opal-CT forms at lower temperature and corresponds to cristobalite/tridymite staking sequences which cristobalite alternates randomly with unidimensionally disordered tridymite layers [10]. After thermal treatment, at 1000°C, opal-CT, transforms mostly to cristobalite (Fig 2) and becomes similar to diatomite industrial filtration aids (Fig. 3). Diatomite industrial filter aid usually treated at a temperature near to 1000°C with flux calcination method as previously reported [14, 15]. Hadjadj-Aoul et al. [16], founded that major elements especially silica increase in Algerian diatomite after thermal treatment at 1000°C. These authors reported that silica amorphous phase disappears after activated thermal treatment and transforms to α-cristobalite [16]. The sharpness of the main reflections of treated sample at 1000°C is an indicator of the increasing crystallinity (Fig. 2, 3).

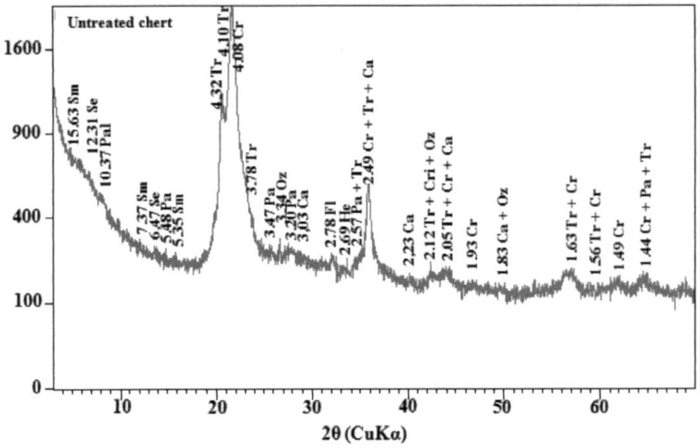

Figure 1. XRD patterns of untreated sample of chert (Ca: Calcite, Cr: Cristobalite, Fl: fluorapatite, He: hematite, Pa: Palygorskite, Se: Sepiolite, Sm: Smectit, Tr: Tridymite, Qz: Quartz)

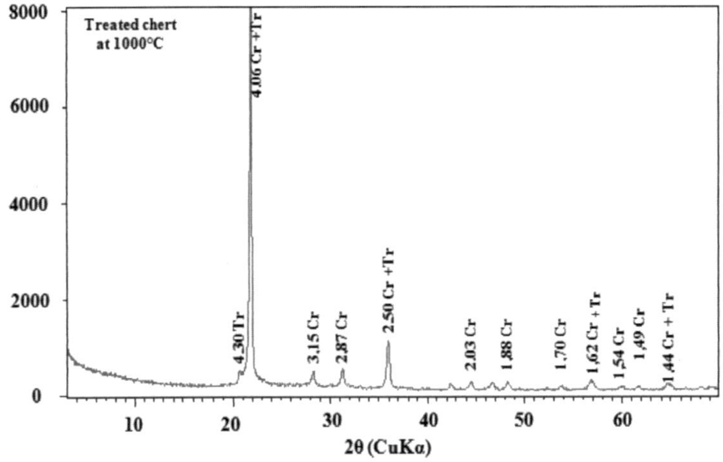

Figure 2. XRD patterns of treated sample of chert at 1000°C with alkaline flux

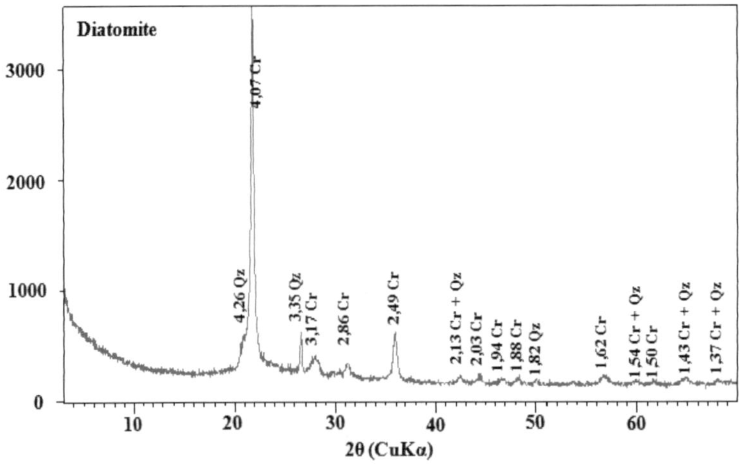

Figure 3. XRD patterns of diatomite.

3.3. Infrared spectra

Infrared spectra show absorption bands of silicates generally observed in the 400 - 1400 cm^{-1} region [9]. The absorption bands at 447; 791; 1095 cm^{-1} appearing in all samples can be assigned to the lattice vibrations of tetrahedral Si-O [9]. Figure 4 shows that infrared spectra of calcined chert and diatomite are closely similar and showing characteristic absorption bands of cristobalite at 619, 795, 1094 cm^{-1} and shoulder at 1202 cm^{-1}. The broad and strong bands of diatomite and calcined chert at 1094 cm^{-1} is asymmetrical by the occurrence of a shoulder at 970 cm^{-1}. The spectrum of raw chert is also similar to the spectra of calcined chert and diatomite excepting the band at 615 cm^{-1} which is absent from the infrared spectrum of raw cherty sample. The absorption bands of hygroscopic water occur at 1645 and 3535 cm^{-1} for the raw chert. These bands are weak or absent from calcined cherty sample and diatomite, which may indicated that chert and diatomite samples become less hygroscopic after thermal treatment. These vibration bands may also camouflaged by sample fluorescence at higher wavenumber region.

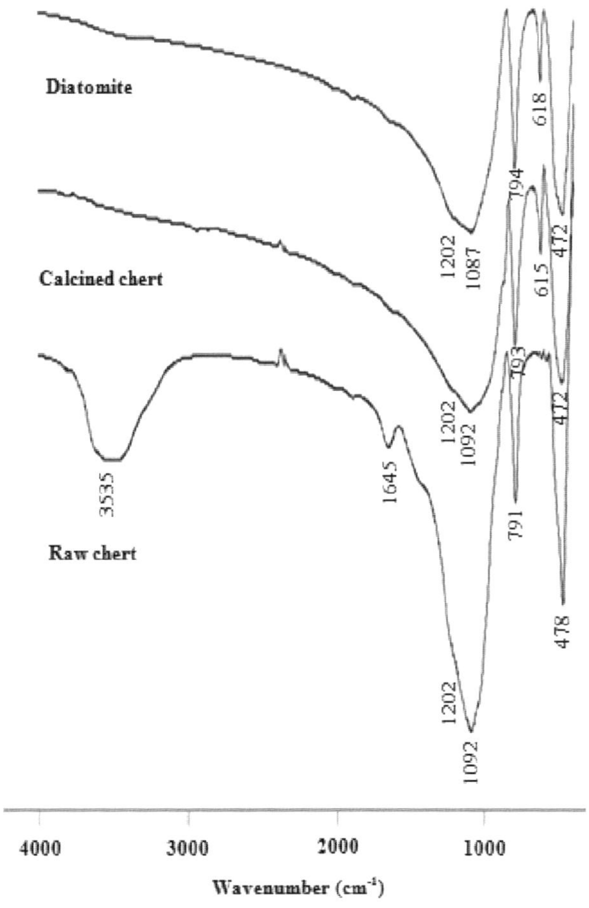

Figure 4. Infrared bands of the untreated and treated chert compared to diatomite.

3.4. Granulometry

The granulometry of chert starting powder was maintained lower than 100 µm and displays unimodal distribution between 1- 100 µm (Fig. 5). D (v; 0.5) have a value of 6.17 µm. After treatment, the value of D (v; 0.5) increases to 32.68µm and the calcined sample at 1000°C displays bimodal distribution (Fig. 6). Thus, a new mode appears systematically, between 0.1 µm and 1 µm, for all thermal treated samples with alkaline flux. This new mode occurs also for diatomite (Fig. 7). It appears that, after treatment, the cristobalite grains may have two different sizes (Fig. 6) as observed for industrial filtration aids like diatomite (Fig. 7).

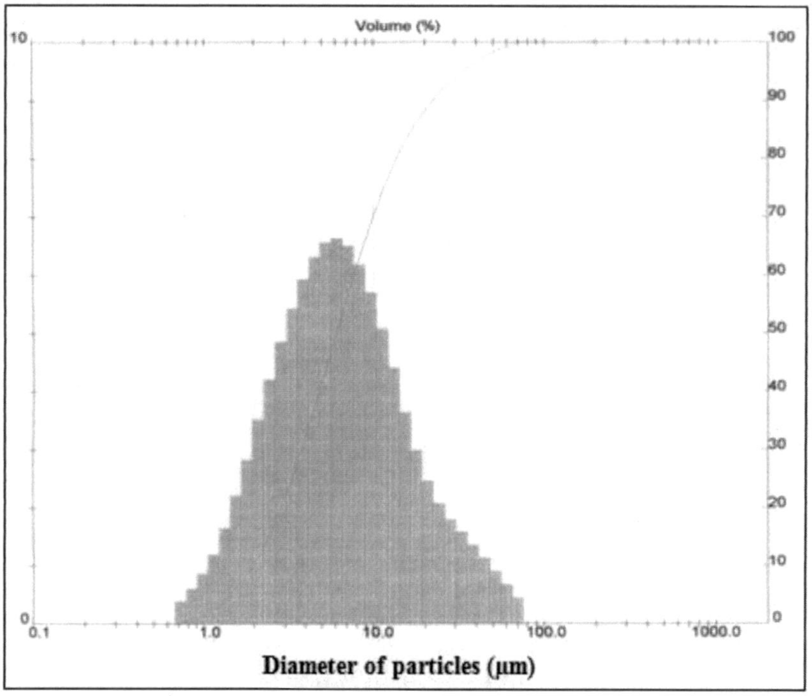

Figure 5. Granulometry distribution histogram of untreated chert.

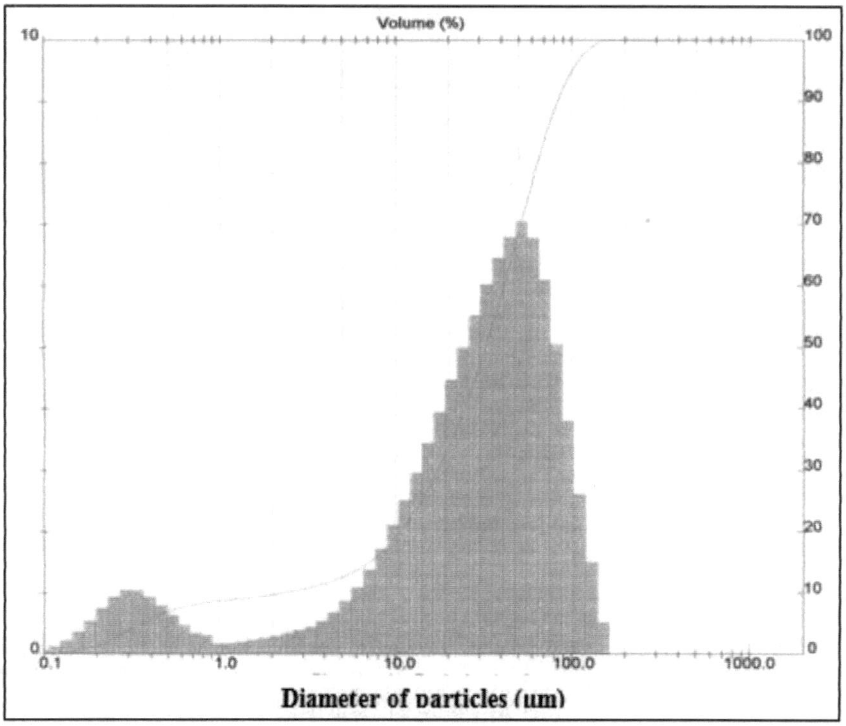

Figure 6. Granulometry distribution histogram of treated chert at 1000°C.

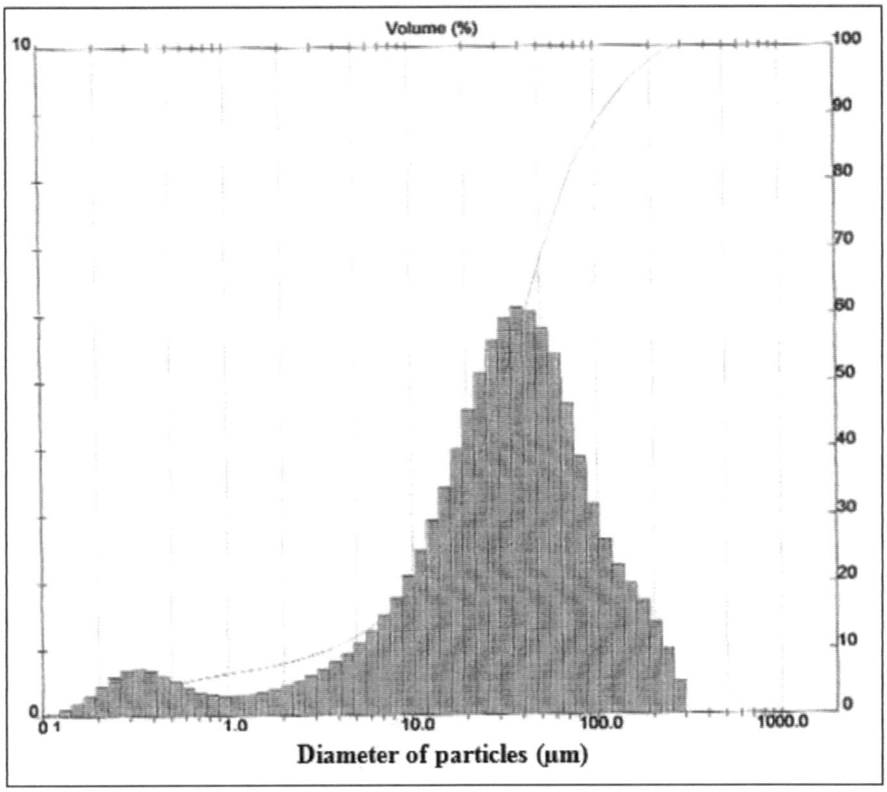

Figure 7. Granulometry distribution histogram of the diatomite.

4.4. SEM analysis

SEM micrograph of untreated sample shows that the particle of chert and their aggregates display rounded shape with different size (Fig. 8A). While, the treated sample displays angular to sub-angular habit (Fig. 8B). Thus flux calcined treatment at 1000°C induces recrystallisation and reorganisation of the chert particles, such that the pore space become better distributed and more homogenous in the sample volume.

Figure 8. A) SEM image of the untreated chert; B) SEM image of the treated chert

4. Conclusion

The high temperature treatment of the chert induces same mineralogical transformation and granulometry distribution. During heating at 1000°C, opal-CT changes in to opal-C. The infrared spectrum of treated chert shows the growing of cristobalite characteristic bands and confirms XRD results. The raw chert displays unimodal distribution between 1- 100 μm. After thermal heating, the granulometry distribution of cert changes and a new mode appears systematically between 0.1 μm and 1 μm, for all thermal treated samples with alkaline flux as same as diatomite industrial filter aids.

References

[1] Sassi S 1974 Thèse de doctorat Es-Sciences Naturelles, Université de Paris Sud, Centre d'Orsay. 292.
[2] Belayoun H Thèse de doctorat d'état. Faculté des Sciences de Tunis 205.
[3] Belayouni H and Trichet J 1980 *Office of Geological Mining Research* France (BRGM). **24** 37.
[4] Chaabani F 1995 Thèse de doctorat Faculté des Sciences de Tunis 428.
[5] Henchiri M 2007 *Journal of African Earth Sciences* **49** 187.
[6] Felhi M, Tlili A and Montacer M 2008 *Resource Geology*. **59** 428.
[7] Tlili A, Felhi M and Montacer M 2010 *Clays and Clay Minerals* **58** 486.
[8] Henchiri M, Slim-S'Himi N 2006 *Sedimentology*. **53** 1135
[9] Farmer V. C 1974 The Infrared Spectra of Minerals. Edited by Farmer. Mineralogical Society.
[10] Hesse R 1989 *Geoscience. Canada.* **15** 227.
[11] Önal M, Kahraman S, Sarıkaya Y 2007 *Applied Clay Science*. **35** 25.
[12] Önal M, Sarıkaya Y 2007 *Journal of Non-Crystalline Solids*. **353** 4195.
[13] Nagase T, Akizuki M 1997 *Canadian Mineralogist*. **35** 947.
[14] Martinovic S, Vlahovic M, Boljanac T, Pavlovic L 2006 *International Journal of Mineral Processing*. **80** 255.
[15] Ediz N, Bentli I, Tatar I 2010 *International Journal of Mineral Processing* **94** 129.
[16] Hadjadj-Aoul O, Belabbes R, Belkadi M and Guermouche M. H 2004 *Applied Surface Science* **240** 131.

IOP Conf. Series: Materials Science and Engineering **28** (2012) 012028 doi:10.1088/1757-899X/28/1/012028

Mechanical properties of tricalcium phosphate–alumina composites

S. Sakka[1], F. Ben Ayed and J. Bouaziz
Laboratory of Industrial Chemistry, Sfax University, National School of Engineering, Box 1173, 3038 Sfax, Tunisia

E-mail: sakka.siwar@yahoo.fr

Abstract. Tricalcium phosphate and alumina powder were mixed in order to elaborate biphasic ceramics composites. This study deals to produce bioceramics composites sintered at various temperatures for differents times. The characterization of samples, before and after the sintering process was investigated, using X-Ray diffraction, scanning electronic microscopy, ^{31}P and ^{27}Al nuclear magnetic resonance and differential thermal analysis. Mechanical properties of biphasic composites were studied using Brazilian test. The tricalcium phosphate - 75 wt% alumina composites mechanical resistance increased with sintered temperature. The mechanical resistance reach it's optimum value (8.6 MPa) at 1550°C for two hours.

1. Introduction

Biomaterials are the subject of biomedical research that has concerned early metals and plastics [1]. Those materials haven't solved all problems of their biocompatibilities and their mechanicals and chemicals properties. For this reason, research was oriented to use bioceramics due to their chemical inertness and wear resistance [2]. Calcium phosphates particularly the tricalcium phosphate (TCP) is resorbable material known by their chemical exchange with human tissue [3-12]. However, the use of bioceramics has faced some problems because of their fragility and their low mechanical resistance [2]. Hence, the idea of strengthening these bioceramics with the addition of inert oxide such as alumina [8, 9]. In the context, this study deals to produce biphasic β-TCP - 75 wt% Al_2O_3 composites sintering at various temperatures for different times. This work propose to study the mechanical properties and the densification of tricalcium phosphate – alumina composites and then to characterize the resulting composites by X-Ray diffraction, scanning electron microscopy and nuclear magnetic resonance (^{31}P and ^{27}Al).

2. Materials and methods

In this study, we used commercial alumina (Riedel-de Haën) and synthesized tricalcium phosphate. The β-TCP powder was synthesized by solid state reaction [12]. A calcium carbonate (Fluka, purity ≥99%) was added to calcium phosphate dibasic anhydrous (Fluka, purity ≥ 98,5%) at 1000°C for one hour according to the following reaction:

$$2\,CaHPO_4(s) + CaCO_3(s) \rightarrow Ca_3(PO_4)_2(s) + H_2O(g) + CO_2(g) \quad (1)$$

[1] To whom any correspondence should be addressed.

The β-TCP and α-alumina powder were mixed in an agate mortar. The powder mixtures were milled in ethanol using ultrasound machine drying 100°C for 24 hours. After drying, the powder mixtures were woulded in a cylinder having a 20 mm diameter and 4 mm thickness and pressed under 150 MPa.

These samples were sintered at various temperatures (1250°C, 1300°C, 1350°C, 1400°C, 1450°C, 1500°C, 1550°C and 1600°C) for different times (0 mn, 15 mn, 30 mn, 60 mn, 120 mn, 180 mn and 300 mn). Heat treatment was carried out in vertical resistance furnace (Pyrox 2408). The heating and cooling rates were, respectively, 10°C mn^{-1} and 20°C mn^{-1}. The relative densities of the sintered discs were calculated by the dimensions and weight. The relative error of densification value was about 1%.

The received powder was analyzed using X-ray diffraction (XRD). The X-ray have used the Seifert XRD 3000 TT diffractometer. The X-ray was produced by using CuKα radiation (λ = 1.54056 Å). The crystalline phases were identified by reference to the ICDD (International Center for Diffraction Data) files. Nuclear Magnetic Resonance (MAS-NMR) ^{31}P and ^{27}Al of spectra were run on a Brucker 300WB spectrometer. The ^{31}P and ^{27}Al observational frequency were 121.49 MHz and 78.2 MHz, respectively. ^{31}P shift are given in parts per million (ppm) referenced to 85 wt% H_3PO_4. The ^{27}Al NMR chemical shifts were attribuated to a static signal obtained from an aqueous aluminium chloride solution. The microstructure of the sintered compacts was investigated by scanning electron microscopy (SEM Phillips XL 30) on sample-fractured surfaces. Differential thermal analysis and thermomechanical analysis are carried out using about 20 mg of powder in Argon (DTA-TGA and TMA; Model Setaram). The heating rate is 10°C mn^{-1}.

Mechanical properties of compacts were measured by Brazilian test [13, 14]. The optimum rupture strength σ_r was offered by equation:

$$\sigma_r = \frac{2 \times P}{\pi \times D \times t} \quad (2)$$

Where P is the maximum applied load, D is the diameter, t is the thickness of the disc and σ_r is the tensile strength. The error of mechanical resistance value was about 1 MPa.

3. Results and discussion

3.1. Characterization of powder

The X-ray diffractogram of the β-TCP synthesized powder was shown in Fig.1a. Referring to ICDD file, all pics of the diffractogram confirms that contains only β-TCP phase (ICDD data file no. 70-2065). The XRD pattern obtained from Al_2O_3 powder illustrated the α phase peaks relative to ICDD data file no. 43-1484 (Fig. 1b).

The ^{31}P MAS-NMR spectra of β-TCP powder synthesized was illustrated in Fig. 2a. There are several peaks of tetrahedral P sites. In particular, an intense peak around 0.36 ppm, a shoulder to 1.46 ppm and another peak around 4.83 ppm more intense. The atoms of phosphorus tricalcium phosphate were located in three different crystallographic sites: $P(1)O_4$, $P(2)O_4$ and $P(3)O_4$ [15]. Other peaks, at -7.43 ppm, -9.09 ppm and -10.35 ppm, reveal low quantity of pyrophosphate [5, 9-11].

The spectra of ^{27}Al MAS NMR of alumina was given in Fig. 2b. It shows two peaks, the most intense located at 7.36 ppm corresponds to aluminum coordination VI (Al^{VI}) and the second located at 37.36 ppm corresponds to the coordination of aluminum V (Al^{V}). These results are similar to those previously obtained by O'Dell et al [16].

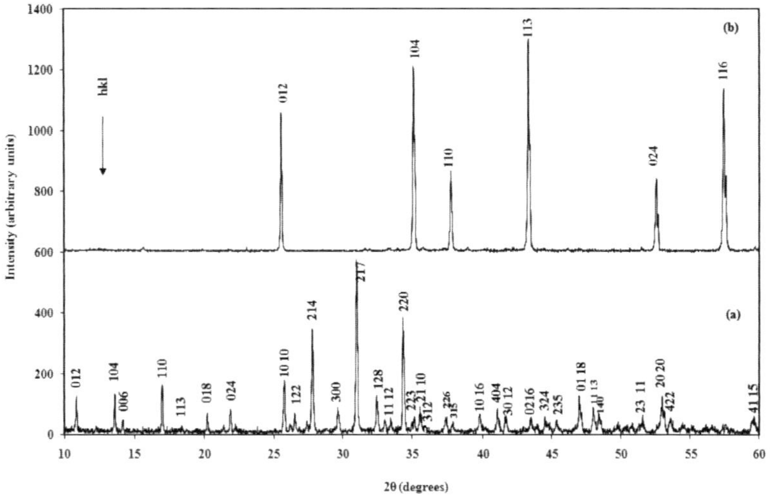

Figure 1. XRD patterns of (a) β-TCP powder, and (b) α-Al_2O_3 powder.

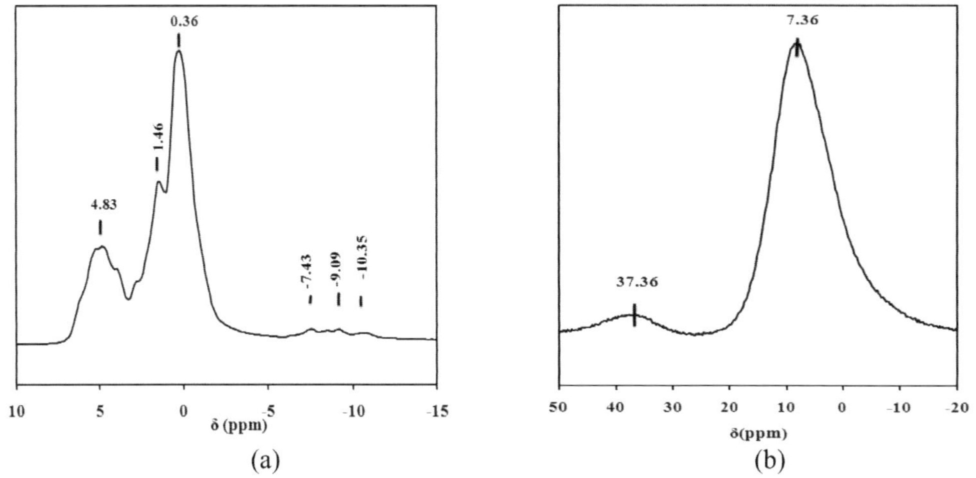

Figure 2. ^{31}P MAS-NMR spectra of (a) β-TCP, and ^{27}Al MAS-NMR spectra of (b) α-Al_2O_3.

Fig. 3 shows the differential thermal analysis (DTA) curves β-TCP, α-Al_2O_3 and β-TCP-75 wt% Al_2O_3 composites. The DTA thermogram of β-TCP (Fig. 3a) shows 2 endothermic peaks:

- A first peak at about 1258 °C characterize the first allotropic transformation of tricalcium phosphate (β to α).
- A second peak around 1469 °C corresponding to the second allotropic transformation of the tricalcium phosphate phase (α to α ').

These results have been observed in several studies [3, 5, 9-11].

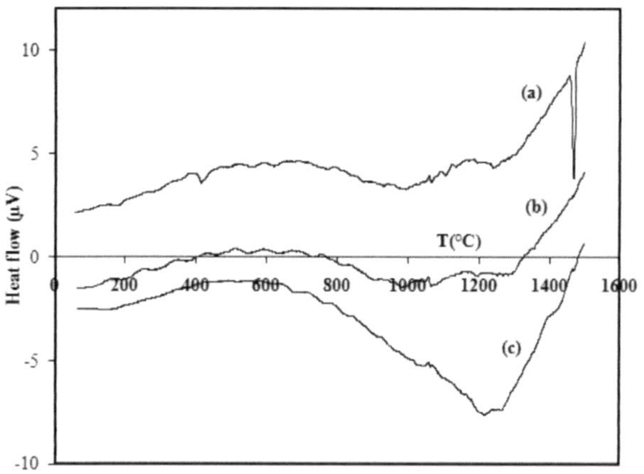

Figure 3. DTA curves of (a) β-TCP, (b) α-Al_2O_3, and (c) TCP - 75 wt% Al_2O_3 composites.

The DTA curve of alumina don't report any process (Fig. 3b). The DTA thermogram on the TCP-75 wt% Al_2O_3 composites shows the presence of endothermic peak located around 1257°C (Fig. 3c). This peak is due to the allotropic transformation of TCP (β→α).

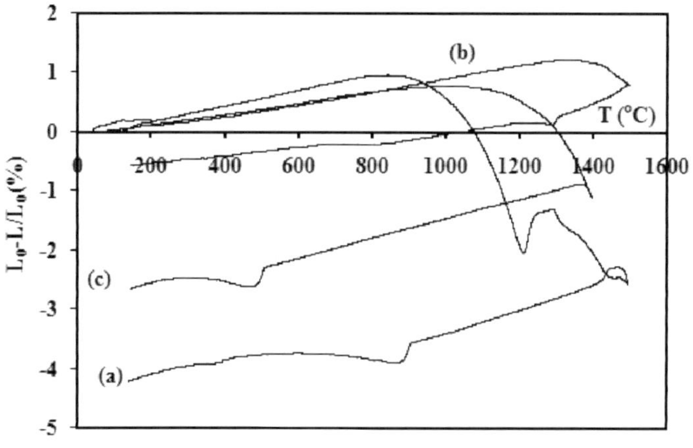

Figure 4. Linear shrinkage versus temperature of (a) β-TCP, (b) α-Al_2O_3, and (c) β-TCP -75wt% Al_2O_3 composites.

Fig. 4 shows the dilatometric measurements of a different samples used in this study. The starting sintering temperature of TCP corresponding at 1080°C whereas, the starting temperature of alumina is observed at 1657°C approximately (Fig. 4a and 4b). The addition of 25 wt% TCP in alumina matrix decreases the sintering temperature of the pure alumina about 377°C (Fig. 4c). Indeed, the starting sintering temperature of TCP – 75wt% Al_2O_3 composites is detected at 1280°C.

3.2. Sintering and mechanical properties of tricalcium phosphate

The heat treatment of tricalcium phosphate was carried out between 1200°C and 1450°C for one hour. Fig. 5 shows the evolution of the densification rate and the rupture strength of TCP as a function of sintering temperature. The densification increases with sintering temperature until 1350°C. At this temperature, the optimum densification reached 78%, whereas, the maximum of mechanical resistance reaches 3.2 MPa. Beyond this temperature, the mechanical resistance decreases rapidly to 1 MPa at 1450°C.

Time dependence of densification and rupture strength were shown in Fig. 6. The optimum densification was obtained for 1 hour (78%). The mechanical strength reaches a maximum value at 1350°C for 3 hours (5.6 MPa).

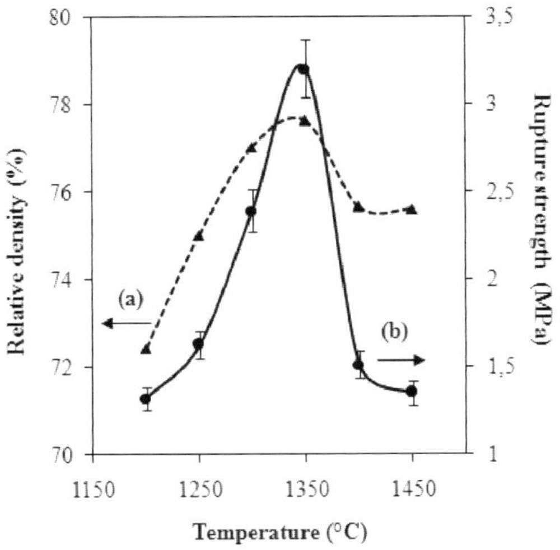

Figure 5. Relative density (a) and rupture strength (b) versus temperature of β-TCP sintered for 1h.

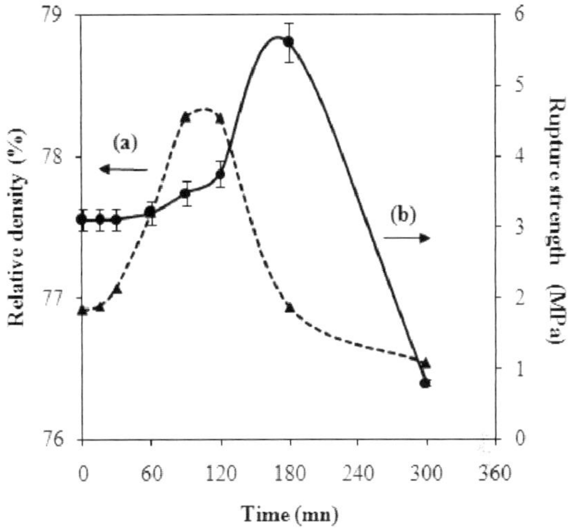

Figure 6. Relative density (a) and rupture strength (b) versus sintering time of β-TCP sintered at 1350°C.

3.3. Sintering and mechanical properties of β-TCP–75 wt% Al_2O_3 composites

In this paper, we study the effect of temperature and sintering duration on the performance of β-TCP–75 wt% Al_2O_3 composites. The rupture strength evolution and densification of the composites sintered for 1 hour depending on temperature as shown on Fig.7. The densification and tensile strength gradually increases with temperature reaching 61% as densification and 6.3 MPa for tensile strength at 1550°C. At 1600°C, the relative density and the mechanical strength decrease.

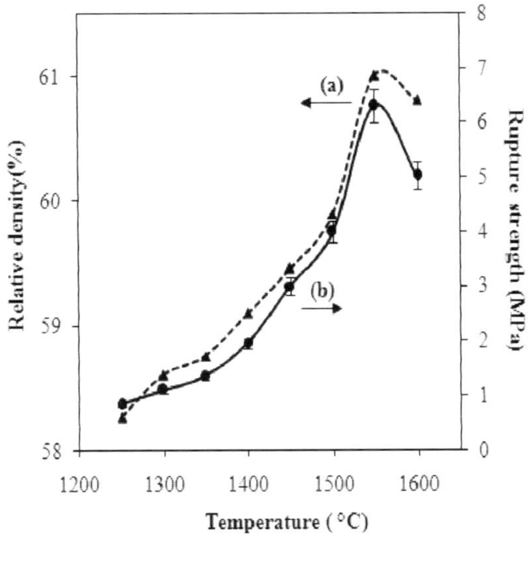

Figure 7. Relative density (a) and rupture strength (b) versus temperature of β-TCP – 75 wt% Al_2O_3 composites sintered for 1h.

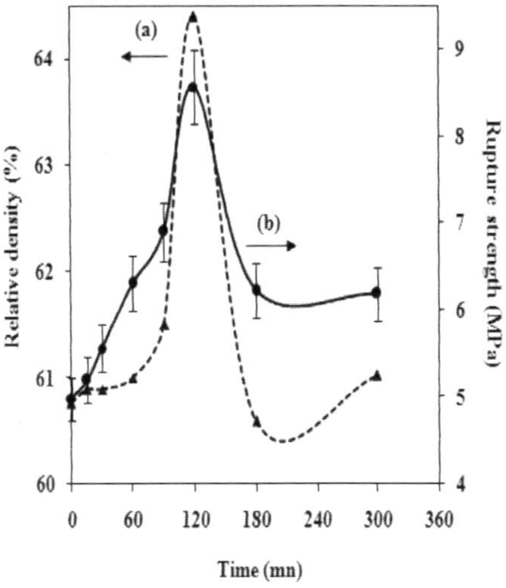

Figure 8. Relative density (a) and rupture strength (b) versus sintering time of β-TCP – 75 wt% Al_2O_3 composites sintered at 1550°C.

Fig. 8 reflects the evolution of the relative density and the rupture strength of the TCP-75% Al_2O_3 composites as a function of sintering time at 1550°C. The optimum value of mechanical resistance was obtained during 2 hours (8.6 MPa).

3.4. Characterization of samples after the sintering process

After sintering, the samples were characterized using: X-ray diffraction (XRD), ^{31}P and ^{27}Al nuclear magnetic resonance (MAS-NMR) and scanning electronic microscopy (SEM).

The XRD diffractogram of β-TCP sintered at various temperatures for one hour were showed in the Fig. 9. Above 1350°C, the diffractogram shows β-TCP and α-TCP (ICDD data file no. 09-0348) (Fig. 9b).

(b)

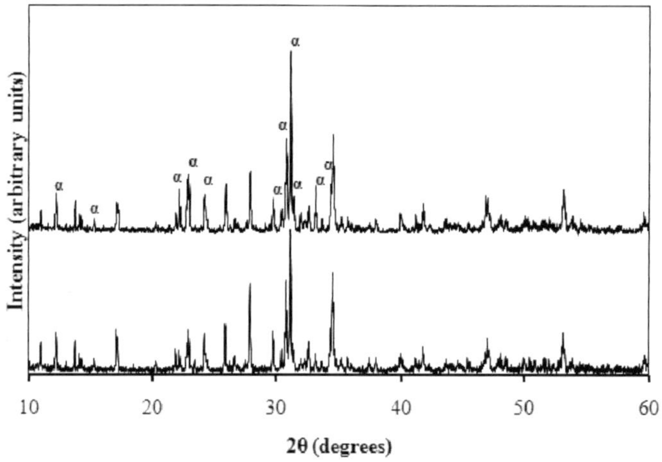

(a)

Figure 9. XRD patterns of β-TCP sintered for 1h at (a) 1350°C, and (b) 1450°C, (α:α-TCP).

Fig 10 shows the XRD diffractograms of TCP - 75wt% Al_2O_3 composites sintered at various temperatures for one hour. The X-ray of the samples show the presence of α-alumina (ICDD data file no. 43-1484) and β-TCP (ICDD data file no. 70-2065).

(b)

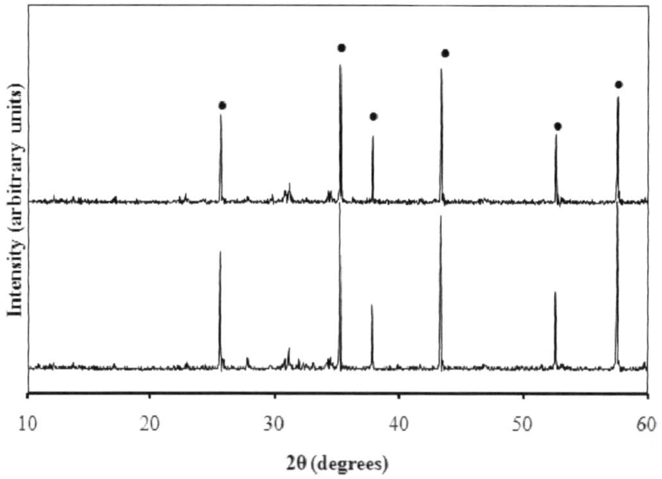

(a)

Figure 10. XRD patterns of β-TCP - 75 wt% Al_2O_3 composites sintered for 1h at (a) 1550°C and (b) 1600°C, (•: α-Al_2O_3).

The ^{31}P MAS-NMR spectra of tricalcium phosphate sintered until 1300°C for one hour show the presence of four peaks which assigned to the tetrahedral environment of P sites (0.12; 1.27; 2.63 and 4.93 ppm) (Fig. 11a and 11b). Above 1300°C, ^{31} P MAS NMR spectra of tricalcium phosphate shows several peaks of tetrahedral environment P sites (-0.006, 0.249, 0.887, 1.397, 1.525, 2.163, 2.801, 4.46 and 5.48 ppm) (Fig. 11c, 11d, 11e and 11f).

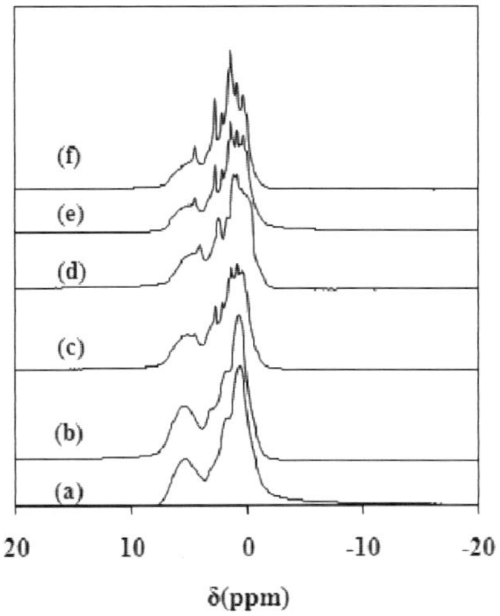

Figure 11. ^{31}P MAS-NMR spectra of β-TCP sintered for 1h at (a) 1200°C, (b) 1250 °C, (c) 1300°C, (d) 1350°C, (e) 1400°C and (f) 1450°C.

The ^{31}P MAS NMR spectra of the TCP-75% Al_2O_3 composites sintered at various temperatures for one hour were illustrated in Fig.12. Four peaks (0.5, 1.9, 3.18 and 4.46 ppm) were observed indicating the presence of tetrahedral P sites.

The ^{27}Al MAS NMR spectra of the TCP -75% Al_2O_3 composites sintered at various temperatures were illustrated in Fig.13. Several peaks detected at approximately 70 ppm, 39.47 ppm and 14.89 ppm assigned to three different aluminium coordination Al^{IV}, Al^{V} and Al^{VI}, respectively. This result was similar to the previous report by Bouslama et al and O'Dell et al [9, 16].

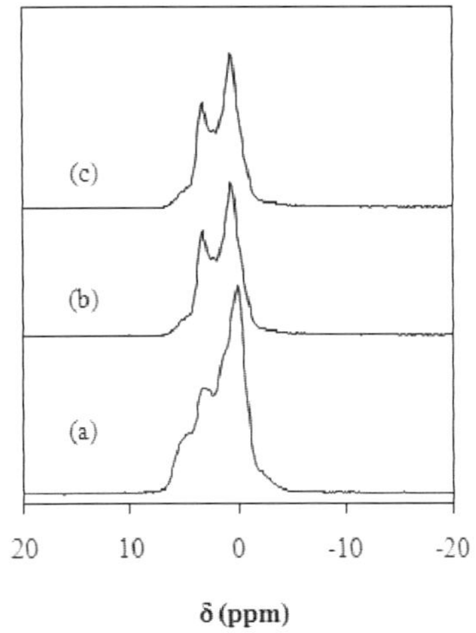

Figure 12. ^{31}P MAS-NMR spectra of TCP - 75wt% Al_2O_3 composites sintered for 1h at (a) 1300°C, (b) 1400 °C and (c) 1500°C.

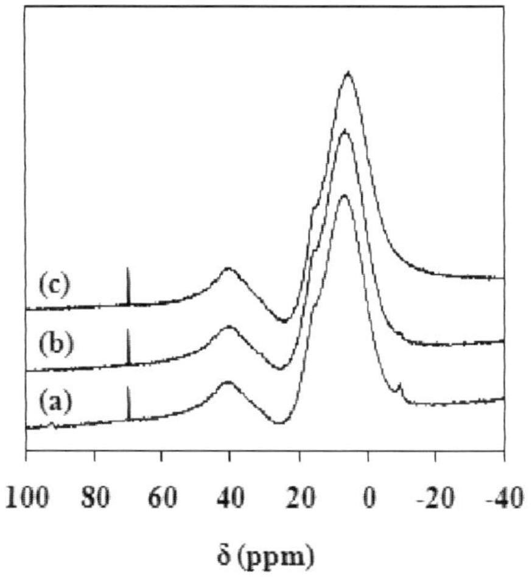

Figure 13. ^{27}Al MAS-NMR spectra of β-TCP - 75 wt% Al_2O_3 composites sintered for 1h at (a) 1300°C, (b) 1400 °C, and (c) 1500°C.

Fig.14 shows SEM micrographs of the β-TCP after the sintering process for 1h at 1350°C and 1450°C. At 1350°C, β-TCP particles went through the stage of partial coalescence (Fig.14a). At 1450°C, the samples present an important intergranular porosity (4µm) (Fig.14b).

The effect of both sintering temperature and time on the morphology of the TCP-75% Al_2O_3 composites was illustrated in Fig.15. The SEM micrographs show the evolution of the microstructure of composites sintered at 1550°C depending on the sintering time variation (Fig.15a, 15c and 15d). Fig.15a shows the presence of intergranular pores and the presence of cracks probably due to allotropic transformations of TCP (β→α). The composites sintered for 2 hours and 5 hours lead to the formation of intragranular pores whose size varies between 3 and 12µm, respectively (Fig.15c and 15d).

In this study, we developed a bioceramic composite containing β-tricalcium phosphate and α-alumina. The analysis by ^{31}P MAS-NMR has demonstrated the formation of a small amount of calcium pyrophosphate in the initial powder of tricalcium phosphate. Calcium pyrophosphate was formed during the synthesis of tricalcium phosphate. Thermal analysis of the powder of tricalcium phosphate shows the presence of two allotropic transformations (β to α and α to ά) at 1258°C and at 1469°C, respectively. These results are similar to those previously obtained by Ben Ayed et al. and Bouslama et al. [5, 9]. The starting temperature of sintering of TCP was close to 1080°C while that of alumina was greater than 1400°C. The maximum relative density of tricalcium phosphate (78.3%) and TCP-75%Al_2O_3 composites (64.4%) were obtained respectively at 1350°C for 3 hours and at 1550°C for 2 hours. The maximum rupture strength for TCP - 75% wt Al_2O_3 composites, was obtained at 1550°C for 2 hours (8.6 MPa). This value is comparable to mechanical resistance of TCP-Fap-alumina composites studied by Bouslama et al. on the study on mechanical properties of TCP- Fap-Al_2O_3 composites [9]. Many other studies such as TCP-Fap-Alumina composites and alumina-Hap-CaF_2 composites prove the significant effect of alumina for enhancing the mechanical properties of bioceramics [8, 17].

Figure 14. SEM micrographs of β-TCP sintered for 1h at (a) 1350°C and (b) 1450°C

Figure 15. SEM micrographs of β-TCP - 75wt% Al_2O_3 composites sintered at (a) 1550°C for 1h, (b) 1600°C for 1h, (c) 1550°C for 2h and (d) 1550°C for 5h.

4. Conclusion

The influence of alumina addition on the tricalcium phosphate was detected in the mechanical properties of β-TCP - Al_2O_3 composites. The mechanical properties of the composites were measured by Brazilian test. The evolution of the rupture strength composites was investigated as a function of sintering temperature and sintering duration. The mechanical resistance of β-TCP increases by the addition 75 wt% of alumina, and it reaches its maximum value of 8.6 MPa at 1550°C after the sintering process for 2 hours.

References

[1] Best S. M, Thian E. S and Huang J 2008 *Jounal of the European Ceramic Society* 1319 1327.
[2] Bouler J. M, Trecant M, Delecrin J, Royer J, Passuti N and Daculsi G 1996 *J Biomed Mater Res* 603 609.
[3] Destainville A, Champion E. D, Assollant B and Laborde E 2003 *Materials Chemistry and Physics* 269 277.
[4] Ryu U, Hong K. S, Lee J. K, Kim D. J, Lee J. H, Chang B. S, Lee C. K and Chung S. S 2004 *Biomaterials* 393 401
[5] Ben Ayed F, Chaari K, Bouaziz J and Bouzouita K 2006 *Comptes Rendus Physique* 825 835.
[6] Ben Ayed F and Bouaziz J 2007 *C R Physique* 101 108.
[7] Ben Ayed F and Bouaziz J 2008 *Journal of the European Ceramic Society* 1995 2002.
[8] Ben Ayed F and Bouaziz J 2008 *Ceramics International* 1885 1892.
[9] Bouslama N, Ben Ayed F and Bouaziz J 2009 *Physics Procedia* 1441 1448.
[10] Bouslama N, Ben Ayed F and Bouaziz 2009 *Ceramics International* 1909 1917.

[11] Bouslama N, Ben Ayed F and Bouaziz J 2010 *Journal of the mechanical behavior of biomedical materials* **2** 13.
[12] Ben Ayed F 2003 *Thesis Faculty Sciences in Sfax, Tunisia.*
[13] ISRM, Suggested methods for determining tensile strength of rock materials, Int J Rock Mech. Min 1978 Sci Geomech Abstr 99 103
[14] ASTM C496, Standard test method for splitting tensile strength of cylindrical concrete specimens 1984 Annual Book of ASTM, Standards vol. **0.042**,ASTM, Philadelphia 336 341
[15] Yashima M, Sakai A, Kamiyama T and Hoshikawa A 2003 *J Solid State Chem* 175 272.
[16] O'Dell L. A, Savin S. L. P, Chadwick A. V and Smith M. E 2007 *Solid State Nuclear Magnetic Resonance* 169 173.
[17] Kim H. W, Koh Y. H, Seo S. B, Kim H. E 2003 *Materials Science and Engineering C* 515 521.

IOP Conf. Series: Materials Science and Engineering **28** (2012) 012029 doi:10.1088/1757-899X/28/1/012029

Effect of fluorapatite additive on the mechanical properties of tricalcium phosphate-zirconia composites

I. Sallemi[1], F. Ben Ayed, J. Bouaziz
Laboratory of Industrial Chemistry, Sfax University, National School of Engineering, Box 1173, 3038 Sfax, Tunisia.

E-mail: imen.sallemi@hotmail.com

Abstract. The effect of fluorapatite addition on the mechanical properties of tricalcium phosphate – 50 wt% zirconia composites was investigated during the sintering process. The Brazilian test was used to measure the mechanical resistance of bioceramics. The mechanical properties of composites increase with the sintering temperature and with fluorapatite additive. At 1400°C, the fluorapatite additive ameliorates the densification and the mechanical resistance of tricalcium phosphate – 50 wt% zirconia composites. The ^{31}P magic angle spinning nuclear magnetic resonance analysis of tricalcium phosphate – zirconia composites sintered with fluorapatite additives reveals the presence of tetrahedral P sites.

1. Introduction

Calcium phosphates are of interest for many biomedical applications due to their good biocompatibility and bioactivity [1-2]. Among materials for medical applications, tricalcium phosphate (TCP) is one of the best candidates in orthopedics, since it exhibits a composition similar to the mineral part of bone [3-11]. In order to enhance the mechanical properties of tricalcium phosphate at high temperature, bioinert ceramic oxides like zirconia (ZrO_2) was used as an amount. Tricalcium phosphate – Zirconia composites have been studied, but its mechanical properties are generally inadequate for many load-carrying applications. Hence, fluorapatite can be used to ameliorate the mechanical properties of tricalcium phosphate – zirconia composites [12]. In fact, Fap is an attractive material due to its similarity in structure and composition to bone [13, 14]. In vitro studies have shown that Fap is biocompatible [2]. It is well-known that the presence of additive makes it possible either to lower the temperature of sintering or to improve the densification and the mechanical properties of material. In our work, we chosed the fluorapatite because of its good thermal stability at high temperature and its better mechanical properties than tricalcium phosphate [2, 5, 9, 15]. So, the aim of our study was to prepare a TCP – 50 wt% ZrO_2 composites with different percentages of fluorapatite and to characterize the resulting composites with density, mechanical resistance, X-ray diffraction, nuclear magnetic resonance (^{31}P) and scanning electron microscopy measurements.

[1] To whom any correspondence should be addressed.

2. Experimental

Tricalcium phosphate powder was synthesized by solid state reaction between calcium carbonate ($CaCO_3$, Fluka) and dicalcium phosphate dihydrate ($CaHPO_4.2H_2O$, Fluka) at 1000°C as the following reaction [16]:

$$2\,CaHPO_4.2H_2O_{(s)}+CaCO_{3(s)} \rightarrow \beta\,Ca_3(PO_4)_{2\,(s)}\ (\beta\text{-TCP}) +5\,H_2O_{(g)}+CO_{2\,(g)} \quad (1)$$

The reaction finish was indicated by phenolphtleine test.

The fluorapatite (Fap) powder was prepared by the precipitation method [13]. A calcium nitrate ($Ca\,(NO_3).4H_2O$, Merck) solution was slowly added to a boiling solution containing diammonium hydrogenophosphate $(NH_4)_2HPO_4$, Merck) and ammonium fluorine (NH_4F, Merck), with continuous magnetic stirring. During the reaction, pH was adjusted to the same level (pH 8-9) by adding ammonia. The obtained precipitate was filtered and washed with deionised water; it is then dried at 70°C for 12 hours.

Calculated quantities of tricalcium phosphate, zirconia and fluorapatite powders were mixed with absolute ethanol in an agate mortar in order to elaborate TCP – ZrO_2 – Fap composites with different percentages of Fap (2 wt%, 5 wt%, 10 wt%, 15 wt%, 20 wt% and 25 wt%). Cylindrical discs were produced by compacting 4 g of composite powder into 20mm diameter and 6mm thichness with uniaxially compacting at pressure of 150 MPa (under a load of 47.1 kN) in the cylindrical mould. Finally, the compacts were sintered, in a vertical resistance furnace (Pyrox 2408), at various temperatures for one hour. The heating and cooling rates of temperature were 10 and 20°C / min, respectively.

The relative density of sintered body was determined by geometrical measurement. To ensure reproducibility of results, six tests were made for every experiment in the same conditions. The relative error of density value was about 1%.

Each powder was identified using X-ray diffraction (XRD) analysis. The X-ray diffraction patterns was obtained using a Seifert XRD 3000 TT diffractometer by a monochromated CuK_α radiation (λ = 1.54056 Å). The obtained crystalline phases were compared with the International Center for Diffraction Data (ICDD) files.

Differential thermal analysis and thermomecanical analysis were carried out using about 30 mg of powder in Helium (DTA-TGA ; Setaram SETSYS Evolution_1750 and TMA; Setaram TMA 92 dilatometer). The heating rate is 10°C min^{-1}.

The NMR spectra were recorded, by high resolution solid state MAS-NMR (BRUKER 300WB) with a ^{31}P frequency of 121.5 MHz, in parts per million (ppm) referenced to 85 wt% H_3PO_4.

The microstructure of the sintered compacts was characterized by scanning electron microscopy (Philips XL 30) on fractured sample surfaces coated with gold layer.

The specific surface area (SSA) was measured by the BET method using azotes (N_2) as an adsorption gas (ASAP 2010) [17]. The primary particle size (D_{BET}) was calculated by assuming the primary particles to be spherical [18]:

$$D_{BET} = \frac{6}{S\rho} \quad (2)$$

Mechanical properties of sintered specimens were measured by Brazilian test [19, 20] using a LLOYD EZ50 device on the cylindrical compacts. The maximum rupture strength (σ_r) was offered by the following equation:

$$\sigma_r = \frac{2\times P}{\Pi\times D\times t}$$

3. Results and discussion

3.1. Characterization of powders

The SSA of β-TCP, Zirconia, Fap and TCP – 50 wt% ZrO_2 composites are 2.25, 2.6, 29 and 1.49 $m^2\,g^{-1}$, respectively. Table 1 summarizes the characteristics of different powder used in this study.

The X-ray diffraction (XRD) pattern of tricalcium phosphate powder reveals only peaks of β–TCP (ICDD data file no. 09-0169) without any other phase (Fig.1a). The phenolphtaleine test was negative. All the X-ray diffraction of zirconia peaks, reported in Fig.1b, corresponds to monoclinic zirconia (ICDD data file no. 37-1484). The XRD pattern of Fap was presented in Fig.1c which assigned the peaks relative to ICDD data file no. 15-876.

Table 1. SSA and average grain size obtained by specific surface area of different powder used in this study.

Powder	SSA (m^2/g) ± 1.0	D_{BET} (μm) ± 0.2	d[(b)]
TCP	2.25	0,86	3,07 (β) 2.86 (α)
m-ZrO_2	2.60	0,40	5.83
Fap [15]	29.00	0.07	3.19
Composites[(a)]	1.49	0,90	4.45

[a]TCP – 50 wt% ZrO_2
[b]theoretical density.

The ^{31}P MAS-NMR spectra of β–TCP powder reveals several peaks (Fig.2a). In particular, a peak towards 4.65 ppm, a shoulder towards 1.12 ppm and another peak, more intense, towards 0.24 ppm. These peaks were assigned to tetrahedral P sites. They show that the phosphorus atoms of tricalcium phosphate are localised in three different crystallographic sites: $P(1)O_4$, $P(2)O_4$ and $P(3)O_4$ [21]. Fig. 2b shows an intense peak at 2.80 ppm relative to the phosphorus of Fap which was assigned to tetrahedral P sites. The ^{31}P MAS-NMR spectrum of composites shows the peaks relative toTCP and Fap (Fig.2c).

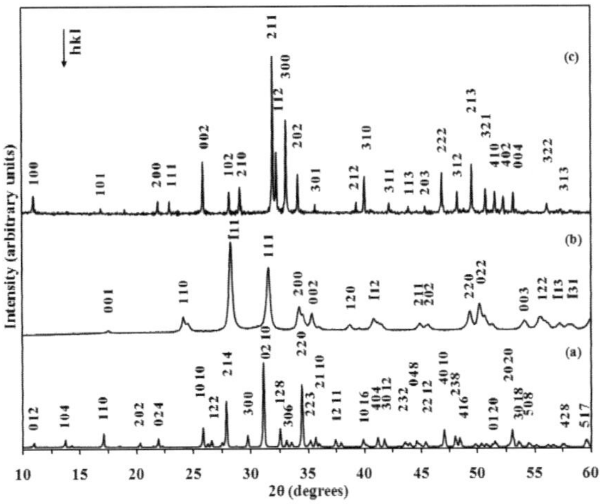

Figure1. XRD patterns of (a) TCP powder, (b) Zirconia powder, and (c) Fap powder.

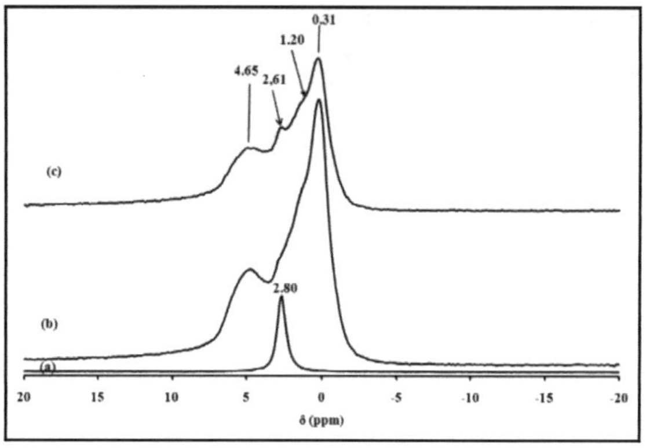

Figure 2. ^{31}P MAS-NMR spectra of (a) Fap powder (b) TCP powder and (c) TCP – 50 wt% ZrO_2 composites.

The DTA analyse proves that TCP, Zirconia and Fap present an endothermic peaks (Fig.3). The differential thermal analysis (DTA) curve of tricalcium phosphate shows 3 peaks (Fig.3a). The first peak, at 186°C, attributed to the hydration water. A second peak, around 1288°C, characterizes the first allotropic transformation of tricalcium phosphate: β to α. The last peak towards 1457°C linked to the second allotropic transformation of TCP : α to α'. Fig.3b presents an endothermic peak, nearly 1170°C, corresponds to the allotropic transformation of zirconia from the monoclinical phase (m) to the tetragonal phase (q). DTA of Fap shows a peak at 1200°C relative to a liquid phases which assigned to the presence of a binary eutectic between CaF_2 and Fap (Fig.3c) [13].

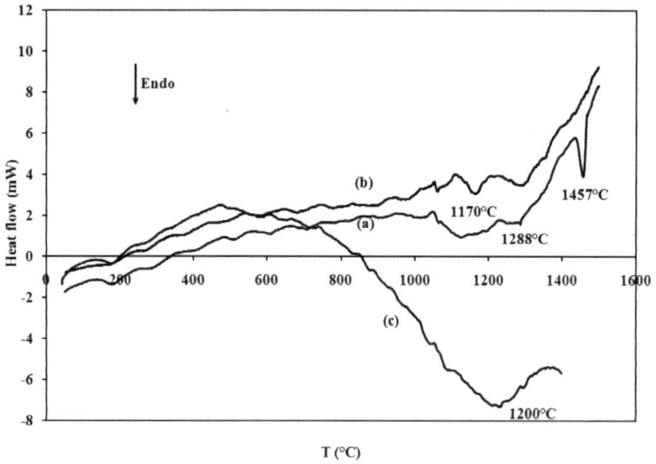

Figure 3. DTA curves of (a) TCP powder, (b) Zirconia powder, and (c) Fap powder.

3.2. Sintering of powders

The evolution of the densification and the mechanical resistance of TCP-50 wt% ZrO_2 composites, sintered between 1200°C and 1600 °C, was reported in Fig. 4. The optimum of composite densification was around 72%, reached at 1600°C (Fig. 4a). Wheras, the rupture strength of the composite reached 5.2 MPa (Fig.4b). Between 1400°C and 1550°C, we registered a small variation of mechanical resistance from 3.4 MPa to 3.8 MPa at 1400°C and 1550°C, respectively (Fig. 4b).

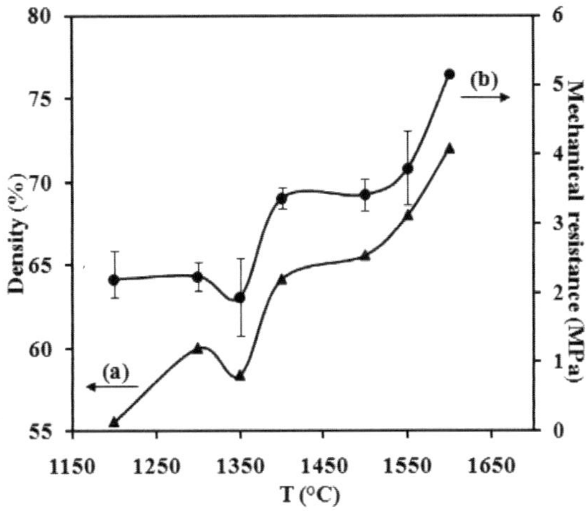

Figure 4. Relative density and mechanical resistance versus temperature of TCP–50 wt% ZrO_2 composites sintered for 1 h (a: relative density, b: mechanical resistance).

Fig.5 illustrates the evolution of relative density and mechanical resistance of TCP – 50 wt% ZrO_2 composites sintered for 1 hour at 1400°C with different percentages of fluorapatite. The composite densification increase with the amount of Fap (Fig.5a). The optimum of composites densification (70%) was reached with 25 wt% of Fluorapatite (Fig.5a). The evolution of TCP – 50 wt% ZrO_2 composites mechanical resistance versus the percentage of Fap was reported in Fig.5b. The addition of 25 wt% Fap ameliorate the rupture strength of composites untill reached 6.2 MPa (Fig.5b).

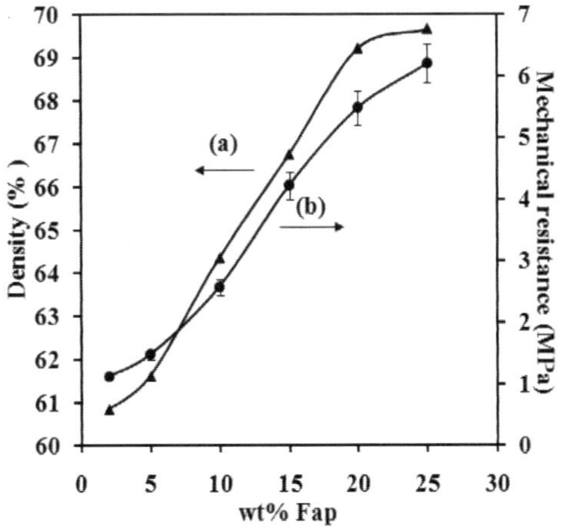

Figure 5. Relative density and mechanical resistance of TCP–50 wt% ZrO_2 composites sintered with different percentages of fluorapatite at 1400°C for 1h (a: relative density, b: mechanical resistance).

3.3. Characterization of samples after the sintering process

After the sintering process, the characterization of sintered samples was carried out using X-ray diffraction, scanning electronic microscopy and ^{31}P nuclear magnetic resonance.

The XRD analysis of TCP – 50 wt% ZrO_2 composites sintered at 1400°C with fluorapatite addition (2 wt%, 5 wt%, 10 wt%, 20 wt% and 25 wt%) were presented in Fig.6. XRD patterns show the presence of β-TCP, m-ZrO_2, Fap and some traces of α-TCP and $Ca_4(PO_4)_2O$. Nevertheless, the Fap peaks increase when the percentage of Fap increase more than 10 wt% (Fig. 6c-6e).

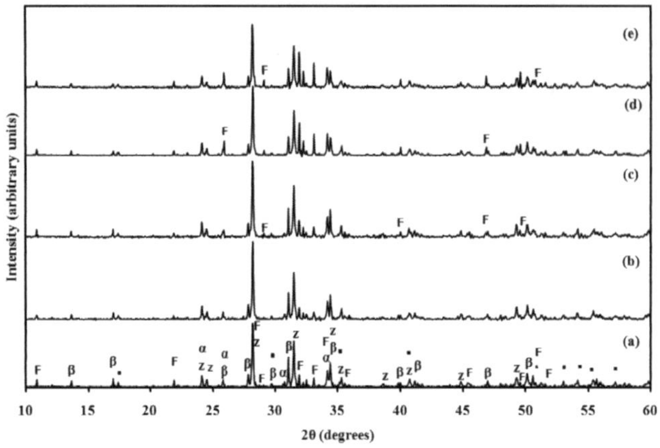

Figure 6. XRD patterns of TCP – 50 wt% ZrO_2 composites sintered at 1400°C with different percentages of fluorapatite (a) 2 wt%, (b) 5 wt%, (c) 10 wt%, (d) 20 wt% and (e) 25 wt% (β : β-TCP ; α : α-TCP ; ▪ : $Ca_4(PO_4)_2O$; Z : m-ZrO_2; F: Fap).

The ^{31}P MAS-NMR solid spectra of TCP – 50 wt% ZrO_2 composites sintered for 1 h at 1400°C with different wt% of Fap addition (2 wt%, 5 wt% and 25 wt%) were presented in Fig.7. The spectra reveal the peaks relatives to β–TCP and Fap. Especially, four peaks correspond to the phosphorous environnement was observed at 0.25 ppm, 1.015 ppm, 2.80 ppm and 4.97 ppm. The Fap's peak was detected at 2.80 ppm, which its intensity increases after adding 25 wt% Fap (Fig.7d).

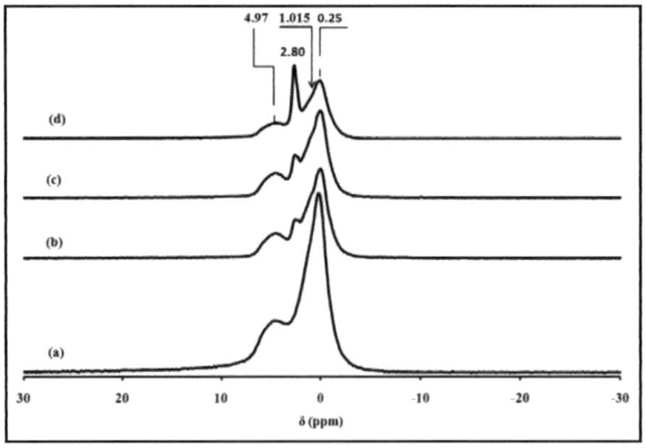

Figure 7. ^{31}P MAS-NMR spectra of TCP – 50 wt% ZrO_2 composites sintered for 1 h at 1400°C with different percentages of fluorapatite (a) 0 wt%, (b) 2 wt%, (c) 5 wt%, (d) 25 wt%.

Fig. 8 shows the SEM micrographs of TCP – 50 wt% ZrO_2 composites sintered at 1400°C for 1h with different amounts of fluorapatite. A few quantitie of Fap (2 wt%) decreases the grain sizes and reduces the composites densification (Fig.8a). However, the presence of 10 wt% Fap increases the densification of TCP – 50 wt% ZrO_2 composites (Fig.8b). In addition, with 20 wt% Fap, the pore slightest size permits the amelioration of composites densification (Fig.8c and d).

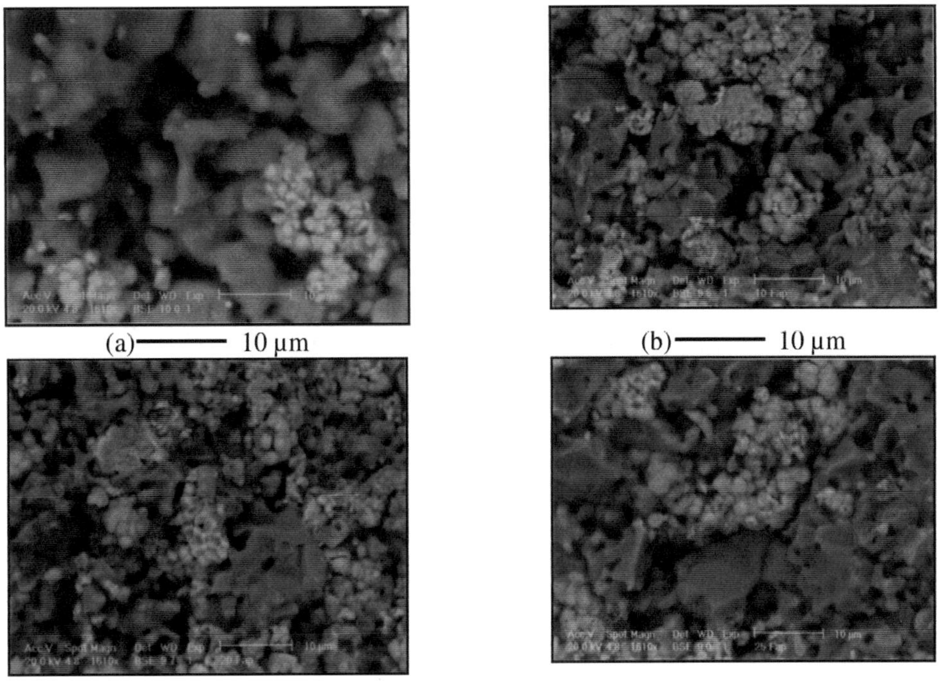

(a) —— 10 µm (b) —— 10 µm

(c) —— 10 µm (d) —— 10 µm

Figure 8. SEM micrographs of TCP – 50 wt% ZrO_2 composites sintered for 1 h at 1400°C with different percentages of fluorapatite (a) 2 wt%, (b) 10 wt%, (c) 20 wt% and (d) 25 wt%.

In this study, fluorapatite has been used in the TCP – 50 wt% ZrO_2 composites as an additive for its important biocompatibility with hard tissue and higher thermal stability. At 1400°C, the mechanical resistance of TCP – 50 wt% ZrO_2 composites reached 3.5 MPa. With 25 wt% Fap, the resistance increases to 2.7 MPa and passes to 6.2 MPa. So, Fap was a good additive in the TCP-ZrO_2 composites during the sintering process. These results are similar to those reported previously by Ben Ayed et al during the study on sintering of tricalcium phosphate – Fluorapatite composites with adition of Zirconia [12].

The preliminary results obtained in this study have shown that the Fluorapatite has a potential to be further developed into an alternative system to produce denser TCP – ZrO_2 bodies. Further investigations are still under way to investigate the influence of Fap on the densification, microstructure and mechanical properties of TCP – ZrO_2 composites bioceramic. This result was similar to pervious report in different studies [2, 5, 8, 9, 11].

4. Conclusion

The effect of fluorapatite additive in TCP – 50 wt% ZrO_2 composites was studied during the sintering process. The evolution of the composite rupture strength was investigated as a function of Fap's amout. The mechanical properties of TCP – 50 wt% ZrO_2 composites increased with the fluorapatite additive. At 1400°C, the composites densification and its mechanical resistance reached its maximum values (70% and 6.2 MPa respectively) after adding 25 wt% of Fap.

References

[1] Aryee S, Imhoff A. B, Rose T, Tischer T 2008 *Biomaterials* **29** 3497-3502.
[2] Elliott J. C 1994 *Elsevier Science B.V, Amsterdam.*
[3] Larry L, Hench L. L 1998 *Bioceramics Journal of the American Ceramic Society* **81** 1705 1728.
[4] Ramachandra Rao R, Kannan T. S 2002 *Materials Science and Engineering C* **20** 187 193.
[5] Wang C. X, Zhou X and Wang M 2004 *Materials Characterization* **52** 301.

[6] Ben Ayed F, Chaari K, Bouaziz J and Bouzouita K, 2006 *C. R. Physique* **7** 825 835.
[7] Ben Ayed F, Bouaziz J, 2007 C. R. Physique **8** 1 101 108.
[8] Bouslama N, Ben Ayed F, Bouaziz J 2009 *Physics Procedia* **2** 1441 1448.
[9] Bouslama N, Ben Ayed F, Bouaziz J 2009 *Ceramics International* **35** 1909 1917.
[10] Chevalier J, Gremillard L 2009 *Journal of the European Ceramic Society* **29** 1245 1255.
[11] Bouslama N, Ben Ayed F and Bouaziz J 2010 *Journal of the mechanical behavior of biomedical materials* **2** 13.
[12] Ben Ayed F, Bouaziz J 2008 *J. Eur. Ceram. Soc.* **28** (10) 1995 2002.
[13] Ben Ayed F, Bouaziz J, Bouzouita K, *J. Eur. Ceram. Soc.* 2000 **20** 8 1069.
[14] Ben Ayed F, Bouaziz J and Bouzouita K, *J. Alloys Compd.* 2001 **322** 1–2 238 245.
[15] Ben Ayed F, Bouaziz J, Bouzouita K, *Ann. Chim. Sci. Mat.* 2006 **31** 4 393 406.
[16] Ben Ayed F, 2003 *Thesis Faculty Sciences in Sfax, Tunisia.*
[17] Brunauer S, Emmet P.H and Teller J 1938 *Amer. Chem. Soc. J.* **60** 310.
[18] Bernache-Assollant D 1993 *Chimie-physique du frittage Editions Hermès 200.*
[19] ISRM. Suggested methods for determining tensile strength of rock materials 1978 Int. J. Rock Mech. Min. Sci. Geomech. Abstr. **15** 99 103.
[20] ASTM C496, Standard test method for splitting tensile strength of cylindrical concrete specimens 1984 Annual Book of ASTM, Standards, ASTM, Philadelphia (24) 336 341.
[21] Yashima M, Sakai A, Kamiyama T, Hoshikawa A 2003 *J. Solid State Chemistry* **175** 272-277.

IOP Conf. Series: Materials Science and Engineering **28** (2012) 012030 doi:10.1088/1757-899X/28/1/012030

The damage of the cutting tools out of carbide metallic during the turning of a soaked and not hardened steel XC38

M. Seghouani[1] [a], A. Tafraoui [ab] and S. Lebaili[a]
[a]Laboratory of Sciences and engineering of Materials, USTHB, BP. 32 El Alia Bab-Ezouar (16000) Alger, Algeria.
[b]University of Bechar BP. 417 (08000) Bechar Algeria.

E-mail: seghouanimohamed@yahoo.fr

Abstract. The purpose of this study widened knowledge on the use of the cutting tools out of metal carbide and to define of it the influence of the elements of the mode of cut on the behaviour of these tools during the machining of treated steel XC38 and untreated. This work aims at evolution determined in experiments of the wear of a cutting tool out of metal carbide with plate reported of P30 nuance for an operation of slide-lathing in turning on soaked and not hardened steel XC38 test-tubes. This research is based on the model of Taylor to determine the lifespan of the cutting tool according to the various parameters of cut, like the cutting speed *Vc*, the advance of cut a, the depth of cutting *P*. In order to express the operational limits of the tool for slide-lathing in a preventive way. The model makes it possible to determine the time of change of the tool and to regard it as constraint for the respect of the roughness of the work piece during a work of series in conventional machining.

1. Introduction

The industrialists need to optimize their production processes in order to increase the productivity, to reduce the wear of the cutting tools [1]. The improvement of the quality of the production will remain always one of the paramount concerns for the industrial development. Technological progress in the machining of metals is related to the improvement of the behaviour of materials to the cut. The increase in the performances on the machine tools influences directly over the lifespan of the cutting tools engaged in production. The selection of material of the tool is a delicate phase, taking into account the conditioning of feasibility of the operation of machining. Consequently the rational choice of a nuance of tool can be done only with experiments quite specific to material given [2].
The materials of cut out of high speed steel, carburizes and cermet are largely used in turning for the majority of machined materials [3], [4]. However their behaviour is limited and their use is sometimes impossible when certain work pieces are characterized by a very great hardness and an abrasion resistance very high [5], [6]. It is the case mainly special steels, pig iron and cast iron soaked, alloys refractory containing nickel and of composite materials with metal matrix. We can also add the sintered materials whose hardness's exceed easily the 70 HRc. Technical progress reveals new materials likely to be used to produce cutting tools answering these requirements. The rational choice of these materials of cut can be done only with experiments specific to each nuance. Thus their implementation requires a sufficient control of the course of the process of cut, in particular the evolution of the wear of the tools and required roughness. During the machining of the materials whose hardness exceeds the 45 HRc, at cutting speeds raised, the mechanical and thermal requests on

[1] To whom any correspondence should be addressed.

the edge of cut of the tool, which constitutes the active element carrying out the work of deformation, become significant. Frictions and the high constraints which are exerted with the interfaces between the part and the tool cause various forms of wear on the active part of tool [7], [8], [9], [10]. Morphologies of wear depend primarily on the nature of the tool, machined material, the cutting conditions and the type of machining. According to standard NF E66-505, the principal forms of wears met are characterized primarily by wears in skin *Vb* and crater K_T. Wear in skin or frontal wear is due to the friction of the part on the face in skin of the tool [11], [12]. From the point of view practises, frontal wear is most significant to consider since it determines the surface quality of the part machined as well as dimensional accuracy [13]. On the other hand, wear in crater is characterized by a basin formed on the cutting face of the tool following frictions of the chip [11], [12]. This form of wear is allotted to the presence of strong pressures and the release of a great quantity of heat in the zone of contact between the chip and the tool. Consequently, the temperature rises and causes a significant thermal diffusion process. The shape of the crater is defined by its depth K_T. Wear notches some occurs under certain cutting conditions following wrenching of particles. Several work [14], [12], [15] proposed mathematical relations which express the lifespan of the tool T according to the parameters of cut [16]. Today hard machining is in full rise and the applications of hard turning include/understand the gears (30%), the bearings (25%), the trees (20%), the casings (5%) and more particularly those intended for the car industry which call upon the semi-completion and the completion of the transmissions, the axles, the discs of brakes, etc, [17], [18]. The turning of the chilled steels by hardening requires the use of component materials having excellent properties of hardness at high temperatures, wear resistance and chemical stability. The latter had relatively high performances in turning for their substitution for the operations of correction [19], [3], and [20]. The more of time, the hard turning products of surfaces of qualities, minimizes the number of operations and the rejected parts [21]. In end of this study is devoted to the comprehension of the damage mechanisms of the cutting tools in service.

2. Materials and methods

2.1. Materials

2.1.1. Machined material

The cylindrical test-tubes of diameter ∅ 80 mm, and length L = 500 mm machined are out of steel XC38, of which a first series of test-tubes not soaked and a second series of test-tubes soaked. This steel is selected as material to be machined because it roughly presents 47% of thermically treated materials used in the manufacture of the parts [22].
The chemical and physical properties of materials used are presented on table 1.

Table 1. Physical and chemical properties of the test tube XC38

% C	% Mn	% Si	Rm (MPa)	Re (MPa)	Rp0.02 (MPa)	% A_{min}
0.35-0.46	0.5-0.8	0.1-0.4	580-670	350	335	21

2.1.2. Materials of the tools

The cutting tool used is a turning tool with removable plate brought back square, fixed on the tool by fastening. The square brought back plates are out of metal carbide of nuance P30 having the following geometry: Xr = 75°, α = 6°, γ = -6°, λ = -6°, whose the radius of the nozzle used is Rε = 0.8 mm. The chemical composition of the plates is: WC = 82%, TiC = 8%, Co = 10%.

2.2. Methods

2.2.1. Models of lifespan and models for wear

The lifespan of a tool is the time of cut total necessary to reach a criterion of specific lifespan. To evaluate the lifespan, we refer to a limiting value of a direct criterion (Vb, K_T). The analysis of the damage is very often done by considering wear on the face of skin because the latter influences directly the surface quality of the part (Figure 1). For a tool out of carbide, the following criterion is recommended by the ISO standard: average height $Vb = 0.3$ mm

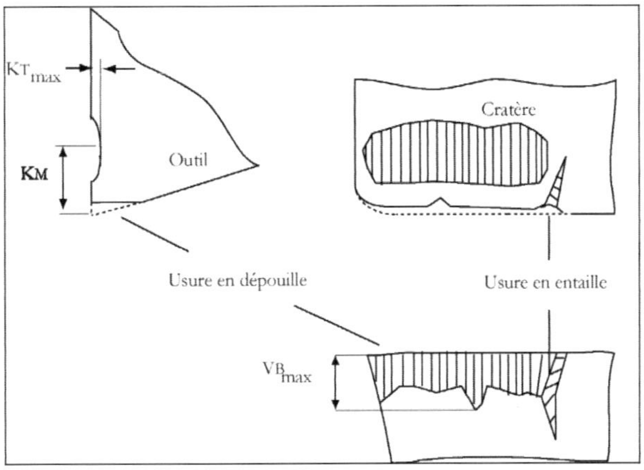

Figure 1. Schematic of wear to the specimen and the cutting tool.

From the criteria of wear, we establish models of lifespan. Oldest and the most used is that of Taylor (equation 1). Taylor was the first to be proposed a mathematical model connecting the effective duration of cut T of a tool to the parameters of modified cut or Taylor (equation 2) [23], [24]:

$$T = C_1 . V_c^n \tag{1}$$

$$T = C_2 . P^x . a^y . V_c^n \tag{2}$$

C_1 : Constant depends on the nuance of the tool and machined material.
V_c : Cutting speed [m/min].
n : Constant.
C_2 : Constant depends on the nuance of the tool and machined material.
P : Depth of cut [mm].
a : Advance [mm/tr].
x, y : Exhibitors depending on the nuance of the tool.

These equations describe the relation between the lifespan T and the parameters of cut like the cutting speed V_C the advance a and the depth of cut P they utilize constants (C_1, C_2, n, x, y) which must be identified in experiments for each couple tool/matter considered and each process of machining. Contrary to the models of lifespan, the models of wear describe the volume of matter lost in the zones of contact of the tool by connecting them to physical parameters like the temperature or the pressures applied to the tool [25], [26], and [27].

3. Presentation of the results

3.1. Case of the operations of slide-lathing of steel XC38 not hardened

3.1.1. Test: Multifactorielle method

The tests carried out allowed the follow-up of the evolution of wear according to time for the various combinations.

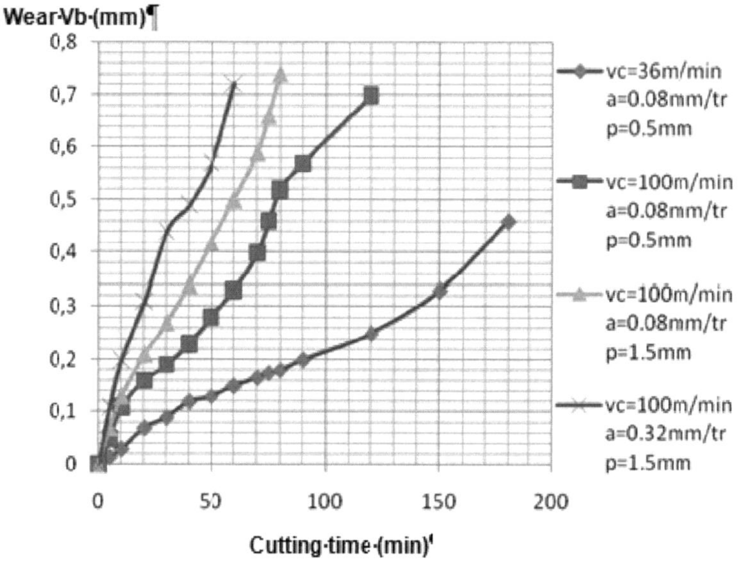

Figure 2. Variation of cutting tool wear (*V*b) according to time for different combinations.

The curves of the Figure 2 have the results of different experimental combinations from the modes of cut (cutting speed, advances, depth of cutting) according to time. For two identical experiments, if we increases the cutting speed (Vc_1 = 36 m/min, Vc_2 = 100 m/min), the wear of the tool increases. Same manner if we increases the advance of cut (a_1 = 0.08 mm/tr, a_2 = 0.32 mm/tr), the wear of the tool increases also even for the increase the depth of cutting (P_1= 0.5 mm, P_2 = 1.5 mm). With the tests of Figure 2, we could deduce the life spans from the cutting tools represented by the Figure 3 according to the model of Taylor. The graphic treatment of these results led to the determination of the life spans corresponding to the various conditions of tests. The statistical processing of the behaviours allowed the deduction of mathematical model of the modified equation of Taylor (2). The latter express the functional relation between the lifespan and the elements of mode of cut in the form of equation 3 [24].

$$T = e^{6.85} \,.\, V_c^{-0.9} \,.\, a^{-0.4} \,.\, P^{-0.41} \quad (3)$$

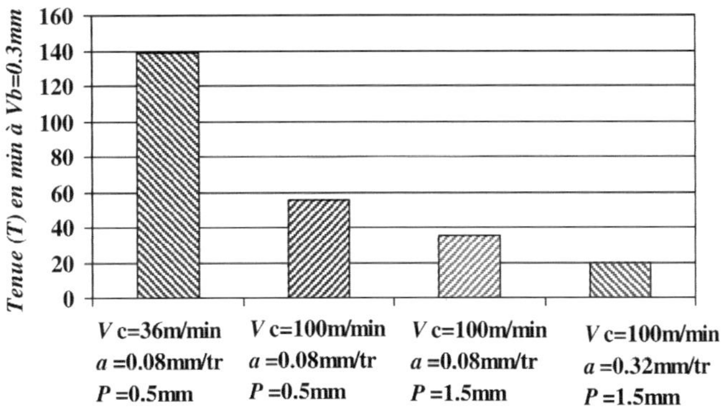

Figure 3. Experimental behaviour for different combinations, *Vb* = 0.3 mm.

The figure 3 is a histogram which recapitulates the various experiments realized previously according to the various combinations of the parameters of cut (cutting speed, advances, depth of cutting). We note that the greatest behaviour (lifespan of the tool) is according to the parameter cutting speed smallest (*Vc* = 36 m/min), then that of the smallest advance (*a* = 0.08 mm/tr) and in end the depth of the smallest cutting (*P* = 0.5 mm).

3.1.2. Micrography of wear Vb of the plates P30 after machining on a steel XC38 not hardened according to time

The Figure 4 shows the various wears Vb recorded on plates out of metal carbide P30 according to time on steel XC38 test-tubes not hardened.

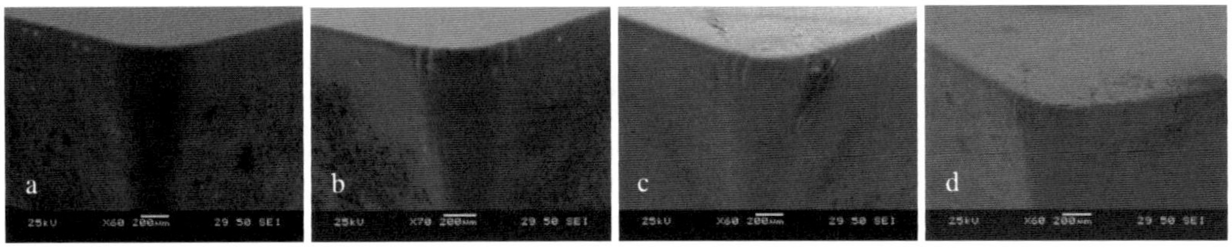

Figure 4. Micrographics of wear *Vb* for metallic carbide layer to P30 nuance at *Vc* = 100 m/min, a = 0.32 mm/tr and *P* = 1.5 mm. (a) : t_c = 10 min, (b) : t_c = 20 min, (c) : t_c = 40 min, (d) : t_c = 60 min

The photographs on the Figure 4 they is images made with the MEB of stop work of the plates of the tools for each test carried out in order to even the degree of wear. The micrographic one of the wear of the plates to pay out of metal carbide according to variation of the time of machining of steel XC38 not soaked shows that the wear of the plate is not significant and that the tools resisted the criteria of wear well.

3.2. Case of the operations of slide-lathing of hardened steel XC38

3.2.1. Heat treatment (Hardening, Returned)

According to the percentage out of carbon and their volume, the steel XC38 test-tubes have were put in a furnace heated until a temperature of 850°C, after 2 hours, the parts are soaked in water. Then, the test-tubes have undergoes an income to know: we gives the test-tubes in the furnace, at the temperature of 250°C left we them during 2 hours with the furnace then cooled with the free air with 20°C. The conditions of the heat treatment are presented in the Figure 5. The average hardness of the test-tubes obtained at the end of the operations is of 112 HRB.

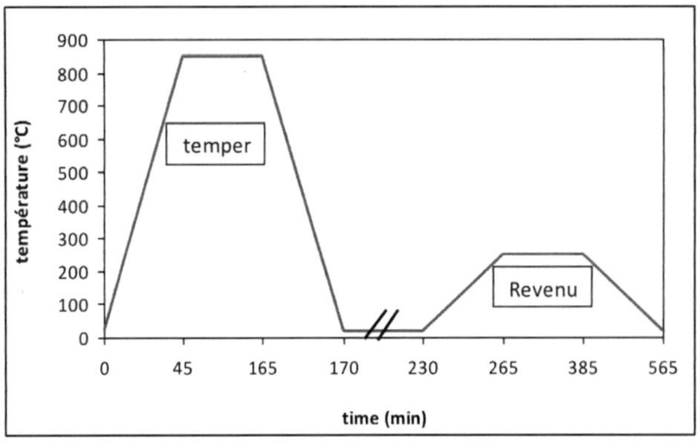

Figure 5. Cycle heat treatment of the test-tubes.

3.2.2. Multifactorielle method

The tests carried out allowed the follow-up of the evolution of wear according to time for the various combinations of the experimental design.

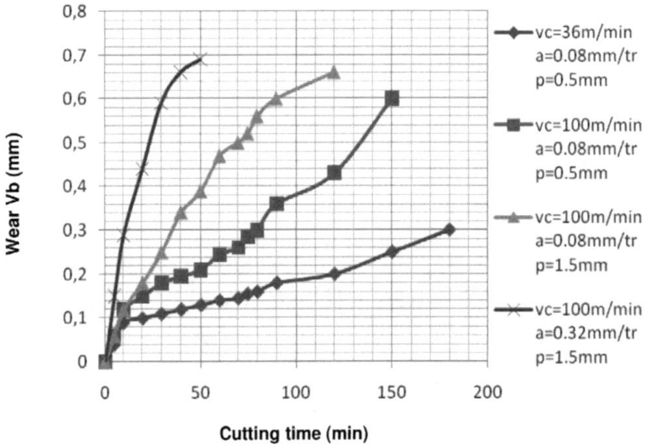

Figure 6. Variation of cutting tool wear (*Vb*) according to time and different combinations.

The functional relation between the lifespan and the elements of mode of cut is in the form:

$$T = e^{4.2} . V_c^{-1.36} . a^{-0.57} . P^{-0.89} \quad (3)$$

The curves of Figure 6 have the results of different combinations experimental from the mode of cut (cutting speed, advances, depth of cutting) according to time. We note the checking of the various experimental results (curved) to find previously with knowing: for two experiments identical, if we increase the cutting speed (example Vc_1 = 36 m/min and Vc_2 = 100 m/min), the wear of the tool increases. Same manner, if we has two experiences identical but with different advances of cut (example a_1 = 0.08 mm/tr and a_2 = 0.32 mm/tr), the wear of the tool Vb_1 is an inferior with Vb_2. It is the same case to note for various depth of cutting (example P_1= 0.5 mm and P_2 = 1.5 mm). In end, for three tests identical relatively, we notes that the value of wear is according to the cutting speed initially, then of the value of the advance in second and end of the value the depth of cutting. By comparison of Figure 2 and Figure 6, the value of wear is different following the increase from the hardness of the soaked test-tube. It is noticed that the values of the wear of the tool are definitely higher than those with the experiments ridges with a test-tube not hardening.

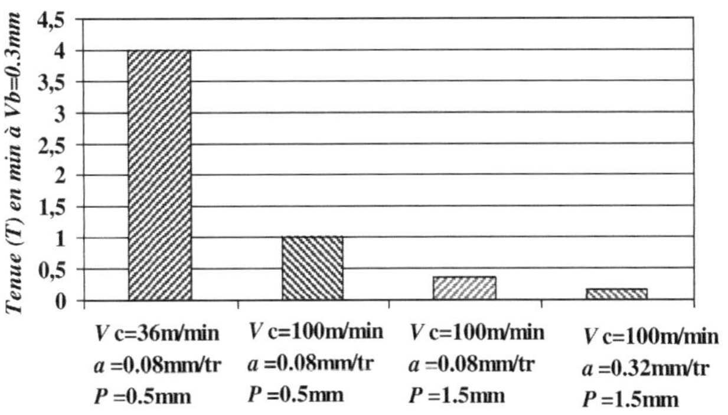

Figure 7. Experimental behaviour for different combinations of experiences plan, *V*b = 0.3 mm.

The behaviour of the tool for the various experiments realized according to the various combinations of the parameters of cut (cutting speed, advances, depth of cutting), shows that the greatest lifespan of the tool is according to the parameter cutting speed smallest, then that of the smallest advance and in end the depth of the smallest cutting. In end, if we compare the results of the behaviour tool presenting it in Figure 7 with those obtained and presenting in Figure 3, we notice that the behaviour of the tools for these experiments is not the same bus the hardness of the test-tubes is not the same one, although the mode of cut is the same one.

3.2.3. Micrography of wear V B of the plates P30 after machining on a steel XC 38 hardened of hardness 112 HRB

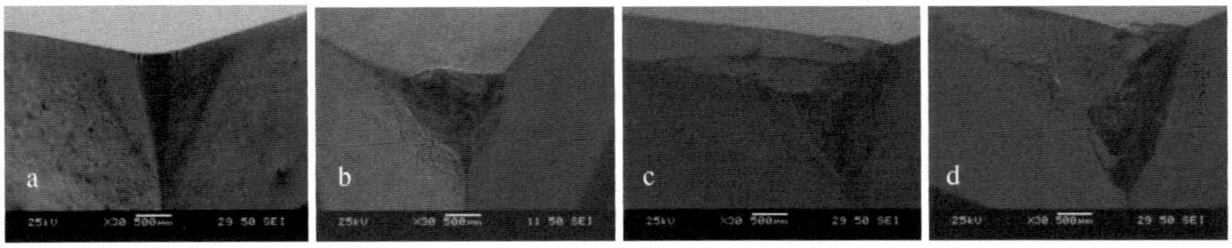

Figure 8. Micrographics of wear *V* B of the plates P30 after machining on a steel XC 38 hardened of hardness 112 HRB. at Vc=100m/min. a = 0.32 mm/tr, and P = 1.5mm (a): Tc = 10s, (b): tc=20s, (c): tc=30s, (d) tc=1min.

In Figure 8, the micrographic variation of the wear of the plate out of metal carbide is according to the variation of the time of machining initially, the mode of cut and the hardness of the work piece. We notice that wear is significant as the time of machining increases is that the hardness of the test-tube increases.

4. Discussion and Conclusion

The analysis of these results shows that the cutting speed has a significant influence on wear, indeed with the increase in the latter; wear increases in a remarkable way see (Figure 2, Figure 6).
It comes out from this analysis that the effect the cutting speed over the lifespan is more pronounced for speeds high Figure 3, Figure 7; it is explained by the fact why when one works with higher cutting speeds:

- The machining system becomes unstable because of the great vibrations recorded during machining.
- The temperature in the zone of cut increases (in particular with the interfaces chip surfaces of attack, and part surfaces of principal skin, which results in a loss of hardness as well as the manifestation of the various mechanisms of wear of the tool.

As the figures Figure 2 and Figure 6 show it that it is in the case of the machining of steel XC38 hardened or not, it is noted that wear increases with the increase in advance and the depth of cut. This rise in wear induced to a degradation of the surface quality of the part machined according to time. The increase in the advance enormously decreases the behaviour of the tool which does not exceed 1 min (Figure 7). Amongst other things, the wear of the tool presented by the figures Figure 4 and Figure 8 show the variation of the damage of the tool according to time with selected modes of cuts.
After having evaluated wear *Vb* for procedures of tests which take into account the variation of the various parameters of cut (*Vc*, *a* and *P*), we noted that:

- The tool out of metal carbide of nuance P30 resists wear for the machining of half-hard steel XC38, to reach acceptable wear *Vb* = 0.3 mm. It him took a time of cut T = 60 min for a cutting speed *Vc* = 100 m/min, and T = 140 min for *Vc* = 36 m/min.
- Hardened steel XC38 of hardness 112 HRB is very difficult to machine, the tool of nuance P30 to reach acceptable wear very quickly (T = 1 min for *Vc* = 100 m/min, and T = 4 min for *Vc* = 36 m/min).
- The surface quality obtained in the case of the machining of hardened steel XC38 is better compared to that obtained during the machining of steel XC38 not hardened.

The modes of wear observed are those of abrasion and diffusion which appears on surface in skin and the surface of attack.
The analysis of the curves of follow-up of wear according to time shows that the pace of the latter obeys the universal law of the wear of any machine element.
The influence of the advance and that the depths of cut on the behaviour are not very significant compared to the cutting speed.
The mathematical model suggested expresses the qualitative and quantitative relation between the wear and the elements of the mode of cut. It is necessary for optimization and the industrial exploitation.
Following the experiments carried out on the soaked test-tubes, we noted and checks the same observations quoted in the references [19], [3], and [20] that the surface quality is better and does not ask for an operation of correction.
This comparative study made it possible to establish the variations of performance of the tool out of metal carbide of nuance P30 during the machining of two series of hardened steel XC38 test-tubes and not soaked with the same modes of cut.

IOP Conf. Series: Materials Science and Engineering **28** (2012) 012030 doi:10.1088/1757-899X/28/1/012030

Notation

L	Length	(mm)
Ø	Diameter	(mm)
% C	Percentage of carbon	
% Mn	Percentage of manganese	
% Si	Percentage in silicon	
Rm	Maximum resistance to traction	(MPa)
Re	Normal load with the elastic limit	(MPa)
$Rp_{,0.02}$	Conventional load with the elastic limit	(MPa)
Amin	Elongation at the fracture of the sample	(%)
α	Clearance angle	(°)
γ	Angle of attack	(°)
λ	Angle of inclination of edge	(°)
$R\varepsilon$	Radius of nozzle	(mm)
WC	Carbon of tungsten	
TiC	Carbon of titanium	
Co	Cobalt	
Vb	Cutting tool wear	(mm)
K_T	Wear in crater	(mm)
T	Behaviour of the tool (lifespan)	(min)
C1	Constant depends on the nuance of the tool and machined material	
n	Constant	
Vc	Cutting speed	($m.min^{-1}$)
C_2	Constant depends on the nuance of the tool and machined material	
P	Depth of cut	(mm)
a	Advance	($mm.tr^{-1}$)
tc	Time of cut	(min)
x, y	Exhibitors depending on the nuance of the tool	
P30	Nuance of carbide of the tool	
XC38	Steel XC38	

References

[1] Ben Salem S, Hadj Sassi B and Moisan M 2003 Performance evaluation of the cutting tools in UGV by the method of the experimental designs. 16th French congress of Nice Mechanics, September 1-5.

[2] Shultz H and Moriwahim T 1992 High speed machining Annals of the CIRP **41**.

[3] Schneider G. Jr 2002 Cutting tool applications Chapter 1 prentice-Hall publication 2-65.

[4] Brandt G, Gerendas A and Mikus M 1990 *Journal of the European Ceramic Society* **6** issue 5 273-290.

[5] Fallböhmer P, Rodriguez C. A, Özel T and Altan T 2000 *Journal of Master Process Tech.* **98** 104-115.

[6] Chou Y. K 2003. Hard turning of M50 steel with different microstructures in continuous and intermittent cutting. Wear **225** 1388-1394.

[7] Bauer C. E, Inspektor A and Oles E. J 2003 A comparative machining study of diamond-coated tools made by plasma torch. Microwave and Hot Filament Techniques **28**, Part 5: 933-944.

[8] Luo S.Y, Lias Y.S and Tsai Y. Y 1999 *Journal of Mater Process Technology* **88** 114-121.

[9] Boulanouar L, Andonov I 1996 Statistical modeling of the workability of the tool steel Z40 CrMo 6-2 The metallurgy-CIT review.

[10] Remadna M 2001. Behavior of the system machining in hard turning, Application to the case of tempered steel machined with plates CBN. Thesis, INSA-Lyon.

[11] List G, Sutteran G and Bi X. F 2009 *Science Direct Wear* **267** 1673-1679.

[12] Kaufeld M and Torbaty S 1999 Rationalization of machining very high speed. SOFETEC.

[13] Amri B 1987 Contribution to the study of the behaviour of modern materials for cutting tools. Thesis, INSA-Lyon.
[14] Roumesy B 1975. Determination of the laws of wear of the tools, Machine tools **322** 319-323.
[15] Boulanouar L, Yallese M, Chaoui K and S. Dominiak S 2006. Comparative study on the behavior with the wear of the ceramic tools and CBN during turning lasts of a bearing steel. 5°day of mechanics EMP (ALGERIE).
[16] Poulachon G, Moison A, Jawahir L. S 2002 Behaviour, properties and workability of chilled steels, hard turning of the bearing steel 100 Cr6. Materials and Techniques.
[17] Defretin, A., and G. Levaillant, 1999 Machining at high speed. Techniques of the engineer. Doc. BM 7180.
[18] Schultz H, 1997 State of the art of high speed machining. First French and German Conference of high speed machining 1-7.
[19] Poulachon G and Moison A 2003. Performance evaluation on hardened steel-PCBN tool pair in high speed turning. Materials and Techniques 1-2: 23-24
[20] Benchiheub S and Boulanouar L 2006. Study of the wear of tools CBN and ceramics during the machining of steel 100 cr6 soaked 5th Day of Mechanics, EMP (ALGERIE).
[21] Poulachon G 1999. Performance evaluation on hardened steel-PCBN tool pair in high speed turning Matériaux et Techniques 1-2: 23-24.
[22] Menegaux F 2004 Method of Taylor: antiquity or topicality. Annals of surgery, flight 129, Issue 10.
[23] Lelong Ferrand J and Arnaudiès J. M 1977 Run of mathematics T2: Analyse, Bordas.
[24] Roumesy B 1975 Determination of the laws of wear of the cutting tools. Machine Tools 301: 34-36.
[25] Roumesy B and Bedrin C 1981 Testing methods fast for the determination of the behaviour of cut of the tools Materials and techniques 6 277-284.
[26] Roumesy B and Bedrin C 1984 Computing program for obtaining optimal cutting conditions in multi tool optimization Annals of the CIRP 33 549-552.

IOP Conf. Series: Materials Science and Engineering **28** (2012) 012031 doi:10.1088/1757-899X/28/1/012031

First-principales calculations on the origins of the gap bowing in $Pb_{1-x}Ca_xS$, $Pb_{1-x}Ca_xSe$ and $Pb_{1-x}Ca_xTe$ alloys

C. Sifi[a1], M. Slimani[a], H. Meradji[b]
[a]Laboratoire LESIMS, Département de Physique, Faculté des Sciences, Université de Annaba, Algeria
[b]Laboratoire de Physique des Rayonnements, Département de Physique, Faculté des Sciences, Université de Annaba, Algeria

E-mail: sifichahra@yahoo.fr

Abstract. The ab initio full potential linearized augmented plane wave (FP-LAPW) method within density functional theory was applied to study the effect of composition on the structural and electronic properties of $Pb_{1-x}Ca_xS$, $Pb_{1-x}Ca_xSe$ and $Pb_{1-x}Ca_xTe$ temary alloys. The effect of composition on lattice parameter, bulk modulus, band gap was investigate. Deviations of the lattice constant from Vegard's law and the bulk modulus from linear concentration dependence were observed for all three alloys. In addition, the microscopic origins of compositional disorder were explained by using the approach of Zunger and co-workers. The disorder parameter (gap bowing) was found to be strong for the three alloys and was mainly caused by the chemical charge transfer effect. The volume deformation contributions for all alloys were also found to be significant, while the structural relaxation contributions to the gap bowing parameter were relatively smaller for all alloys.

1. Introduction

Lead chalcogenide compounds semiconductors are expected to be the tunable laser diodes which laser at mid-infrared region between 3 and 4 μm. In this study band gap IV-VI compounds are promising for wave-length optoelectronic applications in laser diodes and in light emitting diodes. They are considered to be mainly utilized in advanced measurement systems for detecting hydrocarbon pollutants in the air [1] and in a new fiber communication system over super-long distances; which has not yet been developed [2]. In order to use this laser diode, it is required to operate it close to room temperature. So far, many efforts have been made to fabricate laser diodes [3-6].

The compositional variation in $Pb_{1-x}Ca_xS$, $Pb_{1-x}Ca_xSe$ and $Pb_{1-x}Ca_xTe$ alloys induces significant changes in their physical properties such as electronic band structures and lattice parameters. The possible development of heterostructures based on these new material systems needs a detailed investigation of these alloys. Although the structural and electronic properties of $Pb_{1-x}Ca_xS$ and $Pb_{1-x}Ca_xSe$ alloys have been studied experimentally [7,8], to the best of our knowledge no theoretical investigations of $Pb_{1-x}Ca_xS$, $Pb_{1-x}Ca_xSe$ and $Pb_{1-x}Ca_xTe$ alloys have appeared in the literature. Hence, in order to exploit fully these materials for new optical devices, the structural and electronic properties of these alloys need to be investigated in more detail. Motivated by the above considerations, we have

[1] To whom any correspondence should be addressed.

carried out a study of the alloys of interest using the full potential linearized augmented plane wave (FP-LAPW) method. The alloys studied crystallize in the cubic phase over the whole range of composition x ($0 \leq x \leq 1$). The concentration of band gap crossover has also been considered. The optoelectronic properties of the semiconductors alloys are essential in the design and fabrication of devices, the refractive indices and the optical dielectric constants of the materials have been to be known as a function of composition.

After a brief description of the computation al details in Section 2, we present the theoretical results in Section 3. Conclusions are summarized in Section 4.

2. Computational methods

We employed the FP-LAPW method to solve the Kohn Sham equations as implemented in the wien2k code [9]. The exchange-correlation contribution was described within the generalized gradient approximation (GGA) proposed by Perdew et al. [10] to calculate the total energy, while for the electronic properties, in addition to the GGA correction Engel- Vosko's (EVGGA) formalism [11] was also applied. The core states of Pb, Ca, S, Se and Te atoms were treated self-consistently and fully relativistically relaxed in a spherical approximation, while the valence states were treated self-consistently within the semi-relativistic approximation (no spin-orbit effects included). Wave functions, charge density and potential were expanded inside muffin-tin spheres of radius RMT by using spherical harmonies expansions, while in the remaining space of the unit cell a plane wave basis set was chosen. The plane wave cutoff of Kmax = $8.0/R_{MT}$ (R_{MT} is the smallest muffin-tin radius) was used for the expansion of the wave function in the interstitial region for all three alloys and the six binary compounds CaS, CaSe, CaTe, PbS, PbSe and PbTe. Values of R_{MT} were assumed to be 1.9, 2.0, 2.1, 2.2 and 2.4 a.u. for Ca, S, Se, Pb and Te atoms, respectively. The charge density was Fourier expanded up to Gmax = 14 $(Ry)^{1-2}$. The maximum l value for the wave function expansions inside the spheres was confined to l_{max}=10. Meshes of 47 k-points for binary compounds and 125 k-points for alloys were used in the irreducible wedge of the Brillouin zone. Both the plane wave cutoff and the number of k-points were varied to ensure total energy convergence.

3. Results and discussion

3.1 Structural properties

In the present work we analyze the structural properties of the binary compounds PbS, PbSe, PbTe, CaS, CaSe, CaTe and their alloys. A rocksalt structure was assumed. The alloys were modeled at some selected compositions with ordered structures described in terms of periodically repeated supercells with eight atoms per unit cell, for the compositions x = 0.25, x = 0.5, x = 0.75. For the structures considered, the calculated total energies at many different volumes around equilibrium were fitted by the Murna-ghan equation of state [15] in order to obtain the equilibrium lattice constant and the bulk modulus for the binary compounds and their alloys. Our results for the materials of interest are compared with the available experimental and theoretical predictions in Table 1. Considering the general trend that the GGA usually overestimates the lattice parameters, our GGA results for the binary compounds are in reasonable agreement with experimental values and other calculated values. Usually, in the treatment of alloy problems, it is assumed that the atoms are located at ideal lattice sites and the lattice constants of alloys should vary linearly with composition x according to Vegard's law [16], however, violations of Vegard's rule have been reported in semiconductor alloys both experimentally [17] and theoretically [18].

The results obtained for the composition dependence of the calculated equilibrium lattice parameters for $Pb_{1-x}Ca_xS$, $Pb_{1-x}Ca_xSe$ and $Pb_{1-x}Ca_xTe$ alloys, respectively are shown in Fig. 1. (a)-(c). The calculated lattice parameters at different compositions of $Pb_{1-x}Ca_xS$, $Pb_{1-x}Ca_xSe$ and $Pb_{1-x}Ca_xTe$ alloys exhibit tendency to Vegard's law with a marginal upward bowing parameters equal to -0.016, -0.014 and -0.010 Å, .respectively, obtained by fitting the calculated values with a polynomial function.

This small deviation is due to the relaxation of Pb-S and Ca-S bond lengths in $Pb_{1-x}Ca_xS$, Pb-Se andCa-Se bond lengths in $Pb_{1-x}Ca_xSe$ and Pb-Te and Ca-Te bond lengths in $Pb_{1-x}Ca_xTe$.

Table 1. Calculated lattice parameter (a) and bulk modulus (B) of for $Pb_{1-x}Ca_xS$, $Pb_{1-x}Ca_xSe$ and $Pb_{1-x}Ca_xTe$ ternary alloys at equilibrium volume.

	x	Lattice constants a (Å)			Bulk modulus B (GPa)		
		This work	Experiment	Other calculations	This work	Experiment	Other calculations
$Pb_{1-x}Ca_xS$	1	5.722	5.69 [16]	5.717[17]	57.10	64 [16]	57.42 [17]
	0.75	5.807			55.84		
	0.5	5.880			54.33		
	0.25	5.946			52.82		
	0	6.007	5.94[19]	5.936[18]	51.31		
$Pb_{1-x}Ca_xTe$	1	6.396	6.345[16]	6.390 [17]	37.29	41.80 [16]	39.60 [17]
	0.75	6.446			37.47		
	0.5	6.492			37.66		
	0.25	6.532			37.85		
	0	6.567	6.462[19]	6.34 [18]	38.04		
$Pb_{1-x}Ca_xSe$	1	5.964	5.910[16]	5.968[17]	47.52	51[16]	48.75 [17]
	0.75	6.034			48.05		
	0.5	6.105			48.59		
	0.25	6.166			49.12		
	0	6.217	6.123[19]	6.124[18]	49.66		

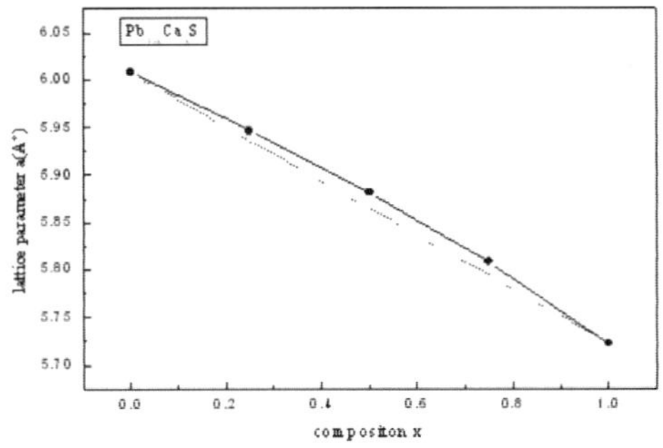

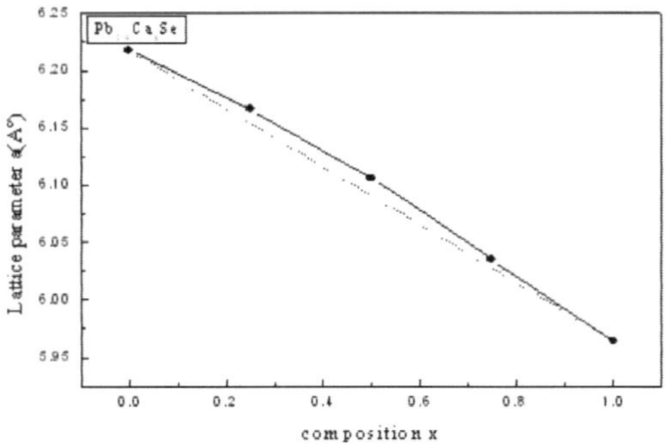

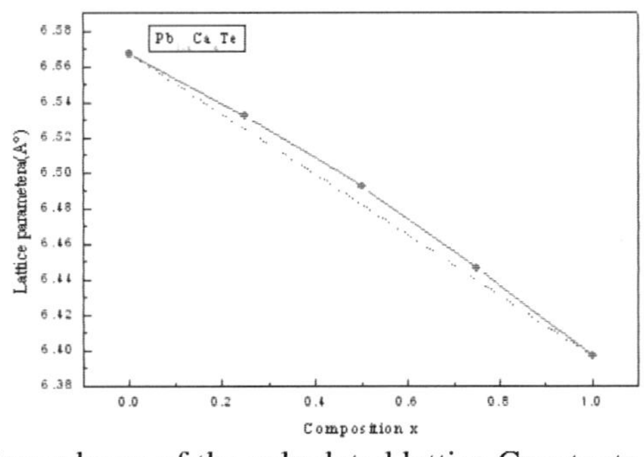

Figure 1. Composition dependence of the calculated lattice Constants (solid squares) of (a) $Pb_{1-x}Ca_xS$, (b) $Pb_{1-x}Ca_xSe$, (c) $Pb_{1-x}Ca_xTe$ with Vegard's prediction (dashed line).

In Fig. 2(a)-(c) the composition dependence of the bulk modulus for $Pb_{1\_x}Ca_xS$, $Pb_{1\_x}Ca_xSe$ and $Pb_{1\_x}Ca_xTe$ alloys is compared with the results predicted by linear concentration dependence (LCD). A small deviation from LCD is observed, with downward bowing equal to -0.54 GPa for $Pb_{1\_x}Ca_xS$ alloy and upward bowing equal to 0.01 GPa for $Pb_{1\_x}Ca_xSe$ $Pb_{1\_x}Ca_xSe$ alloy, while in the case of $Pb_{1\_x}Ca_xTe$ our calculations exhibit a good agreement with LCD with a bowing parameter close to zero. This deviation is mainly due to the mismatch of the bulk modulus of binary compounds which is in our case very small. It is clearly seen that the bulk modulus decreases by increasing the atomic number of the chalcogen atom. Hence, we conclude that PbTe and CaTe are more compressible compared the other calcium and lead chalcogenide compounds, respectively.

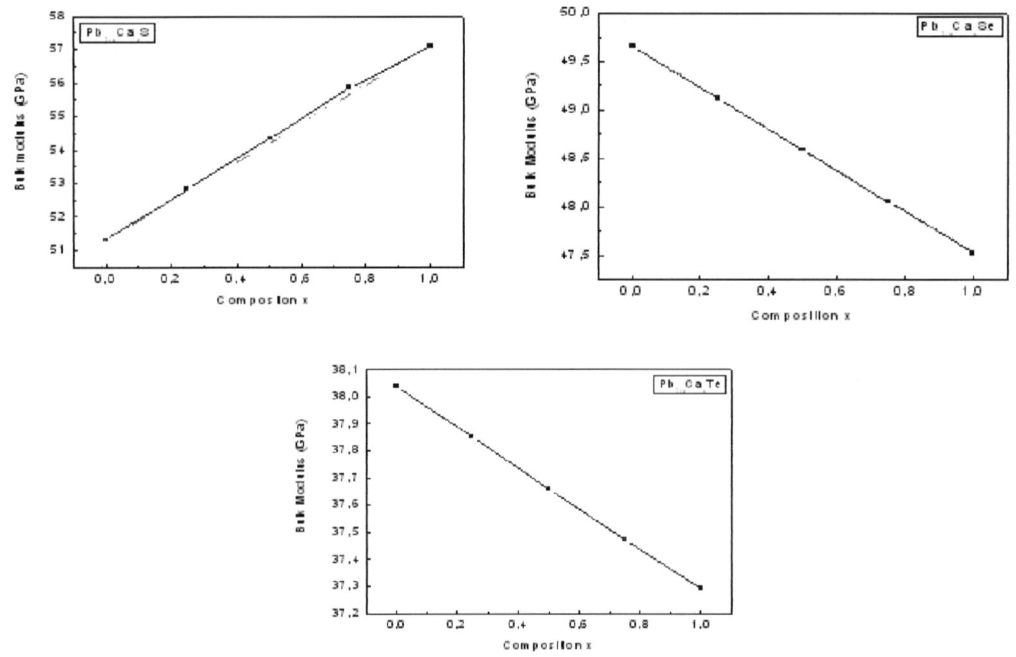

Figure 2. Compositoin dependanceof the calculated bulk modulus (solid squares) of a) $Pb_{1-x}Ca_xS$, (b) $Pb_{1-x}Ca_xSe$, (c) $Pb_{1-x}Ca_xTe$ alloys compared with LCD prediction (dashed line).

3.2. Electronic properties

The self consistent scalar relativistic indirect band gaps of lead calgonate compounds and their alloys were calculated within the GGA and EVGGA schemes. The results for each compound are given in Table 2. The values obtained for the band gap of binary compounds within EVGGA are in better agreement with available experimental results in comparison with the values calculated by GGA. It is well known that the GGA usually underestimates the experimental energy band gap [19-22]. This is mainly due to the fact that the functionals within this approximation have simple forms that are not sufficiently flexible to accurately reproduce both exchange-correlation energy and its charge derivative. Engel and Vosko by considering this underestimation constructed a new functional form of GGA which was able to better reproduce the exchange potential at the expense of less agreement as regards exchange energy when compared to experiment. This approach which is called EVGGA yields a better band splitting and some other properties which mainly depend on the accuracy of exchange-correlation potential. However, in this method, the quantities that depend on an accurate description of exchange energy E_x such as equilibrium volumes and bulk modulus are in poor agreement with experiment.

Table 2. Gap energy E_g of $Pb_{1-x}Ca_xS$, $Pb_{1-x}Ca_xSe$ and $Pb_{1-x}Ca_xTe$ ternary alloys at equilibrium volume.

	x	Band gap E_g	
		This work	
		GGA	EVGGA
$Pb_{1-x}Ca_xS$	1	2.405	3.176
	0.75	1.195	1.883
	0.5	0.764	1.441
	0.25	0.458	1.216
	0	0.486	1.221
$Pb_{1-x}Ca_xSe$	1	2.105	2.815
	0.75	1.162	1.598
	0.5	0.739	1.312
	0.25	0.440	1.092
	0	0.423	1.079
$Pb_{1-x}Ca_xTe$	1	1.582	2.228
	0.75	1.310	1.575
	0.5	1.073	1.484
	0.25	0.871	1.389
	0	0.819	1.331

The band gap E_{gABC} of an alloy such as $A_xB_{1-x}C_x$ is not given by the linear concentration x weighted average of the AC(E_{gAC}) and BC (E_{gBC}) gaps but has a quadratic form:

$$E_{gABC}= xE_{gAC}+(1-x)E_{gBC}-bx(1-x)$$

Where b is known as the bowing parameter. The need to fine tune the band gap of alloys in various device applications has provoked an interest in computing the optical band gap bowing in terms of the constituents elements AC and BC.

In order to better understand the physical origins of the gap bowing in Lead calgonate alloys, we follow the procedure of Bernard and Zunger [18] in which the bowing parameter (b) is decomposed into three physically distinct contributions. By considering the fact that the bowing dependence to the composition is marginal, the authors limited their calculations to x=0.5 (50%-50% alloy). The overall gap bowing coefficient at x = 0.5 measures the change in band gap according to the reaction:

$$AB(a_{AB})+AC(a_{AC})\rightarrow AB_{0.5}C_{0.5}(a_{eq}) \quad (1)$$

where a_{AB} and a_{Ac} are the equilibrium lattice constants of the binary compounds AB and AC, respectively, and a_{eq} is the alloy equilibrium lattice constant. We now decompose reaction (1) into three steps:

$$AB(a_{AB})+AC(a_{AC})\xrightarrow{VD} AB(a)+AC(a), \quad (2)$$

$$AB(a)+AC(a) \xrightarrow{CE} AB_{0.5}C_{0.5}(a), \quad (3)$$

$$AB_{0.5}C_{0.5}(a)\xrightarrow{SR} AB_{0.5}C_{0.5}(a_{eq}). \quad (4)$$

The first step measures the volume deformation (VD) effect on the bowing. The corresponding contribution to the total gap bowing parameter b_{VD} represents the relative response of the band structure of the binary compounds AB and AC to hydrostatic pressure, which here arises from the change of their individual equilibrium lattice constants to the alloy value a=a(x) (from Vegard's rule).

The second contribution, the charge-exchange (CE) contribution b_{CE}, reflects a charge transfer effect which is due to the different (averaged) bonding behavior at the lattice constant a. The final step measures changes upon passing from the unrelaxed to the relaxed alloy by b_{SR}. Consequently, the total gap bowing parameter is defined as:

$$b = b_{VD} + b_{CE} + b_{SR}, \quad (5)$$
$$b_{vD} = 2[\varepsilon_{AB}(a_{AB}) - \varepsilon_{AB}(a) + \varepsilon_{Ac}(a_{Ac}) - \varepsilon_{Ac}(a)], \quad (6)$$
$$b_{CE} = 2[\varepsilon_{AB}(a) + \varepsilon_{AC}(a) - 2\varepsilon_{ABC}(a)], \quad (7)$$
$$b_{SR} = 4[\varepsilon_{ABC}(a) - \varepsilon_{ABC}(a_{eq})], \quad (8)$$

Table 3. Decomposition of the optical bowing into volume deformation (VD), charge exchange (CE), and structural relaxation (SR) contributions compared with that obtained by a quadratic fit (all values are in eV).

		This work (Zunger)		Quadratic equation	
		GGA	EVGGA	GGA	EVGGA
$Pb_{1-x}Ca_xS$	b_{VD}	0.240	0.286		
	b_{CE}	2.577	2.851		
	b_{SR}	-0.093	-0.110		
	b	2.724	3.027	2.972	3.214
$Pb_{1-x}Ca_xSe$	b_{VD}	0.141	0.194		
	b_{CE}	2.046	2.427		
	b_{SR}	-0.089	-0.082		
	b	2.098	2.539	2.256	2.827
$Pb_{1-x}Ca_xTe$	b_{VD}	0.057	0.070		
	b_{CE}	0.535	1.205		
	b_{SR}	-0.084	-0.094		
	b	0.508	1.181	0.541	1.355

where ε is the energy gap that has been calculated for the indicated atomic structures and lattice constants. All terms in Eqs. (6)-(8) are calculated separately via self-consistent band structure calculations within density functional theory and the results are given in Table 3. The total gap bowing for all the three alloys were found to be mainly caused by the charge transfer contribution b_{CE}. It is due to the large electro-negativity difference between atoms. Indeed, the significant role of b_{CE} is correlated with the ionicity factor difference among constitute binary compounds PbS (fi=0.115), CaS (fi=0.430), PbTe (fi=0.022), CaTe (fi=0.260), PbSe (fi=0.086) and CaSe (fi=0.380). These values are calculated using the Pauling's scale[xx]. The volume –deformation term b_{VD} contributes to the bowing parameter at smaller magnitude. The weak contributions of b_{VD} can be correlated to the small mismatch of the lattice constants of the corresponding binary compounds. The contribution of the structural relaxation b_{SR} is small in the three alloys. Finally, it is clearly seen that our EVGGA values for bowing parameters are larger than the corresponding values within GGA.

Alternatively, we calculated the total bowing parameter by fitting the non-linear variation of the calculated band gaps versus concentration with quadratic functions. The results are shown in Fig. 3(a)-(c) and obey the following variations:

$$Pb_{1-x}Ca_xS \Rightarrow \begin{cases} E_g^{GGA} = 0.518 - 1.142x + 2.972x^2 \\ E_g^{EVGGA} = 1.274 - 1.384x + 3.215x^2 \end{cases} \quad (9)$$

$$Pb_{1-x}Ca_xSe \Rightarrow \begin{cases} E_g^{GGA} = 0.439 - 0.624x + 2.258x^2 \\ E_g^{EVGGA} = 1.137 - 1.236x + 2.826x^2 \end{cases} \quad (10)$$

$$Pb_{1-x}Ca_xTe \Rightarrow \begin{cases} E_g^{GGA} = 0.806 + 0.243x + 0.542x^2 \\ E_g^{EVGGA} = 1.375 - 0.563x + 1.355x^2 \end{cases} \quad (11)$$

We note that the calculated quadratic parameters (gap bowing) within GGA and EVGGA are close to their corresponding results obtained by Zunger approach.

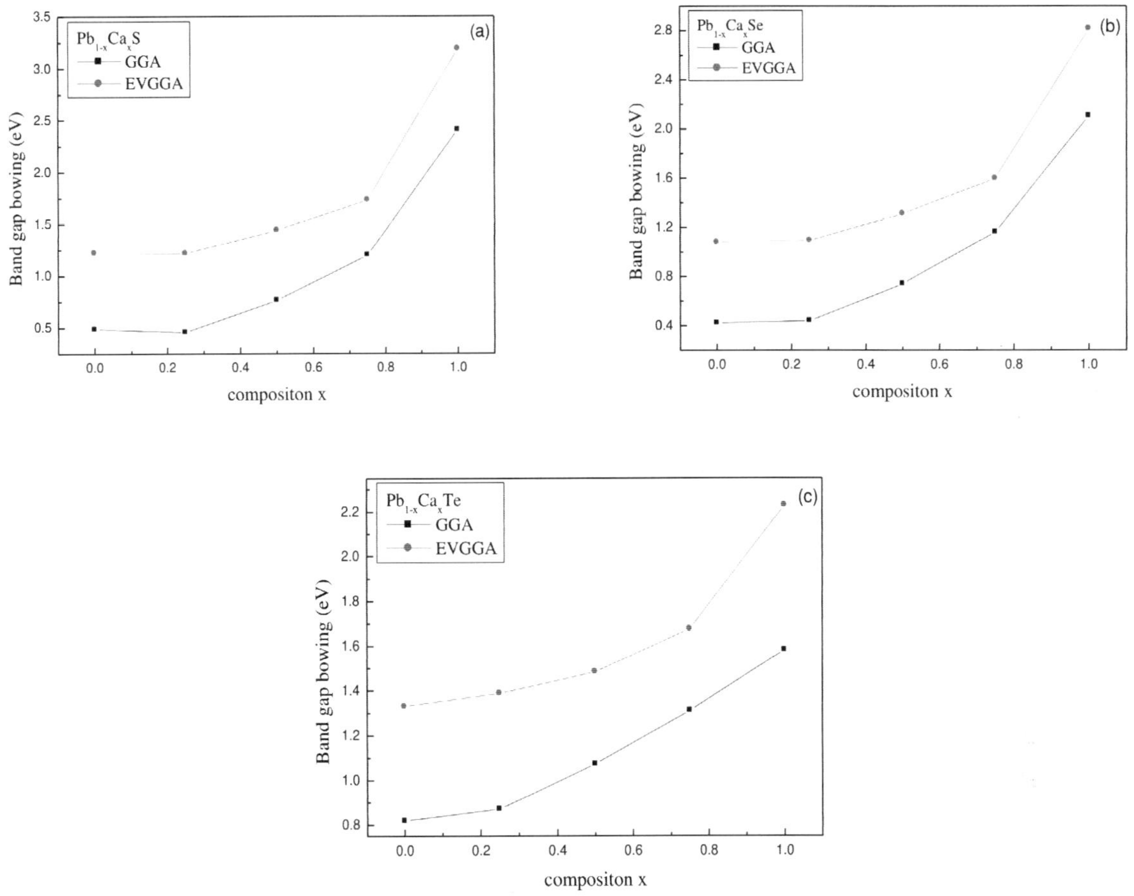

Figure 3. Composition dependence of calculated band gap using GGA (solid squares) and EVGGA (solid circles) for (a) $Pb_{1-x}Ca_xS$, (b) $Pb_{1-x}Ca_xSe$, (c) $Pb_{1-x}Ca_xTe$ alloys.

4. Conclusion

In summary, we have applied the FP-LAPW method in order to study the structural, electronic properties, We have investigated the composition dependence of the lattice constant, bulk modulus, band gap, The calculated lattice parameters for the three alloys exhibit a tendency to Vegard's law with a marginal bowing parameter. A small deviation of bulk modulus from LCD has been observed for all the three alloys. The physical origin of this effect should be mainly due to the weak mismatch of the bulk modulus of the binary compounds. The band structure was calculated by GGA. In addition, the EVGGA scheme has been used to obtain more accuracy. The bowing is found to be mainly caused by the charge-transfer effects, while the volume deformation and the structural relaxation contribute to the bowing parameter with a smaller magnitude.

References

[1] Masumoto K, Koguchi N, Takahashi S, Kiyosawa T and Nakatani I 1985 *Sci. Rept. National Res. Inst. Metals* **6** 101.

[2] Shibata S, Horiguchi M, Jinguji K, Mitachi S, Kanamori T and Manabe T 1981 *Electron Lett.* **17** 775.
[3] Koguchi N, Kiyosawa T and Takahashi S 1987 *J. Crystal Growth* **81** 400.
[4] Ishida A, Muramatsu K, Takashiba H and Eujiyasu H 1989 *Appl. Phys. Lett.* **55** 430.
[5] Koguchi N and Takahashi S 1991 *Appl. Phys. Lett.* **58** 799.
[6] Mochizuki M, Iwata H, Isshiki M and Masumoto K 1991 *J. Crystal Growth* **115** 687.
[7] Jain C, Willis J. R and Bullogh R 1990 *Adv. Phys.* **39** 127.
[8] Albanesi E. A, Peltzer E. L, Blanca Y and Petukhov A. G 2005 *Computational Materials Science,* **32** 85 95.
[9] Cohen M. L, Chelikowsky J. R 1989 *electronic structure and optical properties of semi conductors, second ed. Springer Series I Solid State Sciences* **75** Springer-Verlag, Berlin.
[10] Hinkel V, Hoak H, Mariana C, Sorba L, Horn K, Christensen N. E 1989 *Phys. Rev. B* **40** 5549.
[11] Zunger A, Wei S. H, Ferreira L. G and Bernard J. E 1990 *Phys. Rev. Lett.* **65** 353.
[12] Blaha P, Schwarz K, Madsen G. K. H, Kvasnicka D and Luitz J 2001 WIEN2k, an Augmented Plane Wave Plus Local Orbitals Program for Calculating Crystal Properties (Vienna University of Technology, Vienna, Austria.
[13] Perdew J. P, Burke S and Ernzerhof M 1996 *Phys.Rev.Lett.* **77** 3865.
[14] Engel E and Vosko S. H 1993 *Phys. Rev. B* **47** 13164.
[15] Murnaghan F. D 1944 *Proc. Natl.Acad. Sci. USA* **30** 5390.
[16] Luo H, Greene R.G, Handechari K. G, Li T and Ruoff A. L 1994 *Phys. Rev. B* **50** 16232.
[17] Charifi Z, Baaziz H, El Haj Hassan F and Bouarissa N 2005 *J. Phys. Condens. Matter* **17** 4083.
[18] Mohamed Lach-had, Dimitrios A. Papaconstantopoulos, Michael J. Mehl *2002 Journal of Physics and Chemistry of Solids* **63** 833.
[19] Madelung O, Schulz M, Weiss H (Eds) 1983 Numerical Data and Functional Relationships in science and technology Landolt-Bornstei, *New Series* **17**, Springer, Berlin.
[20] Zaoui A and El Haj Hassan F 2001 *J. Phys: Condens. Matter* **13** 253.
[21] Vegard L 1921 *Z. Phys.***5** 17.
[22] Jobst J, Hommel D, Lunz U, Gerhard T and Landwehr G 1996 *Appl. Phys. Lett.* **69** 97.
[23] El Haj Hassan F 2005 *Phys. Stat. Sol. B* **242** 909.
[24] Dufek P, Blaha P and Schwarz K 1994 *Phys. Rev. B* **50** 7279.
[25] Bachelet G. B and Christensen N. E 1995 *Phy. Rev B* **31** 879.
[26] Bernard J. E and Zunger A 1986 *Phys. Rev. Lett.* **34** 5992.
[27] Pauling L 1932 *Journal of the American Chemical Society* **54** 3570.

IOP Conf. Series: Materials Science and Engineering **28** (2012) 012032 doi:10.1088/1757-899X/28/1/012032

Grafting of 4-aminomethylbenzensulfonamide-lipoic acid conjugate on gold nanoparticles

M. Stiti[a1], H. Bouzit[a], M. Abdaoui[a] and J. Y. Winum[b]
[a]Laboratoire de chimie appliquée; Université 08 Mai 1945, B.P.401- Guelma.
[b]Institut des Biomolécules Max Mousseron (IBMM) UMR 5247 CNRS-UM1-UM2, Ecole Nationale Supérieur e de Chimie, 34296 Montpellier Cedex, France.

E-mail: stiti_maamar@yahoo.fr; bouzit.habiba@Gmail.com; abdaoui_ med@yahoo.fr

Abstract. In this paper, we describe the synthesis of goldnanoparticles bearing aminomethylbenzensulfonamide via a lipoyl moiety.The resulting stable nanoparticles with an average size of 4.0 nm have been achieved by a facile and high-yielding one phase method, by the action of 4-aminomethylbenzensulfonamide-lipoic acid bioconjugate on chloroauric acide, using dimethylsulfoxide (DMSO) as the solvent and sodium tetrahydridoborate ($NaBH_4$) as the reducing agent. UV-vis absorption, transmission electron microscopy (TEM) and X-ray diffraction were used to analyse the morphology and the structure of the obtained nanoparticles. Preliminary study shows that these new nanoparticles are endowed with highly and specific inhibitory activity for the isoform (IX) of carbonic anhydrase over expressed in many cancers ,and are therefore attractive candidate to be used both in diagnosis and in treatment of tumours.

1. Introduction

Carbonic anhydrases (CAs) are zinc enzymes involved in a variety of physiological and physiopathological processes.Up to 16 CA isozymes were found in different tissues in the human body to deliver protons and hydrogen carbonate for physiological reactions [1-3]. Many of these Carbonic anhydrases isozymes are important targets for the design of inhibitors with clinical applications [4-6]. Among a variety of CA inhibitor classes including many non-sulfonamide.

Organic inhibitors, as well as anionic, inorganic inhibitors, a large number of aromatic, heterocyclic, and more recently aliphatic sulfonamides have been synthesized and tested for their CAs inhibitory activities; they are used in the treatment and prevention of glaucoma, gastroduodenal ulcers, certain neurological disorders, altitude sickness, obesity and tumors[7]. However, many sulfonamides possess high affinity for all major isozymes and interfere with multiple isozymes in different tissues, leading to severe side effects due to the widespread distribution of these isozymes (e.g. such in eye, lungs, kidney, liver, stomach,pancreas, brain, and erythrocytes) [8, 9].

Although some progress was done on the development of the specific inhibitors [10], a constant need exists for the design of novel CA inhibitors belonging to different classes of compounds. In recent years, nanomaterials have received considerable attention because of their potential application in a wide spectrum of areas that include biology and medicine [11-14]. Examples of applications, include DNA detection by gold nanoparticles, cell imaging using fluorescent semiconductor nanocrystals, rational assembly of quantum dots using antigen-antibody interactions, and magnetic nanoparticles as contrast enhancement agents in magnetic resonance imaging (MRI).Gold

[1] To whom any correspondence should be addressed.

nanoparticles have recently emerged as an attractive and excellent candidate for delivery of a variety of biomolecules such as proteins, DNA, and drugs into their targets [15].

The tunable shape and size-dependent optical properties of gold nanoparticles have been exploited in various biomedical applications. They are biocompatible, nontoxic, bind readily to a large range of biomolecules such as amino acids, proteins/enzymes, and DNA [16].

The goal of the present study was the design and synthesis of gold nanoparticles as a novel approach for specific carbonic anhydrase targeting. The first step was to choose a suitable ligand that can attach to the gold nanoparticle and would create an effective and selective inhibitor for carbonic anhydrase (IX)overexpressed in many cancers [17].

2. Experimental Section

Chemicals: All chemicals used were of highest available purity, and were supplied by Aldrich, Alfa Aesar and Fluka.

Instruments: ^{1}H NMR and ^{13}C NMR spectra were recorded in DMSO-d6 on a Brüker DRX 400 spectrometer.

A fast atom bombardment mass spectra (FAB-MS) were recorded in positive and negative ion mode on a Joel DX 300 mass spectrometer.

UV-Vis spectra were recorded in a conventional quartz cell on a (Perkin Elmer Lambda 35) spectrophotometer .The powder X-ray diffraction (XRD) patterns were obtained using a Rigaku D/max-2500 diffractometer with Cu Ka radiation.

Transmission electron microscopy (TEM) experiments were performed using Joel 1200 Ex 2 microscope operating at 100 kV.Samples were prepared for TEM analysis by placing a drop of the colloidal solutions of gold nanoparticles on to a carbon-coated copper grid, followed by the evaporation of the solvent at room temperature.

Synthesis: We report in this section the synthetic rout (Scheme1) and the experimental protocols used to obtain gold nanoparticles4.Some characteristics of ligand 3 are also described.

2.1. N-[(4-Sulfonamidophenyl)-methyl]-5-(1, 2-dithiolan-3-yl) pentanamide (3):

To a solution of 0.2 g (1 equiv) of lipoic acid in 2.5 ml of dimethylacetamide (DMA) was added 0.216 g (1equiv) of 4 – aminomethylbenzen- sulfonamide hydrochloride and 0.15 g (1equiv) solution of dimethylaminopyridine(DMAP).

The obtained mixture was treated with 0.14ml (1equiv) of triethylamine (TEA) and 0.186g (1 equiv) of EDCI (1-éthyl-3-(3'-diméthylaminopropyl)- carbodiimide) .After 6h stirring at room temperature, the organic phase is extracted with ethyl acetate, then dried with Na_2SO_4 and concentrated under reduced pressure and the resulting product was purified on silica gel column using methylene chloride/methanol (85-15, v/v)) as eluent system.

Yield: 40 %

Rf = 0,42 (CH_2Cl_2/MeOH: 9,5/0,5).

mp 176°C.

M S (ESI-/MeOH / CH_3CN): m/z (exp.) 359,17 [M-H]- ;394,99 [M+Cl]-^{1}H NMR (DMSO-*d6*, 400 MHz, δ in ppm): 10.23 (s, 1H), 7.74 (s, 4H), 7.24 (s, 2H), 3.63 (s, 1H), 3.16 (m, 3H), 2.37(m,3H), 1.86(m, 1H), 1.61(m,3H), 1.42(d,2H,J= 7.21Hz).

^{13}C NMR (DMSO-*d6*, 400 MHz , δ in ppm): 171.69,142.12,126.61, 118.47 , 56.06 ,39.71 , 39.43 , 38.07 , 36.21 , 34.10 , 28.25 , 24.67.

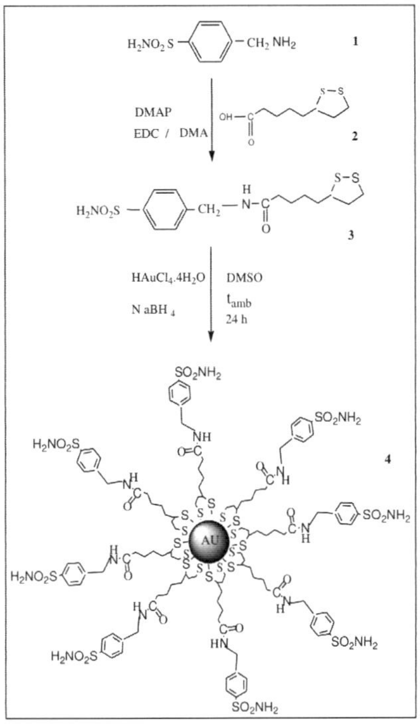

Shema 1. Chemical Synthesis of bioconjugate 3 and goldnanoparticles 4.

2.2. Goldnanoparticles (4):

An amount of HAuCl4.4H_2O (50 mg ,8.12 equiv) was dissolved in 20 ml DMSO, and was rapidly mixed with an equal volume of DMSO (20ml) containing 75.5 mg (133.5 equiv) of NaBH4 and (1 equiv) of ligand 3, The reaction mixture turned deep brown immediately.After stirring for 24h at room temperature, 40ml of CH_3CN are added to give finally a black precipitate, the latter is collected by centrifugation, washed twice with 60ml CH_3CN, DMSO (V / V = 1 / 1) and 60ml of EtOH and then dried under vacuum for 12 h at 80° C to give the desired gold nanoparticles 4.
Elemental analysis, calc (%):Au 71.45, S 6.98,for [Au_{619} $(C_{14}H_{20O3}N_2S_3)_{117}$];
Found: Au 71.68, S 7.56.

3. Results and discussion

One of the emerging goals of nanotechnology is to functionalize inert and biocompatible materials to impart precise biological functions. Several hybrid organic/inorganic nanoparticles have been described for diagnosis or therapeutic use.

In this study, we report the synthesis of gold nanoparticles grafted with arylsulfonamide moiety. It is well documented that the arylsulfonamide compounds have been shown to inhibit carbonic anhydrases even at low concentrations [18, 19]. The gold nanoparticles are in fact objects perfectly adapted not only by their non-toxicity but also by their size range: they gives a good bio-compatibility and bioavailability.Their potential applications are not only related to the intrinsic properties of the metal core but also their enormous capacities of fonctionalization[20].The sulfur-containing molecules, because of their strong affinity for gold, allow functionalization possibilities virtually unlimited gold nanoparticles. This finding already resulted in the development of a large number of functionalized nanoparticles for application areas as diverse as catalysis, optics not linear, electrochemistry and biology. In our study, the first step was to prepare ligand that can attach to the gold surface and can serve as inhibitor of carbonic anhydrases, the easiest manner for grafting arylsulfonamide ligands on the surface of goldnanoparticles requires the development of a disulfide derivatives by coupling these arylsulfonamides with lipoic acid. (Scheme1).

Synthesis of ligand 3

The obtained arylsulfonamide-lipoic acid bioconjugate of general formula 3 is obtained in good yield by conjugation of 4-aminomethyl benzenesulfonamide 1 with lipoic acid 2 in DMA in the presence of EDCI and DMAP at room temperature according to the scheme 1. It should be noted that other alternatives for this purpose are possible but seem much less attractive. The use of carbodiimides such as DCC (1,3-dicyclohexyl carbodiimide) or EDCI to form an amide function is a proven and well known in peptide synthesis; However, this process is often accompanied by side reactions that can complicate the isolation of desired products and significantly decreasing their yields. The optimization of this reaction by the development of new conditions using EDCI as dehydrating agent in the presence of DMAP as a catalyst in DMA, leads to a stable and easily isolated product 3 with satisfactory yield.

Ligand 3 involving peptide bond, is all biodegradable, identifying good candidates for inducing an intense biological activity. For this purpose, lipoic acid is entirely consistent with our strategy.

The spectroscopic characteristics of obtained compound 3 are in agreement with its structure. Indeed, this compound was observed by mass spectrometry which provides molecular peaks in perfect agreement with the masse of the proposed structure (see Experimental protocols for details).The ^{1}H NMR spectra of this compound recorded in DMSO is shown below(figure 1).

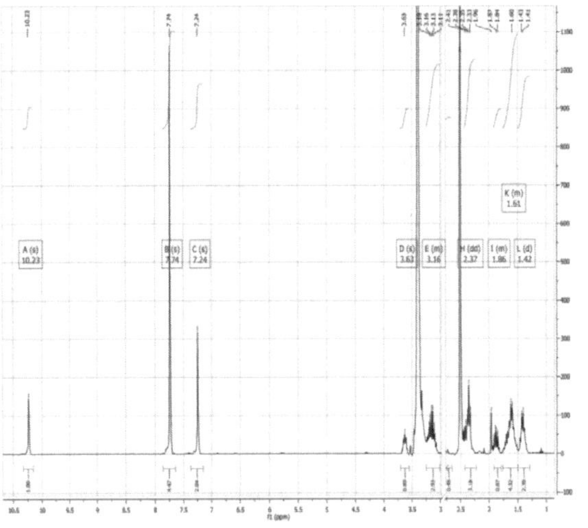

Figure 1. 1H-NMR spectra of ligand 3.

It is particularly noteworthy that the NH proton of the peptide bond is highlighted by the appearance of a signal at 10.23 ppm; the intense signal appearing at 7.24 ppm is attributed to the NH_2 group protons of the sulfonamide. The signals of aromatic protons are observed at 7.74 ppm. Also ,the ^{13}C NMR spectra of this compound recorded in DMSO-d_6(figure 2), reveals the appearance of a characteristic signal at171.69 ppm due to the carbon of amide carbonyl group(C = O), four signals between aromatic carbons and three signals: one at 56.06 ppm corresponding to carbon directly related to carbonyl group and the other two between 39.71 and 39.43 ppm, wich are poorly resolved are attributed to carbon linked to sulphur atoms in heterocyclic ring.Signals located between 38.07 and 24.67 ppm in the area of higher fields are corresponding to aliphatic carbons.

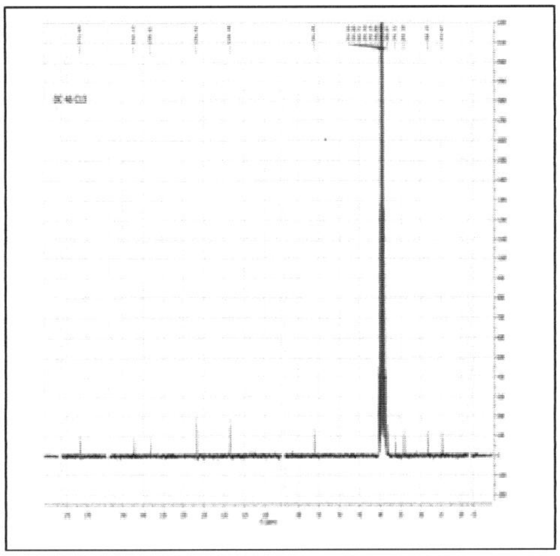

Figure 2. ^{13}C NMR spectra of ligand 3.

Synthesis of goldnanoparticles 4

The second objective of our study has been devoted to the grafting of the obtained arylsulfonamide-lipoic acid bioconjugate on gold nanoparticles.

The preparation of gold nanoparticles can be achieved by a wide range of methods whether physical or chemical which involves the reduction of a gold salt in the presence of a stabilizer.

Two methods have emerged clearly as the most interesting and most widespread, it is the NaBH4 method and the citrate method, leading to nanoparticles with different characteristics according to target applications. In our work, gold nanoparticles 4(Sheme1) were successfully prepared in a single step by a facile and high-yielding one phase method(Brust modified method) by reduction of tetrachloroaurate ($HAuCl_4$) with $NaBH_4$ in DMSO and at room temperature in the presence of the ligand 3.

The average size of the synthesized nanoparticles is much depending on the choice of preparation parameters such as the mole ratio between gold precursors and disulfide species, speed of reducing agent, addition and total reaction time.

Characterization of obtained goldnanoparticles 4

Powder UV-visible spectra:

The Powder UV-vis absorption spectrum of the resulting goldnanoparticles shows an absorption peak centred around 535 nm (Figure 3) which is assigned to the surface plasmon resonance (SPR) band of gold nanoparticles due to the collective oscillation of gold electrons in the conduction band.

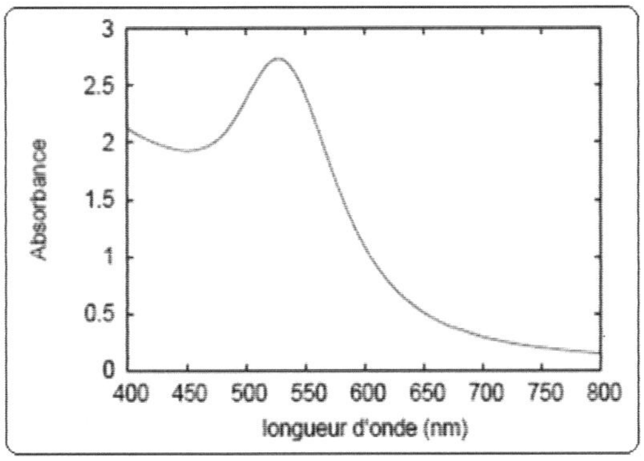

Figure 3. UV–vis spectra of obtained goldnanoparticles

IOP Conf. Series: Materials Science and Engineering **28** (2012) 012032 doi:10.1088/1757-899X/28/1/012032

TEM Characterization:

The transmission electron microscopy high resolution (HR-TEM) is used to obtain information on the morphology of obtained colloidal particles. This technique allowed us to visualize and analyze the shape and size of concerned gold nanoparticles. Figure 4 gives a typical high-resolution TEM image of obtained goldnanoparticles showing numerous discrete nanoparticles.

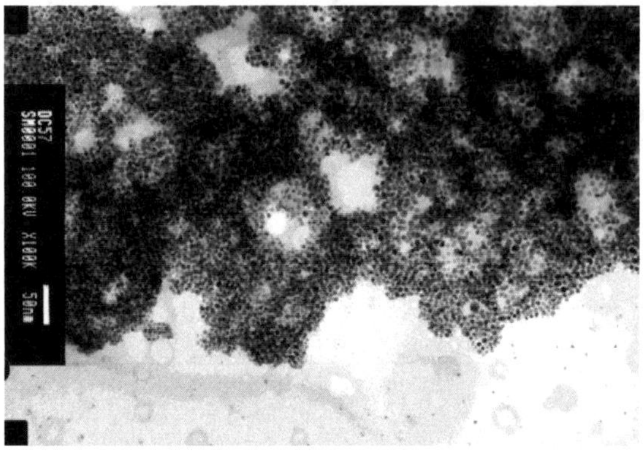

Figure 4. TEM image of synthesised goldnanoparticles 4.

And under our experimental conditions, the average diameter of a typical particle preparation was near the value 4.0 nm;Based on individual measurements on at least 200 particles(figure 5)

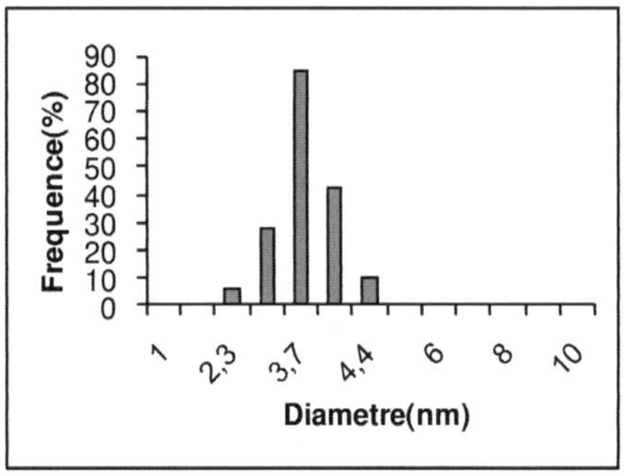

Figure 5. Size distribution histogram of goldnanoparticles 4.

Figure 6 illustrates the Selected-area electron diffraction (SAED) pattern of a sample of these nanoparticles.

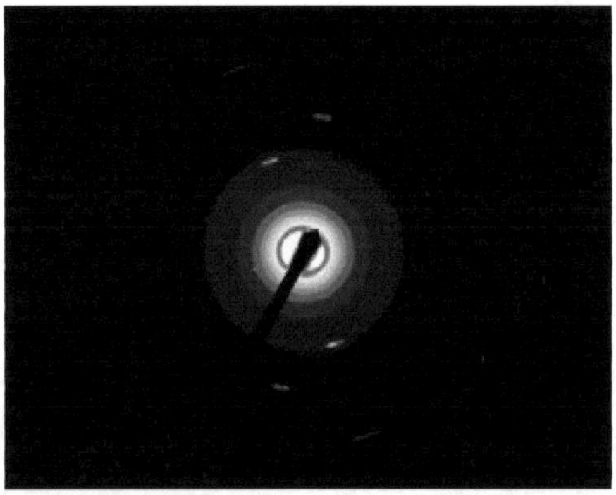

Figure 6. Selected-area electron diffraction (SAED) pattern of the nanoparticles 4.

Powder X-Ray Diffraction: The powder X-ray Diffraction (XRD) pattern of gold nanoparticles 4 in the range 2θ (20°-140°) is given in figure 7.It presents seven of the main diffractions characteristic of the gold cubic phase at 38.6°, 44.9° , 64.9°, 78.1°, 82.2°, 111.2° and 135.5°. The peak broadness may be explained by the small size of the crystalline domains. The peak of the d_{111} reflection has been deconvoluted to Lorentzian curve for the determination of the full-width at half maximum (FMWH) value. The crystalline domain has been calculated from the Debye-Scherrer formula using the FWMH value of respective index peaks, giving an average value of around 3.9 nm .which is concomitant with the value of 4.0 ± 0.8 nm obtained from TEM images.

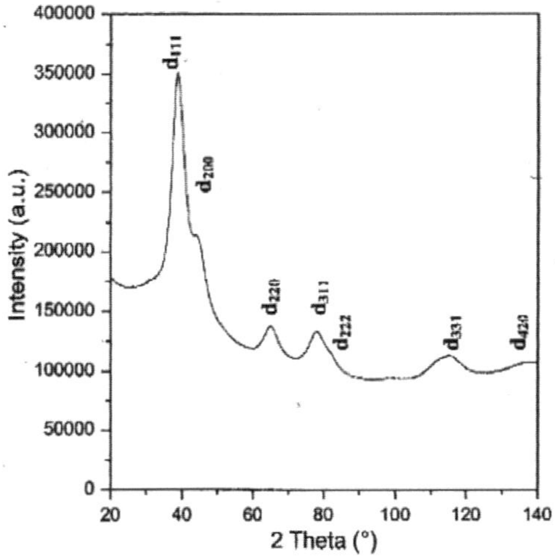

Figure 7. Powder X-ray diffraction spectrum of the synthesized nanoparticles.

4. Conclusion

In conclusion, we have reported the high yield one phase synthesis of N-[(4 Sulfonamidophenyl)-methyl]-5-(1, 2-dithiolan-3-yl) pentanamide coated gold nanoparticle.Due to their nanoscale sizes (4.0nm), these goldnanoparticles are well suited for applications in therapy, either for diagnosis or treatment by targeting specific biological systems such as carbonic anhydrase (IX), a transmembrane isoenzyme involved in several types of cancer [21].

References

[1] Ozensoy O, Puccetti L, Fasolis G, Arslan O, Scozzafava A, Supuran C. T 2005 *Bioorg. Med. Chem. Lett.* ***15*** 4862–4866.
[2] Supuran C. T, Scozzafava A, Casini A 2003 *Med. Res. Re*V. **23** 146-189.
[3] Nishimori I, Minakuchi T, Kohsaki T, Onishi S, Takeuchi H, Vullo D, Scozzafava A, Supuran C. T 2007 *Bioorg. Med. Chem. Lett.* **17** 3585-3591.
[4] Claudiu T. S and Scozzafava A 2007 *Bioorg. Med. Chem.***15** 4336–4350.
[5] Nuti E, Orlandini E, Nencetti S, Rossello A, Innocenti A, Scozzafavab A and Supuran Cl.T 2007 *Med. Chem.* **15** 2298–2311.
[6] Krishnamurthy V. M, Kaufman G. K, Urbach and Whitesides G. M 2008 *Chem. Rev.* **108** 946-1051.
[7] Rami M, Winum J. Y, Innocenti A, Montero J and Supuran C. T 2008 *Med. Chem. Lett.* **18** 836–841.
[8] Smain F. Z, Winum J. Y, Montéro J. L, Regainia Z, Vullo D, Scozzafava A and Supuran C. T 2007 *Bioorg. Med. Chem. Lett.***17** 5096–5100.
[9] Supuran C. T 2008. *Nat. Rev. Drug. Discov.* **7** 168-181.

[10] Supuran C. T, Briganti F, Tilli S, Chegwidden W. R, Scozzafava 2001 A. *Bioorg. Med. Chem.* **9** 703-714.
[11] Daniel M. C and Astruc D 2004 *Chem. Rev.* **104** 293-346.
[12] Byrne,J.D., Betancourt,T., Brannon-Peppas,L.,2008, Advanced Drug Delivery Reviews **60** 1615–1626
[13 Patil S, Reshetinkou S, Haldar M, Seal S and Mallik S 2007 *J. Phys. Chem. C* **111** 8437-8442.
[14] Patra C. R, Bhattacharya R, Wang E, Katarya A, Lau J. S, Dutta S, Muders M, Wang S, Buhrow S. A, Safgren S. L, Yaszemski M. J, Reid J. M, Ames M. M, Mukgerjee P, Mukhopadhyay D 2008 *Cancer. Res.* **68** 1970–1978.
[15] MacDiarmid J. A et al 2009 *Nature Biotech* **27** 643–651.
[16] Haley B, Frenkel E 2008 *Urol. Oncol.* **26** 57–64.
[17] Supuran C. T 2010 *Bioorg. Med. Chem. Lett.* **20** 3467-3474.
[18] Said H. M, Supuran C. T, Hageman C *et al* 2010 *Curr. Pharm. Des.* **9** 3288-3299.
[19] Supuran C. T 2010 *Curr Pharm Des* **16** 3233-3245.
[20] Pfammatter J. M, Siljegovic V, Darbre T and Keese R 2001 *Helvitica Chimica Acta* **84** 678-689.
[21] Stiti M, Cecchi A, Rami M, Abdaoui M, Barragan-Montero V, Scozzafava A, Guari Y, Winum J. Y and Claudiu T. S 2008 *J. Am. Chem. Soc* **130** (48) 16130-16131.

IOP Conf. Series: Materials Science and Engineering **28** (2012) 012033 doi:10.1088/1757-899X/28/1/012033

A Novel poly {(2,5-diyl thiophene) (2-Thiophenyl Methine)} from the Acid-Catalyzed Polycondensation of thiophene and thiophene-2-carbanaldehyde

A. Belmokhtar[1], A. Yahiaoui[1] [1], A. Hachemaoui[1], N. Sahli[2], A. Benyoucef[1], A. Belfedal[2] and M. Belbachir[3]

[1]Laboratoire de chimie Organique, Macromoléculaire et des Matériaux (LCOMM), Université de Mascara, Faculté des Sciences et de la Technologie, BP 763 Mascara 29000, Algérie.

[2]Laboratoire de Chimie des Polymères, Faculté des sciences, Université d'Oran, BP1524 El- Menouer 31000 Oran, Algérie.

[3]Laboratoire de Chimie des Polymères, Faculté des sciences, Université d'Oran, BP1524 El- Menouer 31000 Oran, Algérie.

E-mail: yahmeddz@yahoo.fr

Abstract. Conjugated poly [(thiophene methine) have been synthesized by a simple method: Condensation of thiophene-2-carbaldehyde with thiophene, catalyzed by Maghnite-H+, which is an exchanged montmorillonite sheet silicate with protons to produce Maghnite-H+ non-toxic cationic catalyst, and characterized by FT-IR and UV-Vis. The result reveals that the yield of PTTM increase with time and amount of Mag-H. The results obtained with optical measurements and analyzed by different models, show clearly that our material (PTTM) is a good semi-conductor, when the average gap around 5 eV. The effect of the amount of Mag-H+, and the time of polymer was studied.

1. Introduction

Conjugated polymers have received an increasing interest because of the wide range of their possible applications, which include organic light emitting diodes, field effect transistors, photovoltaic cells, due to their unique optical, electrical and mechanical properties [1]. In this field, polythiophenes and in general thiophene-based functional polymers have been extensively investigated; in particular the synthetic interest have been focused on novel low bandgap and regioregular materials [2, 3]. The reduction of the energy bandgap can facilitate the ease of doping with the possibility of achieving intrinsic metallic conductivity. An important prerequisite for low bandgap materials is the presence of quinonoid contributions in the electronic ground state of the polymer backbone. In this respect, some poly(thienylene methine)s with alternating aromatic and quinonoid segments in the main polymer chain have been prepared [4].

A previous basic inspection of the acid-catalyzed polycondensation of several derivatives of 2-hydroxymethylthiophene [5, 6] was focused on their kinetic and mechanistic features and on the

[1] To whom any correspondence should be addressed.

structure of the ensuing oligomers and polymers. The main differences observed with respect to the homologous furan based systems [7] were: (i) the fact that the 3 and 4 positions on the thiophene ring are as reactive as the 5 position (the only reactive site for furan) and (ii) the lack of cross-linking reactions between different polymer chains due to the very poor dienic character of the thiophene heterocycle. The most relevant common features of these systems were the formation of: (i) irregular sequences of thienylene methylene (-Th-CH2-) units and (ii) conjugated moieties containing exounsatured 2,5-dihydrothiophene structures, due to a sequence of hydride ion and proton losses leading to an electronic reorganization. These synthetic approaches were based on a two-step process involving first the formation of a non-conjugated polymer precursor, by reaction between thiophene derivatives and aromatic aldehydes, and then the conversion to the conjugated polymer via oxidative elimination of bridge hydrogens.

Recently, an Algerian proton exchanged montmorillonite clay called Maghnite-H+ (Mag-H+), a new non toxic cationic initiator, was used as a catalyst for cationic polymerization of a number of vinylic and heterocyclic monomers [8, 10].

Here we report the synthesis of a novel soluble obtained in a single step polymerization through a very simple acid-catalyzed condensation of thiophene and thiophene-2-carbanaldehyde in the presence of an exchanged clay montmorillonite called Mag-H as catalyst (scheme1).

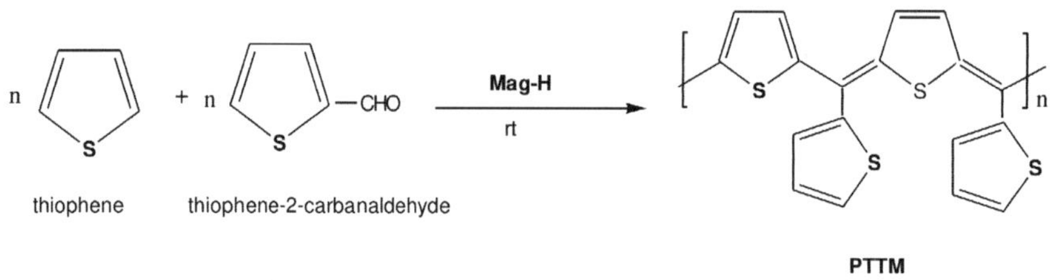

Scheme 1. Synthetic route for PTTM polymer.

2. Experimental

2.1. Reagents

(THF) tetrahydrofurane (99%), dichloromethane, and methanol (98%) were purchased from Aldrich and used as received. Thiophene: was purchased from Aldrich Chemical Co. and distilled under reduced pressure. Thiophene-2-carbaldehyde: was purchased from Aldrich Chemical. Raw-Maghnite (Algerian montmorillonite clay) was procured from BENTAL (Algerian Society of Bentonite).

IR spectra of the polymers as thin films were recorded with a Brucker ALPHA FT-IR A220/D-01 spectrometer. The UV-vis absorption spectra were recorded with an OPTIZEN UV-2120 spectrophotometer. For the measurement of the electrical conductivity of the products, thin pellets were prepared by compaction of the polymer powders under 10 tons of weight. Typical sample diameter was 13 mm and the thickness was 0.41 mm. The conductivity measurements were performed with the four-point probe technique.

2.2. Polymer preparation

The condensation of thiophene-2-carbaldehyde with thiophene in the presence of Mag-H+ as a catalyst was carried out by condensation in bulk under nitrogen for 6 h.

Each mixture was prepared with a weighted quantity of Mag-H+ dried just before use for 1 h in a drying oven at 100°C. Thiophene-2-carbaldehyde (6x10-3mol) and thiophene (6x10-3mol) were mixed, and various amounts of Mag-H+ (1, 2,3, 5, 10 and 15 % by weight) were added. The reaction was carried out at 25°C for 6 h. The solid product was filtered under vacuum and washed with hydrochloric acid and distilled water to remove all traces of oligomers, and finally dried under vacuum at room temperature for 24h. The resulting polymers were characterized by FTIR and UV-Visible spectroscopy.

3. Results and discussion

3.1. Effect of the amount of Mag-H

The effect of the amount of Mag-H on the synthesized of PTTM was examined. Figure 1 shows the effect of the amount of Mag-H, expressed by using various weight ratios Mag-H/monomer, on the polymerization rate. The polymerization was carried out in bulk. As shown in Figure 1, a yield of 38 wt % was reached for 5 wt % of Mag-H, and the use of a higher amount of clay caused an increase of the yield of the reaction. Similar results are obtained by Boutaleb and al [11] in the polymerization of l'o-anisidine, aniline by Mag-H. This phenomenon is probably the result of the number of "initiating active sites" responsible of inducing polymerization, this number is prorating to the catalyst amount used in reaction. Up to 20 wt % of Mag-H, The polymerization yield became constant; this is probably the result of an increase in the medium viscosity.

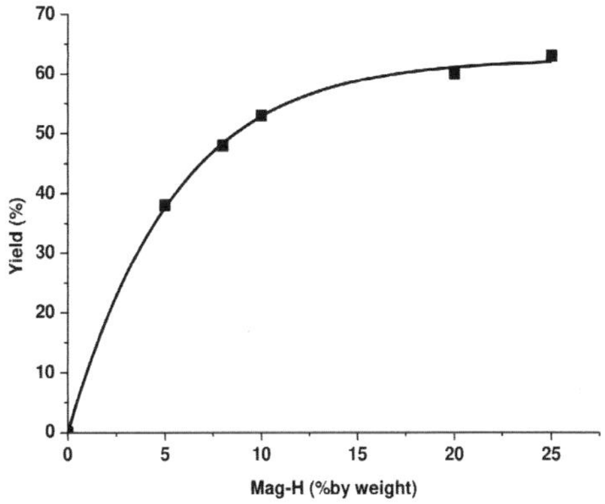

Figure 1. Effect of the amount of Mag-H+ on the yield of poly[(thiophene methine)] ,
Mag-H+ =10%, t = 6h.

3.2. Effect of time on condensation

Figure 2 shows the yield of polymer versus time for polymerization of thiophene using Mag-H as catalyst. As the figure shows, polymerization takes place slowly and smoothly, reaching a yield of 63 % after 9h at 25°C in the presence of 10 % of Mag-H.

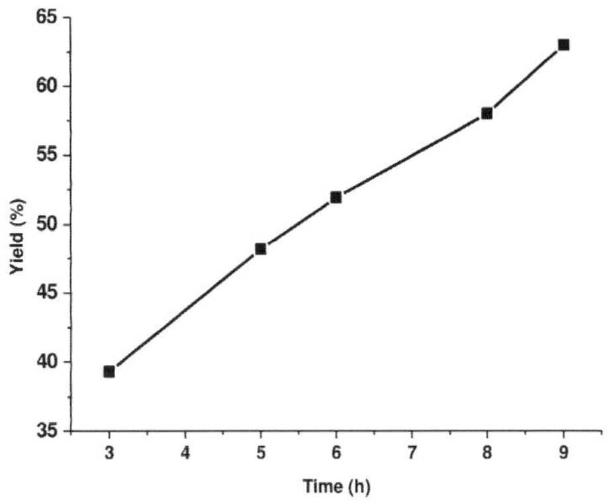

Figure 2. Effect of the time on the yield of poly[(thiophene methine)] ,Mag-H+ =10%, t = 6h

4. Spectroscopic characterization

The electronic spectrum of PTTM in CH_2Cl_2, which an example is provided in Figure 3, shows two major absorption bands. The first band in range of 280–300 nm is assigned to the π -π* transition of the aromatic heterocyclic. Since un-substituted thiophenes absorb only around 220 nm, the high-wavelength maximum strongly suggested the presence of conjugated sequences along the chains. The second broad peak centered between 620 and 650 nm and tailing into the visible is assigned to the π - π* band gap transition [12].

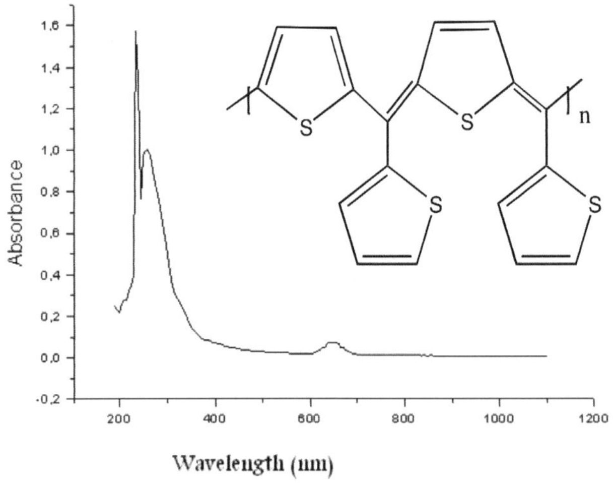

Figure 3. Optical absorption spectra of PTTM in CH_2Cl_2

The figure 4 presents the FTIR spectrum of PTTM and shows the appearance of a strong absorption at 1640 cm-1 which is attributed to the stretching vibration of conjugated C=C and the stretching vibration of aromatic in thiophene. The absorption band of the C=C stretching vibration of quinoidring and methine conjugated with quinoid ring of PTTM is at 1690 cm-1, respectively, which proved the partially dehydrogenation of these polymers. The bands in the 1000-1100 cm-1 region are the C-S stretching vibration bands of the thiophene ring in the main chain. The presence of differently disubstituted thiophene rings as indicated by the appearance of a band at about 850 cm-1, corresponding to =C-H out-of-plane vibrations in 2,5-disubstituted thiophenes. A distinct peack near 737,37 cm-1 is due the out of plane vibration Cβ-H characteristic of the α-linkage in thiophene rings [12].

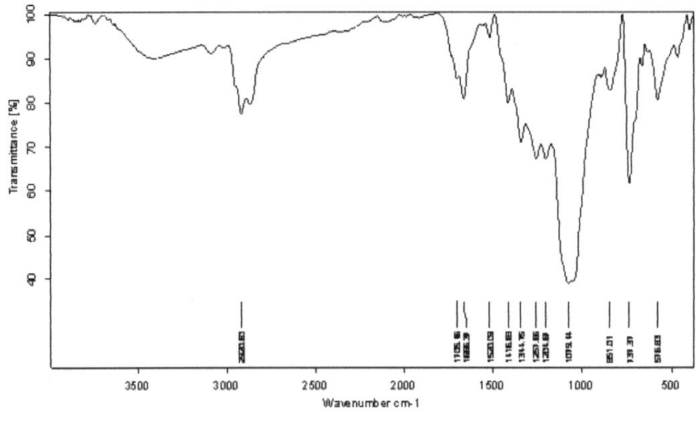

Figure 4. FTIR spectrum of PTTM

5. Conductivity measurements

For the measurement of the electrical conductivity of the products, thin pellets were prepared by compaction of the polymer powders under 10 tons of weight. Typical sample diameter was 13 mm and the thickness was 0.41 mm. The conductivity measurements were performed with the four-point probe technique and calculated with the following equation:

$$\sigma = V^{-1} \cdot I(\text{In}2/\pi d_n)$$

Where V is the potential in volts, I is the current in ampere and dn is the thickness of the samples in cm. The electrical conductivity of the synthesised polymer was σ = 0.02253 Ω . cm-1.

6. Optical results

To determine the optical properties, the average gap Em, dispersion energy E_D and static refraction index n_0 of our material, we use the UV analysis simulations of PTTM by the Wemple Didominico model

When light passes through a medium, its behavior will depend on the proprieties of the medium and of the light (reflection, absorption, refraction and scattering). The refractive index in its real and imaginary parts or dielectric function is the response to interaction between the material and light. To well study these parameters it is necessary to make the optical measurements (transmission and/or reflection) in the UV wavelength range [13]. We present in Fig. 5 the transmittance T and absorptance A spectra normalized with regard to the basic line in the range wavelength 190 to 1100 nm. We distinguish clearly tow regions in these spectra:

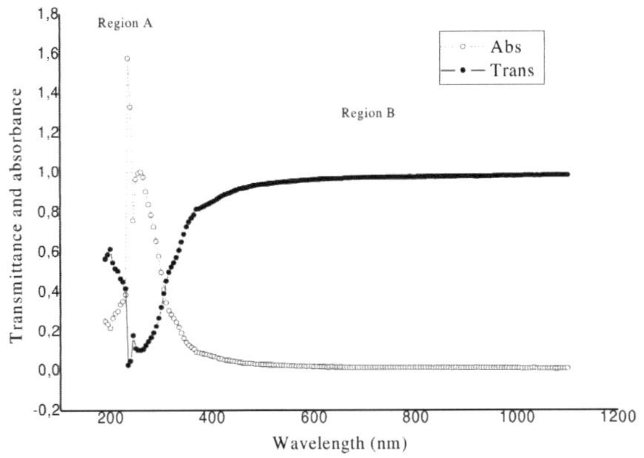

Figure 5. Transmittance and absorptance variation in UV-range.

i- Region A or absorption region: in this range the transmittance and absorptance spectra show a strong variation with diminution the transmittance spectrum and increase the level of absorptance. In this range we can deduce from the transmittance spectra the absorption coefficient $\alpha(\omega)$ at frequency ω, by the following relation [14]:

$$\alpha(\omega) = -\frac{1}{d}\ln[T] \quad (1)$$

Where d is the thickness of the film and T its transmittance normalized by background.

For the material no dispersive, the reflectance spectrum R is calculated by:

$$R = 1 - A - T \quad (2)$$

ii- Region B or transparency region: in this range of wavelength the optical transmittance remains practically constant at 0.97 (97%) and the absorptance is typically equal to zero. In this region we can determined the refractive index n. Where R is related to refractive index n for the normally incidence by relation:

$$R = \left(\frac{n-1}{n+1} \right)^2 \tag{3}$$

From this expression we can deduce the refractive index in transparency region:

$$n = \frac{1+\sqrt{R}}{1-\sqrt{R}} \tag{4}$$

The values found for the refractive index at room temperature for unpolarized light are represented in the Fig. 6. From the results shown in this figure it clearly appears that the refractive index decrease rapidly (2.6 at 1.3) in the wavelength range 190 to 500 nm and remains constant in the region between 500 and 1100 nm. This behavior is a good agreement with results found for semi-conductors such as silicon (Si) and germanium (Ge) [15, 16]. For determined the refractive index in all wavelength range studied we simulated this parameter by Lorentz model [17]. We notice that both spectra are perfectly superposed in the high wavelengths range from 400 to 1100 nm. We can by this model generalize the refractive index variation on the low wavelengths range. The Lorentz model works surprisingly well, provided we remember that real materials correspond to a collection of Lorentz oscillators with different frequencies.

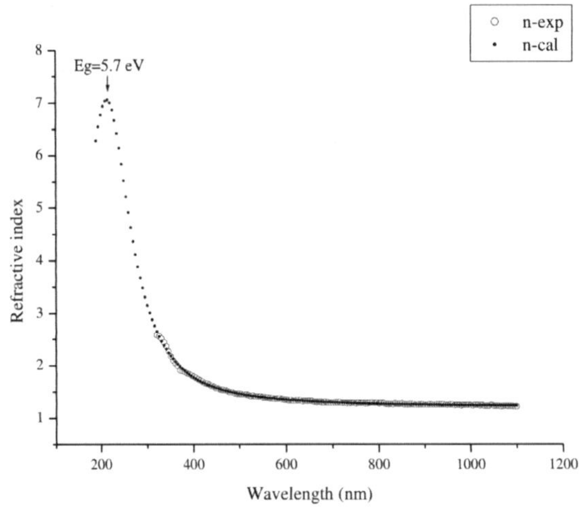

Figure 6. refractive index variation in UV-range and simulated curve by Lorentz model.

The outer, or valence, electrons predominantly determine the characteristics of the optical properties a material. Semiconductors are covalently bonded materials where the electrons are evenly shared between neighboring atoms. (Some insulators are covalently bonded, too.). This means that the electrons are smeared out into broader bands and that their resonance frequencies are lower than for ionically bonded materials. Usually these materials can be described by a single energy gap and single broad absorption band above the energy gap (example Silicon see work [18]). From these simulations by Lorentz model, we can determined the average gap EM centred at 216 nm equivalent to energy EM=5.3 eV. The average gap usually considered as the energy separation between the centers of both the conduction and valence bands. This gap such as defined by this method is relatively high than the optical gap which separate localized and delocalized states defined by Tauc [19]. In order to understand this phenomenon well, the values of the static refractive index n_0 [refractive index at zero energy: i.e $n_0 = n(\hbar\omega = 0)$] is determined from the analysis of the Wemple and Didominico one oscillator model [20]. According to this model, the dispersion relation of the refractive index is given by:

$$n^2(\hbar\omega) = 1 + \frac{E_D.E_m}{Em^2 - (\hbar\omega)^2} \tag{5}$$

Where Em denotes the average gap, E_D denotes the dispersion energy related to the coordination number of the atoms. The static refractive index n_0, average gap Em and dispersion energy E_D are deduced by considering the variation of $\left[n^2(\hbar\omega)-1\right]^{-1}$ as function of $(\hbar\omega)^2$ (figure 7). The values obtained for these parameters by this analysis are:
$Em = 3.62\,eV$, $E_D \approx 2\,eV$ and $n_0 = 1.22$

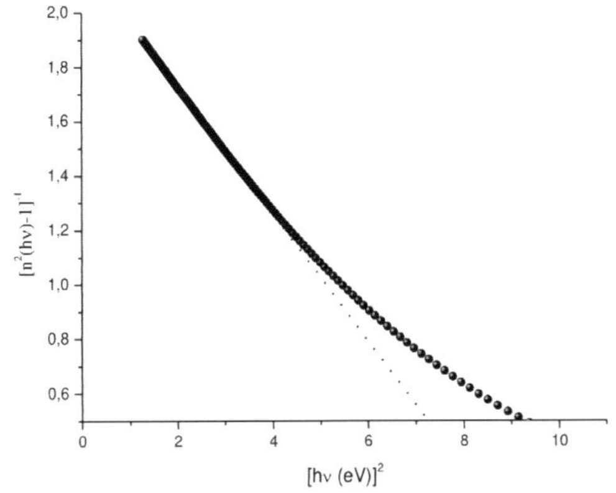

Figure 7. Show the Wemple Didominico model for determined the average gap Em, dispersion energy E_D and static refraction index n_0.

The difference obtained in values for average gap Em by these models due to principally to prediction of each. Nevertheless, the values found for Em show clearly that our material is a good semi-conductor and can be compared to silicon or germanium. Although the value of E_D is relatively lower.

In the figure 8, we represent the variation of the imaginary part, ε_2, of dielectric function deduced from refractive index n and excitation coefficient k by:

$$\varepsilon_2 = 2.n.k \tag{6}$$

Where k related to absorption coefficient α by relation:

$$\alpha = \frac{4\pi.k}{\lambda} \tag{7}$$

λ is the wavelength studied.

Experimentally the dielectric function also can be obtained by spectroscopic ellipsometry (SE) when employed to study thin film growth or modification of the surface in various liquid environnement though in-situ or real-time measurement [21]. It can be seen clearly that this spectra present the look of a semiconductor a microstructure. In fact, in the high energy range, two peaks are observable. One mean peak around 4.8 eV and one intense shoulder around 5.3 eV. They correspond to a convolution of direct electronic transitions in UV region [22]. These also features are an obvious signature of the presence of crystallites structure in this film.

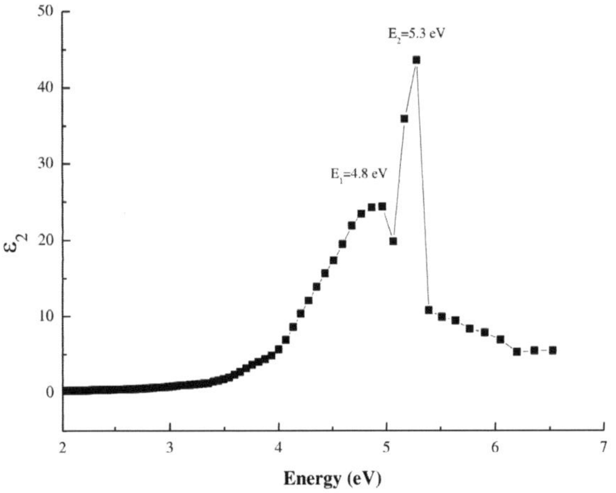

Figure 8. Imaginary part of dielectric function, ε_2.

Our material studied here is a mixture of poly {(2,5-diyl thiophene) (2-Thiophenyl Methine)}and dichloromethane, then two or more liquids are mixed together physically, we may apply the Bruggeman effective medium theory (EMT) to deduce the optical function of the mixture [23] by the formula:

$$f_A \frac{\varepsilon_A - \varepsilon}{\varepsilon_A + 2\varepsilon} + (1 - f_A)\frac{\varepsilon_B - \varepsilon}{\varepsilon_B + 2\varepsilon} = 0 \tag{8}$$

Here, ε_A and ε_B are the dielectric function of liquid A and liquid B respectively. f_A is the volume fraction of liquid A and ε is the resultant dielectric function of mixture. We are going to return on this study in detail in our next works when we employed the spectroscopy ellipsometry for these materials.

6. Conclusion

In conclusion, we have prepared some conjgugated poly {(2,5-diyl thiophene) (2-Thiophenyl Methine)} (PTTM), which has a π- conjugated chain, by polycondensation of thiophene-2-carbaldehyde and thiophene catalyzed by Maghnite-H. Such results may serve primarily to illustrate a new strategy to increase the solubility of low band gap polymers through the arrangement of different aromatic heterocycles in conjugated polymer backbones.

The results obtained with optical measurements and analyzed by different models, show clearly that our material (PTTM) is a good semi-conductor, when the average gap around 5 eV. In fact, the analysis the imaginary part of dielectric function, ε_2, show also the structure of the semiconductor present a crystallites by presence the two peaks of Van Haove

Acknowledgements: This work was supported by The National Agency for the Development of University Research (ANDRU) and the Directorate General of Scientific Research and Technological Development (DGRSDT) of Algeria.

References

[1] Mac Diarmid A. G 2001 *Angew Chem Int* **40** 2581 2590.
[2] Chan H. S. O, Ng S. C 1998 *Prog Poly Sci* **23** 1167 1231.
[3] Bolognesi A, Botta C, Porzio W 2001 *In Molecular Materials and Functional Polymers*, Blau W J, Lianos P, Schubert U, *Eds, Springer: Wien, New York* 121–128.
[4] Chen W. C, and Jenekhe S. A 1995 *Macromolecules* **28** 2 465 480.
[5] Stagnaro P, Costa G and Gandini A 2001 *Macromolecules* **34** 26 32.
[6] Stagnaro P, Costa G, Pioli F and Gandini A 2003 *Polymer* **44** 1359 1365.
[7] Choura M, Belgacem N. M and Gandini A 1996 *Macromolecules* **29** 3839 3850.
[8] Belbachir M and Bensaoula A 2006 *US patent (Houston)* **7** 094 823 B2.
[9] Hachemaoui A, Yahiaoui A and Belbachir M 2010 *J Appl Polym Sci* **118** 3445 3452.
[10] Chabani M,Yahiaoui A, Hachemaoui A and Belbachir M 2011 *J Appl Polym Sci* (in press)
[11] Boutaleb N, Benyoucef A, Salavagione H. J, Belbachir M and Morallón E 2006 *European Polymer Journal* **42** 4 733 739.
[12] Le Floch F, Ho H. A, Harding-Lepage P, Bedard M, Neagu-Plesu R and Leclerc M 2005 *Adv Mater* **17** 1251 1254.
[13] Monfredotti C, Fizzoti F, Borco M, Pastorino P, Polessello P and Vittone E 1994 *Phys. Rev* **50** 24 18046 18053.
[14] Brodsky M. H, Cardona M and Cuomo 1977 *J Phys. Rev.* **16** 8 3556 3571.
[15] Bouizem Y, Belfedal A, Sib J. D, Kebab A and Chahed L 2007 *J. Phys. Condens. Matter* **19** 35 356215.
[16] Bouizem Y, Belfedal A, Sib J. D, Kebab A and Chahed L 2005 *J. Phys. Condens. Matter* **17** 33 5149 5158.
[17] Marquardt D. W 1963 *J. Soc. Indus. Appl. Math.* **I1** 431 441.
[18] Belfedal A, Bouizem Y, Benlakehal D, Baghdad R, Sib J. D, Kebab A, Chahed L and Zellama K 2010 *Phys. Status Solidi* **7** 3-4 565 570.
[19] Tauc J and Menth A 1972 *J. Non-Cryst. Solids* **8-10** 569 585.
[20] Wemple S. H and Didomenico M 1971 *Phys. Rev.* ***B*** **3** 1338 1351.
[21] Arwin H 1998 *Thin Solid Films* **313-314** 764 774.
[22] Van Hove L 1953 *Phys. Rev.* **89** 1189 1193.
[23] Brugemann A. G 1936 *Ann. Phys (leiptzig)* **24** 636 665.

IOP Conf. Series: Materials Science and Engineering **27** (2012) 012034 doi:10.1088/1757-899X/27/1/012034

Optimization of the synthesis based on Resorcine and formaldehyde to obtain carbon nanoxerogels

A. Zweg[a1] , E. Elaloui[a], S. Nouri[a], Y. Moussaoui[b]
[a]Materials, Environment and Energy Laboratory, Sciences Faculty of Gafsa, Gafsa University, Zarroug city, 2112, Gafsa, Tunisia.
[b]Physical Organic Chemistry Laboratory, Science Faculty of Sfax, Sfax University, 3000, Sfax, Tunisia

E-mail: amel.zweg@gmail.com

Abstract. The sol-gel method consists on obtaining very pure products from selected precursors. Indeed, synthesizing very pure activated carbon (> 99% carbon) using organic precursors containing only carbon and oxygen as hetero atoms to form a polymeric resin by pyrolysis. The proposed synthesis reaction is in the border of two sciences: the Sol-gel and polymerization. The polymerization by mechanism followed by the initiation and propagation, the sol-gel by evolution of the media environment starting with a solution of various ingredients to have the particle's sol in the emulsion solvent.
The sol turns into a gel which is the result of three-dimensional crosslinking of the material trapping the solvent.
In this work we propose to introduce a new method of synthesis of these nanoxérogels by polycondensation of resorcinol with formaldehyde in two different media (water and acetone) in the presence of two catalysts (acetic acid and sodium carbonate). By varying the following parameters: solvent, nature and quantity of catalyst, 16 samples were obtained. By the method of factorial design (2^k): an optimization is realized by following three characteristics of nanoxerogels prepared: the density, the gel time and performance. The results indicate the best conditions for the synthesis of a carbon nanoxerogel: the lowest quantity of acid catalyst in aqueous media (A5).

1. Introduction

It is known that by reacting formaldehyde on a phenolic compound (phenol, resorcine, pyrogallol, cresol etc...) a copolymerization may be induced giving a thermosetting resin, such as Bakelite ®, which is used in very Various (electrical insulation, production of moldings such as handsets) [3]. Investigation on the Resorcine - Formaldehyde in water in the presence of Na_2CO_3 [3,4] are reported in the literature, a recent study on carbon xerogels derived from the polycondensation of pyrogallol with formaldehyde to show the feasibility in acidic catalyst [5]. Indeed, the team of RW Pekala [6] exploits in the recent years a method of synthesis of this resin using the achievements of the sol-gel applicated to silica [7]. This solid which has the polymer and nanoxerogel proprieties is used in various sciences and technologies (Electronic, Film, supercapacitor ...)[8,9]

[1] To whom any correspondence should be addressed.

The synthesis reaction proposed to study is in border of two sciences: The Sol-gel and polymerization. The polymerization mechanism followed by the initiation and propagation and the sol-gel by changing the media from a solution of various ingredients giving a sol of in emulision within the solvent. The ground turns into a gel [10] which is the result of three-dimensional crosslinking of the material trapping the solvent in these pores.

2. Experimental

The initial mixture is formed by Resorcine, formaldhyde, solvent and catalyst. The proportions are given in concentration ratio: Formaldehyde / Resorcine = F / R = 2 is a variable catalyst volume. The volume of media which is the solvent is 30 mL. The solvents are distilled water and acetone. The sol is evolving into gel after gel time which depends mainly on the following factors: type of catalysis, the catalyst quantity and nature of the solvent.

The amounts of the precursors are selected due to coefficients stochiometry of the polymerization reaction.

The polymerization reaction is activated by acid-base catalysis. The acid catalyst used is the acetic acid and the basic one is Na2CO3. Sixteen samples are prepared, as shown in figure 1:

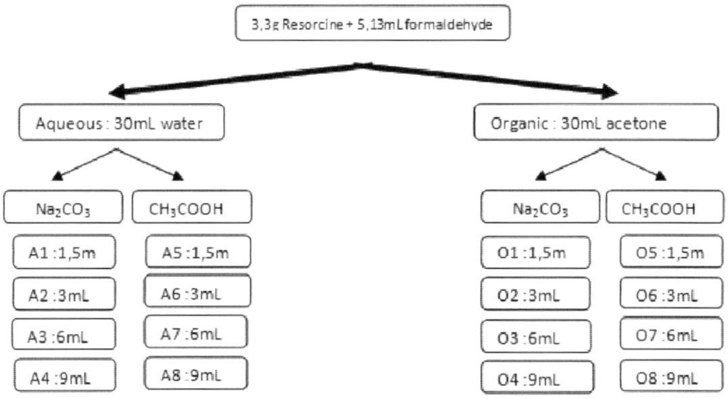

Figure 1. Summary of the preparations.

3. Results and discussion

3.1. Gel time and mass loss

Table 1. Variation of tg and mass of the samples.

Sample	tg (h)	m (25°C)	m (50°C)	m (80°C)	m (110°C)
A1	30.5	35,55	15,24	7,09	4,28
A2	16	36,79	16,49	7,21	4,45
A3	8	38,85	17,65	7,53	4,74
A4	5	41,66	18,76	7,78	4,92
A5	35.5	35,63	15,51	6,91	4,17
A6	45.5	37,3	17,32	7,11	4,19
A7	49	39,16	18,21	7,86	4,21
A8	50	40,99	19,01	8,09	4,23
01	157	7,34	5,21	4,39	4,07
02	166.5	8,45	5,54	4,48	4,14
03	60	10,82	5,65	4,71	4,36
04	81	14,19	5,78	4,98	4,48
05	205	7,25	5,31	4,52	4,15
06	215	8,32	5,65	4,59	4,17
07	194	10,12	5,79	4,71	4,19
08	181	13,09	5,89	4,75	4,21

The following table summarizes the results obtained during the synthesis of the resin R/F: (Tg: gelation time) and in (T°C) witch is the change masse of the temperature at various nanoxerogels formed after drying (mass change).

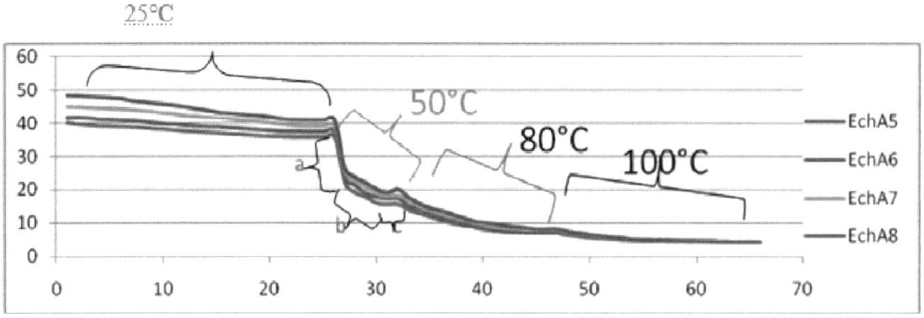

Figure 2. Loss of mass corresponding to samples prepared in an aqueous media catalyzed by acid.

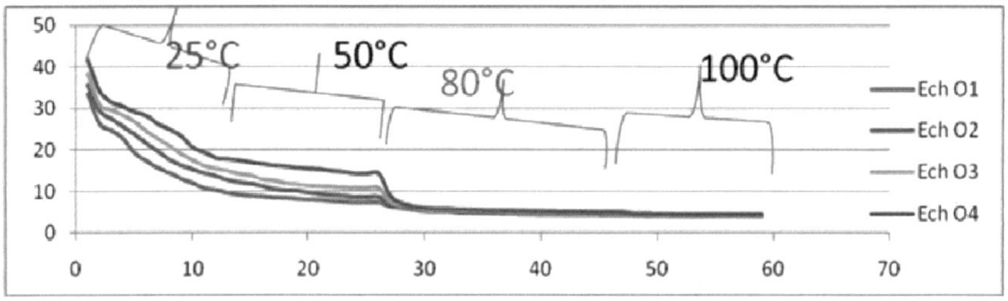

Figure 3. Loss of mass corresponding to samples prepared in an organic media catalyzed by a basic salt.

3.1.1. Kinetics of gels drying

The mass measurements during drying were made for samples of the series prepared in the case of acid and base catalysis. We note that each variation of m = f (t) consists to three parts: fast (a), slow (b) and steady(c) (figure 2).

The first part is attributed to the solvent evacuation from the large pores, the second part is corresponding to the solvent evacuation from the pores of smaller sizes and the third part is corresponding to the mass stabilization of the samples.

During the first and the second part (slower drying step), the properties of the nanoxerogel are kept intact or modified. In this time interval, the trapped solvent is removed to the gel network. Therefore, stresses appear, they are due to the liquid-vapor equilibrium in the network and the gel pores, causing the collapse and shrinkage of the gels.

It should be noted that the sample with a spongy morphology lose more solvent during the first part and the one with more compact morphology loses in the second part the largest quantity of solvent. We can stop the drying of the gels prepared when the sample mass remains constant.

3.1.2. Gelation time (tg)

Reading the table (Table1) highlights the following results:

- Samples prepared in aqueous media, the aquagels (A), give gel faster than those synthesized in organic solved, acetogels (O). This may be due to the presence of other reactions in the second case (polymerization reaction of acetone).
- For aquagels, gelation time decreased when the quantity of catalyst increases whatever its nature, whereas the variation of tg for acetoel is not regular depending catalyst quantity.

3.2. Variation of yield and density

The yield is calculated as follows:

$$R(\%) = \frac{Msample}{Mtotal} \times 100$$

The total mass is the mass of precursor: 3.3 g of resorcine and 5.13 mL (1.923 g) of formaldehyde which gives: $M_{total} = 5.223$ g

Table 2. Variation in yield and density of samples.

Sample	Yield (%)	Density (g cm^{-3})
A1	81,93	1.07
A2	85,18	1.39
A3	90,73	1.18
A4	94,18	1.4
A5	79,44	0.2
A6	79,82	0.23
A7	80,21	0.52
A8	80 ,59	0.84
O1	77,91	1.35
O2	79,25	1.42
O3	83,46	1.09
O4	85,76	1.49
O5	79,82	0.99
O6	80,21	1.04
O7	80,59	1.05
O8	80,97	1.052

From Table 2 (Tab2), we can conclude that the polymerization reaction is not total, because the maximum yield is 94.18% for sample A4, it can reach the minimum value 77, 91% for sample O1. This variation is related to changes in operating conditions.

The density of dry solids Obtained varies between 0.2 and 1.49 through spongy samples and compact ones.

The table above shows clearly the dependence of the density value on the operating conditions.

Indeed, the molar Res / Cat seem to be more important in the case of catalysis by the acid (CH_3COOH) in an aqueous media.

The density increases with the quantity of catalyst, for samples A5, A6, A7 and A8, the density is low because there is no loss in terms of quantity. It increases 0,2 - 0,23 - 0,52 and 0.84. In this case, we obtain a spongy xerogel.

In conclusion, the use of acetic acid as catalyst in aqueous media will produce compounds with different structures by simple variation of the molar rapport Res / Cat (varying quantity of catalyst) during the preparation of gels.

3.3. Experiment design

We chose the factorial plan as a plan for this optimization. It is important to assign to each factor two levels: One will be called "low level" and the other" high level".

If the factor is qualitative, the low level and the high levels will correspond to two forms of the factor, for example, two types of solvent.

In practice, the low level will be coded with the number (-1) and the high level with the number (+1). A plan with each k factors has only 2 levels is named Plan 2^k [11] for k = 3.

with k: number of variables as shown in the following table .

Table 3. Table of variable levels chosen.

Variables	**Catalyst C**	**Solvent S**	**Amount of catalyst Q**
High level +1	Acid	Organic	9 mL
Low level -1	Base	Aqueous	1.5 mL

The matrix of experiments is the table showing the number of experiments to realize by varying factors and order [11] with which experiments must be realized

Table 4. Table showing the number of experiments to be realized and their answers.

		Factor 1	Factor 2	Factor 3	Factor 4	Factor 5	Factor 6	Factor 7	
	medium	C	S	Q	CS	CQ	SQ	CSQ	$Y_{Rép}$
Exp 1	+	-	-	-	+	+	+	-	Y1 (éch A1)
Exp 2	+	+	-	-	-	-	+	+	Y2 (éch A5)
Exp 3	+	-	+	-	-	+	-	+	Y3 (éch O1)
Exp 4	+	+	+	-	+	-	-	-	Y4 (éch O5)
Exp 5	+	-	-	+	+	-	-	+	Y5 (éch A4)
Exp 6	+	+	-	+	-	+	-	-	Y6 (éch A8)
Exp 7	+	-	+	+	-	-	+	-	Y7 (éch O4)
Exp 8	+	+	+	+	+	+	+	+	Y8 (éch O8)
Divider effects	8	8	8	8	8	8	8	8	

$$\begin{pmatrix} +1 & -1 & -1 & -1 & +1 & +1 & +1 & -1 \\ +1 & +1 & -1 & -1 & -1 & -1 & +1 & +1 \\ +1 & -1 & +1 & -1 & -1 & +1 & -1 & +1 \\ +1 & +1 & +1 & -1 & +1 & -1 & -1 & -1 \\ +1 & -1 & -1 & +1 & +1 & -1 & -1 & +1 \\ +1 & +1 & -1 & +1 & -1 & +1 & -1 & -1 \\ +1 & -1 & +1 & +1 & -1 & -1 & +1 & -1 \\ +1 & +1 & +1 & +1 & +1 & +1 & +1 & +1 \end{pmatrix} * \begin{pmatrix} C \\ S \\ Q \\ CS \\ CQ \\ SQ \\ CSQ \\ \square \end{pmatrix} = \begin{pmatrix} Y1 \\ Y2 \\ Y3 \\ Y4 \\ Y5 \\ Y6 \\ Y7 \\ Y8 \end{pmatrix}$$

Matrix of the works plan

Each estimate of a model coefficient is equal to the algebraic sum of the experimental responses Y_i assigned signs in the matrix X corresponding to Factor X_i divised by the number of experiments [12]. This result allows to affranchise every matrix calculation and provides the model coefficients with a simple calculator or better with a spreadsheet. These plans are applied to experimental results found in this study: the gel time, loss of weight, performance and density.

3.3.1. Gelation time (hours)

The results of the effects are:

Table 5. The effects values and their interactions calculated for the gel time.

Cste	C	S	Q	CS	CQ	SQ	CSQ
92,75	24,37	63,25	-14,25	13,87	12,37	-10,75	1,87

Y(h) = 92,75 + 24,37 × C + 63,25 × S – 14,25 × Q + 13,87 × CS + 12,37 × CQ – 10,75 × SQ + 1,87 × CSQ (Equation 1)

From Equation 1, the effect corresponding of solvent type is the closest to the average effect so we have to choose the type of solvent firstly. So we can say that the best response is:

Y (h) = 92.75 + 24.37 × C + 63.25 × S – 14.25 × Q + 13,87 × CS+12,37×CQ-10,75×SQ. (Equation2)

The type of catalyst has a greater influence than the quantity of catalyst. Interactions play also a role in achieving rapid gelation.

The experimental results show that the acid catalyst in an aqueous media increases with the gelation time while the base catalyst in an aqueous media, is inversely proportional to the time of gelation [13].

3.3.2. Loss of mass (g)

Table 6. The values of calculated effects and their interactions for mass loss.

Cste	C	S	Q	CS	CQ	SQ	CSQ
4,31	-0,12	-0,086	0,146	-0,116	-0,116	-0,031	0,29

$Y(g) = 4{,}31 - 0{,}21 \times C - 0{,}086 \times S + 0{,}146 \times Q - 0{,}116 \times CS - 0{,}116 \times CQ - 0{,}031 \times SQ + 0{,}29 \times CSQ$. (Equation 3)

All effects in absolute value are very close to zero and very low to the average effect. By comparing the effects between them, we note that the closest value corresponds to the most influential factor which is the interaction between the type of catalyst, the quantity of catalyst and solvent type. In general, all factors have a secondary effect on mass loss, except the interaction between catalyst and solvent. The best response is:

$Y(g) = 4{,}31 - 0{,}21 \times C + 0{,}146 \times Q - 0{,}116 \times CS + 0{,}29 \times CSQ$. (Equation 4)

3.3.3. Yeild (%)

Table 7. The values of calculated effects and their interactions for performance.

Cste	C	S	Q	CS	CQ	SQ	CSQ
82,55	-2,37	-1,46	2,8	1,65	-2,22	-0,55	0,55

$Y(\%) = 82{,}55 - 2{,}37 \times C - 1{,}46 \times S + 2{,}8 \times Q + 1.65 \times CS - 2{,}22 \times CQ - 0{,}55 \times SQ + 0{,}55 \times CSQ$ (Equation 5)

From Equation 5, we can neglect the interaction of third order and we can say that to get good yield we must choose. The type of basic catalyst (C), the type of aqueous solvent (S), the quantity of catalyst (Q) maximum 9 mL. So:

$Y(\%) = 82{,}55 - 2{,}37 \times C - 1{,}46 \times S + 2{,}8 \times Q + 1.65 \times CS - 2{,}22 \times CQ$. (Equation 6)

From the experimental results, we found that performance in a media either aqueous or organic, with basic catalyst, increases with the quantity of catalyst [14].

It concludes that the performance depends to the type of catalysts, their quantity and the nature of the solvent.

3.3.4. Density (g.cm-3)

Table 8. The values of effects and their interactions calculated for the density.

Cste	C	S	Q	CS	CQ	SQ	CSQ
1,05	-0,28	0,17	0,15	0,08	0,03	0,08	-0,05

$Y (g.cm^{-3}) = 1{,}05 - 0{,}28 \times C + 0{,}17 \times S + 0{,}15 \times Q + 0.08 \times CS + 0{,}03 \times CQ + 0{,}08 \times SQ - 0{,}05 \times CSQ$. (Equation 7)

From Equation 7 obtained, the effects values corresponding to the interaction of second and third order are neglected because they are in the order of zero.

$Y (g.cm^{-3}) = 1{,}05 - 0{,}28 \times C + 0{,}17 \times S + 0{,}15 \times Q$. (Equation 8)

So the density depends on the type of catalysts, solvent type and quantity of catalyst. The experimental results show that the use of acid in an aqueous media as a catalyst will provide compounds having different structures, by simple variation of catalyst concentration. So the equation is in agreement with the experimental results.

3.4. Characterization by Scanning Electron Microscope S.E.M

Analyses by scanning electron microscope show the morphology of a porous media and allow realizing elemental microanalysis by electron microprobe. The results of these analyses are represented by the following pictures:

Magnification 1000 times Magnification 4000 times

All samples are analyzed at the state powder. These initial investigations by SEM showed an apparent porosity. Indeed, the powder is constitute by spherical particles interconnected forming an aggregate with a radius in the average of a micrometer.

Electron probe microanalysis is a fixed probe elemental analysis performed by SEM while taking the snapshot, it shows the elements present on some surface (qualitative analysis) and their percentage (quantitative analysis).

The analysis is performed on the particle. It shows the existence of two elements carbon and oxygen (Figure 4).

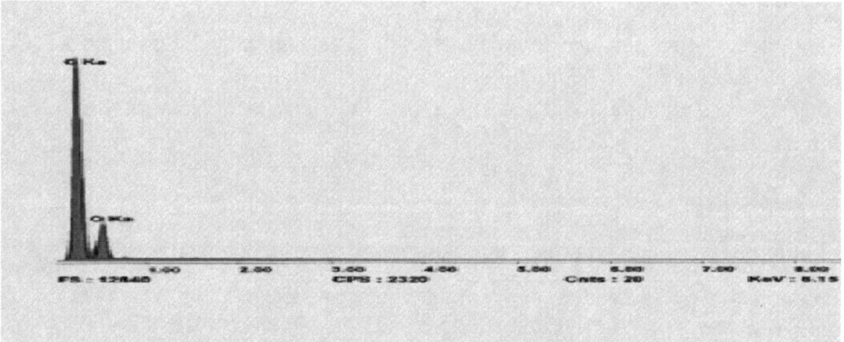

Figure 4. Snapshot of samples prepared in a media catalyzed by a basic salt showing the presence of some element over a surface.

4. Conclusion

The main purpose of this work is the preparation by sol-gel process of xerogel and its physicochemical characterization. A new method of synthesis of the gel is made by varying the operating conditions (solvent and catalyst). In fact, the copolymerization of resorcine and formaldehyde is conducted in aqueous and organic media catalyzed by acid and base followed by a conventional drying to obtain a xerogel. Indeed, the preparation involves the synthesis of a gel for different types of catalysts and different molar rapports Res / Cat. It appears that the catalysis by acid CH_3COOH is the best in an aqueous media. The kinetics of the reaction is inversely proportional to the Res / Cat., this parameter shows the importance of the preparation and its effects on the properties of xerogels.

The drying method chosen is called conventional. The drying is carried out at different temperatures from 25° C to 110° C for several days to obtain xerogels. An important shrinkage appears after drying, it is about 80% for some samples. This shrinkage has an influence on the density and texture of the gel. The method colled experience's plan design allowed us to highlight the influence of parameters and identify the optimal synthesis procedure.

The best sample in our work is the A5 due to its nanostructure.

medium: aqueous	Catalyst: 1.5 ml de CH_3COOH	yield = 79.44%
Density = 0.2	gel time=35.5 h	Mass = 4.17g

*Application: This type of sample: adsorbing pollutions studied in the field of water treatment, it is doing a new experimental plan 2^k in what quantities of the catalyst studied vary between 0.5 mL and 1.5 mL in seeking the best conditions to obtain a sample with increased porosity.

Temperature = 25°C

Temperature =100°C

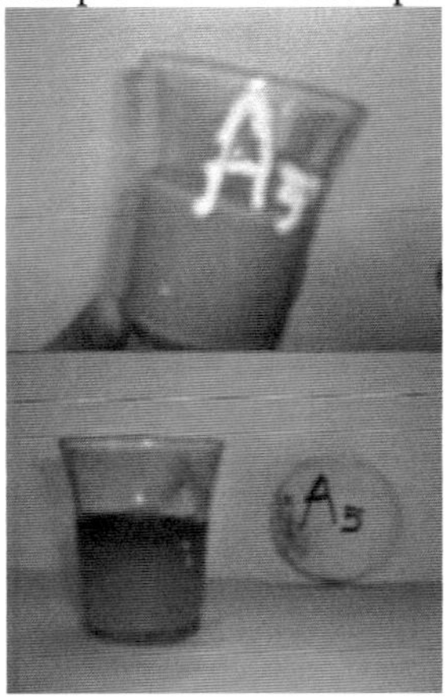

References

[1] Elaloui E and Pajonk G. M 2000 *J. Soc. Chim. Tunisie* **7** 256-263.
[2] Pekala R. W, Kong In F. M, Vacher R, Phalippou J, Pelous J and Woignier T 1989 *Rev. Phys. Appl.* **24 C4** 33.
[3] Pekala R. W, Alviso C. T and LeMay J. D *1990 J. NonCryst. Solids* **125** 67.
[4] Pekala R. W. and Alviso C. T. In: Novel Forms of Carbon 1992, Renschler C.L., Pouch J. J. and Cox D. M., *Editors, Mater. Res. Soc. Symp. Proc.* **270** 3.
[5] Pekala R. W, Alviso C. T, Lu X, Gross J, Fricke J 1995 *J. Non-Cryst. Solids* **188** 34-40.
[6] Pierre A. C 1992 Introduction aux procédés sol-gel, Forcera, Editions, Septima, Paris.
[7] Ller R. K 1979 The chemistry of Silica, Wiley, New York.
[8] El Mir L, Kraiem S, Bengagi M, Elaloui E, Ouederni A and Alaya S 2007 *Physica B* **395** 104.
[9] Glora M, Petricevic R, Moginger A and Fricke J 2001 *J. Non-Cryst. Solids* **285** 272.
[10] Glora M, Wiener M, Petricevic R, Probstle H and Fricke J 2001 *J. Non-Cryst. Solids* **285** 283.
[11] Pekala R. W 1989 US Patent N°4 **873** 218.
[12] Bolte M and Troquet M 1990 *Europ. Polym. J.* **26** 177.
[13] Silou T, Malanda M, Loubaki L 2004 *Journal of Food Engineering* **65** 219.
[14] Pauthe M, Phalippou J 1989 *Revue phys. Appl.* **24** C4-215.

Synthesis, crystal structure and spectroscopic investigations of a new tetrachloridozincate $NH_3(C_6H_4)_2NH_3ZnCl_4$

W. Wamani [1]**, M. Belhouchet and T. Mhiri**
Laboratoire de l'Etat Solide, Département de Chimie, Université de Sfax, Faculté des Sciences de Sfax, B. P.1171, 3000 Sfax, Université de Sfax, Tunisia

E-mail: wafawamani@yahoo.fr; belhouchet2002@yahoo.fr

Abstract. Chemical preparation, crystal structure and spectroscopic investigation are given for a new layered organic-inorganic hybrid, 4,4'-biphenylenediammonuim tetrachloridozincate. This compound is triclinic $P\bar{1}$ with the following unit-cell parameters: a = 6.4052(1) Å, b = 9.5026(2) Å, c = 14.5807(4) Å, α = 100.547(2)°, β = 92.101(1)°, γ = 109.032(2)°, Z = 2, and V = 820.34(3) Å^3. The crystal structure has been solved and refined to R1 = 0.031 and wR2 = 0.104 using 4733 independent reflections. The structure of this compound consists of inorganic layers built by isolated $ZnCl_4^{2-}$ tetrahedral. The organic entities, located between these layers, perform 3D cohesion through N-H...Cl hydrogen bonds. The vibrational properties of this compound were studied using IR absorption. The observed bands were identified by comparison with the vibrational modes and frequencies observed in similar compounds.

1. Introduction

In recent years organic-inorganic hybrid materials have been synthesized in large quantities, their structures, performances and applications were studied [1, 2]. The design and synthesis of organic-inorganic hybrid perovskite materials have aroused great interest due to their unusual topological properties and potential applications in the fields of catalysis, biochemistry, magnetism, material science, [3]. It was reported mostly that organic amine and inorganic divalent metal chlorides were used, respectively, as organic and inorganic components [4]. These optical, electrical, magnetic and thermodynamic properties are found to be useful in many application fields, which have aroused widespread attention [5]. For example, the transitional metal layered perovskite-like hybrid compounds can be used as a kind of magnetic material owing to the inherent magnetic property of a transition metal [6]. Most of layered organic-inorganic hybrid materials have pervovskite-type structures where inorganic sheets consist of corner-sharing MX_6 octahedrons. Few compounds show layered structure of isolated tetrahedral MX_4 instead of corner-sharing octahedral sheets observed in perovskite-type materials. In this context we report synthesis, crystal structure, thermal behavior and infrared spectroscopy of a new organic-inorganic hybrid compound $NH_3(C_6H_4)_2NH_3ZnCl_4$.

[1] To whom any correspondence should be addressed.

2. Experimental

1.1. Synthesis
Benzidine (98%), zinc oxide (99, 99%) were purchased from Aldrich chemical company and used without further purification. HCl (36%) was obtained from Tunisian Chemical Industry. Ethanol refers to absolute ethanol unless otherwise specified. The title compound was obtained by a reaction of two aqueous solutions containing respectively 1 mmol of zinc oxide dissolved in dilute hydrochloric acid and 1 mmol of 4,4-biphenylene diamine dissolved in 10 ml distilled water/ethanol. The mixture was stirred and then allowed to stand at room temperature until the formation of single crystals suitable for X-ray structure analysis. The obtained crystals remained stable under normal conditions of temperature and humidity.

1.2. X-ray crystallography
Table 1. Crystal data and structure refinement parameters for $NH_3C_{12}H_8NH_3ZnCl_4$.

Formula	$NH_3(C_{12}H_8)NH_3ZnCl_4$
Formula weight	393.42 g.mol^{-1}
Crystal system / Space group	Triclinic / $P\bar{1}$
a (Å)	6.4052(1)
b (Å)	9.5026(2)
c (Å)	14.5807(4)
α (°)	100.547(2)
β (°)	92.101(1)
γ (°)	109.032(2)
V (Å^3)	820.34(3)
Z	2
Crystal form / color	Block / light brown
Crystal size (mm^3)	0.3 * 0.2 * 0.1
hkl ranges	$0 \leq h \leq 8$
	$-13 \leq k \leq 12$
	$-20 \leq l \leq 20$
θ Range (°)	1.4 - 30.0°
Measured reflections	16092
Independent reflections	4733
Reflections with $I > 2\sigma(I)$ / R_{int}	4080 / 0.031
Absorption correction	Multi-scan SADABS
T_{min} / T_{max}	0.607 / 0.808
Goodness-of-fit on F^2	1.11
Final R indices	R1 = 0.031; wR2 = 0.104
Number of refined parameters	174
$\Delta\rho_{min}$ / $\Delta\rho_{max}$(e Å^{-3})	−0.64 / 0.93

Table 2. The final atomic coordinates and the equivalent temperature factors for $NH_3C_{12}H_8NH_3ZnCl_4$.

	x	y	z	U_{eq}
Zn1	0.43489 (4)	0.27333 (3)	1.066083 (16)	0.03136 (9)
Cl1	0.80981 (8)	0.35878 (6)	1.06733 (4)	0.03479 (12)
Cl2	0.31920 (10)	0.02734 (6)	1.09328 (4)	0.03902 (13)
Cl3	0.34159 (10)	0.40860 (6)	1.18998 (4)	0.04099 (14)
Cl4	0.29982 (10)	0.25759 (7)	0.91732 (4)	0.03969 (13)
N1	0.8265 (3)	0.3176 (2)	0.85128 (12)	0.0343 (4)
N2	0.8806 (3)	0.1039 (2)	0.15958 (12)	0.0365 (4)
C1	0.8376 (4)	0.2841 (2)	0.74969 (14)	0.0315 (4)
C2	1.0408 (4)	0.3284 (3)	0.71531 (16)	0.0400 (5)
C3	1.0491 (4)	0.2983 (3)	0.61875 (16)	0.0429 (5)
C4	0.6442 (4)	0.2130 (3)	0.69164 (16)	0.0414 (5)
C5	0.6542 (4)	0.1850 (3)	0.59534 (16)	0.0449 (6)
C6	0.8562 (4)	0.2270 (3)	0.55783 (15)	0.0366 (5)
C7	0.8640 (4)	0.1959 (3)	0.45439 (15)	0.0368 (5)
C8	1.0141 (5)	0.1319 (3)	0.41408 (17)	0.0469 (6)
C9	1.0190 (5)	0.1007 (3)	0.31770 (17)	0.0447 (6)
C10	0.7213 (5)	0.2309 (4)	0.39542 (17)	0.0485 (6)
C11	0.7272 (5)	0.2019 (3)	0.29898 (16)	0.0467 (6)
C12	0.8751 (4)	0.1358 (3)	0.26153 (14)	0.0343 (4)

A single crystal (light brown Block, 0.3*0.2*0.1 mm) was carefully selected under polarizing microscope in order to perform its structural analysis by X-ray diffraction. Intensity data were collected on Bruker APEX-II area detector diffractometer (chromated Mo-Kα radiation, λ=0.71073 Å). The structure was solved by means of the heavy-atom methods, using the SHELXS-97 [7] program, and was refined by full matrix least-squares methods on F^2 with SHELXL-97 [8]. The crystal of $NH_3C_{12}H_8NH_3ZnCl_4$ is triclinic with $P\bar{1}$ space group. A total of 16092 reflexions were collected (θ: 1.43-30.0; index ranges: $0 \leq h \leq 8$, $-20 \leq k \leq 20$, $-13 \leq l \leq 12$); 4080 of them were unique reflexions (Rint = 0.031). Absorption correction based on the crystal shape was applied using SADABS (Tmin = 0.607, Tmax = 0.808) [9]. The hydrogen atoms were placed in geometrically calculated positions and refined as riding on their bonded atom. The refinement converged with R1 = 0.031 and wR2 = 0.104 for 4080 reflections with I>2(I) (174 refined parameters, Goof S = 1.11). The experimental conditions of collection, the strategy followed for the structure determination and the final results are given in Table 1. All non–hydrogen atoms of $NH_3C_{12}H_8NH_3ZnCl_4$ were refined. The final atomic coordinates of the non-hydrogen atoms and their equivalent isotropic displacement parameters, Ueq, are given in Table 2.

3. Results and discussion

3.1. Structure description

Table 3. Main interatomic distances (Å) and angles (°) of $NH_3C_{12}H_8NH_3ZnCl_4$.

Zn1—Cl1	2.2685 (6)	Cl1—Zn1—Cl2	109.15 (2)
Zn1—Cl2	2.3241 (6)	Cl1—Zn1—Cl3	108.84 (2)
Zn1—Cl3	2.2367 (6)	Cl1—Zn1—Cl4	105.55 (2)
Zn1—Cl4	2.2661 (6)	Cl2—Zn1—Cl3	103.95 (2)
		Cl2—Zn1—Cl4	106.87 (2)
		Cl3—Zn1—Cl4	122.06 (2)
N1—C1	1.466 (2)	C4—C1—C2	121.9 (2)
N2—C12	1.466 (3)	C6—C3—C2	120.9 (2)
C1—C4	1.369 (3)	C1—C4—C5	118.9 (2)
C1—C2	1.375 (3)	C6—C5—C4	121.0 (2)
C2—C3	1.391 (3)	C5—C6—C3	118.6 (2)
C3—C6	1.389 (3)	C5—C6—C7	120.3 (2)
C4—C5	1.388 (3)	C3—C6—C7	121.1 (2)
C5—C6	1.388 (3)	C10—C7—C8	118.1 (2)
C6—C7	1.489 (3)	C10—C7—C6	120.7 (2)
C7—C8	1.390 (3)	C8—C7—C6	121.2 (2)
C7—C10	1.389 (3)	C9—C8—C7	121.2 (2)
C8—C9	1.386 (3)	C9—C12—C11	121.3 (2)
C9—C12	1.370 (3)	C9—C12—N2	119.4 (2)
C10—C11	1.387 (3)	C11—C12—N2	119.3 (2)
C11—C12	1.376 (3)	C12—C9—C8	119.2 (2)
C4—C1—N1	118.9 (2)	C11—C10—C7	121.0 (2)
C2—C1—N1	119.2 (2)	C12—C11—C10	119.2 (2)
C1—C2—C3	118.7 (2)		

The main geometrical features of different entities are reported in Table 3. Fig. 1 shows the X-SEED [10] plot of the asymmetric unit of $NH_3C_{12}H_8NH_3ZnCl_4$, containing one protonated, 4,4-biphenylene diamine $[NH_3C_{12}H_8NH_3]^{2+}$ cation and one tetrachloridozincate $[ZnCl_4]^{2-}$ anion. A view of the structure projected along a axis is given in Fig. 2. This projection shows that the crystal structure of the title compound consists of alternating organic layers and inorganic sheets stacked along [100] direction. The inorganic and organic species are interconnected through six N-H...Cl hydrogen bonds (Table 4). Each organic cation is linked to four $[ZnCl_4]^{2-}$ tetrahedra. The corresponding N...Cl distances vary between 3.121(2) and 3.438(2) Å, which can be considered relatively weak [11]. In the organic layers, there are no π–π interactions between antiparallel cations, with face to face distances of 4.983 and

8.249 Å, up than 3.8 Å, the maximum value accepted for $\pi - \pi$ interactions [12]. The dihedral angle between the two phenyl rings of the organic cation is 47.2(1).

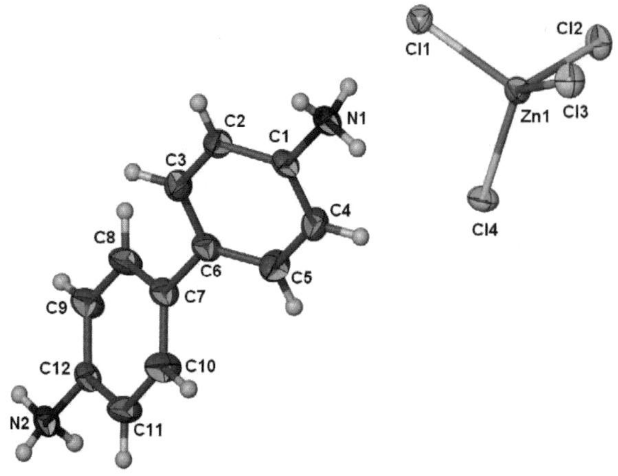

Figure 1. Asymmetric unit of $NH_3C_{12}H_8NH_3$ $ZnCl_4$ with atom labels and 50 % probability displacement ellipsoids for non-H atoms

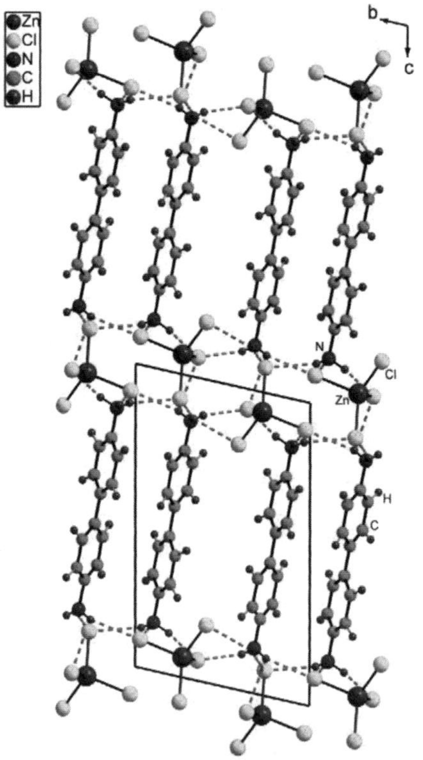

Figure 2. View down the a axis of the $NH_3C_{12}H_8NH_3$ $ZnCl_4$ showing alternating organic inorganic layers, anions and cations are connected by N-H...Cl hydrogen bonds (dashed lines)

As expected, in $[ZnCl_4]^{2-}$ anion, the Zn atom has a tetrahedral coordination surrounded by four Cl atoms. In the $[ZnCl_4]^{2-}$ tetrahedra, the main values of Zn-Cl distances and Cl-Zn-Cl angles are 2.2738(6) Å and 109.4(2)°, respectively, which are in agreement with those found in other organic tetrachloridozincate [13-16]. The slight deviation from the perfect tetrahedron around Zn(II) can be explained by the involvent of the chlorine ions in the hydrogen bonding. However, the Zn-Cl bond lengths do not differ appreciably with the shortest and longest bond lengths of 2.2367 (6) and 2.3241 (6) Å associated to Zn-Cl2 and Zn-Cl3, respectively. In addition, the Cl-Zn-Cl bond angles are also comparable, varying between 105.55 (2) and 122.06 (2)°. All of these data indicate a slight deviation from a perfect tetrahedral environment around the zinc cation according to the literature results [13-16].

Table 4. Bond length (Å) and angles (°) in the hydrogen-bonding scheme in $NH_3(C_{12}H_8)NH_3ZnCl_4$.

D—H···*A*	*D*—H	H···*A*	*D*···*A*	*D*—H···*A*
N1—H1a···Cl2^{i}	0.86	2.58	3.361 (2)	152
N1—H1b···Cl4	0.86	2.63	3.438 (2)	156
N1—H1c···Cl1^{ii}	0.86	2.32	3.175 (2)	171
N2—H2a···Cl4^{iii}	0.86	2.56	3.209 (2)	134
N2—H2b···Cl2^{iv}	0.86	2.40	3.261 (2)	174
N2—H2c···Cl1^{v}	0.86	2.27	3.121 (2)	169

An examination of the organic moiety geometrical features shows that, the carbon atoms building the phenyl rings of the title compound have a good coplanarity, and they form conjugated planes with average deviations of 0.0031 and 0.0055 Å for phenyl 1 (from C1 to C6) and phenyl 2 (from C7 to C12), respectively. The main value of the C-C bond lengths in the two phenyl rings is 1.383 Å, which is between a single bond and a double bond, agree with that observed in 4,4'-biphenylenediammonuim tetrachloridocuprate and 4,4'-biphenylenediammonuim tetrachloridomercurate [17].

3.2. IR spectroscopy

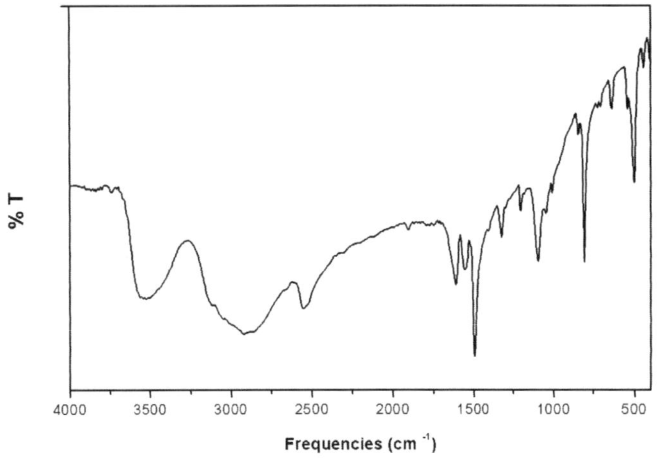

Figure 3. IR spectrum of $NH_3C_{12}H_8NH_3$ $ZnCl_4$.

Because IR spectroscopy is one of the major physical methods for the investigation of molecular structures, we have studied the IR spectrum of crystalline organic-inorganic hybrid compound $NH_3C_{12}H_8NH_3ZnCl_4$ which is shown in Fig. 3. To assign the IR peaks to vibrational modes, we examined the modes and frequencies observed in similar compounds [18-19]. The broad bands between 3600 and 2700 cm^{-1} are attributed to stretching vibrations of CH_2 and NH_3 groups (ν(C-H) and ν(N-H)). Frequencies in the range 2600-1700cm^{-1} are assigned to combination bands. The intense band at 1494 cm^{-1} accompanied by a low intensity peak at 1326 cm^{-1} is attributed to the phenyl ring vibrations. Bands of the symmetric and antisymmetric deformation vibrations of the NH_3^+ groups ($\delta_s(NH_3)$ and $\delta_{as}(NH_3)$) are observed in 1650-1500 cm^{-1} region indicate the presence of the cation in the doubly protonated base in the crystal structure. All peaks located in the frequency range 800-1250 cm^{-1} are attributed to stretching vibrations of C-N and C-C bonds (ν(C-N) ν(C-C)) and deformation in the plane δ(CH). The bands of low intensities at 726 and 636 cm^{-1} correspond to the twist out of plane t(CH). Deformation δ(NCC) and the twist out of plane twisting type of NH_3 observed in the range 400-550 cm^{-1}.

4. Conclusion

$[NH_3C_{12}H_8NH_3]ZnCl_4$ was prepared as single crystals at room temperature and characterized by various physico chemical methods. On the structural level, the atomic arrangement of the title compound can be described by inorganic layers organization. The inter layer spacing is occupied by the organic molecules. The structural of the title material is stabilized by interesting H-bond. The vibrational proprieties of this structure were studied by infrared spectroscopy. The assignment of the vibrationnal bands was performed by comparison with the vibration modes frequencies of homologous compounds.

References

[1] Yan B, Xu Y, Goh N. K, Chia L. S 2000 *Inorg. Chem. Commun* **3** 379.
[2] Ferrey G 2001 *Chem. Mater.* **13** 3084.
[3] Berrocal T, Mesa J. L, Pizarro J. L, Lezama L, Bazan B, Arriortua M. I, Rojo T 2008 *J. Solid State Chem.* **181** 884.
[4] Willet R. D, Gomez-Garcia C. J, Twamley B 2005 *Polyhedron* **24** 2293.
[5] Jarboui A, Louati B, Hlel F, Guidara K 2010 *J. Alloys Compd.* **503** 340.
[6] Xiao Z. L, Chen H. Z, Wang M, Tang B. Z 2005 *Gongneng Gailiao/Journal of Functional Materials,* **36** 1217
[7] Sheldrick G. M 1997 *SHELXS-97: Program for Crystal Structure solution, University of Gottingen, Germany.*
[8] Sheldrick G. M 1997 *SHELXL-97: Program for Crystal Structure solution, University of Gottingen, Germany* 1997.
[9] Sheldrick G. M 1996 *SADABS. University of Göttingen, Germany* 1996.
[10] Barbour L. J 2001 *J. Supramol. Chem. 1* **189.**
[11] Brown I. D 1976 *Acta crystallogr. A* **32** 24.
[12] Janiak C 2000 *Journal of the Chemical Society, Dalton Transactions* **21** 3885.
[13] Ben Gharbia I, Kefi R, Ben Nasr C 2007 *Anal. Sci. Xray struct. Anal. Online,* **23** 243.
[14] Jin Z. M, Shun N, Lu Y. P, Hu M. L, Shen L 2005 *Acta Crystallogr. C* **61** m43.
[15] Hlel F, Rhaim A, Guerfel T, Guidara K, Naturforsch Z 2006 **61** B 1002.
[16] Guo N, Yi J, Chen Y, Liao S, Fu Z 2007 *Acta Crystallogr. E* **63** m 2571.
[17] Belhouchet M, Wamani W, Mhiri T 2010 IOP *Conf. Series: Materials Science and Engineering* **13** 12039.
[18] Topacli C, Topacli A 2007 *J. Molecular Structure* **658** 9.
[19] Akalin E, Akyüz S 2003 *J. Molecular Structure* **571** 650.

IOP Conf. Series: Materials Science and Engineering **28** (2012) 012036 doi:10.1088/1757-899X/28/1/012036

Quantification of surface state effects in GaAs MESFETs

Z. Hadjoub[1], K. Cheikh, A. Zouyed, A. Khoualdia and A. Doghmane
Laboratoire des Semi-Conducteurs, Département de Physique, Faculté des Sciences, Université Badji-Mokhtar, BP 12, Annaba, DZ-23000, Algérie

E-mail: E-mail: hadjoub.zahia@univ-annaba.org

Abstract. The operation of gallium arsenide schottky barrier field effect transistor is greatly affected by several anomalies such as frequency dispersion of output impedance, z_{ds}; this low frequency behaviour is related to the presence of capture centres at the channel/substrate interface. In this context, we investigate the influence of such defects via a circuit consisting of a capacitance in series with a resistance placed in parallel to the output of a transistor, between drain and source. Then, using PSPICE program we made some simulations at several bias of the device and different values of elements characterising traps. It was found that the dispersion increases when the value of the drain bias increases. It was also shown that the dispersion increases with increasing trap resistance. Whereas, the trap capacitance greatly affects the initial and final saturation of the output impedance

1. Introduction

Despite the great evolution of modern technology in order to achieve high performance devices, surface and interface phenomena still play undesirable effects [1-3]. In particular, for gallium arsenide metal semiconductor field effect transistor, GaAs MESFET, the conducting channel, limited by two depleted zones, is highly perturbed by uncontrollable defects and capture centres. Thus, there always exist some persistent limitations such as low-frequency anomalies which affect the performance of broadband systems [4-6]. In fact, it was noted that the output impedance of GaAs MESFET at high frequencies can be substantially different than its dc value [7, 8]. This behavior, although of relatively minor importance in microwave and digital circuits, can profoundly affect the performance of analogical integrated circuits.

The origin of frequency dispersion phenomenon is still a subject of some controversy. Although it usually attributed [9-11] to the influence of deep traps localized at the interface between active layer and semi-insulating substrate, it was also reported that its origin could be due to short channel effect and/or to the diminution of the square resistance of n^+ layer beneath ohmic contacts. Other interpretations were related to device structure with buried channel or p-well structure [12, 13]. Consequently, the presence of localized surface states was modeled by an equivalent circuit consisting of a surface state capacitance, C_{ss}, and a resistance, R_{ss}, [14].

In this work, to put into evidence the dispersive phenomenon, we first observe experimental frequency dependence of output impedance, $Z_{ds}(f)$ in GaAs MESFETs. Then, we use SPICE program to simulate $Z_{ds}(f)$ at different bias values via an equivalent circuit in low frequency that takes into

[1] To whom any correspondence should be addressed.

account trapping effects. Finally, comparing experimental and simulated results, we determine the influence of R_{ss} and C_{ss} values on $Z_{ds}(f)$.

2. Dispersion phenomena of output impedance

2.1 Experimental observations

Since the output impedance, Z_{ds}, is related to drain current, I_d, as well as to drain-source voltage, V_{ds}, it strongly depends on polarization intensities. Hence, measurements were carried out, at room temperature, at V_{gs} = 0 V and several V_{ds} polarizations ranging from 0.1 V to 0.9 V. The frequency interval was varied from 10 Hz to 1 MHz.

In order to illustrate both linear and saturated regimes, the experimentally measured magnitude of the output impedance of a commercial GaAs MESFET is plotted in Fig. 1, at a constant gate source potential (V_{gs} = 0 V) and for different drain-source polarizations (0.3 V, 0.5 V, 0.7 V and 0.9 V). According to V_{ds} values, two situations of $Z_{ds}(f)$ can clearly be observed: (i) dispersive behavior for $V_{ds} > 0.4$ V and (ii) constancy for $V_{ds} \leq 0.4$ V.

In the linear region ($V_{ds} \leq$ 0.4 V), the output impedance remains almost constant in the whole frequency range. Therefore, in this low biasing interval, the output impedance is not dispersive. This could be attributed to the fact that the conducting channel is uniform. Hence, transport phenomena are not influenced by the channel/substrate interface. However, when the device is operated in the saturated regime for higher V_{ds}, it is easy to note that the output impedance of a GaAs MESFET undergoes some variations at low frequencies (f < 1 KHz). It can be seen that the curves show an initial decrease, starting from a maximal value obtained for f = 10 Hz, to attain the lowest values, for each V_{ds}, at higher frequencies. This dispersion behavior obtained for V_{ds} = 0.7 V and 0.9 V is usually observed with most GaAs MESFET devices. Such frequency dispersions are considered as unwanted anomalies.

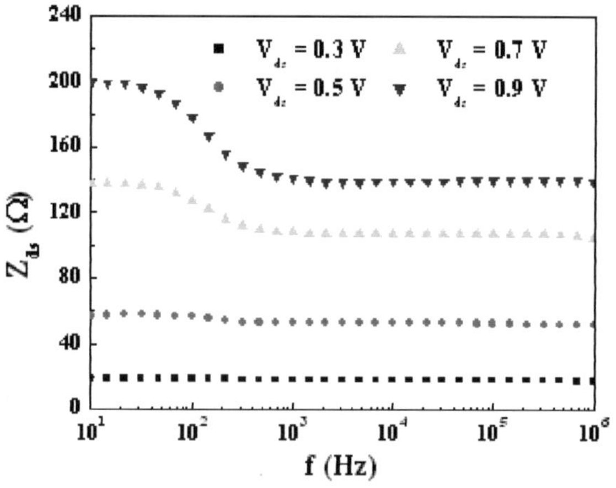

Figure 1. Output impedance variations with frequency at gate-source bias V_{gs} = 0 V.

2.2 Phenomenon simulation

Frequency dispersion phenomena of output impedance are attributed to the presence of localized surface states at the channel/substrate interface of the FET device. The gate biasing via a negative voltage creates a space charge region in the channel. Drain polarisation by a positive tension generates a non homogeneous width of the space charge region that is larger in the drain region than that near the gate (as schematically illustrated by Fig. 2a). Consequently, the depletion region under the gate extends towards the channel/substrate interface. This mechanism influences trap states situated at this interface at which an electron accumulation region appears. Hence, the semi-insulating substrate would acquire an n-type semiconductor behavior. Thus, this conducting region can be represented by a resistance R_{SS}. Whereas, below the source, the space charge region is too far from the channel/substrate interface so that electron traps are inactive. Therefore, this region can be considered as a capacitance, C_{ss}. Hence, due to the position of R_{SS} and C_{ss} elements, representing trap electrical activity, they can be associated in parallel to drain and source (Fig. 2a). Thus, in this investigation, we consider the initially [14] proposed circuit shown in Fig. 2b; it includes a branch consisting of a capacitance, C_{ss}, in series with a resistance, R_{ss}. Both C_{ss}, and R_{ss}, are then branched in parallel to the output impedance of the device.

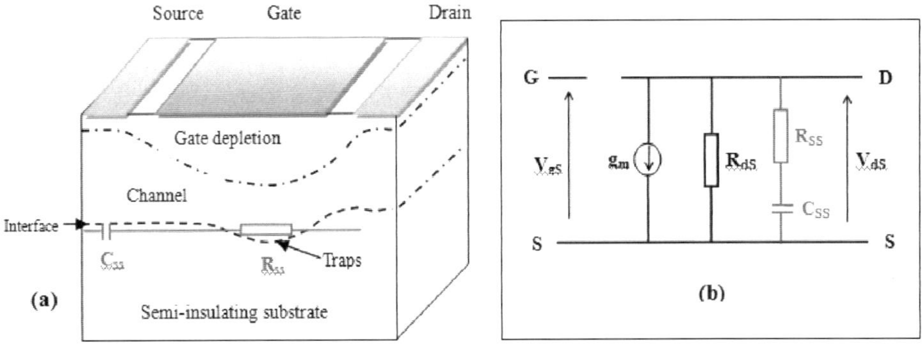

Figure 2. (a) Cross section of a GaAs MESFET, (b) Equivalent circuit model at low frequency; R_{SS} and C_{SS} represent traps characteristics.

The present methodology consists of several steps, for every polarization point: (i) measurements of the output impedance, $Z_{ds}]_m$, calculation of $Z_{ds}]_c$ in the same frequency range and (iii) comparison between measured and calculated values, $Z_{ds}]_m$ - $Z_{ds}]_c$ in order to determine, with high precision, trap characteristics, i.e., R_{ss} and C_{ss}.

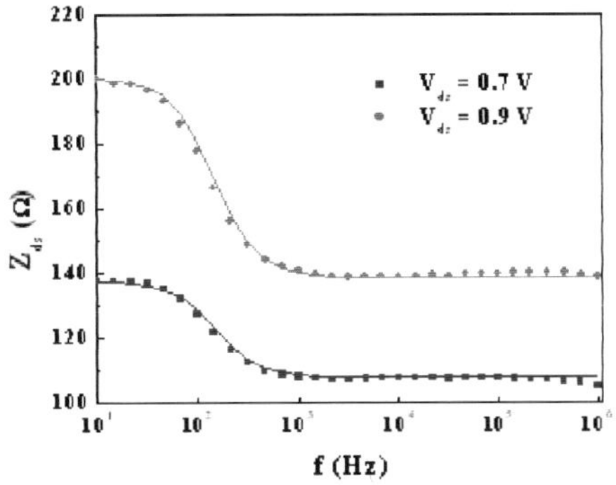

Figure 3. Frequency dispersion of output impedance (V_{GS}=0V) considering trapping effect: calculated (———) and measured results at V_{DS} = 0.7 V (■) AND 0.9 V (●).

The simulated results, thus obtained on a commercial GaAs MESFET, are plotted in Fig. 3 (continuous line) in terms of output impedance as a function of frequency at V_{gs} = 0 V. For comparison, we also included experimental results (from Fig. 1) for V_{ds} = 0.7 V (■) and 0.9 V (●).
It can clearly be seen that the agreement between experimental and calculated values is very good. Moreover, using an optimization method we were able to determine both C_{ss} and R_{ss}; they were found to be: (i) C_{ss} = 1.8 µF and R_{ss} = 0.45 kΩ for V_{ds} = 0.7 V and (ii) C_{ss} = 1.7 µF and R_{ss} = 0.5 kΩ for V_{ds} = 0.9V.

3. Quantification of surface state effects

3.1 Effects of trap resistance
In order to quantify the effects of the resistance R_{SS} on the frequency behavior of the output impedance, we fixed polarizations at constant values (V_{gs} = 0 V and V_{ds} = 0.7V) but varied R_{ss} values from 0.1 kΩ to 1.50 kΩ.

The obtained results are plotted in Fig. 4a in terms of Z_{ds} versus frequency at different R_{ss}. Several behaviors can clearly be observed. At the lowest frequency, f = 10 Hz, all the curves of different R_{ss} values start at the same Z_{ds} value, i.e., Z_{ds} = 137.5 Ω. However, as the frequency increases, Z_{ds} decreases initially then saturates. The extension of this decrease in Z_{ds} depends on the values of R_{ss}: it is larger for lower R_{ss}.

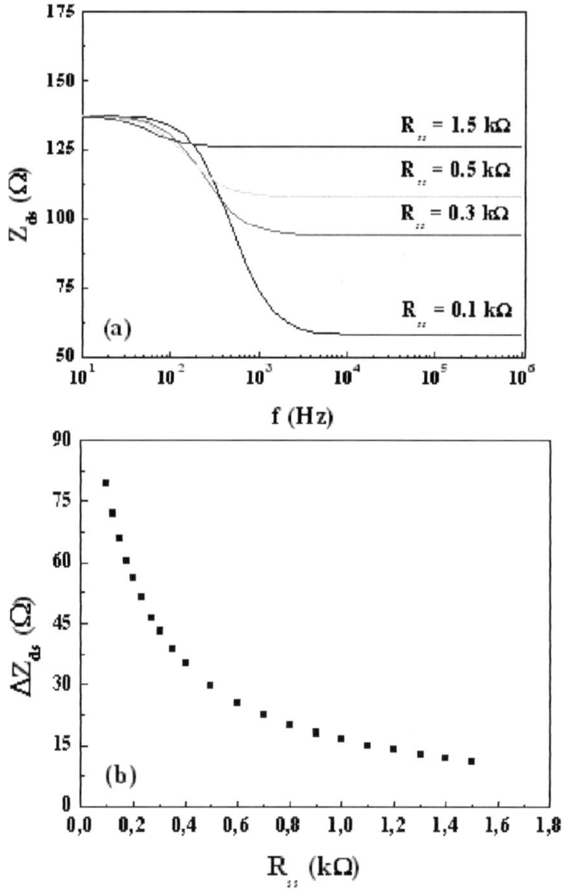

Figure 4. Frequency dispersion of output impedance (V_{GS}= 0V) considering trapping effect: calculated (———) and measured results at V_{DS} = 0.7 V (■) and 0.9 V (●).

To better quantify R_{ss} effects, we define the dispersion extension, ΔZ_{ds}, in terms of the difference between the highest and the lowest Z_{ds} values obtained for 10 Hz and 1 MHz, respectively, i.e.,:

$$\Delta Z_{ds} = Z_{ds}]_{(f = 10\ Hz)} - Z_{ds}]_{(f = 1\ MHz)} \tag{1}$$

Thus, the dependence of ΔZ_{ds} parameter on R_{ss} is better illustrated in Fig. 4b. It is clear that the dispersion becomes less important as R_{ss} increases. To quantify this phenomenon, we used limiting conditions: $Z_{ds}]_{(f=10Hz)} = Z_{ds}]_{(f\to 0)}$ and $Z_{ds}]_{(f=1MHz)} = Z_{ds}]_{(f\to\infty)}$. Hence, making use of the equivalent circuit in Fig. 2, the calculation of output impedance, consisting of a resistance R_{ds} placed in parallel to the characteristic impedance of traps ($Z_{ss} = R_{ss} + 1/j\omega C_{ss}$) leads to the following relation:

$$Z_{ds} = R_{ds}\left(\frac{1 + J\omega R_{SS}C_{ss}}{1 + J\omega C(R_{ss} + R_{ds})}\right) \tag{2}$$

By applying the limiting conditions to relation (2), it can readily be shown that ΔZ_{ds} takes the form:

$$\Delta Z_{ds} = \frac{{R_{ds}}^2}{R_{ss} + R_{ds}} \tag{3}$$

Therefore, under the present polarization conditions, $V_{gs} = 0$ V and $V_{ds} = 0.7$V, one can easily find that $R_{ds} = 137.5\ \Omega$ and C_{ss} $C_{ss} = 1.7\mu F$. Hence, relation (3) becomes:

$$\Delta Z_{ds})_{\Omega} = \frac{19.10^3}{140 + R_{ss}} \tag{4}$$

3.2 Effects of trap capacitance

Following the same steps, as for R_{ss}, we investigate the effects of trap capacitance, C_{ss}, on frequency variation of Z_{ds} by varying C_{ss} from 0.1 µF to 3 µF. The obtained results are shown in Fig. 5a. Unlike previous R_{ss} results, the C_{ss} values do not seem to influence ΔZ_{ds} that remains constant. We define the frequency band variations, Δf, as:

$$\Delta f = f_2 - f_1 \tag{5}$$

where f_1 and f_2 are cutoff frequencies deduced from relation (2); they are given by:

$$f_1 = \frac{1}{2\pi C_{ss}(R_{ss} + R_{ds})} \qquad f_2 = \frac{1}{2\pi C_{ss}R_{ss}} \tag{6}$$

Hence, replacing f_1 and f_2 into (5) one gets:

$$\Delta f = \frac{R_{ds}}{2\pi C_{ss}R_{ss}(R_{ds} + R_{ss})} \tag{7}$$

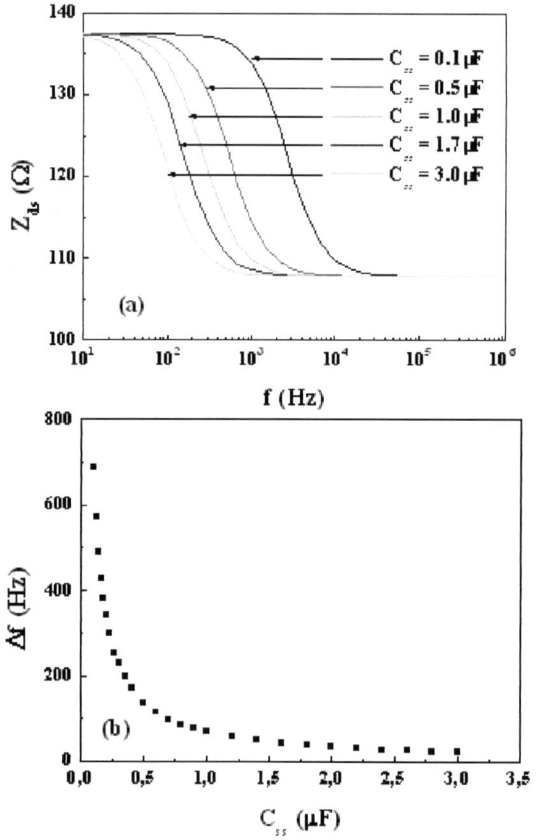

Figure 5. (a) output impedance as function of frequency with C_{SS} parameter and (b) overall band frequency variation with C_{SS}.

The dependence of Δf as a function of C_{ss} is better illustrated in Fig. 5b. It is clear that Δf is inversely proportional to C_{ss}. Using R_{ds} = 137.5 Ω and R_{ss} = 0.5 kΩ, this relation could be simplified to take the form:

$$\Delta f)_{Hz} = \frac{6{,}87.10^{-5}}{C_{ss}} \quad (8)$$

3.3 Interdependence of Δf and ΔZ_{ds}

Figure 6 illustrates output impedance variations, ΔZ_{ds}, as a function of frequency variations, Δf; it is clear that ΔZ_{ds} increases with increasing Δf. Such dependence could be expressed analytically via curve fitting (—) to obtain the following expression:

$$\Delta Z_{ds} = 120 - 116\, e^{-11.7.10^{-4}\Delta f} \quad (9)$$

If f is given in Hz, relation (9) can be re-written as:

$$\Delta Z_{ds)\Omega} = 120 - 116 \exp\left(-\frac{\Delta f}{856}\right) \quad (10)$$

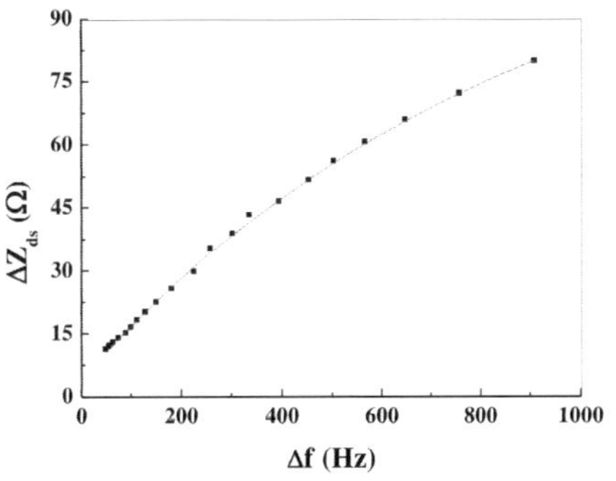

Figure 6. Maximum dispersion of output impedance as function of band frequency variation

This relation is of great importance because the maximum dispersion of output impedance could be determined by just knowing the frequency band width and vice versa. Consequently, it becomes possible to evaluate the influence of trapping characteristics via R_{SS} and C_{SS} values.

4. Conclusions

Output conductance of a GaAs MESFET was measured, at room temperature, under different bias condition in the frequency range [10 Hz – 1MHz]. It was found that, at low polarizations ($V_{ds} < 0.4V$) no frequency variations was noticed. Whereas, for $V_{ds} > 0.4$ V a negative frequency dispersion behavior was obtained. These phenomena are attributed to deep traps situated at the active zone/semi-insulating substrate interface. The comparison of the obtained results with measured ones permitted the deduction of the deeps characteristics represented with a circuit composed of capacitance, C_{ss}, in series with resistance, R_{ss}. Subsequently the influence of the R_{ss} and C_{ss} values on the $Z_{ds}(f)$ is quantified. The maximum dispersion is larger for lower R_{ss}. Whereas, C_{ss}, controls the frequency band width that is limited by cutoff frequencies. Finally, a relation between variations of output impedance and frequency was established.

References

[1] Lu H, Zhang Yi, Zhang Yu, and Zhang T 2009 *Solid-State Electronics* **53** 285
[2] Hadjoub Z, Khoualdia A, Cheikh K and Doghmane A 2007 *ICTON'MW'07* IEEE, FRP.8.
[3] Kunihiro K and Ohno Y 2001 *Solid-State Electronics* **45**, 1763.
[4] Hadjoub Z, Cheikh K and Doghmane A 2009 *Optoelectronics and Advanced Materials – Rapid Communications* **3** 360.
[5] Johnson D. A 1990 *IEEE Transactions on Electron Devices* **37** 1217.
[6] Graffeuil J, Hadjoub Z, Fortea J. P and Pouysegur M 1986 *IEEE Solid-State Electronics,* **29**, 1087.
[7] Khoualdia A Hadjoub Z and Doghmane A 2010 *International Journal of Nanoelectronics and Materials,* **3**, 9.
[8] Reynoso-Hernandez J A and Graffeuil J 1989 *IEEE. Transactions on Microwave Theory and Techniques,* **37** 1478.
[9] Hasumi Y, Matsunaga N, Oshima T and Kodera H 2003 *IEEE Transactions on Electron Devices* **50** 2032.
[10] Nakajima S, Yanagisawa M, Tsumura E and Sakurada T 2000 *IEEE Transactions on Electron Devices* **47** 2255.
[11] Reynoso-Hernandez J. A, Escotte L, Plana R. and Graffeuil J 1995 *Electronics Letters* **31** 677.
[12] Canfield P, Medinger J and Forbes L 1987, *IEEE Electron Device Letters,* **3** 88
[13] Choi S and Das M. B 1994, *IEEE Transactions on Electron Devices* **41** 1725.
[14] Camacho-Penalosa C and Aitchison C. S 1985, *Electronics Letters,* **21** 528.

IOP Conf. Series: Materials Science and Engineering **28** (2012) 012037 doi:10.1088/1757-899X/28/1/012037

Acoustic microscopy evaluation of anisotropy effects on elastic properties of Na_2CoGeO_4 single crystals

A. Doghmane[1a], I. Al-Suraihy[a,b], I. Hadjoub[a,c], and Z. Hadjoub[a]
[a] Laboratoire des Semi-Conducteurs, Département de Physique, Faculté des Sciences, Université Badji-Mokhtar, BP 12, Annaba, DZ-23000, Algérie
[b] Department of Physics, Faculty of Education, AL-Mahweet, Sana'a University, Al-Mahweet, Yemen
[c] Ecole Préparatoire aux Sciences et Techniques, EPST, Annaba, DZ-23000 Algérie,

E-mail: E-mail: doghmane.abdellaziz@univ-annaba.org

Abstract. In this paper, we investigate anisotropy effects on the elastic acoustic parameters (reflection coefficients R(θ), critical angles of wave excitation, acoustic signatures V(z), spatial periods, velocities) of sodium cobalt germanate Na_2CoGeO_4 single crystals with six crystallographic orientations [100], [001], [010], [110], [101], and [011]. Simulations were carried out in the case of a scanning acoustic microscope at an operating frequency of 142 MHz, a half opening angle of 50° and water couplant. Interesting results were found. Shear modes were first excited at θ_S ~ 26.8° on [100] direction. However, at other propagation directions [001], [010], [110], [101], and [011] the shear mode appears at higher angles: θ_S ~27.1°, 31°, 33.9°, 40.5°, 43° respectively. Propagation of acoustic waves showed a decrease of longitudinal waves velocities from 5687 m/s for [101] direction to 4471 m/s for [011] direction. Similar behaviors were obtained with R(θ) and V(z) curves. These variations are due to either low or high atomic occupation of different orientations.

1. Introduction

Materials elastic properties are of great importance in industrial applications as well in basic research. In industry, elastic data are very much useful to determine the strength of the materials under various strained conditions while in basic research, the data are useful in obtaining an insight into the structure of the inter-atomic and inter-ionic forces in solids especially of the long-range type forces. These data can be obtained via static or dynamic methods; the latter case based on acoustic wave propagation is non destructive. However, elastic wave propagations strongly depend on the nature of propagating media, in particular in solids that could be homogeneous or inhomogeneous, isotropic or anisotropic [1-4].

Therefore, it is our aim to investigate anisotropy effects on several acoustic parameters: reflection coefficients R(θ), critical angles of wave excitation, acoustic signatures V(z), spatial periods Δz, velocities of propagating modes). This investigation is carried out, in the case of a scanning acoustic microscope, on sodium cobalt germanate Na_2CoGeO_4 single crystals that are playing very important roles in many modern applications.

[1] To whom any correspondence should be addressed.

2. Materials and computing procedure

2.1 Materials

Sodium-cobalt germanate single crystals, Na_2CoGeO_4, possess orthorhombic structure. These crystals are isomorphous to Na_2ZnGe0_4, whose monoclinic unit cell is strongly pseudo-orthorhombic; its lattice parameters are a = 7.175, b=5.605 and c =5.325 and the angles α, βand γ are equal to 90°. The characteristic equation of orthorhombic structure, as in case of Na_2CoGeO_4, is of the same form as for the hexagonal classes, but with more general stiffness coefficients [5, 6]. The idealized form [7] of Na_2CoGeO_4 single crystals and crystallographic orientations are schematically illustrated in Fig. 1.

The single crystals considered in this work were prepared by direct synthesis of Na_2CoGeO_4 from the oxides as well as recrystallization of Na_2CoGeO_4 and seeded growth [8].The spontaneous crystallization technique does not permit the stable production of large crystals with a high degree of perfection. It may be reasonable of seed growth or by the modified spontaneous crystallization method with the aim to suppress the nucleation and thus to produce somewhat larger single crystals. It was in these directions that work was carried out in search of the technique of growing single crystals [8].

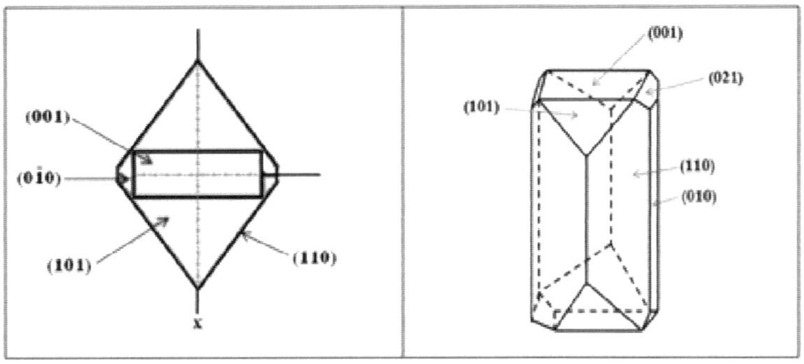

Figure 1. Schematic representation of crystallographic orientations of Na_2CoGeO_4 single crystals

By applying the pulse ultrasonic method in the frequency of 14 MHz and the synchronous ring method, it was reported [8, 9] that the velocities of longitudinal (V_L) and transverse (V_S) elastic waves can be experimentally measured (Table 1) along definite crystallographic directions: [100], [001], [010], [110], [101], and [011]. These published data are to be used in the present simulations to investigate all other acoustic parameters.

Table 1. Experimental data [8] of longitudinal and shear Velocities in Na_2CoGeO_4 directions.

Velocities	**Crystallographic directions**					
	[100]	[010]	[011]	[101]	[001]	[110]
V_L (m/s)	5540	5260	4480	5790	5430	5690
V_S (m/s)	3250	3230	2810	2610	2240	2130

2.2 Calculating procedure

The calculation procedure, described in detail elsewhere [10-14], consists of several steps:

a) calculating the reflection coefficient, R(θ) from [14]:

$$R(\theta)=\frac{(Z_L Cos^2 2\theta_T + Z_T Sin^2 2\theta_T - \rho_{Liq} V_{Liq} / Cos\theta)}{(Z_L Cos^2 2\theta_T + Z_T Sin^2 2\theta_T + \rho_{Liq} V_{Liq} / Cos\theta)} \tag{1}$$

where Z_L, Z_S, ρ_{Liq} and V_{Liq} are liquid impedance, solid impedance, coupling liquid density and the propagating wave velocity in the liquid, respectively.

b) computing the V(z) curves of the whole specimen-lens system from the angular spectrum model [13] given by the expression:

$$V(z)=\int P^2(\theta)R(\theta)\exp(2jk_o z\cos\theta)\sin\theta\cos\theta d\theta \quad (2)$$

where $P^2(\theta)$ is the pupil function, θ is the half-opening angle of the lens, z is the defocusing distance and $k_o = 2\pi/\lambda$ is the wave number in the coupling liquid, $j = \sqrt{-1}$.

c) deducing V(z) of the sample by subtracting the response of the lens $V_l(z)$ from V(z),

d) applying fast Fourier transform, FFT, a spectral technique to the obtained signal $V(z)-V_l(z)$, in order to determine spatial periods Δz,

e) deducing the phase velocity of the corresponding mode from the following relation [12]:

$$V_R = V_{Liq.}/[1 - (V_{Liq.}/2f\Delta z)^2]^{1/2} \quad (3)$$

where f is the operating frequency.

2.3 Simulation conditions

Theoretical simulations were carried out in the case of a scanning acoustic microscope in the reflection mode under normal operating conditions: an operating frequency of 142 MHz, a half opening angle of 50° and water as the coupling liquid whose density, ρ = 1000 kg/m^3 and longitudinal velocity V_L = 1500 m/s.

3. Results and discussions

3.1 Anisotropy effects on R(θ)

Reflection coefficients of sodium germanate Na_2CoGeO_4 along definite crystallographic directions: [100], [001], [010], [110], [101], and [011] were calculated. Figure 2 illustrates typical results of the amplitude (Fig. 2a) and the phase (Fig. 2b) as a function of incidence angle. It is clear that, as the angle of incidence increases we notice important changes in both amplitude and phase.

In amplitude, the change occurs at the critical angle, θ_S, at which the shear mode is excited for every crystallographic direction. Beyond these critical angles all the energy is reflected. The modulus of R(θ) becomes equal to unity as a result of total reflection and a null transmission. It can readily be deduced that shear modes are excited at θ_S ~ 26.8°, 27.1°, 31°, 33.9°, 40.5° and 43° for [100], [001], [010], [110], [101], and [011], respectively.

In phase, the most important fluctuation, with $\Delta\varphi = 2\pi$, occurs at the critical angle, θ_R, at which Rayleigh modes are excited. Thus, one can easily determine that θ_R ~ 30.5°, 31°, 36.2°, 38.2°, 45.4° and 48.2° for wave propagation in the directions [100], [001], [010], [110], [101], and [011], respectively. It should be noted that the onset of the 2π phase change corresponds to the shear critical angle θ_S, (as indicated by the arrow in Fig. 2b). The slight shift between θ_S and θ_R is due to the fact that the Rayleigh velocity varies slightly from 87 to 95 per cent of the shear velocity [14]. The degree of fluctuations in phase and amplitude in R(θ) in the different crystallographic directions of each critical angle are indicative of the generation efficiency of different modes.

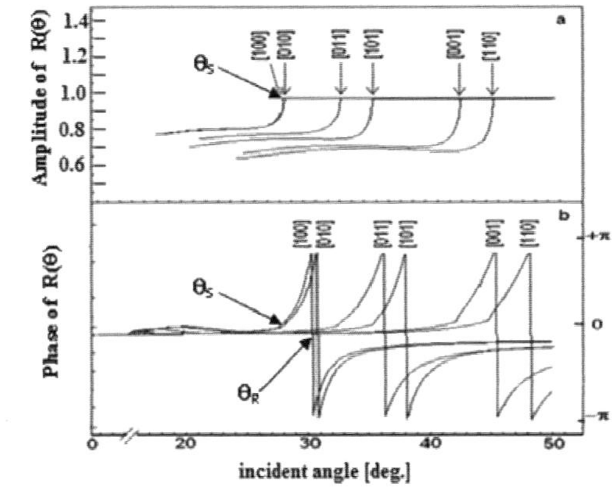

Figure 2. Amplitude (a) and phase (b) of refection coefficient of Na_2CoGeO_4 at different crystallographic directions.

3.2. Anisotropy effects on V(z) curves

Acoustic materials signatures, V(z), are the most important parameter that can be experimentally measured by a scanning acoustic microscope. These signatures represent the output signal, V, as a function of the defocusing distance, z, when the sample is moved vertically in the z axis towards the acoustic lens. The V(z) curves can also be calculated from equation 2 and previously determined reflection coefficients (Fig. 2). Typical obtained curves are illustrated in Fig. 3a for Na_2CoGeO_4 at six crystallographic directions [110], [001], [101], [011], [010] and [100].

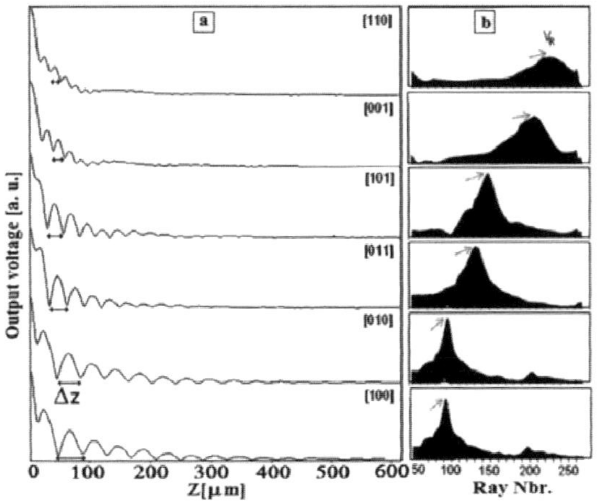

Figure 3. Acoustic materials signatures (a) and their FFT spectra (b) at different crystallographic directions on Na_2CoGeO_4.

It can clearly be seen that all the curves show oscillatory behavior due to constructive and destructive interference between different propagating modes in different direction. The difference between two successive minima (or maxima) is known as the spatial period, Δz, related to the velocity of the propagating mode by equation 3. Thus, it is obvious (Fig. 3a) that the spatial period Δz decreases gradually from direction [100] to [110]. The deduced values are assembled in Table 2.

3.3. Anisotropy effects on wave velocities

The variations in Δz is better put into evidence by the spectral treatment of V(z) periodic signals (Fig. 3a) via fast Fourier transform analysis. Figure 3b illustrates FFT spectra of the corresponding V(z) curves of Na_2CoGeO_4 at different orientations. The principal peak corresponds to the most dominant propagating mode. It is well established that under the simulating conditions the propagating mode corresponds to that of leaky Rayleigh waves. It can be noticed that the height as well as the position of the peak change with crystallographic orientations. The height indicates that generation efficiency of the mode; whereas, the position corresponds to the value of Rayleigh wave velocity, V_R. The determined values (Eq. 3), regrouped in Table 2, put clearly into evidence the anisotropy effects on such acoustic parameters.

Table 2: Calculated elastic parameters of Na_2CoGeO_4 in different crystallographic directions

Modes	Parameters	Crystallographic directions					
		[100]	**[010]**	**[011]**	**[101]**	**[001]**	**[110]**
Rayleigh	**θ_R(deg.)**	30.2	30.7	36.1	37.8	45.3	48.2
	Δz (μm)	39.7	38.9	28.7	25.1	18.9	17.3
	V_R(m/s)	2991	2962	2580	2464	2151	2077
Longitudinal	**θ_L(deg.)**	15.6	16.7	19.6	15.1	16.1	15.3
	Δz (μm)	141.3	128.5	92.4	151.3	133.3	147.5
	V_L(m/s)	5500	5251	4471	5687	5354	5613

To enrich this investigation, we considered longitudinal acoustic waves which are a type of acoustic modes that can propagate in materials faster than shear and Rayleigh modes. To put into evidence anisotropy effects on such mode, we first calculated R(θ) in the vicinity of longitudinal critical angles, then the corresponding V(z) curves and finally their FFT spectra. Typical obtained results are displayed in Fig. 4. Similar behaviours were obtained as those of Rayleigh modes for different crystallographic orientations:

(i) variations in longitudinal critical angles,
(ii) (ii) changes of spatial periods,
(iii) changes in positions of FFT principal rays,
(iv) variations in peak heights and
(v) variations in the values of longitudinal velocities.

All these observations are regrouped in Table 2. This behaviour could be explained by the atomic occupation of different planes. For denser directions the velocity is higher.

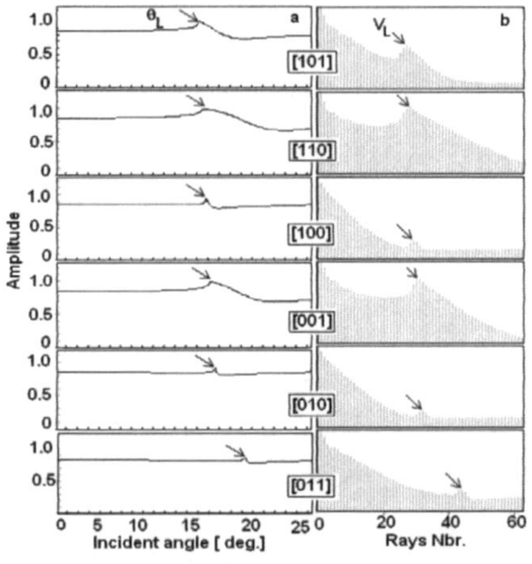

Figure 4. Effects of anisotropy on (a) amplitude of reflection function in the vicinity of the longitudinal mode and (b) FFT spectra extracted from the corresponding V(z) curves.

3. Conclusions

The crystallographic direction effects in Na_2CoGeO_4 crystals are put into evidence for reflections coefficients, acoustic signatures, critical angles of mode excitation, spatial periods as well as velocities of different modes. For example, the propagation of acoustic waves in six crystallographic directions [101], [110], [100], [001], [010] and [011] of Na_2CoGeO_4 shows a decrease of the velocity of longitudinal waves from 5687 m/s to 4471 m/s due to the decreasing the atomic occupations of these planes.

References

[1] Maev R. G 2008 *Acoustic Microscopy: Fundamentals and Applications* (Berlin: Wiley-VCH).
[2] Doghmane A, Bouziane N and Hadjoub Z 2009 *J. Mater. Sci. Eng.* **3**(8) 8.
[3] Briggs A (ed.) 1995 *Advances in Acoustic Microscopy* (New York: Plenum Press,)
[4] Zinin P. V 2001 *Handbook of Elastic Properties of Solids, Liquids and Gases.* **1***: edited by: Levy et al.* 187 (New York: Academic Press).
[5] Al-Surayhi I, Doghmane A and Hadjoub Z 2009 *Damage and Fracture Mechanics, edited by Boukharouba et. al.* 415 (Springer science).
[6] Seetawan T, Amornkitbamrung V, Burinprakhon T, Maensiri S, Kurosaki K, Muta H, Uno M and Yamanaka S 2006 *J. Alloys Compd.* **407** 314.
[7] Lobachev A. N, Demianets L. N, Kuzm1na I. P and Emelianova E. N 1972 *J. Cryst. Growth* **13-14** 540.
[8] Duderov N. G, Demianets L. N, Lobachev A. N, Pisarevsky Y. V and Sil'vestrova I. M 1978 *J. Cryst. Growth* **44** 483.
[9] Wintenberger M and Andron B. L 1973 *Phys. Status Solidi (a)* **18** 209.
[10] Yu Z and Boseck S 1995 *Rev. Mod. Phys.* **67** 863.
[11] Hadjoub Z, Touati I, Doghmane M, and Doghmane A 2008 *C. R. Phys.* **9** 903.
[12] Kushibiki J and Chubachi N 1985 *IEEE Sonics Ultrason.* **32** 189.
[13] Sheppard C. G. R and Wilson T 1981 *Appl. Phys. Lett.* **38** 858.
[14] Briggs A 1992 *Acoustic Microscopy* (Oxford: Clarendon Press).

IOP Conf. Series: Materials Science and Engineering **28** (2012) 012038 doi:10.1088/1757-899X/28/1/012038

Bulk metallic glass matrix composite for good biocompatibility

F. Hadjoub[1] [a,b], W. Metiri[a], A. Doghmane[a] and Z. Hadjoub[a]

[a] Laboratoire des Semi-Conducteurs, Département de Physique, Faculté des Sciences, Université Badji-Mokhtar, BP 12, Annaba, DZ-23000, Algérie

[b] Département de Médecine, Faculté de Médecine, Université Badji Mokhtar, BP 205, Annaba DZ-23000, Algeria

E-mail: fati_hadjoub@yahoo.fr

Abstract. Reinforcement volume fraction effects on acoustical parameters of $Zr_{41.2}Ti_{13.8}Cu_{12.5}Ni_{10}Be_{22.5}$ matrix composites reinforced by Mg, Ag and Cd metals have been studied via a simulation program based on acoustic microscopy technique. Moreover, acoustical parameters of human bone were compared to those of BMGs in both monolithic and reinforced case. It was found that elastic behavior of BMGs matrix composites in high reinforcement volume fraction is similar of that of human bone. This behavior leads to high biocompatibility and good transfer of stress between composite material and human system.

1. Introduction

Bulk metallic glass, BMGs, have attracted a lot of attention due to their remarkable properties like high strength, elasticity, corrosion resistance and unique processing capability [1-4]. Because of their flexible chemistry, atomic structure and high biocompatibility, they offer potential opportunity to fabricate complex implants and devices. They are used in the development of new medical devices, prostheses, tissue repair and replacement technologies. However, the uses of BMGs materials in biomedical applications are limited by their catastrophic failure at room temperature and their high value of elastic constants [5]. To overcome this failure mode and to enhance their elastic properties, BMGs matrix composites have been produced [6].

The aim of our work is to study acoustical behaviors of $Zr_{41.2}Ti_{13.8}Cu_{12.5}Ni_{10}Be_{22.5}$ BMG composites for different fibers in various volume fractions. This behavior is then compared to human bone.

2. Materials and methodology

Matrix ($Zr_{41.2}Ti_{13.8}Cu_{12.5}Ni_{10}Be_{22.5}$) and fibers (Mg, Ag and Cd) densities and elastic constants [7, 8] represented in table 1, were used to calculate densities, ρ_c, Young's modulus, E_c, and shear modulus, G_c of BMGs composites by employing the following relationships [9]:

$$\rho_c = 1/[(\square_f/\rho_f)+(\square_m/\rho_m)] \quad (1)$$

$$E_c = E_f\square_f+E_m\square_m \quad (2)$$

$$G_c = (G_m\, G_f)/[G_m\square_f + G_f\,(1-\square_f)] \quad (3)$$

[1] To whom any correspondence should be addressed.

where ρ_f, E_f and G_f are the density, Young's and shear modulus of fibers, P_m, E_m and G_m are those of matrix, $\square_f$ and $\square_m$ are the volume fraction of fiber and matrix composite, respectively.

Table 1. Densities and elastic constants of Vit.1 BMG's & Mg, Ag, Cd metals

Materials [Ref]	ρ [Kg/m^3]	E [GPa]	G [GPa]
Vit.1 [7]	6130	101.2	37.4
Mg [8]	1738	44	17
Ag [8]	10600	81.2	29.7
Cd [8]	8600	66.6	25.1

The calculated ρ_c, E_c and G_c from relations (1), (2) & (3) are used to deduce longitudinal, V_L, and transversal waves velocities, V_T, from the following familiar relations [10]:

$$E_c = \rho_c V_T^2(3V_L^2 - 4V_T^2)/(V_L^2\ V_T^2) \quad (4)$$

$$G_c = \rho_c\ V_T^2 \quad (5)$$

Using the calculated values of V_L and V_T for such materials in the simulation program, we can deduce the acoustical reflection coefficient, R(θ), which expressed as [9]:

$$R(\theta) = (Z_{sol}-Z_{liq})/(Z_{sol}+Z_{liq}) \quad (6)$$

with Z_{liq}, Z_{sol} being acoustical impedances of coupling liquid and total impedance of the solid material, respectively; they are given by:

$$Z_{liq} = \rho_{liq}.V_{liq}/\cos\theta \quad (7)$$

$$Z_{sol} = Z_L\cos^2 2\theta_L + Z_T\sin^2 2\theta_T \quad (8)$$

where θ_L and θ_T are the critical angles at which the excitation of longitudinal and transverse mode occurs. The simulation program uses R(θ) to calculate acoustic materials signatures, V(z), using the following relation:

$$V(z) = \int_0^{\pi/2} P^2(\theta)R(\theta)[\exp(2jk_0 z\cos\theta)]\sin\theta.\cos\theta.d\theta \quad (9)$$

with θ is the lens opening angle, P(θ) is the lens pupil function, k_0 is the liquid wave number and z is the sensor defocusing distance. The V(z) is obtained by recording the reflected acoustic signal variation as a function of the sample defocus z, when the latter is moved along the axis from the focal plane towards the acoustic lens. Under these conditions, the output voltage, V, produced by the transducer presents pseudo-oscillations which the defocusing distances, z [11-14]. Moreover, the periodic V(z) can be analyzed through fast Fourier transform, FFT, a spectral method used in numerical signal processing.

3. Results and Discussions

3.1 Simulation conditions

In this investigation we considered the usual operating conditions of a conventional scanning acoustic microscope, SAM, in the reflection mode with point focus beam: a lens opening angle of 50°, an operating frequency of 142 MHz. However, instead of using water as a coupling liquid, we chose oil-silicon whose density of 980 kg/m^3 and a velocity of 968 m/s [15, 16] in order to detect propagating modes of materials characterized with low surface acoustic wave, SAW, velocities. Furthermore, we studied reinforcement volume fraction effect on the acoustical properties of ($Zr_{41.2}Ti_{13.8}Cu_{12.5}Ni_{10}Be_{22.5}$) BMG's composites. For such fiber, several volume fractions were chosen varied from 0%, 20%, 40%, 60%, 80% and 100% and fibers were unidirectional.

In the present simulation we considered the isotropic cases of these materials (known to be anisotropic) due to the fact that the SAM experimental investigations [12, 14] with point focus beam could also be applied to specific planes by taking an average values of determined acoustic parameters (SAW velocities, elastic moduli, etc.). For a more precise data, one may use line focus beam acoustic microscopes [13].

3.2 Reflection coefficient variations

The propagation of SAW in different media is greatly affected by any discontinuities or defects in such media. This phenomenon could be deduced be measuring the changes in acoustic parameters such as critical angles at which the modes are excited and SAW velocities. In this context, the reflection coefficient (eqs. 6, 7 & 8), which is a complex function that admits a phase and amplitude, would be an interesting parameter that leads to the detection of different modes.

The deduced curves of reflection coefficient phase of Vit.1/Cd composites using their acoustical velocities are plotted in Fig. 1 in terms of incidence angle for different reinforcement volume fraction. It can clearly be seen that with increasing the incidence angle, we notice an important fluctuation with a 2π transition for each volume fraction. Under the present simulation conditions, these 2π changes correspond to the critical angle for the excitation of acoustical Rayleigh waves, θ_R. Moreover, we notice that θ_R of Vit.1/Cd composite shift to higher values with increasing $\square_f$. It should be noted similar variations were also observed with Vit.1/Ag & Vit.1/Mg composites.

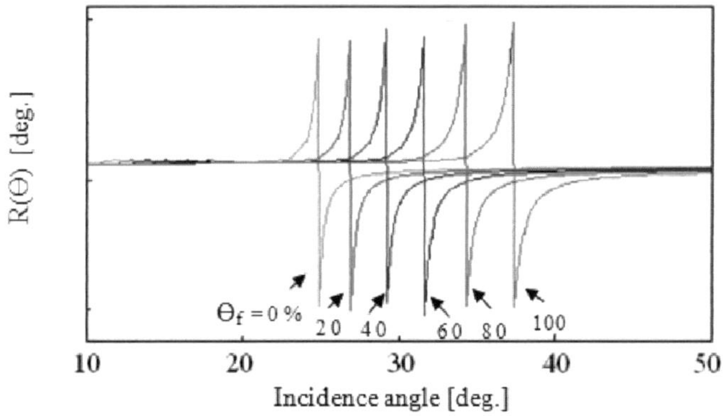

Figure 1. Vit.1/Cd reflection coefficient as a function of reinforcement volume fraction.

3.3. Acoustic signatures and spectral analysis

Acoustical signatures of Vit.1 composites were calculated from expression (9) and the calculated $R(\theta)$. The obtained curves of Vit.1/Cd composites for different volume fractions are illustrated in figure 2. It is clear that the deduced V(z) curves contain two distinct regions: (i) a region near the focal zone (z = 0) that depends mostly on the lens and (ii) a region characterized by periodic variations that contains information on the surface waves velocities. The oscillatory behavior is due to constructive and destructive interference between different propagating modes; it is characterized by a period Δz (between two successive maxima or minima) that is related to SAW velocity values [13]. Moreover, as can be seen, V(z) curves of Vit.1/Cd attenuate and their periods Δz narrowed with the increase of reinforcement volume fraction.

The spectral treatment of periodic V(z) curves via fast Fourrier trasfom, FFT, techniques of vit.1/Cr led to the determination of the spectra shown next to their corresponding curves (Fig 2). These spectra exhibit various peaks that differ in height and position. Each peak corresponds to an acoustical mode. The longitudinal mode appears near the coordinate axis. Rayleigh is the most dominate mode. Transversal mode sited near and below the Rayleigh mode. Acoustical modes velocities can be determined from the spectrum ray number. A high value of ray number corresponds to small acoustic velocity. Therefore, the corresponding velocities, V_L, V_T and/or V_R are determined. The displacements of peak positions are indicative of variations of the velocities of propagating modes with fibers volume percentage._It can be concluded that acoustical velocities decrease as function of fibers volume percentage.

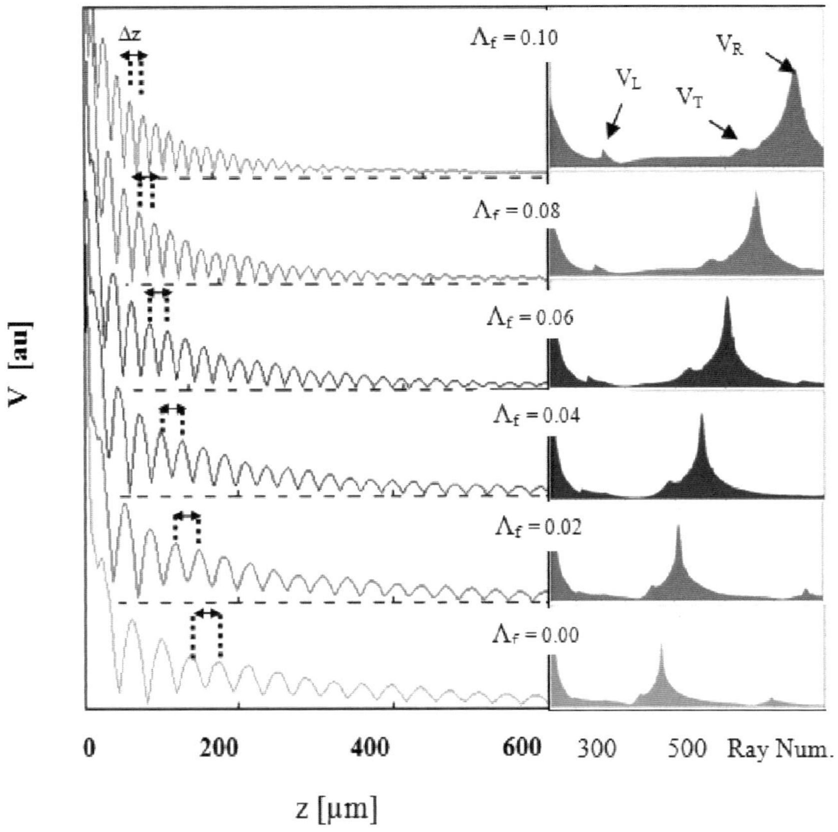

Figure 2. Reinforcement volume fraction effect on V(z) & FFT spectrum of Vit.1/Cd composite

3.4 Acoustic properties of human tibia bone

In ordered to enrich this investigation, we considered acoustical parameters of human tibia bone. Figure 3 illustrates the obtained curves of R(θ), V(Z) & FFT spectra. Rigorous analysis of FFT spectra of human tibia bone reveal that ray numbers of longitudinal, transversal and Rayleigh modes are comparable of those of Vit.1/Cd composite's in 60 % of Λ_f. Hence, their acoustical velocities are similar to the studied composites. Same results have been obtained for Vit.1/Ag and Vit.1/Mg BMGs composites. Comparison between acoustical parameters of BMGs composites and human bone show that they are similar when reinforcement volume fraction closes to 60 %. These results can be used in the choice of materials for biomedical applications. These curves are similar to those of Vit.1/Cd, Vit.1/Ag & Vit.1/Mg composites in high value of reinforcement volume fraction. This ensures a good transfer of stress between BMGs composites and the human bone.

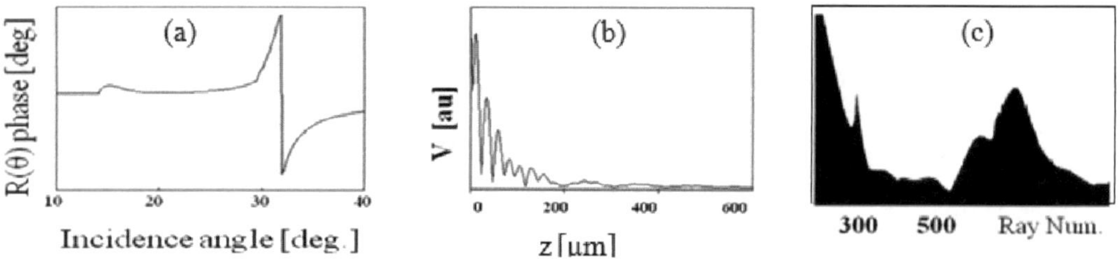

Figure 3. Acoustic parameters of human bone tibia (a) reflection coefficient, (b) acoustic signature and (c) FFT spectrum.

4. Conclusion

The effects of fiber volume fractions on reflection coefficient, acoustic signature & their FFT spectra of vit.1 composites were studied. It was found that BMG's composites acoustical parameters are close to those of human bone at high value of reinforcement volume fraction, in particular when reinforcement volume fraction closes to 60 %. These results can be used in the chose materials for biomedical applications. This behavior permits a good biocompatibility between these composites and the human system.

References

[1] Balch D. K, Ustandag E and Dunand D. C 2003 *Metallurg. Mater. Trans.* **34** 1787.
[2] Johnson W. L and Jom J 2002 *Miner. Met. Mater. Soc.* **54** 40.
[3] Horton J. A and Parsell D. E 2003 *Mater. Res. Soc. Symp. Proc.* **754**.
[4] Bruck H. A, Christman T, Rosakis A. J and Johnson W. L 1994 *Scripta Metall.* **30** 429.
[5] Cornner R, Dandlaiker R. B and Johnson W. L 1998 *Acta Mater.* **46** 6089.
[6] Telford M 2004 *Mater. Today* **7** 36.
[7] Zhang Y, Zhao D. Q, Wei B. C, Wen P, Pan M. X and Wang W. H 2001 *J. Mater. Res.* **16** 1675.
[8] Briggs A. 1993 *Acoustic Microscopy* (Oxford: Clarendon).
[9] Berthelot J 1996 *Matériaux Composites* (Paris: Masson).
[10] Al-Surayhi I, Doghmane A, Hadjoub Z 2009 *Damage and Fracture Mechanics*, ed. by Boukharouba *et. all.* p. 415 (Berlin: Springer Verlag).
[11] Cros B, Gigot V and Despaux G 1997 *Appl. Surf. Sci.* **119** 242.
[12] Doghmane A, Hadjoub Z, Alami K, Saurel J. M and Attal J 1992 *J. Acoust. Soc. Am.* **92** 1545.
[13] Kushibiki J and Chubachi N 1985 *IEEE Sonics Ultrasound* **SU-32** 189.
[14] Gomez-Moralez J, Rodrigues R, Durand J, Ferdj-Allah H, Hadjoub Z, Attal J and Doghmane A 1991 *J. Mater. Res.* **6** 2484.
[15] Schaaff W 1967 *New Series Group II: Atomic and Molecular Physics*, **5**: *Molecular Acoustics*, Eds: Hellwege K. H and Hellwege A. M 1967 (Berlin: Springer-Verlag).
[16] Metiri W, Hadjoub F, Doghmane A and Hadjoub Z 2009 *Physics Procedia* **2** 1421.

IOP Conf. Series: Materials Science and Engineering **28** (2012) 012039 doi:10.1088/1757-899X/28/1/012039

Evolution of microstructural and mechanical properties of an industrial 2017 Aluminium alloy processed by equal channel angular extrusion

M. A. Rekik [1], **T. Makhlouf, N. Njah**
Laboratory of Applied Metallurgy, Faculty of Sciences, BP 1171, 3000, University of Sfax-Tunisia

E-mail: rekik_medali@yahoo.fr;

Abstract. The evolution of mechanical properties, microstructure and their stability after different thermal treatments of an industrial Al-4wt% Cu alloy processed by Equal Channel Angular Extrusion (ECAE) were studied by Hardness measurements, tensile test, X-ray diffraction (XRD), transmission electron microscopy (TEM) and differential scanning calorimetry (DSC) techniques. A strong increase in the flow-stress and in the Vickers Hardness of the alloy was observed after ECAE. An important conclusion is that the strength level achieved by ECAE is higher than that obtained by ageing. X-ray diffraction analysis showed that the microstructure after ECAE consisted of low-sized crystallites (≈ 100 nm) and high values of root mean squared strain (rms-strain), almost 1%. In Transmission Electron Microscopy (TEM), it was found that sub-micrometric grains are formed by severe plastic deformation. The study of thermal stability showed that for N=1 and N=3, the alloy does not recover after annealing at 150°C for one hour. A substantial recovery is observed only at 250°C. At lower annealing temperatures, an increase in hardness is remarked, due to a reprecipitation enhanced by ECAE. The full recrystallized alloy exhibits a coherency length as low as 110 nm, revealing the effects of precipitates in dragging grain growth.

1. Introduction

The increase of the mechanical performances of conventional alloys is one of the attractive fields in material science. Aeronautic materials are known to have an enhanced E/ρ or Re/ρ ratios (E is the Young modulus, Re the flow stress and ρ is the material density) in addition to a high corrosion resistance. Aluminium alloys mainly heat treatable ones are of relatively high mechanical properties. Despite, the importance of the age hardening due to the formation of precipitates as the Guinier-Preston zones (GP zones), which may be regarded as fully coherent metastable precipitates, or the more stable Al-Cu based phase (θ”, θ’ and Al_2Cu (θ)) and the Al-Cu-Mg based precipitates as the Guinier-Preston Bagaryatsky zones (GPB zones) , S”,S’ and S (Al_2 CuMg) in the Al-Cu-Mg alloys [1, 2]. The increase of strengthening through the reduction of grain size to the nanometre range has attracted much attention several years ago. In fact, ultrafine-grained materials exhibit superior mechanical properties [3] and several techniques are now available for this purpose. Among others, procedures based on severe plastic deformation such as Equal Channel Angular Pressing (ECAP),

[1] To whom any correspondence should be addressed.

Torsion under High Pressure (THP) and Mechanical Milling (MM) are used. ECAP is the most attractive one since THP can be applied only on small specimens and MM requires a subsequent consolidation of the powder at high temperatures. Invented by Segal et al. [4, 5], ECAP can apply large cumulative strains to material without a change of the shape; through a specially designed die having two equally sized channels connected at a finite angle [6, 7].

It is well established that physical and mechanical properties are strongly related to the microstructure [8] so that, it is important to control the evolution of the latter via the amount of deformation introduced. In the present work, an industrial Al-4%Cu (2017) alloy was processed by ECAP. Transmission Electron Microscopy (TEM) was used to control the microstructure. X-ray diffraction (XRD) peak profile analysis was used to determine the coherency length D and the lattice strain ε. Tensile tests and Vickers Hardness (VH) were used to characterize the evolution of mechanical properties. Thermal stability of the microstructure was controlled by DSC and hardness measurements in addition to XRD.

2. Experimental Procedure

An industrial 2017 aluminium-copper alloy was received in form of rods (12x12x5000) mm^3 in size. Chemical composition is shown in Table 1. Specimens of 10×10×50 mm^3 were annealed at 550°C for 24 hours for homogenisation, quenched in icy water and over-aged at 350°C for one hour. After this treatment, the alloy exhibits the minimum hardness level corresponding to an overaged state [9]. ECAE was performed at room temperature. The die is formed by two square channels intersecting at an angle Φ=90°. The angle Ψ representing the outer arc of curvature is 90°. The total strain, ε_N, accumulated by N passages through the die is given by [10]:

$$\varepsilon_N = [N (2\cot(\Phi/2+\Psi/2) + \Psi \mathrm{cosec}(\Phi/2+\Psi/2))] /\sqrt{3} \quad (1)$$

In the present experiments where Φ = 90° and Ψ = 90°, ε_N = 0.906N.

Table 1. Chemical composition of the investigated alloy (weight %)

Al	Mg	Si	Mn	Ni	Fe	Cu	Cr	Zn	Pb
Bal.	0.29	0.71	0.93	< 0.1	0.64	4.44	0.05	< 0.15	0.12

X-ray diffraction patterns were recorded on a wide angle diffractometer in the θ-2θ step scan mode by using CoK_α radiation. Scans were collected over a range of 20°-120° in 2θ. To measure the intensity of each Bragg reflections an angle steps of 0.016° and fixed counting times of 4s were taken. Integral breadth (β), defined by the ratio between the integral intensity and the height of each peak was determined by using WinPLOTR software. For each diffraction peak, the reciprocal broadening β^* is related to the finite coherency length D and the rms-strain ε by [11, 12]:

$$(\beta^*/d^*)^2 = D^{-1}\beta^*/(d^*)^2 + (\varepsilon/2)^2 \quad (2)$$

where $\beta^*= \beta\cos\theta/\lambda$ and $d^*=2\sin\theta/\lambda$, θ is the peak position and λ is the wavelength of the X-rays.

The experimental breadth β_e was corrected for the peak breadth β_i due to the instrumental effects to obtain β using the relation given for a Voightian profile by [11]:

$$\beta = [(\beta_e^2 - \beta_i^2)^{1/2} \times (\beta_e - \beta_i)]^{1/2} \quad (3)$$

β_i was taken as the integral breadth corresponding to the same 2θ value for NaCl powder chosen as a standard reference material. Differential Scanning Calorimetry (DSC) was performed at a constant heating rate of 10°C/min in sealed crucibles using a NETZSCHDSC 204 calorimeter.

3. Results and discussion

3.1. Effects of ECAP

Figure 1a shows two selected TEM micrographs of the alloy before extrusion (N=0) and after extrusion three times (N=3) with the corresponding diffraction patterns.

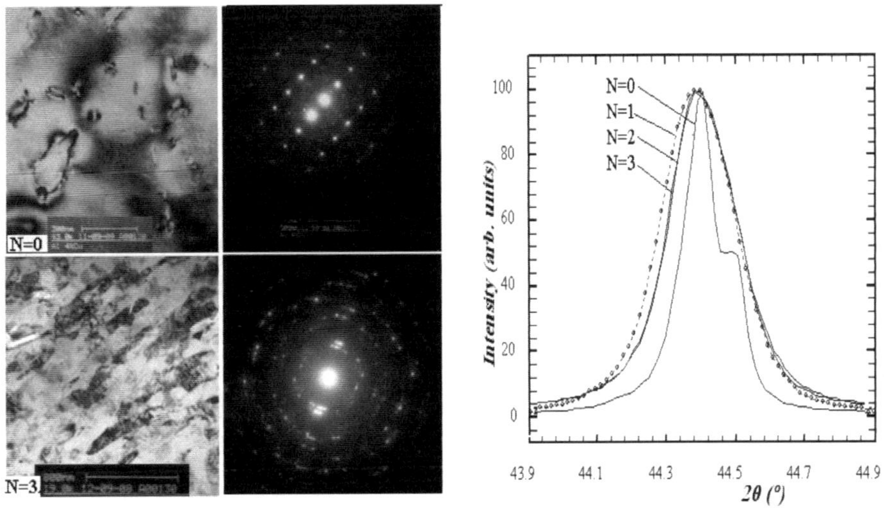

Figure 1. (a) Bright field images and corresponding diffraction patterns of the alloy Al-4%Cu for N=0 and N=3. (b) (111) XRD peak of the same alloy before and after ECAE.

TEM examination showed that equiaxed grains and the well defined and individualized spots on the diffraction pattern observed for N=0 are replaced by substantially elongated grains and rings spots for N=3 indicating a high refinement of grains after ECAE. An example of the broadening of the (111) reflection is shown in Figure 1b. An important broadening was observed after the first pass through the die, which causes the absence of the $K\alpha_2$ shoulder observed for N = 0.Taking into account that the broadening is due to finite coherency length D and to an enhanced rms-strain ε. The experimental values of D and ε were obtained from a linear representation referring to equation (2). The obtained results are plotted in figure 2. The average value of D obtained since the first passage is around 54 nm whereas the rms-strain ε increases from 0.4% for N= 1 to 0.56% for N=3.

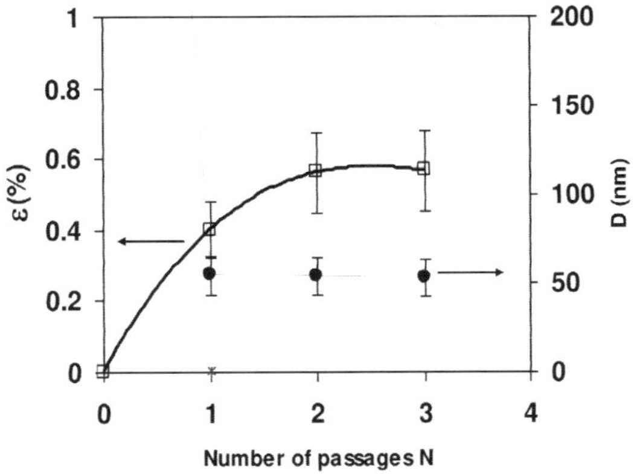

Figure 2. Mean values of the coherency length D and of the rms- strain ε as a function of the number of passages; N. Error bars represent calculation accuracy, which does not exceed 20% of the value

Figure 3 shows a part of XRD patterns before and after ECAP. In addition to the fundamental peaks of aluminium matrix, peaks relative to Al_2Cu and $Al_2(Cu,Mg)$ precipitates[13-14], are also detected.

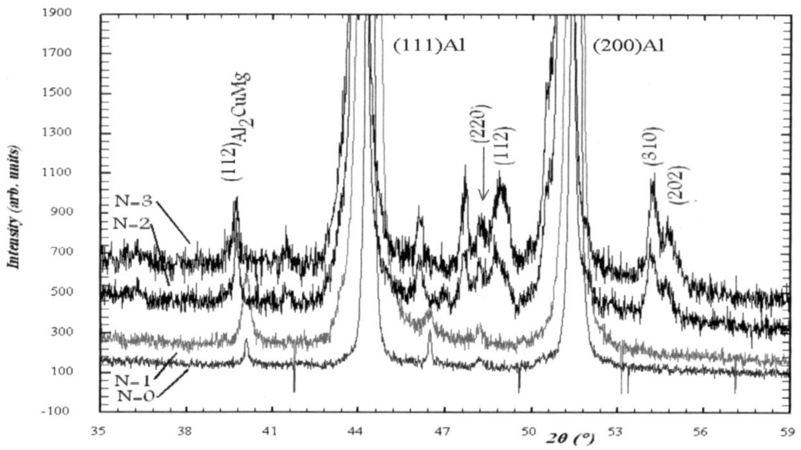

Figure 3. XRD patterns recorded on the Yo plane before and after ECA- pressings revealing broadening peaks of Al_2Cu and Al_2CuMg Precipitates. Non-subscribed peaks are relative to Al_2Cu.

A slight shifting of the diffraction peaks to lower θ values is remarked; it is due to the compression filed generated by high deformation resulting in higher lattice spacing. On the other hand, a careful examination of the peaks relative to the precipitates shows that the latter are significantly broadened and shifted to low values of θ after extrusion; this is due probably to the creation of a high defect density within them during deformation. In fact, after the first passage, the work-hardening of the matrix may cause a transition from by-passing mechanism of deformation to shearing one as it is often the case in many two-phase alloys [15].

Stress-strain curves of the alloy before and after extrusion were given in a previous paper [9]. Figure 4 shows the evolution of the tensile properties as a function of the number of passes. It can be seen that the flow stress $\sigma_{0.2}$ and the ultimate tensile strength (UTS) σ_m are improved by ECAE after the first passage; when the number of passes increases, the dependence of these strength properties on the number of passages slow-down. On the other hand, the homogeneous elongation ε_R decreases from pass to pass; this can be explained by the habitual competition between hardening and plasticity.

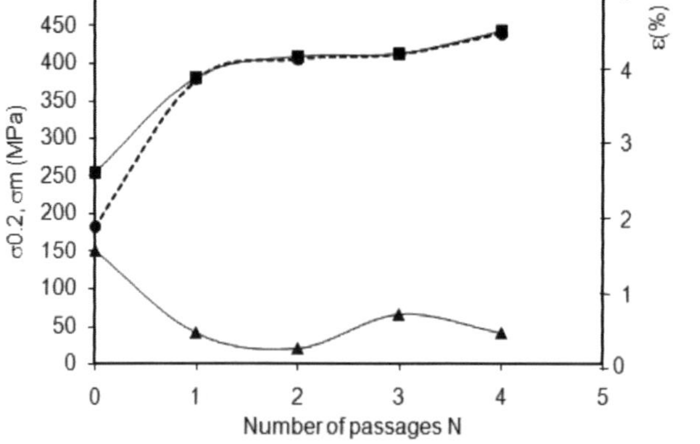

Figure 4. Influence of ECAE on the 0.2% flow stress (circles), on the ultimate tensile strength (squares) and on the elongation (triangles) of the alloy.

The values of 0.2% flow stress after extrusion is significantly larger than the values obtained for the quenched and overaged states (291± 9) and (182 ± 14) MPa respectively [9]. Since the first passages through the die, the values of $\sigma_{0.2}$ merge with those of the UTS. $\sigma_{0.2}$ depends on the mechanisms of deformation in its early stages, i.e. on the total amount of defects pinning dislocations; it increases when the defect density increases, which is the case after ECAE. σm and more exactly σm/σ0.2 ratio is related to the total amount of deformation which can be accumulated in the material during tensile tests or other forming process. The decrease of this ratio from about 1.4 before ECAE to a constant value of 1 reflects that the alloy has been deformed to saturation after the first passage. Therefore, as it's the case of conventional modes of deformation, ECAE can lead to a higher resistance level than aging but the disadvantage is a loss of elongation, which drops dramatically. Similar variation of the tensile properties was also observed in casting Al-0.63 wt. % Cu and Al-3,9 wt. % Cu model alloys processed by ECAE [16]. In Al-3.9 wt. % Cu alloy, the UTS reached 207 MPa after one pass and 290 MPa after four passes. These values are much lower than those obtained in the alloy of the present investigation (381MPa for N=1 and 442 for N=4). The difference may be related to the presence of other elements in our alloy (industrial alloy) and to the age-hardening which is more efficient when using controlled heat treatments and so, controlled microstructure.

3.2 Effects of subsequent heating

DSC thermograms of the alloy before and after ECAE are plotted in figure 5. The values of Vickers hardness after subsequent annealing of the extruded alloy are also plotted on the same figure for N=1 and N=3. In the present alloy, either GP→θ"→θ'→θ Al-Cu based precipitates or GPB→S"→S'→S Al-Cu-Mg based precipitates are expected [17-19]. This agrees with the results of figure 3, which reveal the presence of peaks related to these precipitates.

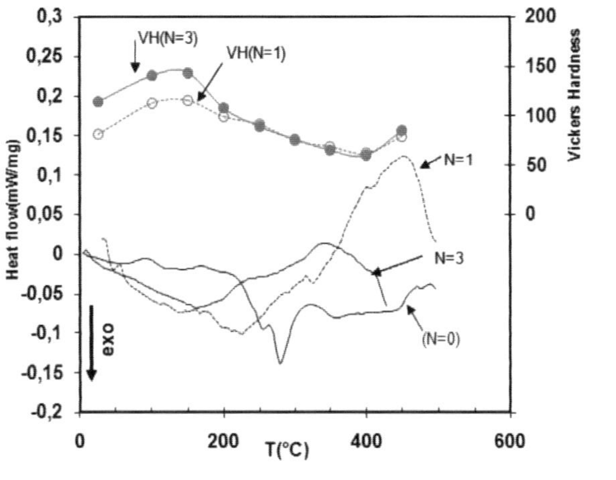

Figure 5. DSC thermograms of the alloy overaged (N=0) and ECAE processed. Hardness variations for N=1 and N=3 are also plotted

The increase of Vickers Hardness (VH) between 25°C and 175°C indicates that wide exothermic peaks around 225°C for N=1 and 175°C for N=3 are due to the formation of GP /GPB zones and of Al-Cu and Al-Cu-Mg co-clusters rather to an eventual recovery. The formation of these phases leads to a substantial increase of hardness in this temperature range. However, the threshold value of hardness corresponding to GP/GPB and/or coherent precipitates θ"/S" is observed between 100°C and 150°C; the difference in temperatures is due to a relatively long ageing time (1hour) for hardness specimens and to the usual lateness of DSC peaks. The intense peaks at 250°C and 275°C for N=0 are ascribed to the formation of θ'/S' and θ/S precipitates [17]; the equivalent peak in the extruded samples may be screened and, consequently only the corresponding decrease of hardness was observed. It should be noted that all of the peaks are shifted toward lower temperatures with respect to those of N=0. The intense peak relative to θ'/S' and θ/S precipitation is shifted by about 50°C and broadened.

XRD shows no reduction of the peak breadth below 250°C (figure 6a). The values of D and ε obtained from the HW representations are plotted on figure 6b. We note that heating above 250°C leads to a reduction of ε reaching then an equilibrium value close to zero. A substantial increase of D is also obtained but the final value at 350°C is still measurable and by XRD does not exceed 110 nm. The limited value of coherency length obtained after annealing, which is widely larger before extrusion reflects clearly the role of precipitates in dragging grain growth. The effect is however important mainly in the low-precipitate size range.

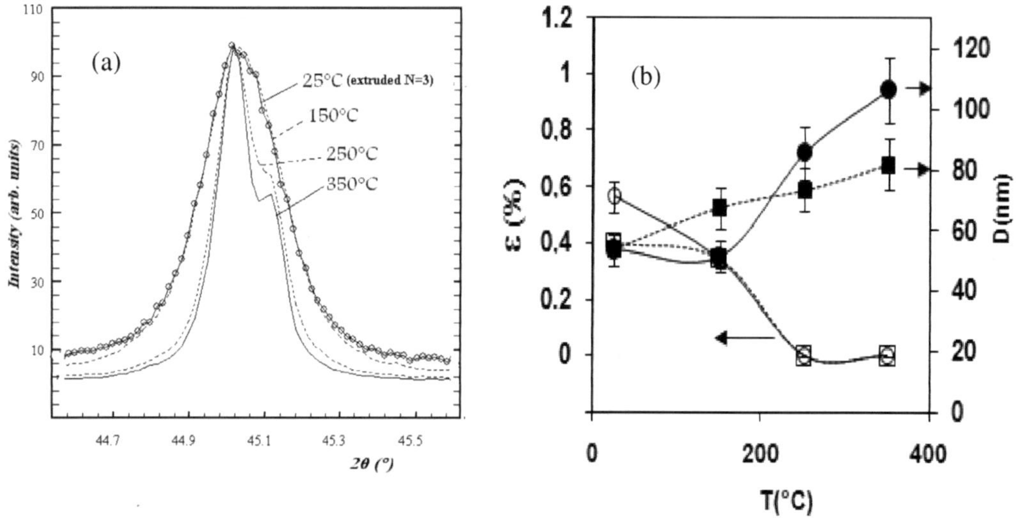

Figure 6. Normalized intensity of (111) reflection of the pressed (N=3) before and after annealing. (b) Changes in coherency length D (full marks) and in the rms- strain ε (open marks) as a function of annealing temperature for the ECAE processed material up to N=1 (squares) and N=3 (circles).

4. Conclusion

High strength of heat treatable Al-Cu alloys can be improved by plastic deformation. Equal channel angular extrusion (ECAE) is known to lead to a high strength via a reduction of grain size without a change of the specimen shape. In the present work, this technique was applied to an overaged Al-4%Cu alloy up to three passes using a die characterized by the angles Φ=90° and ψ =90°. XRD analysis revealed the presence of Al_2Cu and $Al_2(Cu,Mg)$ precipitates. Change of peak breadth was observed after ECAE; it was related to a decrease of the coherency length D and to an increase of the rms- strain ε. The average values obtained from the Halder-Wagner representation show that D is reduced to about 54 nm whereas the ε is increased up to 0.56%. At the same time, precipitate peaks are also broadened and shifted indicating that they are partially destroyed by extrusion. An increase of strength properties is obtained since the first passage. A subsequent heating leads to a modification of the microstructure: The combination of DSC, XRD and VH measurements show that the microstructure is maintained unchanged up to 150°C. At higher temperatures, the rms-strain falls to zero whereas coherency length does not exceed 110nm at a temperature as high as 350°C; this reveals clearly the effects of precipitates in dragging recovery.

References

[1] Verlinden B, Zahra A-M 2003 Precipitation hardenableAl-Mg-Cu alloys: mechanical properties and hardening mechanisms *Mater Sci Forum* 426-432 423-8.

[2] Zoqui E. J and Robert M. H 1998 *J. Mater, Process. Technol.* **78** 1.
[3] Valiev R. Z, Korznikov A. V, Mulyukov R. R 1993 *Mater Sci Eng A* **168** 141.
[4] Segal M, Reznikov V. I, Dorbyshevskity A. E, Kopylov V. I 1981 *Russ Metall 1* **99**.
[5] Segal V. M 2003 *Mater Sci Eng A* **345** 36.
[6] Segal V. M 1995 *Mater Sci Eng* A **197** 157.
[7] Aida T, Matsuki K, Horita Z, Langdon T. G 2001 *Scripta Mater* **44** 575.
[8] Hansen N 2001 *J Mater Sci Technol* **17** 409.
[9] Rekik M.A, Njah N 2009 *Ann. Chim. Sci. Mat.* **34** 99.
[10] Iwahashi Y, Wang J, Horita M, Nemoto M and Langdon T. G 1996 *Scripta Mater.* **35** 143.
[11] Langford J. I 1992 *Proceedings of the international conference Accuracy in powder Diffraction II, held at NIST, Gaithersburg, MD,* May 26.
[12] Halder N. C, Wagner C. N 1996 *Adv X-ray Anal* **9** 91.
[13] Perlitz H, Westgren A, Kemi Ark 1943 *Mineral.Geol. B* **16** 13.
[14] Meetsma A, De-Boer J. L, Van-Smaalen S. J 1989 *Solid State Chem* **83** 370.
[15] Njah N, Dimitrov O, Chamberaud A 1990 *J Nucl.Mater* **172** 228.
[16] Fang D. R, Zhang Z. F, Wu S. D 2006 *Mater. Sci Eng. A* **426** 305.
[17] Wang S. C, Starink M. J 2007 *Acta Mater* **55** 933.
[18] Starink M. J, Cerezo A, Yan J. L. Gao N 2006 *Phil Mag Lett* **86** 243.
[19] Starink M. J, Gao N, Davin L, Yan J. L, Cerezo A 2005 *Phil Mag* **85** 1395.

Crystal structure and spectroscopic investigations of a new benzidinuim dichloride $NH_3(C_6H_4)_2NH_3Cl_2$

W. Wamani[1], M. Belhouchet, T. Mhiri
Laboratoire de l'Etat Solide, Département de Chimie, Faculté des Sciences de Sfax, B.P.1171, 3000 Sfax, Université de Sfax, Tunisia

E-mail: belhouchet2002@yahoo.fr; wafawamani@yahoo.fr

Abstract. Chemical preparation, crystal structure and spectroscopic investigations for a new 4,4'-biphenylenediammonium dichloride $NH_3(C_6H_4)_2NH_3Cl_2$ are presented. This compound crystallizes in the triclinic space group $P\bar{1}$, with flowing unit cell parameters a = 6.665(5) Å, b = 7.722(5) Å, c = 12.581(5) Å, α = 84.526(3)°, β = 76.312(5)°, γ = 74.008(5)° , V = 604.6 (6) $Å^3$ and Z = 2. The crystal structure has been determined and refined to R1 = 0.038 and wR2 = 0.130 using 6845 independent reflections. The structural arrangement can be described as inorganic layers of Cl^- anions parallel to (a, b) plane alternating with layers of organic $[NH_3(C_6H_4)_2NH_3]^{2+}$ cations. Cations and anions are connected through multiple N-H...Cl hydrogen bonds. The vibrational properties of this compound were studied using IR absorption. The observed bands were identified by comparison with the vibrational modes and frequencies observed in similar compounds.

1. Introduction

The coordination chemistry of anions has found widespread applications in supramolecular chemistry [1] and biochemical processes [2]. Moreover, the halide anion can be useful for such applications because of the high flexibility of their coordination [3]. Here, a new member of this family, the title compound $NH_3(C_6H_4)_2NH_3Cl_2$, is presented, which was obtained during our studies on the preparation of new organic chloride compounds and organic-inorganic hybrid materials.

2. Experimental

2.1. Synthesis

An aqueous HCl solution and 4,4-biphenylene diamine (Aldrich, Purity (98%)) in a 2:1 molar ratio were mixed and dissolved in sufficient water/ethanol. Crystals for single-crystal X ray diffraction structural analysis were grown by slow evaporation at room temperature, and then set aside for two weeks to obtain colorless needle crystals. The obtained crystals remained stable under normal conditions of temperature and humidity

---

[1] To whom any correspondence should be addressed.

2.2. X-ray crystallography

Table 1 Crystal data and structure refinement parameters for $NH_3C_{12}H_8NH_3Cl_2$

Formula	$NH_3C_{12}H_8NH_3Cl_2$
Formula weight	M_r = 257.15 g.mol^{-1}
Crystal system / Space group	Triclinic / $P\bar{1}$
a (Å)	6.666 (1)
b (Å)	7.722 (1)
c (Å)	12.581 (1)
α (°)	84.577 (2)
β (°)	76.312 (1)
γ (°)	74. 008 (2)
V (Å^3)	604.56 (13)
Z	2
Crystal form / color	needle / colorless
Crystal size (mm^3)	0.35 × 0.24 × 0.2
hkl ranges	$-11 \le h \le 11$ $-13 \le k \le 13$ $-22 \le l \le 22$
θ Range (°)	2.8 - 38.7°
Measured reflections	20809
Independent reflections	6845
Reflections with $I > 2\sigma(I)$ / R_{int}	4112 / 0.027
Absorption correction	Multi-scan SADABS
T_{min} / T_{max}	0.863 / 0.903
Goodness-of-fit on F^2	1.02
Final R indices	R1 = 0.038; wR2 = 0.108
Number of refined parameters	201
$\Delta\rho_{min}$ / $\Delta\rho_{max}$(e Å^{-3})	−0.27 / 0.56

A single crystal (colorless needle, 0.35*0.24*0.2 mm) was carefully selected under polarizing microscope in order to perform its structural analysis by X-ray diffraction. Intensity data were collected on Bruker APEX-II area detector diffractometer (chromated Mo-Kα radiation, λ=0.71073 Å). The structure was solved by means of the heavy-atom methods, using the SHELXS-97 [4] program, and was refined by full matrix least-squares methods on F^2 with SHELXL-97 [5]. The crystal of $NH_3C_{12}H_8NH_3Cl_2$ is triclinic with space group $P\bar{1}$. A total of 20809 reflexions were collected (θ: 2.8-38.7; index ranges: $-11 \le h \le 11$, $-13 \le k \le 13$, $-22 \le l \le 22$); 4112 of them were unique reflexions (Rint = 0.027). Absorption correction based on the crystal shape was applied using SADABS (Tmin = 0.863, Tmax = 0.903) [6]. The hydrogen atoms were located by difference-Fourier synthesis and refined. The refinement converged with R1 = 0.038 and wR2 = 0.108 for 4112 reflections with I>2(I) (201 refined parameters, Goof S = 1.02). The experimental conditions of collection, the strategy followed for the structure determination and the final results are given in Table 1. The final atomic coordinates of all atoms and their equivalent isotropic displacement parameters, Ueq (U_{iso} for hydrogen atoms), are given in Table 2.

Table 2 The final atomic coordinates and the equivalent temperature factors for $NH_3C_{12}H_8NH_3Cl_2$

	x	*y*	*z*	U_{iso}*/U_{eq}
Cl1	0.53499 (4)	0.17738 (3)	−0.411906 (19)	0.03528 (8)
Cl2	0.96331 (4)	0.32194 (3)	−0.581387 (18)	0.03343 (7)
N1	1.10440 (17)	0.06359 (13)	−0.38417 (6)	0.03066 (17)
N2	0.40822 (17)	0.42787 (13)	0.38539 (6)	0.03177 (17)
C1	0.99564 (15)	0.11760 (12)	−0.27272 (6)	0.02506 (16)
C2	0.92912 (18)	−0.00737 (13)	−0.19650 (7)	0.03090 (19)
C3	0.83175 (18)	0.04522 (13)	−0.08992 (7)	0.03090 (19)
C4	0.96554 (18)	0.29285 (13)	−0.24467 (7)	0.0312 (2)
C5	0.86453 (18)	0.34422 (13)	−0.13856 (7)	0.0320 (2)
C6	0.79635 (15)	0.22224 (12)	−0.05842 (7)	0.02540 (16)
C7	0.69464 (15)	0.27618 (12)	0.05601 (7)	0.02583 (16)
C8	0.57670 (18)	0.17259 (13)	0.12821 (8)	0.0327 (2)
C9	0.48507 (18)	0.22008 (13)	0.23606 (7)	0.0324 (2)
C10	0.71716 (19)	0.43092 (14)	0.09679 (8)	0.0339 (2)
C11	0.62615 (18)	0.48059 (14)	0.20414 (8)	0.0341 (2)
C12	0.51015 (15)	0.37468 (12)	0.27281 (7)	0.02703 (17)
H1N1	1.040 (3)	0.150 (2)	−0.4287 (14)	0.050 (4)*
H2N1	1.231 (3)	0.071 (2)	−0.3959 (13)	0.042 (4)*
H3N1	1.105 (3)	−0.049 (2)	−0.3943 (14)	0.047 (4)*
H1N2	0.473 (2)	0.341 (2)	0.4324 (13)	0.044 (4)*
H2N2	0.282 (3)	0.4233 (19)	0.3985 (13)	0.040 (4)*
H3N2	0.408 (3)	0.545 (2)	0.3942 (15)	0.056 (5)*
H2	0.950 (3)	−0.127 (2)	−0.2163 (13)	0.064 (5)*
H3	0.793 (2)	−0.0426 (18)	−0.0381 (11)	0.041 (3)*
H4	1.011 (2)	0.3761 (18)	−0.2971 (11)	0.049 (4)*
H5	0.837 (2)	0.462 (2)	−0.1239 (12)	0.057 (4)*
H8	0.556 (2)	0.0661 (19)	0.1025 (11)	0.054 (4)*
H9	0.400 (2)	0.1467 (18)	0.2831 (11)	0.045 (4)*
H10	0.796 (2)	0.5050 (18)	0.0502 (11)	0.041 (3)*
H11	0.645 (3)	0.587 (2)	0.2329 (13)	0.068 (5)*

3. Results and discussion

3.1. Structure description

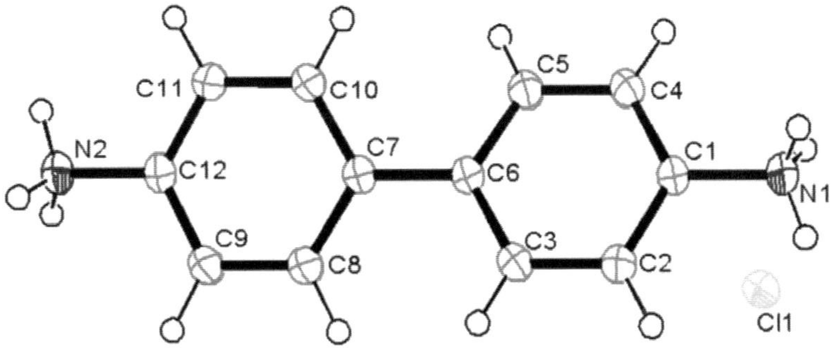

Figure 1. Asymmetric unit of $NH_3C_{12}H_8NH_3\ Cl_2$ with atom labels and 50 % probability displacement ellipsoids for non-H atoms.

Table 3. Main interatomic distances (Å) and angles (°) of $NH_3C_{12}H_8NH_3Cl_2$

N1—C1	1.4588 (11)	C6—C7	1.4838 (13)
N2—C12	1.4619 (11)	C7—C8	1.3979 (13)
C1—C4	1.3796 (13)	C7—C10	1.4017 (13)
C1—C2	1.3810 (12)	C8—C9	1.3882 (12)
C2—C3	1.3892 (12)	C9—C12	1.3848 (13)
C3—C6	1.4006 (13)	C10—C11	1.3861 (13)
C4—C5	1.3857 (12)	C11—C12	1.3827 (13)
C5—C6	1.3964 (12)		
C4—C1—C2	121.04 (8)	C8—C7—C10	117.31 (8)
C4—C1—N1	119.06 (8)	C8—C7—C6	121.38 (8)
C2—C1—N1	119.88 (8)	C10—C7—C6	121.29 (8)
C1—C2—C3	119.04 (8)	C9—C8—C7	121.70 (9)
C2—C3—C6	121.67 (8)	C12—C9—C8	119.11 (8)
C1—C4—C5	119.26 (8)	C11—C10—C7	121.73 (9)
C4—C5—C6	121.75 (8)	C12—C11—C10	119.13 (9)
C5—C6—C3	117.22 (8)	C11—C12—C9	121.02 (8)
C5—C6—C7	121.61 (8)	C11—C12—N2	119.80 (8)
C3—C6—C7	121.15 (8)	C9—C12—N2	119.16 (8)

The main geometrical features of the organic entities are reported in Table 3. Fig. 1 shows the ORTEP [7] plot of the asymmetric unit of $NH_3C_{12}H_8NH_3Cl_2$, containing one protonated, 4,4-biphenylene diamine $[NH_3C_{12}H_8NH_3]^{2+}$ cation and two chloride anions. A view of the structure projected along a axis is given in Fig. 2. This projection shows that the crystal structure of the title compound consists of alternating organic layers and inorganic sheets stacked along [100] direction. The inorganic and organic species are interconnected through six N-H...Cl hydrogen bonds (Table 4). The corresponding N...Cl distances vary between 3.1607(11) and 3.2302(10) Å, which can be considered relatively weak [8]. In the organic layers, there are no π–π interactions between antiparallel cations, with face to face distances of 4.361 and 5.70 Å, up than 3.8 Å, the maximum value accepted for $\pi - \pi$ interactions [9]. The dihedral angle between the two phenyl rings of the organic cation is 18.6.

Table 4 Bond length (Å) and angles (°) in the hydrogen-bonding scheme in $NH_3(C_{12}H_8)NH_3Cl_2$

D—H···*A*	*D*—H	H···*A*	*D*···*A*	*D*—H···*A*
N1—H1N1···Cl2	0.911 (17)	2.306 (18)	3.1868 (10)	162.5 (14)
N1—H2N1···Cl1^{i}	0.837 (17)	2.350 (17)	3.1607 (11)	163.2 (13)
N1—H3N1···Cl2^{ii}	0.889 (17)	2.336 (17)	3.2143 (10)	169.3 (15)
N2—H1N2···Cl1^{iii}	0.932 (16)	2.289 (17)	3.1853 (10)	161.1 (13)
N2—H2N2···Cl2^{iv}	0.829 (17)	2.410 (17)	3.2133 (12)	163.2 (13)
N2—H3N2···Cl1^{v}	0.917 (18)	2.321 (18)	3.2302 (10)	171.0 (16)

Symmetry codes: (i) $x+1, y, z$; (ii) $-x+2, -y, -z-1$; (iii) $x, y, z+1$; (iv) $x-1, y, z+1$; (v) $-x+1, -y+1, -z$.

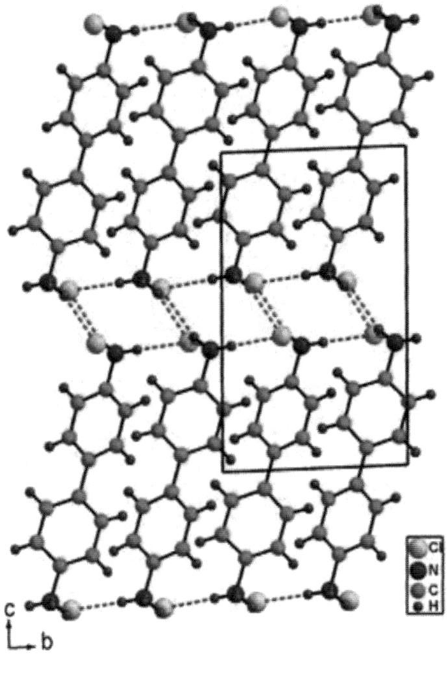

Figure 2. View down the a axis of the $NH_3C_{12}H_8NH_3\ Cl_2$ showing alternating organic inorganic layers, anions and cations are connected by N-H...Cl hydrogen bonds (dashed lines).

An examination of the organic moiety geometrical features shows that, the carbon atoms building the phenyl rings of the title compound have a good coplanarity, and they form conjugated planes with average deviations of 0.0053 and 0.0018 Å for phenyl 1 (from C1 to C6) and phenyl 2 (from C7 to C12), respectively. The main value of the C-C bond lengths in the two phenyl rings is 1.390 Å, which is between a single bond and a double bond, agree with that observed in 4,4'-biphenylenediammonuim tetrachloridocuprate and 4,4'-biphenylenediammonuim tetrachloridomercurate [10].

3.2. IR spectroscopy

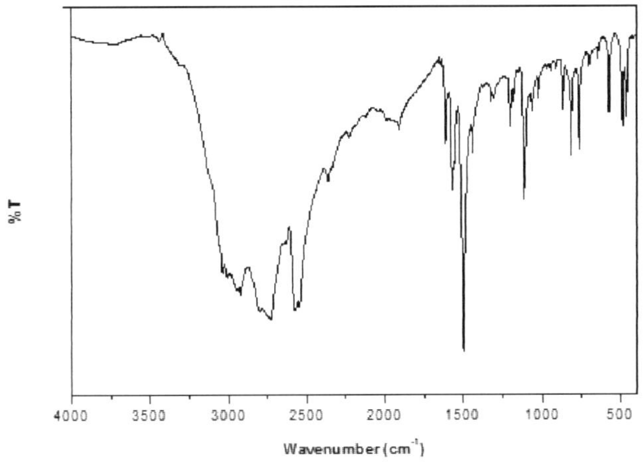

Figure 3. IR spectrum of $NH_3C_{12}H_8NH_3\ Cl_2$.

Because IR spectroscopy is one of the major physical methods for the investigation of molecular structures, we have studied the IR spectrum of crystalline organic-inorganic hybrid compound $NH_3C_{12}H_8NH_3Cl_2$ which is shown in Fig. 3. To assign the IR peaks to vibrational modes, we examined the modes and frequencies observed in similar compounds [11-12]. The broad bands between 3200 and 2700 cm^{-1} are attributed to stretching vibrations of CH_2 and NH_3 groups (ν(C-H) and ν(N-H)). Frequencies in the range 2600-1700 cm^{-1} are assigned to combination bands. The intense band at 1498 cm^{-1} accompanied by a low intensity peak at 1324 cm^{-1} is attributed to the phenyl ring vibrations. Bands of the symmetric and antisymmetric deformation vibrations of the NH_3^+ groups ($\delta_s(NH_3)$ and $\delta_{as}(NH_3)$) are observed at 1572 and 1614 cm^{-1}, respectively, indicate the presence of the cation in the doubly protonated base in the crystal structure. All peaks located in the frequency range 800-1250 cm^{-1} are attributed to stretching vibrations of C-N and C-C bonds (ν(C-N) ν(C-C)) and deformation in the plane δ(CH). The bands of low intensities at 800 and 600 cm^{-1} correspond to the twist out of plane t(CH). Deformation δ(NCC) and the twist out of plane twisting type of NH_3 observed in the range 400-580 cm^{-1}.

4. Conclusion

$NH_3C_{12}H_8NH_3Cl_2$ was prepared as single crystals at room temperature and characterized by various physico chemical methods. On the structural level, the atomic arrangement of the title compound can be described by inorganic layers organization. The inter-layer spacing is occupied by the organic molecules. The structural of the title material is stabilized by interesting H-bond. The vibrational proprieties of this structure were studied by infrared spectroscopy. The assignment of the vibrationnal bands was performed by comparison with the vibration modes frequencies of homologous compounds.

References

[1] Schmidtchen F. P, Berger M 1997 *Chem. Rev.* **97** 1609.
[2] Pajewski R, Ferdani R, Schlesinger P. H, Gokel G. W 2004 *Chem. Commun.* 160.
[3] Ilioudis C. A, Hancock K. S. B, Georganopoulou D. G, Steed J. W 2000 *New J. Chem.* **24** 787.
[4] SHELXS-97, Sheldrick G. M, 1997 Program for Crystal Structure solution, University of Gottingen, Germany.

[5] SHELXL-97, Sheldrick G. M 1997 Program for Crystal Structure Refinement, University of Gottingen, Germany
[6] Sheldrick G.M 1996, SADABS. University of Göttingen, Germany
[7] Farrugia J. L 1997 *J. Appl. Cryst.* **30** 565.
[8] Brown I. D 1976 *Acta crystallogr. A* **32** 24.
[9] Janiak J 2000 *J. Chem. Soc., Dalton Trans.* 3875.
[10] Belhouchet M, Wamani W, Mhiri T 2010 *IOP Conf. Series: Materials Science and Engineering* **13** 12039.
[11] Civan M, Topacli C, Topacli, Gürol I, Ahsen V 2007 *J. Molecular Structure* 842.
[12] Akalin E, Akyüz S 2003 *J. Molecular Structure* **571** 650.

Synthesis, Calorimetric and X ray diffraction studies of the solid Solution $[Rb_{0.92}(NH_4)_{0.08}]_2TeCl_4Br_2$

W. Ben Aribia[a], M. Abdelmouleh[a], A. Van Der Lee[b] and A. Kabadou[a1]
[a]Laboratoire des Sciences des Matériaux et d'Environnement, Faculté des Sciences de Sfax BP 1171, 3000, Université de Sfax, Tunisia.
[b]Institut Européen des Membranes (UMR 5635), Université de Montpellier II, cc 047, Place E. Montpellier- France

E-mail: kabadouahlem@yahoo.fr

Abstract. The structure of $[Rb_{0.92}(NH_4)_{0.08}]_2TeCl_4Br_2$ has been determined from conventional X-ray powder diffraction data, by Rietveld method, in P4/mnc tetragonal space group (a = 7.315 *(2)* Å, *c* = 10.354*(4)* Å, *Z* = 2). The refinement of the structure led to final factor χ_2 =4.20%. The structure is considered as isolated octahedred $TeCl_4Br_2^{2-}$. These anions show a 4° rotation around the fourfold axis against the cubic arrangement of the K_2PtCl_6 type structure. The monovalent cations (Rb^+/NH_4^+) are located between the octahedra ensuring the stability of the structure by ionic and hydrogen bonding contacts: Rb....Cl/Br and N-H...Cl/Br. At low temperature a second order transition phase was detected at 213K by DSC.

1. Introduction

Antifluorites A_2BX_6 (A being an alkali, B a tetravalent and X a halogen atom) belong to typical high symmetry crystal structures. Within the antifluorite family, $(NH_4)_2BX_6$ [1-3] is of particular interest: the ammonium group distortion and orientation differ from one compound to another.

For compounds A_2TeBr_6 (with A = K, NH_4, Rb, Cs) [4-6] the high-temperature phases show the cubic K_2PtCl_6 type structure. Lowering the temperature, second order phase transitions can be observed, especially when the anions are comparatively large. These phase transitions can be explained by a softening of rotary phonons. Generally, in A_2MX_6 compounds this instability is a result of a double minimum in the Coulomb potential with respect to the octahedra tilt angle φ around (0 0 1); the depth of this minimum, depends on the interaction study [7], will give us a hint for understanding the lattice dynamics of A_2MX_6 phases.

In $(A)_2TeX_6$ the partial of both cation and anion substitutions on the structure and especially on the bonding within the $(TeX_6)^{2-}$ anions may give a wide range of substitutional solid solutions, which opens up possibilities for preparing new materials with gradually varying properties.

This work is a part of our systematic physico-chemical investigation of mixed-alkali hexachlorotellurates [8-9]. We present here the Rietveld crystal structure analysis of $[Rb_{0.92}(NH_4)_{0.08}]_2TeCl_4Br_2$ obtained from powder diffraction data.

[1] To whom any correspondence should be addressed.

2. Experimental

The aforementioned mixed compound was synthesized by adding the stochiometric mixtures of RbBr, NH_4Br, and TeO_2 (molar 1:1:1 ratio) hot saturated solutions in hydrochloride acid. The precipitated powder was thoroughly dried and sealed under vacuum. All starting components were of a purity of 99.99 wt %. The sample was mounted in the toploaded sample holder and investigated by X-ray powder diffraction (XRD). Powder X-ray diffraction data were collected at room temperature using a PANalytical X'pert-Pro diffractometer equipped with an ultrafast X'celerator detector. Ni-filtered Cu K_α X-rays (40 kV/40 mA) were used. In order to limit the axial divergence, 0.02 rad primary and secondary Soller slits were inserted. Data were collected by continuous scan from 10 to 120° 2θ with an effective step size of 0.016° 2θ and 20 s integration time per step. It is noted that the position sensitive X'celerator detector has an opening angle of 2.122° with 127 channels. Prior to data analysis the experimental data were recalculated to step sizes of 0.016° using a cubic spline interpolation method. Indexing was performed using DICVOL04 software [10] with standard options confidence factors were M(20)=41 and F(20)=33.8 [11]. Our assumption was subsequently confirmed by successful Rietveld refinement of the investigated structure.

The X-ray powder diffraction data were refined by the Rietveld method [12] using the FULLPROF code [13]. The Rietveld refinement was carried out using the pseudo-Voigt function [14] for the simulation of the peak shapes. Le Bail profile refinement was first used to find the profile and instrumental parameters such as the zero-angle displacement error and asymmetry effects. The background was fitted with a linear interpolation between the 30 given points [15]. The refinement with halfwidth parameters *U*, *V*, *W*, unit cell parameters mixing coefficient and the symmetry parameters converged to χ^2=0.042. Then the assumed structure was refined while fixing the profile and instrumental parameters. The refinement of Br and Cl in the same positions diverges and is not stable. For this, the Cl and Br atoms are fixed in two different positions and then they are ordered.

Table 1. Crystal structure data and experimental conditions of structure determination of $[Rb_{0.92}(NH_4)_{0.08}]_2TeCl_4Br_2$.

Formula	$\mathbf{[Rb_{0.92}(NH_4)_{0.08}]_2TeCl_4Br_2}$
Temperature (K)	293
Space group	*P4/mnc*
a (Å)	7.315(2)
c (Å)	10.3540(4)
V(Å^3)	554.204(4)
Z	2
D_x(gcm^{-3})	3.53
Radiation (Å)	$CuK_{\alpha1}$ 1.54060, $CuK_{\alpha2}$ 1.54433
Diffractometer	PANalytical X'pert-Pro
2θ range of refinement(°)	10 - 120
Number of structural variables[a]	7
Number of global parameters[b]	38
R_P	0.010
R_{wp}	0.090
χ^2	0.042
R_{bragg}	0.016

[a]Including the scale factor.

[b]Including background coefficients, cell parameters, Pseudo-Voigt function parameters, and asymmetry parameters.

A final refinement taking into account the overall isotropic displacement factor (*B*ov) converged to R_P=0.010, R_{wp}=0.090, and R_{bragg}=0.016 (Table 1). Low values of the profile residuals, *R*p and *R*wp, indicate a high background level rather than the real profile agreement, but low R_B values (which are not affected by the background level) show the reasonable agreement between the experimental and theoretical structural models. The occupancy probability of the Rb and N atoms was refined in such a way that the total occupancy was equal to 1 (Table. 2).

Table 2. Positional equivalent and isotropic thermal parameters for the structure of $[Rb_{0.92}(NH_4)_{0.08}]_2TeCl_4Br_2$.

	Atoms	**Site**	x	y	z	B_{iso} (Å^2)	**Occ. factor**
$[Rb_{0.92}(NH_4)_{0.08}]_2TeCl_4Br_2$	**Te**	2b	0.5	0.5	0	0.2424	1.000
	Rb	4d	0	0.5	0.25	0.0226(9)	0.920(3)
	N	4d	0	0.5	0.25	0.0226(9)	0.080(3)
	Br	4e	0	0	0.2478(2)	0.9313(8)	1.000
	Cl	8h	0.7619 (3)	0.2631(3)	0	0.6399(7)	1.000
	H	16i	0	0.4380	0.1800	0.9274	0.02000

Crystal structure data and experimental conditions are collected in Table 1. Positional and isotropic thermal parameters of the ternary halides $[Rb_{0.92}(NH_4)_{0.08}]_2TeCl_4Br_2$ are summarized in Table 2. Fig. 1 illustrates the final Rietveld plots for $[Rb_{0.92}(NH_4)_{0.08}]_2TeCl_4Br_2$.

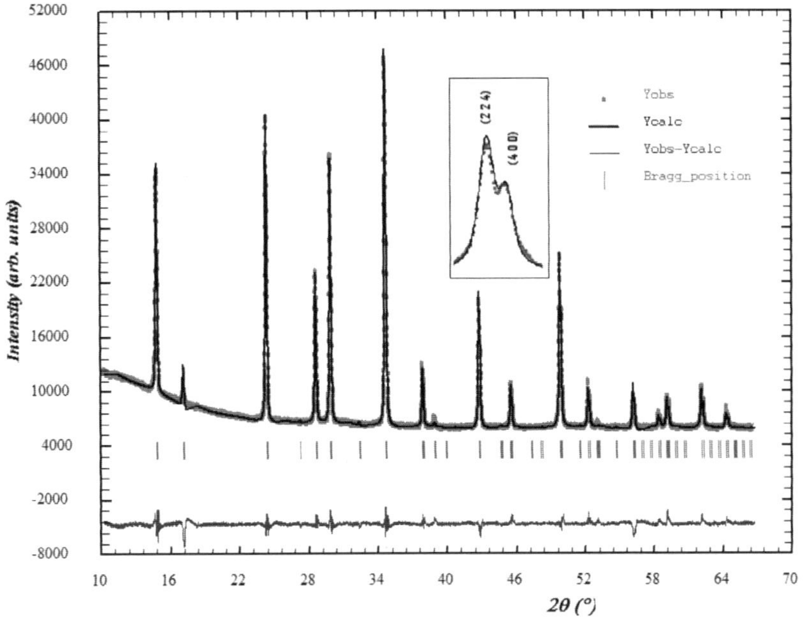

Figure 1. X-ray Powder diffraction pattern and refinement of $[Rb_{0.92}(NH_4)_{0.08}]_2TeCl_4Br_2$.

Differential scanning calorimetry (DSC) measurements at low temperature were carried out using a Perkin-Elmer DSC7 calorimeter in the temperature range from 120K to 300K at a heating rate of 10 $K.min^{-1}$ using a polycrystalline sample in a flowing nitrogen atmosphere.

3. Results and discussion

3.1. Structure description

The crystal structures of the A_2BX_6 usually belong to the large family of K_2PtCl_6 (the so-called antifluorite structure). For a tetragonally distorted K_2PtCl_6 type structure the space group P4/mnc was first proposed by Das and Brown, reporting a low temperature phase of $(NH_4)_2TeBr_6$ [16]. In the present work, $[Rb_{0.92}(NH_4)_{0.08}]_2TeCl_4Br_2$ crystallizes in the tetragonal space group (P4/mnc) with a = 7.315 *(2)* Å, *c* = 10.354*(4)* Å.

For $[Rb_{0.92}(NH_4)_{0.08}]_2TeCl_4Br_2$, the Te atoms lie on the 2(b) sites of the P4/mnc space group. Each Te atom is surrounded by four chlorine and two bromine atoms forming a slightly distorted octahedron with distances Te-Cl is 2.584 (12) Å and Te-Br is 2.615 (15)Å.

In the tetragonal structure each monovalent cation is coordinated by 12 anions: 8 Cl atoms at distances ranging from 3.569 (5) Å to 3.752(4) Å and 4 Br atoms at distances equal to 3.558(4) Å for $[Rb_{0.92}(NH_4)_{0.08}]_2TeCl_4Br_2$.

The Rb/NH_4^+ groups are located between the $TeCl_4Br_2$ octahedra ensuring the stability of the structure by ionic and hydrogen bonding contacts (Rb^+)–Cl^-/Br^- and N–H…Cl/Br, respectively (Fig.2).

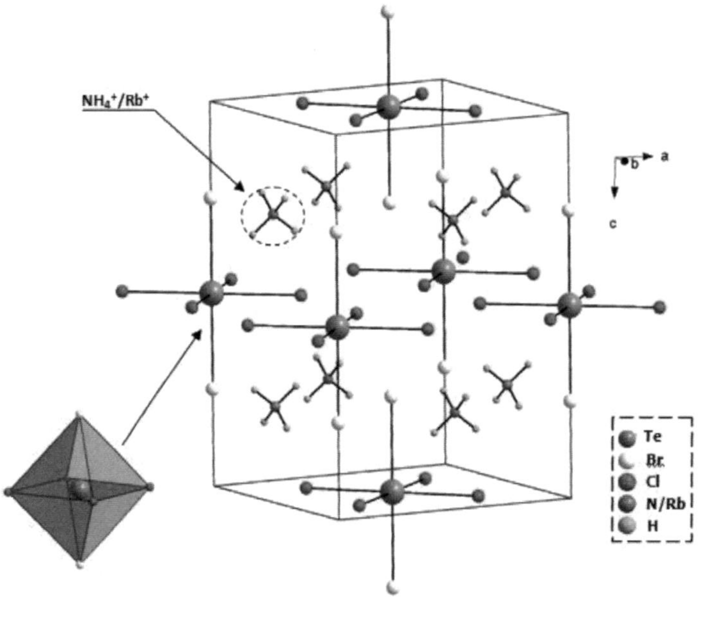

Figure 2. The structure of $[Rb_{0.92}(NH_4)_{0.08}]_2TeCl_4Br_2$.

The hydrogen atoms were located geometrically in the site symmetry 4/m. All the hydrogen atoms of NH_4^+ are involved with the atoms of bromine and chlorine octahedra ($TeCl_4Br_2$) to form a three-dimensional network of hydrogen bonding ensuring the cohesion of the crystal structure.

The Br atom enters into hydrogen bond H....Br with distance equal to 3.100 (6) Å. But, the H bond length H...Cl is in the order of 2.944(18) Å. The difference in length of bond is the origin of the strong interaction H...Cl compared to that H...Br due to the difference of the electronegativity between chlorine and bromine atoms.

To examine the influence of cation and anion substitutions on the bonding within the TeX_6^{2-} ion, we have now refined the structure of $[Rb_{0.92}(NH_4)_{0.08}]_2TeCl_4Br_2$ which possess the distorted K_2PtCl_6 structure at room temperature.

For $[Rb_{0.92}(NH_4)_{0.08}]_2TeCl_4Br_2$, the Te–Br distances vary, at most 0.066Å in $(NH_4)_2TeBr_6$ structure to 0.074 Å in Rb_2TeBr_6 structure and that the Te–Cl distances vary, at least 0.006Å in $(NH_4)_2TeCl_6$ structure to 0.058 Å in Rb_2TeCl_6 structure.

From this comparison it follows that small concentration of NH_4^+ ions really act as a "chemical pressure" along the [0 0 1] direction leading to the contraction of the octahedral $TeCl_4Br_2$. The same effect is observed for $[Rb_{0.92}(NH_4)_{0.08}]_2TeCl_4Br_2$.

The combined substitution leads to a change from cubic to tetragonal structure, the octahedral $TeCl_4Br_2^{2-}$ anions are rotated of about 4° around the [0 0 1] direction (Fig.3).

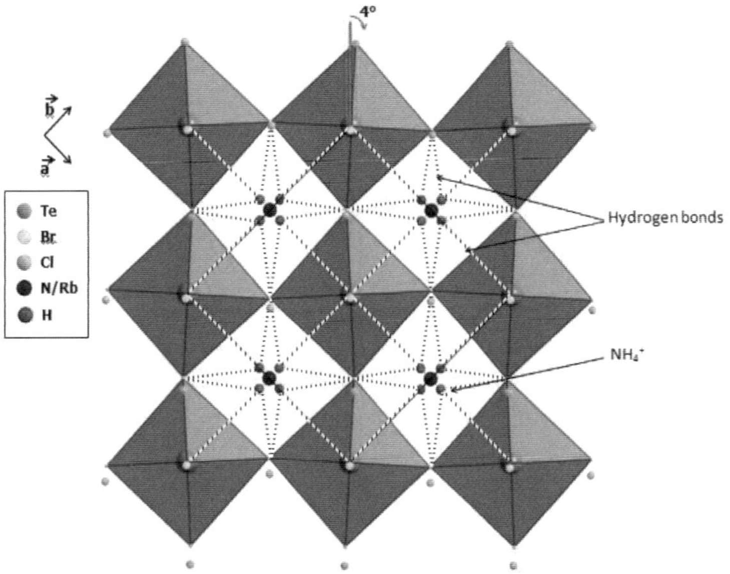

Figure 3. Projection along the fourfold axis with an anti-ferrorotative displacement of the $[TeCl_4Br_2]^{2-}$ octahedral: Show the Hydrogen bond in compound $[Rb_{0.92}(NH_4)_{0.08}]_2TeCl_4Br_2$.

3.2. Calorimetric study

At low temperature the DSC curves show one endothermic peak at 213K with corresponding enthalpie change ΔH= 12.21 $Jmol^{-1}$ (Fig.4). This temperature can be attributed to a second order phase transition.

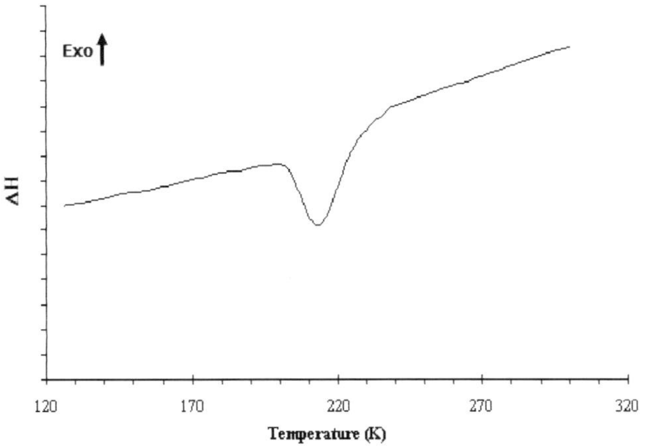

Figure 4. DSC thermogram of $[Rb_{0.92}(NH_4)_{0.08}]_2TeCl_4Br_2$ in the temperature range 120 to 320K.

4. Conclusion

The present work is the first presentation of the results of detailed XRD, low temperature DSC studies of $[Rb_{0.92}(NH_4)_{0.08}]_2TeCl_4Br_2$. The title compound crystallises in P4/mnc tetragonal space group. The main feature of the structure of this material is based on isolated TeX_6 (X = Cl or Br) octahedra connected by ionic and hydrogen bonding through the Rb or N atoms which are located between the octahedral ensuring the stability of the structure. Moreover, DSC measurement in the temperature range (120-300K) was also presented. The occurrence of a low-temperature phase transition near 213K is well evidenced.

References

[1] Lisher E. J, Cowlam N and Gillott L 1979 *Sect. B: Struct. Crystallogr. Cryst. Chem.* **35** 1033 1038.
[2] Dimitropoulos C, Pelzl J and Borsa F 1990 *Phys. Rev. B* **41** 3914-3919.
[3] Prager M, Raaen A. M and Svare I 1983 *J. Phys. C* **16** L181 L186.
[4] Brown I. D 1964 *Canad. J. Chem.* **42** 2758
[5] Das A. K and Brown I. D 1966 *Canad. J. Chem.* **44** 939.
[6] Berg R. W and Nielsen K 1979 *Acta Chem. Stand. A* **33** 157.
[7] Abriel W 1982 *Kristallogr Z.* **1** 159.
[8] Ben Aribia W, Abdelmouleh M, Karray R, Van Der Lee A, Kabadou A and Ben Salah A 2011 *Journal of Molecular Structure* **986** 86-91.
[9] Belgaroui H, Loukil M, Karray R, Ben Salah A, Kabadou A 2010 *Journal of Alloys and Compounds* **499** L5 L8.
[10] Louër D, Louër M 1972 *J. Appl. Crystallogr.* **5** 271-275.
[11] Boultif A, Louër D 2004 *J. Appl. Crystallogr.* **37** 724-731.
[12] Rietveld H. M 1969 *J. Appl. Crystallogr* **2** 65-71.
[13] Rodriguez-Carjaval J 1990 *Proceedings of the Satellite Meeting on PowderDiffraction* 127.
[14] Young R. A, Wiles D. B 1982 *J. Appl. Crystallogr.* **15** 430-438
[15] Roisnel T 2001 *WinPLOTR Program, Laboratoire de Chimie du Solide et Inorganique Moléculaire, UMR 6511 CNRS-Université de Rennes 1, Institut de Chimie de Rennes, Avenue du Général Leclerc, 35042 Rennes Cedex, France.*
[16] Das A. K, Brown I. D 1969 *Canad. J. Chem.* **47** 4288.

IOP Conf. Series: Materials Science and Engineering **28** (2012) 012042 doi:10.1088/1757-899X/28/1/012042

An investigation into the effects of conventional heat treatments on mechanical characteristics of new hot working tool steel

M. L Fares[1a,b], M. Athmani[b], Y. Khelfaoui[c], A. Khettache[b]
[a]Mechanics of Materials and Plant Maintenance Research Laboratory (LR3MI), Metallurgy and Material Engineering Dept., Badji Mokhtar University, BP 12, Annaba 23000, Algeria.
[b]Metallurgy and Material Engineering Dept., Badji Mokhtar University, BP 12, Annaba 23000, Algeria.
[c]Material Technology and Engineering process Laboratory, A/Mira University, Bejaïa Algeria.

E-mail: fares.lamine@univ-annaba.org

Abstract. The effects of conventional heat treatments, i.e. quenching and tempering, on the mechanical characteristics of non standard hot work tool steel, close to either AISI-H11/H13 are investigated. The major elemental composition differences are in carbon, silicon and vanadium. The objective of the carried heat treatments is to obtain an efficient tool performance in terms of hardness, wear resistance and mechanical strength. Experimental results allow an explanation of the surface properties depending mainly on both chemical composition and optimised preheating parameters. After austenitizing at 1050 °C for 15 min, the as-quenched steel in oil bath exhibited the fully martensitic structure (without bainite) connected to a small fraction of retained austenite and complex carbides mainly of $M_{23}C_6$ type. Twice tempering at 500 °C and 600 °C resulted in initiating the precipitation processes and the secondary hardness effect. As a result, carbide content amounted to 3% while the retained austenite content decreased to 0%. Accordingly, the required mechanical properties in terms of hardness and wear are fulfilled and are adequately favourable in handling both shocks and pressures for the expected tool life. Induced microstructures are revealed using optical and scanning electron microscopes. Phase compositions are assessed by means of X-ray diffraction technique while mechanical characteristics are investigated based on hardness and abrasive wear standard tests.

1. Introduction

The term 'hot work tools' is applied to different kinds of tools for the working or forming of metals at high temperatures, i.e. tools for die-casting; melt extrusion, hot pressing and plastic parts moulding. Usually, hot working steels used for die-forging tools are subjected to high thermal effect and severe conditions of both mechanical loads and wear. Consequently, such effects cause an increased susceptibility to wear leading to considerable service costs. The wear mechanisms depend widely on

[1] To whom any correspondence should be addressed.

chemical composition, microstructure, load level and the material surface properties as premature wear is connected to a combination of heat shocking, erosion and corrosion phenomena [1]. The prevention of such damage is often associated with a critical hardness level that must not be exceeded for a specific application. For instance, the recommended tool steel hardness lies in the ranges 43-52 HRC for extrusion dies and 44-50 HRC for die-casting respectively while it is in the interval 40-55 HRC for forging dies [2].

Several standard steels are used for hot working tools, e.g. AISI H10-H19, and together with a series of commercial special ones. The main sought properties are toughness, resistance to shocks, high hardenability, dimensional stability, excellent wear resistance and thermal resistance under heat treatments. None of the studied steels can satisfy all these requirements from tools for all the different application areas. Steel grades such as AISI H11 and H13 are the most widely used in industry and this is confirmed by the availability of an extensive body of literature, comprising aspects such as chemical composition, operating parameters and tool life, related to conventional heat treatments [3-8]. Focusing on all these factors and their role on tool performance, the slightest modification of chemical composition can have a major impact on mechanical properties of working tools. These conditions can be satisfied by steel alloys having a basic composition in order to yield a microstructure with an even carbide distribution in a ferritic matrix. Since a variety of steels with standard and non-standard chemical composition exists, the problem of choosing the right steel for a given application can become rather confusing if chemical comparisons are used alone. That is why, it is necessary to also consider heat treatment parameters. For instance, to quench drastically enough in order to harden completely, tool steel may be subject to cracking risk while a milder quench may not develop microstructures with the required strength. Therefore, the properties that can be obtained after heat treatments must be a result of a compromise between various operating parameters. This fact supports continued research activities and offers several ways of improving the mechanical resistance. If surface properties such as hardness as well as wear resistance are the primary concerns, other properties, e.g. toughness and thermal fatigue resistance [7] as well as corrosion resistance [9], must not be neglected in order to produce an effective tool with acceptable extended lifetime.

The objective of this work is the characterisation of the structure and main mechanical properties of a modified steel grade ((industrial steel with high carbon content, low silicon and Ni additive), depending on its optimized heat treatments conditions compared to standard AISI H13/H11 hot working tool steels, commonly used for hot working tool manufacturing. This investigation is focused primarily on the searching for competitive materials with better tool life under severe service conditions. The hardening procedure consisted of three steps i.e. austenitising, quenching and tempering to achieve the martensitic phase transformations which increase wear and thermal fatigue steels resistances [5,7]. Microstructural characterisations in terms of chemical composition and formed phases are carried out by means of both optical and scanning electron microscopes (SEM) and X-ray diffraction technique. Additionally, mechanical properties are assessed using hardness and abrasive standard tests.

2. Materials and experimental procedures

2.1. Materials

Table 1. Composition of studied steels (wt %), Fe balance

	C	Mn	Si	Al	Ni	Cr	V	Ti	Mo
As-Received steel	0.52	0.42	0.55	0.03	0.22	5.20	0.34	0.004	1.10
Standard 1.2343/H11	0.38	0.40	1.00	-	-	5.10	0.40	-	1.25
Standard 1.2344/H13	0.40	0.40	1.05	-	-	5.15	1.00	-	1.35

For the present investigation, modified alloy hot-working tool steel is considered. Its chemical composition, identified by the Thermo Scientific ARL 4460 optical emission spectrometer (OES) is reported in Table 1. Compared to AISI H11 and AISI H13 steels, it discloses some difference in carbon, vanadium, nickel and silicone contents. The heat treatments as well as structural and mechanical characterisation are carried out at a laboratory scale.

2.2. Specimens preparation and characterizations

Beforehand, cubic samples (10x10x10) mm^3 in size are subjected to an annealing operation in order to eliminate residual stresses that resulted from pre-machining and to give an appropriate microstructure for hardening. Afterwards, specimens are successively austenitised, quenched and twice tempered in a furnace chamber according to the heat treatment cycle shown in Fig.1. Plan of experience and specimen designations are illustrated in Table 2.

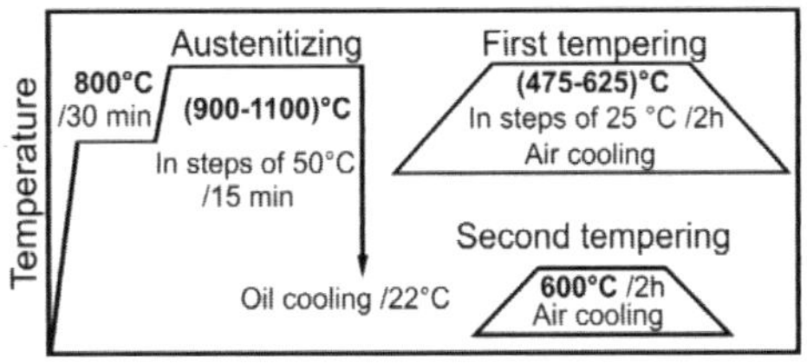

Figure 1. Conventional heat thermal cycle for as-received steel.

Table 2. Treatments parameters and specimen designations of conventional heat treatments

Process	Cycles	Specimen designations
Soft Annealing	550 °C/1h + 880 °C /3h /furnace cooling	A1
Quenching	(900÷1100) °C /11min /Oil cooling/22°C; in 50 °C gradation	Q1
First Tempering	(475÷625) °C /2 h /air cooling; in 25 °C gradation	QT1
Second tempering	600 °C /2h /air cooling	QT2

Prior to optical observations, surface preparation consisted of mechanical grinding using 1200-grit mesh and polishing down to a final step of 1μm diamond pastes. After ultrasonic cleaning in ethanol, samples are etched in a 3% nital solution at ambient temperature. Metallographic investigations are achieved using both optical and scanning electron microscopes types GX.51 Olympus and JEOL SJM-6500F with Energy Dispersive X-ray Spectrometry attachment (EDS) respectively. To reduce the cutting strain effect induced by the mechanical polishing operation, all samples underwent an electropolishing followed by a specific etching. Then, the volume fraction of formed phases (retained austenite and carbides) are determined by the Guinier's method [10], using a Philips PW1730/10 diffractometer operated at 40 kV and 200 mA, with a cobalt anticathode of λ Kα = 0.179026 nm. A 2θ-scan is recorded within the range 35° to 135° with a constant step wise increase of 0.036°/35 s.

Hardness tests of the quenched as well as twice tempered samples are achieved by means of the Rockwell method using a universal machine of Wolper type machine.

Abrasion criterion measurements adopted in this study is based on the weight loss method [11] and the resistance is determined under dry conditions, in air and at room temperature. The abrasive wear results are obtained by means of a pin-on-disc test where specimens are forced to slide on a fresh 240-mesh abrasive paper with an applied normal load of 354 g (~3.5 N) [12]. The abrasive paper is fitted on a steel disc of a polishing machine rotating at a speed of 50 cycles /min. After a prescribed sliding distance, samples are removed and completely rinsed. An analytical microbalance accurate to 10^{-3} g is used to measure weight changes for each 180 seconds interval.

3. Results and discussion

According to the suggested thermal cycle treatment shown in Fig.1, the conventional heat treatments for steels are performed in order to achieve a tempered martensitic microstructure with maximum hardness that should lead to an increased wear resistance [13]. The right choice of austenitising temperature is the basis in getting optimal properties such as hardness and toughness after tempering. Fig. 2(a) shows hardness results versus austenitisation temperatures in the range of 900 to 1100 °C with a step of 50 °C for a holding time of 15 min. It is observed that a maximum hardness value close to 57 HRC (Q1 sample) is obtained at 1050 °C since it can be changed by the only subsequent tempering treatment. This result agrees with earlier reported studies in H11 and H13 steels [6,8]. Besides, it is asserted that quenching H13 steel at 1050 °C allows obtaining optimum high toughness and relevant thermal fatigue resistance [14]. It also noted that, beyond 1050 °C, further increase of the austenitising temperature causes an important drop of hardness close to 42 HRC. This hardness decline is probably caused by the sensitive coagulation of the alloying elements in the austenite and the relative degree of austenite grain coarsening.

Then, the sample Q1 was subjected to the first tempering operation for 2 h in the temperature range of 450-625 °C with gradation of 25 °C. Accordingly, Fig. 2(b) exhibits the temper resistance assessed from the tempering curve for hardness vs. tempering temperatures. It can be seen that in the range of 475 to 500 °C (*broken line*), an increase of the hardness occurs from 51 to 54 HRC corresponding to the maximum value which is owed normally to a homogeneous precipitation of very fine carbides. This increase is connected to three distinct transformations occurring during the tempering operation; (i) destabilization of retained austenite which is sensitized by tempering and changes at the time of the cooling by modification of its chemical composition; (ii) transformation in tempered martensite; (iii) fine carbides precipitation which may be associated to the secondary hardness; the later bounds to the exchange of the carbon between cementite and carbides [15]. An earlier work carried out on H13 steel [16] led to an optimal hardness of 53.5 HRC after quenching and tempering at 1060 °C and 510 °C respectively. Beyond the temperature of 525 °C, this value exhibits a noticeable fall to 43 HRC at tempering temperature of 625 °C. As a result, increasing the tempering temperature above 500 °C may be provoking an advanced recovery in examined specimens. Also, initiation of the recrystallization of the matrix as well as separation and coagulation of newly created alloy carbides can take place and would explain the considerable hardness drop. At this stage, the lower silicon content (~0.55%) can have a favourable effect on the properties of hardened and tempered as-studied tool steel. In view of that, it is known that higher silicon has a detrimental effect on the AISI H11 steel [17], as it shifts the secondary hardening peak towards lower tempering temperatures. As a consequence, reducing the silicon content may result in a possible use of the tool for higher service temperature (an excess temperature from 20 to 30 °C).

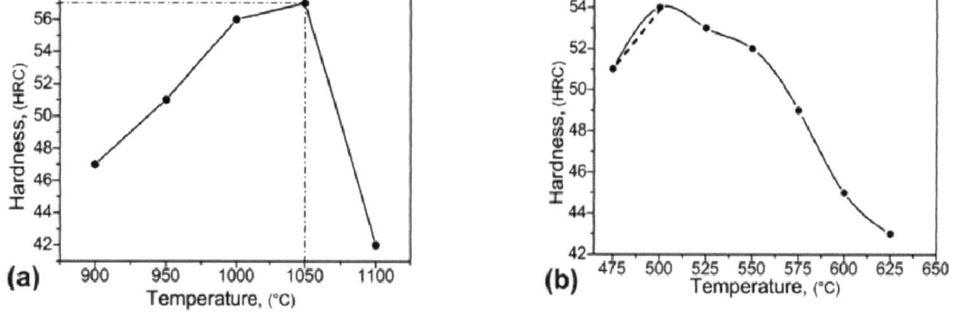

Figure 2. Hardness results versus treatment temperature of the original steel; (a) Quenching (Q1 sample); (b) First tempering (QT1 sample).

Fig. 3 allows a better view of the details associated with microstructure transition under the effect of the suggested heat treatment cycle. Commonly, a balanced chemical composition of carbide forming elements such as chromium, molybdenum and vanadium aimed at providing an excellent soft-annealed initial structure for the subsequent tool hardening; thereby achieving acceptable hardenability, tempering resistance and temperature strength properties. Fig. 3(a) related to an annealed microstructure of as-received steel, so-called A1 sample, shows a homogeneous precipitation of globular carbides embedded in a ferritic matrix. Such configuration corresponds to the optimal sought structure. The corresponding measured hardness value is ~(180–210) HB which is concomitant with a good machinability. XRD results of Fig. 4(a), determining the sequence of induced precipitation, revealed the presence of M_3C (M: Fe mainly), $M_{23}C_6$ (M: Cr mainly) and a small ratio of M_7C_3 (M: V mainly).

Fig. 3(b) exhibits the optical microstructure of the sample Q1 after quenching at 1050 °C for 15 min. It comprises lath martensitic structure, partially twinned with a small fraction of retained austenite. The average diameter of the primary austenite grain size being about 15 μm, which is numbered ASTM No. 8, allows obtaining advantageous mechanical properties [16]. Consequently, it can be avowed that both optimum austenitising temperature (1050 °C) and quenching time (15 min) have a positive effect on the primary grain size despite the lower vanadium content which remained sufficient to prevent grain coarsening. Fig. 4(b) displays X-ray spectra of treated samples. It points up that retained austenite, due to its structural difference from other phases in steel, produced a diffraction peak at different locations than ferrite. The amount of a phase in steel is proportional to the integrated intensity of its diffraction peak [18]. No distinction has been made between ferrite and martensite phases due to the fact that theoretical X-ray diffraction intensities are nearly the same and the term ferrite can also apply to martensite [19].

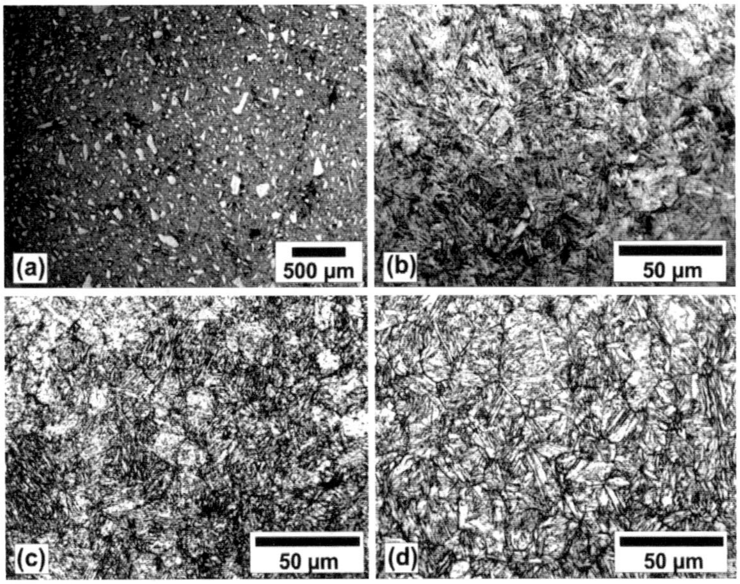

Figure 3. Optical micrographs of studied steel at various treatment states for; (a) Annealed sample: A1; (b) quenched sample: Q1; (c) first tempered sample: QT1 and (d) twice tempered sample: QT2.

Commonly, the higher austenitizing temperature (1050 °C) gives rise to a better dissolution of the primary alloy carbides present in the annealed state, especially vanadium rich carbides [20,21]. In addition, because of the chemical composition of the studied steel, including more carbon in solution, and the preexisting carbide-forming element such as vanadium as well as molybdenum, theoretically it is expected to have the amount of MC and/or M_7C_3 carbides as dominant compared to preexisting $M_{23}C_6$ carbides. It is clear that the pre-cited precipitates during tempering play a key role in improving the wear resistance by increasing the surface hardness since $M_{23}C_6$ are supposed to be less resistant. However, for the sample Q1, as depicted by relevant X-ray spectrum (Fig. 4(b)), only the chromium carbides $M_{23}C_6$ which may have molybdenum dissolved in them are found while the vanadium carbides are not revealed. The disappearance of such carbides precipitation, caused by both higher quenching temperature (1050 °C) and higher cooling rate (oil bath) is probably not as harmful as first implied since during the subsequent temper, the maximal precipitation occurs thereby causing the secondary hardening. In addition, at this stage, a fewer vanadium content of the studied steel (~0.35%) may also not be much of a problem, since it avoids the formation of undesirable indissolved carbides which may remain along prior to austenite grain boundaries throughout the thermal processing of the material.

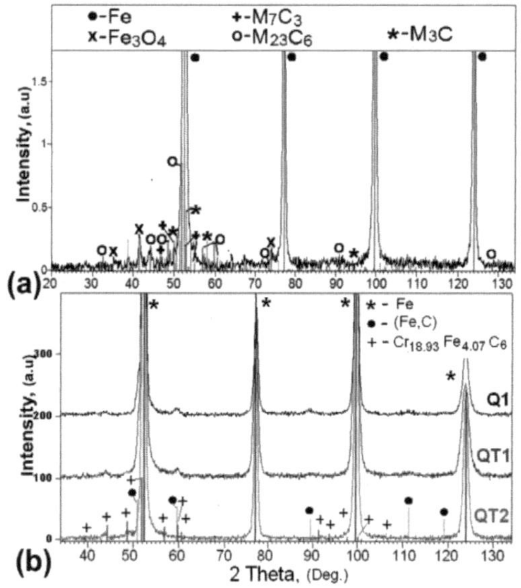

Figure 4. XRD patterns of studied steel at various treatment states; (a) A1 sample; (b) Q1, QT1 and QT2 samples.

Table 3 is a summary of X-ray and electron diffraction data for volume fractions identification of ferrite (martensite), carbides and retained austenite formed by quenching at 1050 °C and twice tempering at 500 °C and 600 °C respectively. The results for Q1 sample, obtained by means of Guinier's method [10] indicate that for the sample Q1, the volume fractions of $M_{23}C_6$ carbides and retained austenite are 2 % and 4% respectively while that of martensite is 96%.

The structural hardening, caused by the martensitic transformation and the better dissolution of carbides, resulted in a decrease of transformation points *Ms* and *Mf*. It is well known that the temperature at which austenite starts to transform martensite on quenching (M_S) is a function of steel chemical composition. Generally, the assessment of *Ms* Temperature is completed by means of the CCT Diagram; however, because of the inaccessibility to such data, several sources can give *Ms* Temperature. Among, these sources, the Steven and Haynes' formula [22] expressed below (Eq. 1), is the most indicated because it can be used for steels containing up to maximum 0.6% of carbon content:

$$Ms(°C) = 561 - 474C\% - 33Mn\% - 17Ni\% - 7Cr\% - 21Mo\% \tag{1}$$

Besides, the relation defined by Koistinen and Marburger [22] (Eq. 2) provides the untransformed austenite fraction related to both *Ms* and *Tq* (quenching temperature), as follows:

$$V_\gamma(\%) = \exp\left(-1.10x10^{-2}(Ms - Tq)\right) \tag{2}$$

Thus; the amount of martensite is deduced as:

$$V_m(\%) = 1 - V_\gamma \tag{3}$$

It is worth mentioning that in general, *Mf* may not correspond to 100% martensite, because some retained austenite might be present [23]. Consequently, the accuracy of equation (1) is corroborated here by the calculated value that gives an austenite amount of ~4.67% which agrees with measured value (~4%) and according to equation (3), the martensite amount is ~95.33%. This feature can be linked to the rapid quenching rate which has a critical significance on microstructure shaping, final working tool properties and its extended service life. Accordingly, it is stated that pure martensitic microstructure resulted in the highest toughness value and that a rapid quench may avoid the formation of bainite thus, achieving a relevant maximum hardness [24]. This result can be asserted by the ratio value c/a (2.98 A/2.85 A) of tetragonal martensite structure, determined from Fig. 4(b) using the current issue of the Powder Diffraction File (PDF, ICDD).

Table 3. X-ray diffraction data using Guinier's method for identification of formed phases (martensite, carbides and retained austenite) for Q1, QT1 and QT2 samples.

ηκλ	2θ	I* χαρβ.	I* α	I* γ	I (α, γ, χαρβ).		
420	22.05	7.95	-	-	2.0	4.5	4.8
200	59.52	-	-	22.49	9.0	8.9	0.0
200	77.37	-	08.88	-	101.4	167.5	173.8
220	89.23	-	-	11.49	7.0	4.8	0.0
211	99.90	-	17.78	-	218.3	365.4	368.2
311	110.85	-	-	15.95	7.5	6.6	0.0
222	118.47	-	-	5.19	4.3	3.1	0.0
220	124.10	-	07.72	-	100.7	164.9	167.2
Σ I* α	–	–	34.38	–	420.4	697.8	709.3
Σ I* γ	–	–	–	55.12	27.8	23.3	0.0
Iχαρβ/ι*χαρβ					0.2	0.6	0.6
Iα/I*α					12.2	20.3	20.6
Iγ/I*γ					0.5	0.4	0.0
Σ I/I*					13.0	21.3	21.2
Samples					**Q1**	**QT1**	**QT2**
V % Carbides					**2**	**3**	**3**
V % Feγ					**4**	**2**	**0**
V % Feα					**94**	**95**	**97**

However, it is important to point out a major difficulty that consisted in noticeable appearance of hardening cracks and distortions in the quenched samples due to rapid quench (one step) leading to material rubbish. In some cases, regarding to the earlier quoted problems, a martempering process is advisable to reduce these risks. In this process, the material is quenched in two steps. First, it is cooled from the hardening temperature in a salt bath having a temperature just above the *Ms* Temperature. It is kept there until the temperature has equalized between the surface and the core, after which the tool can be allowed to cool down freely in air within the martensite transformation range.

A steel through-hardened is not a satisfactory material for most application. Thus, a subsequent tempering treatment is required for obtaining a tempered martensite microstructure with appropriate properties for industrial purposes. At this stage, some transformations occur in one part of the martensite and the second in the retained austenite.

After one tempering, the sample Q1 within the range of temperature 475 °C to 500 °C, resulted in starting of the precipitation processes in martensite and the secondary hardening effect. Fig. 3(c) reveals the optical microstructure of the sample QT1 after tempering at 500 °C for 2 h. It is noted the disappearance of martensite needles which are changed in a newly tempered martensite composed of the wider laths than those of martensite formed in the sample Q1. Besides, $M_{23}C_6$ carbides became larger since it is observed a sensitive decrease in retained austenite content from 4% to 2% related to a slight development of $M_{23}C_6$ carbides from 2% to 3%. The precipitates are the most relevant microstructural constituents with respect to hardness and thermal stability of the material. Since the mechanical properties of such materials strongly depend on type, particle number, density size and lattice structure of the prevailing precipitates, an extensive characterisation is required. Thus, a high resolution method such as transmission electron microscopy (TEM) and atom probe tomography (APT) [25] can be applied in perspective.

Secondary tempering is used to further improve both toughness and thermal stability of the structure. The tool materials should not be tempered at a temperature lower than the desired operating temperature. Fig. 3(d) shows the optical microstructure of the sample QT1 subjected to a secondary tempering (so-called QT2 sample) at 600 °C for 2 h. The second tempering takes care of the secondary martensite formed after the first tempering. The hardness value is 53 HRC which is favourable to shock resistance or high pressures. The laths of newly-formed martensite having 0.1-2μm of width and 0.5-15μm of length are wider than those of martensite formed in the sample QT1. Theses values agree with those obtained on H11 steel by Delagnes in a previous study [26]. At this stage, volume fraction of retained austenite decreased from 2 to 0%, while the one of carbides stayed at a level of 3% and remained the same as for the QT1 sample. According to diffractograms of Fig. 4(b), the intensity of the Fe-peak of QT2 sample is greater than those of both Q1 and QT1 samples; this future can be explained by the presence of tempered martensite which is the result of the austenite transformation leading to slighter network distortions. Thus, the value of the lattice parameter c is expected to tend toward a.

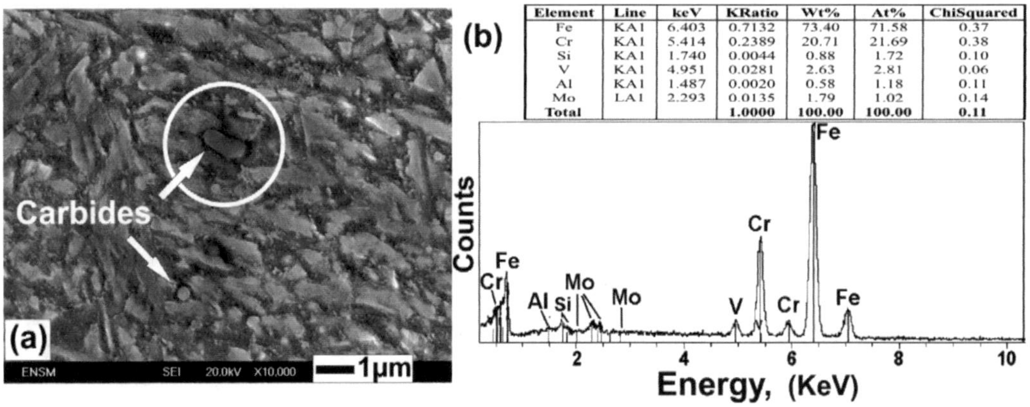

Element	Line	keV	KRatio	Wt%	At%	ChiSquared
Fe	KA1	6.403	0.7132	73.40	71.58	0.37
Cr	KA1	5.414	0.2389	20.71	21.69	0.38
Si	KA1	1.740	0.0044	0.88	1.72	0.10
V	KA1	4.951	0.0281	2.63	2.81	0.06
Al	KA1	1.487	0.0020	0.58	1.18	0.11
Mo	LA1	2.293	0.0135	1.79	1.02	0.14
Total			**1.0000**	**100.00**	**100.00**	**0.11**

Figure 5. SEM analysis of QT2 sample; (a) microstructure; (b) EDS analysis.

As it has been already observed on Fig. 3(d), optical microscopy does not allow the precise determination of the carbides presence in martensitic matrix due to limited resolution. Therefore, the individual microstructural features of the tempered samples, relevant for the mechanical properties, are characterized by EDS analysis. Fig. 5 shows SEM microstructure of the sample QT2 containing carbides with 0.2 to 1μm in size. According to the EDS spectroscopy results (Fig. 5(b)), conducted on the marked area in Fig. 5(a), carbides correspond undoubtedly to $M_{23}C_6$, where M is chromium (~20.71 wt. %) mainly. These carbides may have both molybdenum (~1.79 wt. %) and vanadium (~2.63 wt. %) dissolved in them. If the secondary tempering is performed at a temperature above 600 °C, it then allows the coalescence of $M_{23}C_6$ precipitates departure after the first tempering [27].

The effect of optimal conventional heat treatment, as suggested in this study (quenching for 15 min at 1050 °C and twice tempering for 2 h at 500 °C and at 600 °C) on abrasive wear resistance have been additionally investigated by the pin-on-disc test. Wear test results showing weight loss vs. sliding distance (1690 m) are plotted in Fig. 6. It is observed that at a sliding distance of 113 m, a high initial weight loss of 29.10^{-3} g occurred giving a wear rate of $25.6.10^{-5}$g.m^{-1}. This can be ascribed to the subsistence of a thin oxide layer on the surface [28]. Next, for a global sliding distance of 1690 m, this wear behaviour is followed by a linear steady-state weight loss. The global weight loss is 131.10^{-3} g corresponding to a wear rate of $7.75.10^{-5}$ g.m^{-1}. It is expected that a similar wear mechanism will be observed for Q1 and QT1 samples. Besides, Q1 sample exhibited a lower weight loss due to higher hardness level and a relatively larger amount of residual austenite which may promote wear resistance associated with a decreasing friction coefficient.

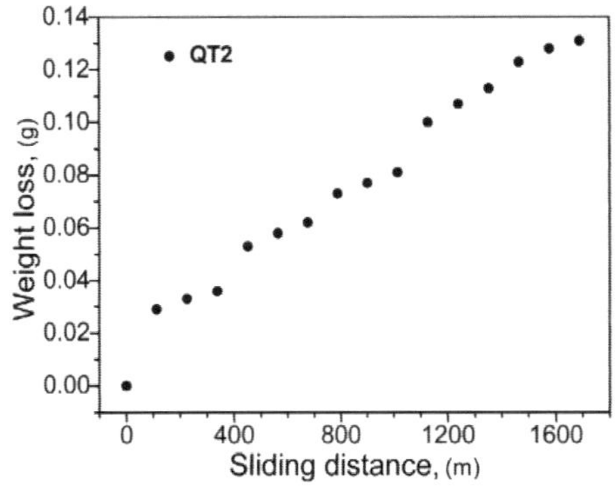

Figure 6. Wear test results of QT2 sample showing weight loss vs. sliding distance.

It becomes clear that the proposed heat treatment can be successfully implemented for this new hot working tool steel which is subjected to severe wear conditions. The obtained results suggest higher both strength properties and tool productivity, especially for hot industrial applications. From this standpoint, these results indicate that further improvements of earlier properties such as higher hardness, better wear resistance and corrosion behaviour can be realized using thermochemical treatments (i.e. gas carburizing, salt bath nitrocarburizing and gas carbonitriding [29]).

4. Conclusions

The experimental results obtained on this non standard tool steel and the suggested preheated parameters (i.e. austenitising temperature: 1050 °C; holding time: 15 min; and rapid quench: oil bath) allow to draw the following conclusions:

The proposed heat treatments give rise to a better dissolution of the primary alloy carbides present in the annealed state, especially vanadium rich carbides. In addition, they show a positive effect on the primary grain size despite the lower vanadium content which remained low enough to prevent grain coarsening.

A reliable martensitic microstructure (without bainite) is obtained and in which is embedded a small fraction of retained austenite (4%) combined with a fewer $M_{23}C_6$ carbides (2%). Such microstructure is responsible for the relevant maximum hardness values and the suitable tribological observed behaviour. After twice tempering at 500 °C and 600 °C, tool steel shows some transformation processes in the martensite phase which led to a decrease in retained austenite content (down to 0%) and to a slight increase in $M_{23}C_6$ carbide content (up to 3%). Consequently, the hardness value reached a satisfactory level of 53 HRC with all the positive results in term of ameliorated wear resistance.

The as-received low vanadium content (~0.35%) is sufficient to avoid the formation of undesirable indissolved carbides which may remain along prior to austenite grain boundaries throughout the thermal processing of the material. However, it is found that high level carbon content (~0.52%) and the chosen cooling medium (oil bath) give unsatisfactory results and led to the formation of cracks. As a remedy, a martempering process is most advisable to avoid cracking and to attain superior finish, high hardness with a lower excessive distortion risk.

According to the reported results above mentioned, it ought to be noted that low silicon content (~0.55%) might have a favourable effect in tempering as it contributes to a possible use of the tool steel in higher in-service temperature.

Acknowledgments

The authors are grateful to the technical staff of Ecole des Mines de Saint-Etienne, Centre SMS France, for their support and for making available the facilities for various characterization tests.

References

[1] Ueda M, Uchino K and Kobayashi A 2002 *Wear* **253** 107.

[2] Shivpury R and Semiatin SL 1988 Wear of Dies and Molds in Net Shape Manufacturing Report ERCNSM-88-05, Ohio State University: Colombus.
[3] Barrau O, Boher C, Gras R and Rezai-Aria F 2007 *Wear* **263** 160.
[4] Rodrıguez J, Martinez D, Perez A, Gonzalez R, Rodrıguez E and Valtierra S 2007 *Wear* **263** 301.
[5] Klobcar D, Tusek J and Taljat B 2008 *Thermal Sciences & Engineering* **A472** 198.
[6] Leskovsek V, Sustarsic B and Jutrisa GJ 2006 *J. of Mat. Processing and Technology* **178** 328.
[7] Mebarki N, Delagnes D, Lamesle P, Delmas F and Levaillant C 2004 *Materials Science Engineering* **A387-389** 171.
[8] Bahrami A, Mousavi Anijdan SH, Golozar MA, Shamanian M and Varahram N 2005 *Wear* **258** 846.
[9] Fares M. L, Chaoui K and Le Coze J 2009 *Surface & Interface Analysis* **41** 549.
[10] Guinier A 1956 Théorie et technique de la radiocristallographie Ed. Dunod, Paris.
[11] Zum-Gahr KH 1989 *Wear* **133** 1.
[12] Fares M. L, Touhami M. Z, Belaid M and Bruyas H 2009 *Surface & Interface Analysis* **41(3)** 179.
[13] Lin Y. C, Wang S. W and Chen T. M 2002 *J. of Materials Processing and Technology* **120** 126.
[14] Yikuan L 1990 *Heat Treatments of Metals* **2** 51.
[15] Bonek M, Dobrzanski L. A, Hajduczek E and A Klimpel 2006 *J. of Mat. Processing and Technology* **175** 45.
[16] Dobrzanski L. A, Mazurkiewicz J, Hajduczek E and Madejski J 2001 *J. of Mat. Processing and Technology* **113** 527.
[17] Delagnes D, Lamesle P, Mathon M. H, Mebarki N and Levaillant C 2005 *Materials Science Engineering* **A394** 435.
[18] Asi O, Can A. C, Pineault J and Belassel M 2009 *Material & Design* **30** 1792.
[19] ASTM E 975–84 1989 *Standard Practice for X-ray Determination of Retained Austenite in Steel with Near Random Crystallographic Orientation ASTM*.
[20] Schmidt L 1987 Tool Materials for Molds and Dies. Colorado School of Mines Press 118–164.
[21] Branco JRT 1989 *Development of Nb Modified AISI H13/H11 Type Tool steels* Thesis Colorado School of Mines.
[22] Parrish G Carburizing, 1999 Microstructures and Properties. ASM International**,** Metals Park Ohio.
[23] Porter D. A and Easterling K. E 2001 Phase transformations in metals and alloys Thornes Cheltenham.
[24] Mayer S, Scheu C, Leitner H, Siller I and Clemens H 2009 *Inter. J. of Mat. Research* **100** 86.
[25] Killer M. K, Cerezo A, Hetherington A. G and Smith GDW 1996 Atom Probe Fields Ion Microsc. Clarendon Press, Oxford.
[26] Delagnes D 1998 *Comportement et tenue en fatigue isotherme d'aciers à outil Z38CDV5 autour de la transition oligocyclique-endurance* Thesis Ecole Nationale Supérieure des Mines de Paris France.
[27] Mebarki N 2003 *Relation microstructure propriétés mécaniques d'aciers martensitiques revenus destinés aux outillages de mise en forme d'alliages légers* Thesis Ecole Nationale Supérieure des Mines de Paris France.
[28] Mohan S, Prakash V and Pathak J. P 2002 Wear characteristics of HSLA steel *Wear* **252** 16.
[29] Fares M. L, Talhi A, Chaoui K and Touhami M. Z 2011 *Surface Engineering* **27** 595.

IOP Conf. Series: Materials Science and Engineering **28** (2012) 012043 doi:10.1088/1757-899X/28/1/012043

X-ray studies of structure defects in recycled aluminum alloy during equal channel angular extrusion: Dislocation density and stored energy

T. Makhlouf [1], **N. Njah**
Applied Metallurgy Laboratory, Sciences Faculty, BP.1171 (3000) Sfax University, Tunisia.

E-mail: thabet.makhlouf@yahoo.fr

Abstract. A recycled aluminum alloy was severely deformed by Equal Channel Angular Extrusion (ECAE) using the processing route B_C, to study the evolution of the microstructure and associated changes of mechanical properties. XRD patterns were analyzed to determine crystallite size, lattice strain and lattice parameter. As a result, ultrafine grains material has been obtained with microstructure showing a mixture of highly strained crystallites. High density of dislocations is achieved as a result of severe plastic deformation through the die. This can explain an important decrease of the lattice parameter observed after the first passage. Changes in mechanical behavior are also reveled after ECAE. This is due to strain hardening. Differential Scanning Calorimetry (DSC) results show overlapped exothermic peaks indicating the succession of the recovery, recrystallization and grain growth phenomena's. Moreover, we demonstrate that the energy stored during ECAE can be related to the dislocation density introduced by severe plastic deformation.

1. Introduction

Ultrafine grain can be introduced into bulk metallic alloys through use of equal channel angular extrusion (ECAE) [1-3]. During the process a metal billet is pressed through a die consisting of two channels, equal in cross section and intersecting at an angle Φ. By multiple passing, very large effective deformation can be developed in bulk products [4-7]. Extensive research has been carried out on the microstructure development and mechanical property evolution during ECAE of several materials. It is well known that ECAE of commercial aluminum alloys leads to ultrafine grained structures with improved mechanical properties [8-11]. The grain sizes obtained by ECAE are generally in the submicrometer/nanometer range [12]. Therefore, ECAE processed materials may be attractive for different applications, mainly those needing superplastic behavior and/or enhanced mechanical properties, i.e. fine-grained microstructure [13, 14]. One of the most common severe plastic deformation (SPD) methods is ECAE, a technique that results in a homogeneous ultrafine-grained microstructure of the work piece [15, 16]. It has been shown that an imposed equivalent strain of~1% is introduced on one passage of the sample through the ECAE die [15].

For understanding the mechanical behavior of materials produced by ECAE, it is necessary to characterize their microstructure. The crystallite size and the lattice strain in nanocrystalline materials

[1] To whom any correspondence should be addressed.

can be determined by X-ray line profile analysis [17, 18]. In SPD processed materials where the lattice distortions are primarily caused by dislocations, the characteristic parameters of the dislocation structure can be obtained by the evaluation of the strain broadening of X-ray line profiles [19, 20]. The nanostructured materials produced by ECAE have a very high strength due to their low grain size and high dislocation density [16].

For the extruded alloy, most of the energy is believed to come from the strain energy stored by severe plastic deformation. For example, a small energy less than 0.5 J/g was reported in the extruded Cu at a strain of 1%, through the shear modulus of Cu is larger than that of Al by about 1.7 times. The energy was no more than 1 J/g even after 4 passes [21]. On the other hand, this measured stored enthalpy can also corresponds to the full precipitation which is expected to have already taken place during the slow cooling in a furnace before being subjected to ECAE [22].

In the present study, we have used X-ray diffraction (XRD), the differential scanning calorimetry (DSC) and Vickers hardness tests (Hv) techniques to control the evolution of the microstructure and the mechanical properties of a recycled aluminium alloy during ECAE via routes B_C. In parallel, we demonstrate that the energy stored during ECAE can be related to the dislocation density introduced by severe plastic deformation.

2. Experimental

The experiments were carried out using a recycled aluminum alloy received in the form of cast ingots. It was analyzed by inductively coupled plasma optical emission spectrometry (ICPOES).The chemical composition is given in Table 1. The material was annealed for 24 h at 500 °C and quenched in iced water.

Table 1. Chemical composition of the investigated material (ppm weight).

Al	Fe	Si	Zn	Mg	Mn	Cu	Ti	Cr	Ni	Ca	Co	Pb
Balance	2500	2190	1660	1071	600	558	60	56	36	14	2.6	<1

The die used for ECAE was made of two tool steel blocks bolted together. Two square channels of (10x10) mm2 cross sectional area intersecting in at an angle $\Phi = 90°$ were machined in one of the two pieces. The arc of curvature at the outer point of the intersection of the two channels delineated an angle $\Psi = 90°$ (Fig. 1). In this case, according to Iwahashi and al. [23], the total equivalent strain, ε_N accumulated by N passes was equal to 0.906 N. In the present work, repeated pressing was conducted up to N = 4 passes using route B_C [24, 25]. Samples with dimensions of ~10mm x 10mm x 70mm were cut for ECA pressing.

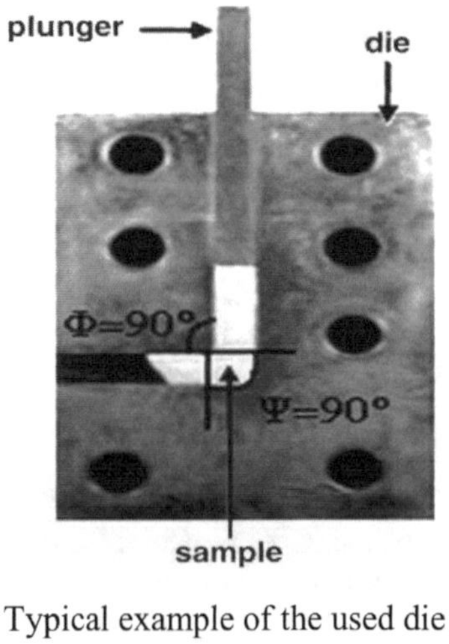

Figure 1. Typical example of the used die for ECAE

IOP Conf. Series: Materials Science and Engineering **28** (2012) 012043 doi:10.1088/1757-899X/28/1/012043

X-ray diffraction was performed on a wide angle diffractometer in the θ–2θ step scan mode by using Co-Kα radiation. Scans were collected over a 2θ range of 20–140° with a step of 0.017°. A 4s acquisition time was used at each step to obtain good statistics. The crystallite size (D) and the equivalent lattice strain (ε) were determined from the broadening (β) of peaks reflections using the Halder–Wagner approach [17]. The lattice parameter of the material before and after ECA pressings was obtained from a linear regression analysis of the measured lattice parameter, a_{app}, obtained from the Nilson–Reley method [26].

Vickers hardness (Hv) was measured with a load of 1Kg applied for 5sec. DSC experiments were performed using a "TA instruments Q100" apparatus under a low pressure of argon. The rise of temperature is programmed to 10°C/mn between 25°C and 550°C.

3. Results and discussion

Fig. 2a depicts the XRD patterns recorded on used material before and after ECAE. Microstructural changes occurring in the extruded materials could be seen with increasing the number of passes in ECAE process. One can see, in addition to characteristic reflections of the FCC structure of aluminum, many peaks at 2θ=36.5, 40.81, 47.23°, 49.09° and 50.65° of lower intensities (Fig. 2b). These peaks were identified to be characteristic of the Al_8Fe_2Si fine precipitates of the hexagonal structure. These precipitates persist even after four passes through the die. The broadening of the width of the reflections shown in Fig. 2c are due to the contribution of the effective crystallite size and an increase of the atomic level strain because of heavy plastic deformation.

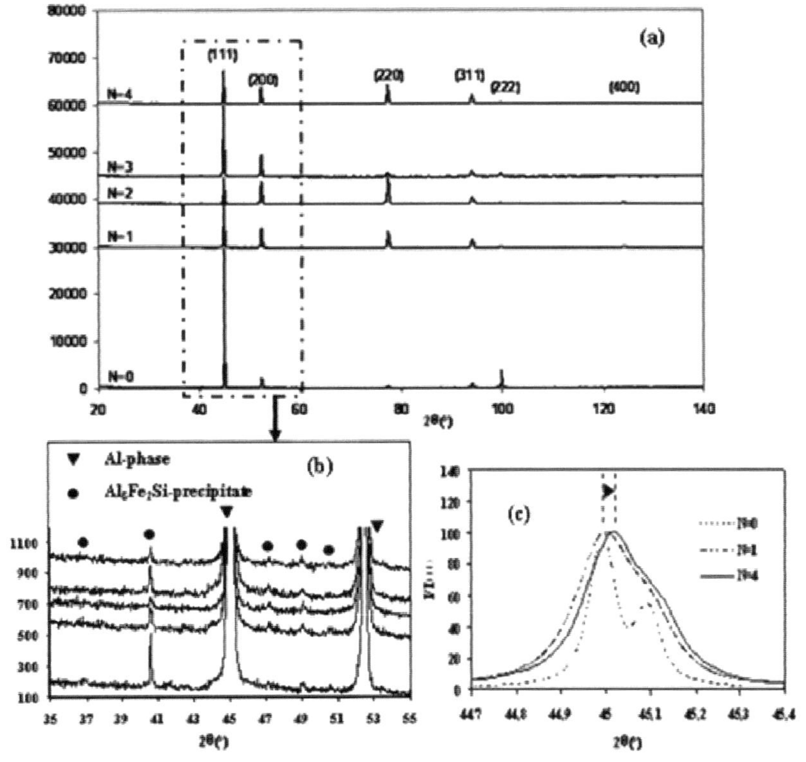

Figure 2: (a) XRD patterns before and after ECAE (Route Bc), (b) precipitates peaks and (c) typical example of (111) line broadening.

The calculated values, of these parameters, as a function of the number of passages through the die, are given in Table 2. These parameters were determinate from the Halder-Wagner plots. A strong decrease of <D> is observed since the first passage accompanied by a substantial increase of ε; this is due to the high amount of deformation introduced. After the second passages, no systematic variation is observed; the means of the values for crystallite size and lattice strains are 80.5 nm and 0.16 % for D and ε, respectively. Moreover, if we consider the bars variations 5 nm and 0.06 % for D and ε, respectively, we can see a variation (up/down) of these parameters. This might be characteristic to the processing mechanics. It is well known that the severe plastic deformation may lead several modifications: atomic site interchange, vacancies and dislocation activities. The atomic site interchange can occur here due to the ECAE of supersaturated solid solution Al alloy. During the whole ECAE process of materials, the lattice expansion is 0.014 %, therefore the formation of vacancies contradictorily be used as a proof to explain the present results. Hence dislocations are considered to be responsible for the final ECAE state. Moreover, lattice strain caused by ECAE is commonly attributed to plastic deformation, i.e. the generation and movement of dislocations. On the other hand, Fig. 2c demonstrates the shift of (111)-peak to higher values of theta after fourth passes in the die. The dependence of lattice parameter on the number of passages is somewhat different from the monotonic variations generally observed. It is not easy to draw a clear conclusion about this behavior since many contributions should operate simultaneously. It is known that an increase in defect concentration (mainly dislocations) leads to an increase in the lattice parameter, whereas a reduction of grain size is the main explanation of its decrease generally observed [27, 28]. Many investigations have revealed that the apparent lattice parameter depends on the defect density and on the coherency length: Ungár et al. [29] reported a decrease of lattice parameter in nanocrystalline copper produced by inert gas condensation and hot compaction as compared with that in coarse grained copper. Before ECAE, the lattice parameter had a value of 4.0545 Å. A significant decrease of the lattice parameter was observed after the first pass through the die ($\Delta a/a_0$=0.056 %). In fact, for N=1, a high amount of deformation is located mainly in the outer part of the specimen. Consequently, deformed grains in this region may apply a compressive stress on the grains of the inner part [30]. On the other hand, the present results show a clear relation between ε and a in the manner that an increase of ε for N=1 to 4 (almost twice) leads to an increase of a. So, the increase of lattice parameter can be due to the increase of the strain field in the coherent domains and then an increase of a.

Table 2. Evolution of microstructural parameters as a function of the number of passes through the die (Route Bc).

N (passes)	<D> nm)	$(\varepsilon^2)^{1/2}$ (%)	a (Å)	ρ (10^{15} m^{-2})	Hv (MPa)	Esto (MPa)
1	86	0.10	4.0522	2.33	540	2.054
2B	73	0.17	4.0522	3.92	570	3.245
3B	89	0.20	4.0524	3.75	610	3.121
4B	74	0.20	4.0528	4.92	690	3.957

For ECAE samples subjected to severe plastic deformation, if we consider that dislocations are the main defects, their densities, ρ, can be calculated by [31]:

$$\rho = \frac{1}{D(\frac{b}{\gamma} + y)} \tag{1}$$

where ρ is the dislocation density, b is the Burgers vector of dislocations and equals $\frac{a\sqrt{2}}{2}$ for the FCC structure with a is the lattice parameter, γ the plastic strain introduced and equals 0.906×*N* with *N* is the number of passes through the die [23] and *y* is the annihilation distance of dislocations and equals to 2 nm [31]. The calculated dislocation densities of the ECAE samples are given in table 2. It is clearly seen that ρ increases from about 2.33×10^{15} to 4.92×10^{15} m^{-2} with increasing the number of passes from 1 to 4. Usually, in conventional polycrystalline materials, the grain boundaries GBs are thought to be barriers to the dislocation motion, therefore the slight increase in the dislocation density within further passages in the die indicates a hardening of the GBs. When the GBs have turned hard or contracted, the amount of the dislocations piled up near the GBs will be increased. The measured stored enthalpy corresponds mainly to the recovery of defects of the ECAE samples which includes two components simply: non-equilibrium lattice defects in grains and the GBs. The former contribution is mainly of the elastic energy of dislocations situated in the grains, and the latter contribution originates mainly from the structural and compositional deviation in the GB region from the inner part of the grains.

The Vickers hardness values after each ECAE pass are given as a function of the number of passes in table 2. Before ECAE, the material had a hardness of 250 MPa. After the first pass, one can see that the hardness increases remarkably. Indeed, it becomes twice larger for N=1 than for the undeformed sample (N=0). This improvement can be allotted to the strain hardening generated by the creation of dislocations. Between the second and the fourth pass, the hardness increase slowly. This can be allotted to the phenomenon of dynamic recovery. This recovery is explained by the high dislocation density which makes possible an annihilation process during ECAE. Generally, the results show that the hardness of the recycled alloy increases with increasing amount of strain for the two processing routes.

The thermal behavior of ECAE samples has been investigated by DSC measurements as a function of number of passes through the die. The results are shown in Fig. 3. Each trace of ECAE samples contains overlapping exothermic peaks, became more clearly as third peaks, in the case of the ECAE sample after four passes in the die. In the case of the four passes extruded sample, the first exothermic peak was rather small and its onset and peak temperatures were approximately 160 and 210 °C, respectively. The second exothermic peak became stronger, and its onset and peak temperatures were approximately 245 and 310 °C, respectively. The third exothermic peak was rather small and large and its onset and peak temperatures were approximately 320 and 355 °C, respectively. For the ECAE samples, besides the energy strain release, the energy release during grain growth of the severely deformed microstructure should be taken into consideration. In fact, the first exothermic reaction can be attributed to the cells formation and recovery process. The others exothermic peaks, in more elevated temperature, can probably due to the recrystallization and grain growth of deformed samples, respectively.

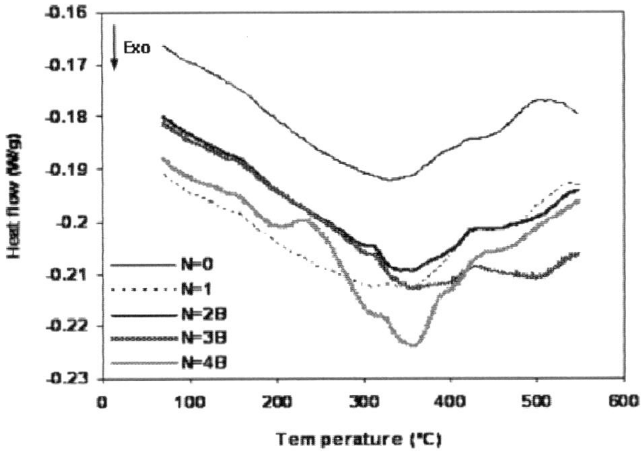

Figure 3. DSC curves for undeformed sample (N=0) and samples processed by ECAE via Route Bc (N=1 to 4).

The stored energy (E_{stor}), during ECAE, can be related to the dislocations density, ρ as follows [32]:

$$E_{Stor} = Gb^2 \frac{\rho}{4\pi\kappa} \ln((b\sqrt{\rho})^{-1}) \quad (2)$$

where G is the shear modulus, b is the Burger vector and κ denotes the arithmetic average of 1 and (1−ν), with ν is the Poisson ratio, assuming equal parts of edge and screw dislocations.

The calculated stored energy as a function of number of passes in the die is given in table 2. We note an increase of E_{stor} from a value of 2.054 MPa after the first passage through the die to a value of 3.957 MPa after fourth passes. Moreover, one can see a decrease of E_{stor} after the third passes indicating a partial dynamic recovery of the material in the advanced stages of deformation. The recovery is explained by the high dislocation density which makes possible an annihilation process during ECAE.

4. Conclusion

A recycled aluminum alloy was severely deformed by ECAE up to four passes using the processing route B_C. The evolution of the microstructure and associated changes of mechanical properties such as Vickers hardness were studied. The following conclusions were drawn:

1. The analysis of microstructure before and after ECAE showed a significant grain refinement and a high level of lattice strains.
2. After the first passes through the die, the lattice parameter decreases indicating that high density of dislocations take place during ECAE. Then, it increases because of the raised strain field in the coherent domains.
3. Changes in mechanical behavior are observed after ECAE: the Vickers hardness increases when the number of the passes through the die increases. This is due to strain hardening.
4. DSC results confirm the succession of strain release, recrystallization and grain growth phenomena after extruded material heating up to 550°C.

Reference

[1] Valiev R. Z, Korznikov A. V, Mulyuko R. R 1993 *Mater. Sci. Eng. A* **168A** 141.
[2] Stolyarov V. V, Zhu Y. T, Alexandrov I. V, Lowe T. C, Valiev R. Z 2001 *Mater Sci. Eng. A* **299** 59 67.
[3] Horita Z, Fujinami T, Nemoto M, Langdon T. G 2001 *J. Mater. Process. Technol.* **117** 288.
[4] Segal V. M 1995 *Mater. Sci. Eng. A* **197** 157.
[5] Valiev R. Z 1997 *Mater. Sci. Eng. A* **59** 234.
[6] Langdon T. G, Furukawa M, Nemoto M, Horita Z 2000 *J. Met.* **52** 30.
[7] Segal V. M 1999 *Mater. Sci. Eng. A* **271** 322.
[8] Zhao Y. H, Liao X. Z, Jin Z, Valiev R. Z, Zhu Y. T 2004 *Acta. Mater.* **52** 4589.
[9] Lee J. C, Seok H. K, Suh J. Y 2002 *Acta. Mater.* **50** 4005.
[10] Nagarajan D, Chakkingal U, Venugopal P 2007 *J. Mater. Proces. Tech.* **182** 363.
[11] Miyamoto H, Ota K and Mimaki T 2006 *Scripta Mater.* **54** 1721.
[12] Chang C. P, Sun P. L, Kao P. W 2000 *Acta Mater* **48** 3377-3385.
[13] Vinogradov A, Ishida T, Kitagawa K, Kopylov V I 2005 *Acta Mater.* **53** 2181-2192.
[14] Urrutia I. G, Morris M. A. M, Morris D. G 2005 *Mater Sci Eng. A* **394** 399-410.
[15] Valiev R. Z, Islamgaliev R. K, Alexandrov I.V 2000 *Prog. Mater. Sci.* **45** 103.
[16] Dalla Torre F, Lapovok R, Sandlin J, Thomson P. F, Davies C. H. J, Pereloma E. V 2004 *Acta Mater.* **52** 4819.
[17] Langford J. I 1978 *J. Appl. Crystallogr.* **11** 4 10.
[18] Langford J. I, Snyder in: R. L, Fiala J, Bunge H. J (Eds.) 1999 Defect and microstructure analysis by diffraction, IUCr Monographs on Crystallography 10, International Union of Crystallography – Oxford Science Publications, New York **59**.
[19] Ungàr T, Borbély A 1996 *Appl. Phys. Lett.* **69** 3173.

[20] Dragomir I. C and Ungàr T 2002 *J. Appl. Cryst.* **35** 556.
[21] Schafler E, Steiner G, Korznikova E, Kerber M, Zehetbauer M. J 2005 *Mater. Sci. Eng. A* **169** 410-411
[22] Kim W. J, Wang J. Y 2007 *Mater. Sci. Eng. A* **464** 23 27
[23] Iwahashi Y, Wang J, Horita Z, Nemoto M, Langdon T. G 1996 *Scripta Mater.* **35** 143.
[24] Iwahashi Y, Horita Z, Nemoto M, Langdon T. G 1997 *Acta. Mater.* **45** 4733.
[25] Zhu Y. T, Lowe T. C 2000 *Mater. Sci. Eng. A* **291** 46.
[26] Azàroff L.V, Buerger M. J 1958 *The powder method in X-Ray cristallography, McGrow-Hill Book company, New York* 342.
[27] Gubicza J, Nam N. H, Balogh L, Hellmig R. J, Stolyarov V. V, Estrin Y, Ungár T 2004 *J. All. Comp.* **378** 248 252.
[28] Gubicza J, Chinh N. Q, Krállics G. Y, Schiller I, Ungár T 2006 *Cur. App. Phy.* **6** 194 9.
[29] Ungàr T, Ott S, Sanders P. G, Borbély A, Weertman J. R 1998 *Acta Mater.* **46** 3693.
[30] Zhang K, Alexandrov I. V, Lu K 1997 *Nanostruct. Mater.* **9** 347.
[31] Li L, Ungár T, Wang Y. D, Fan G. J, Yang Y. L, Jia N, Ren Y, Tichy G, Lendvai J, Choo H, Liaw P. K 2009 *Scripta Mater.* **60** 317-320.
[32] Beaver M. B, Holt D. L, Titchener A. L 1973 *Prog. Mater. Sci.* **17** 5.

IOP Conf. Series: Materials Science and Engineering **28** (2012) 012044 doi:10.1088/1757-899X/28/1/012044

A study of the phase transition behaviour of $[(NH_4)_{0.63}Li_{0.37}]_2TeBr_6$

R. Karray[a 1], D. Linda[a], A. Van Der Lee[b], A. Ben Salah[a] and A. Kabadou[a]
[a]Laboratoire des Sciences des Matériaux et d'Environnement, Faculté des Sciences de Sfax BP 1171, 3000, Université de Sfax, Tunisia.
[b]Institut Européen des Membranes (UMR 5635), Université de Montpellier II, cc 047, Place E. Montpellier- France

E-mail: rimrebai@lycos.com

Abstract. The mixed hexabromotellurate $[(NH_4)_{0.63}Li_{0.37}]_2TeBr_6$, presenting at room temperature a K_2PtCl_6-type structure with space group $Fm\bar{3}m$, exhibits three anomalies at 195, 395 and 498 K in the differential scanning calorimetry diagram. Different techniques: dielectric investigation, High-temperature X-ray powder diffraction and infrared spectroscopic study, in the range temperature (300-470) K are applied to explore the phase transition around 395 K.
Combining XRD, dielectric and differential scanning calorimetry (DSC) results, no phase transition leading to a super-ionic conductivity phase is found. At high temperature, $[(NH_4)_{0.63}Li_{0.37}]_2TeBr_6$ is characterized by a medium conductivity $\sigma_{453} \approx 10^{-4}\ \Omega^{-1}m^{-1}$.

1. Introduction

A number of salts with formula $[(NH_4)_{1-x}A_x]_2TeCl_6$ [1-4] comprise a family of isomorphous compounds having a cubic space group ($Fm\bar{3}m$) and a large unit cell: a ≈ 10 Å. Well investigated example is the $[Li_{0.2}(NH_4)_{0.8}]_2TeCl_6$ compound [5], presenting a high temperature super ionic phase transition which is considered to originate from the mobility of both cations Li^+ and NH_4^+.
A search of further similar properties was undertaken by structural investigation on Lithium-ammonium mixed hexabromotellurate $[(NH_4)_{0.63}Li_{0.37}]_2TeBr_6$. The structure at room temperature of the later compound has been determined by single X-ray diffraction [6]. The space group obtained is $Fm\bar{3}m$, with a= 10.7200(12) Å.
The Differential scanning calorimetry showed three anomalies at 195, 395 and 498K.
The aim of the present work is to characterize the high temperature transition of the $[(NH_4)_{0.63}Li_{0.37}]_2TeBr_6$ solid solution.
In this paper we report and discuss the results of dielectric, X-ray diffraction as well as I.R. spectroscopy measurements, at various temperatures.

[1] To whom any correspondence should be addressed.

2. Experimental

2.1. Synthesis of crystal sample

The title compound was synthesized by adding the stoechiometric mixtures (molar 1:1:1 ratio) of LiBr and NH_4Br and TeO_2 hot saturated solutions in hydrobromide acid. Starting materials were of a purity of 99.99 wt%.

2.2. Thermal and spectroscopic analysis

In order to characterize phase transitions of the title compound we have performed thermal and spectroscopy measurements. The temperature behaviour was examined using a DCS METTLER TA4000 at heating speed of 10 K/mn between 123 and 523K.

Infrared absorption spectra of suspensions of crystalline powders in KBr were recorded on a PERKIN-ELMER 1750 spectrophotometer in the 400-4000cm^{-1} range. The heating of the sample was performed in an air-atmosphere SPECTAC heating cell. A thermocouple EUROTHERM REGLER was used for the temperature measurements.

2.3. Dielectric Measurement

Dielectric measurements were carried out on pellets with dimensions 8 mm in diameter and 1 mm in thickness. Graphite electrodes were deposited on the two sample face. The complex impedance was measured in the frequency range 0.01-13MHz with a Hewlett-Packard 4192 ALF automatic bridge monitored by a HP vectra microcomputer.

2.4. X-ray diffraction and data collection at different temperatures

The diffractograms between 303K and 423K temperatures were collected using a Panalytical X'pert-Pro diffractometer equipped with an ultrafast X'celerator detector; Ni-filtered Cu-$L_{3,2}$ X-rays were employed (λ_1=1.54051 Å and λ_2=1.54443 Å) at 40 kV/40 mA using a power generator.

Indexing was performed using DICVOL04 software [7]. Confidence factors were M(17)=37.6 and F(17)=28.8 [8].It is noted that only 17 diffraction peaks were observed due to the high crystal symmetry. It was assumed that the mixed compound is isotypic with K_2PtCl_6-type compounds [9]

3. Results and discussion

3.1. DSC measurements

The DSC (differential scanning calorimetry) measurements were used to identify the phase transition temperatures. The DSC curves of the $[(NH_4)_{0.63}Li_{0.37}]$2TeBr6 crystals for the heating-up are shown in Fig. 1. We observe three endothermic picks, at 195, 395 and 498 K. The corresponding enthalpy changes are ΔH_1=0,24 J/g, ΔH_2=6,24 J/g et ΔH_3=22,18 J/g, respectively.

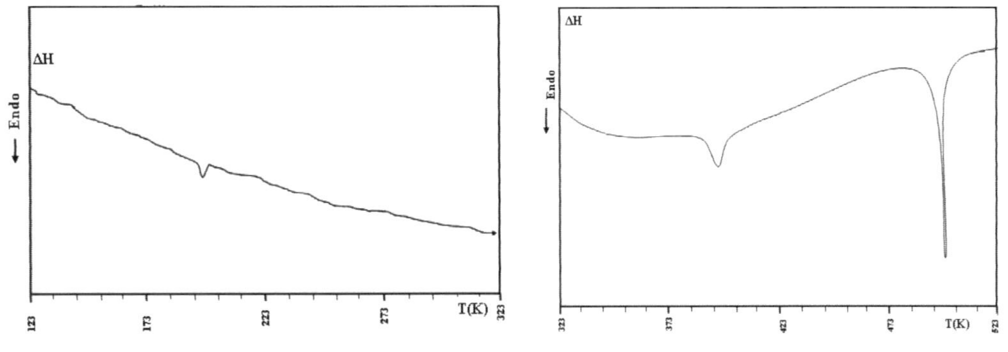

Figure 1. The DSC curves of $[(NH_4)_{0.63}Li_{0.37}]_2TeBr_6$ for the heating-up runs.

The third signal corresponds to the decomposition of the material as final solid residue color changes to coal black.

The first peak at low temperature corresponds to a structural phase transition, attributed to the transition from cubic to tetragonal symmetry, as confirmed by X-ray diffraction study [6], which showed the existence of low temperature tetragonal phase of space group P4/mnc (D_{4h}^{6}).
In what follows, we will try to characterize the second transition appeared around 395 K.

3.2. IR spectroscopy investigation
Infrared spectroscopic study between 300 and 403 K has been undertaken in order to gain more information on the crystal dynamics and to determine the nature of the high temperature phase transitions. The IR study was restricted to the frequency range (400 to 4000cm^{-1}). IR spectra are presented in fig. 2. The major observed bands associated with NH_4^+ were assigned [10].

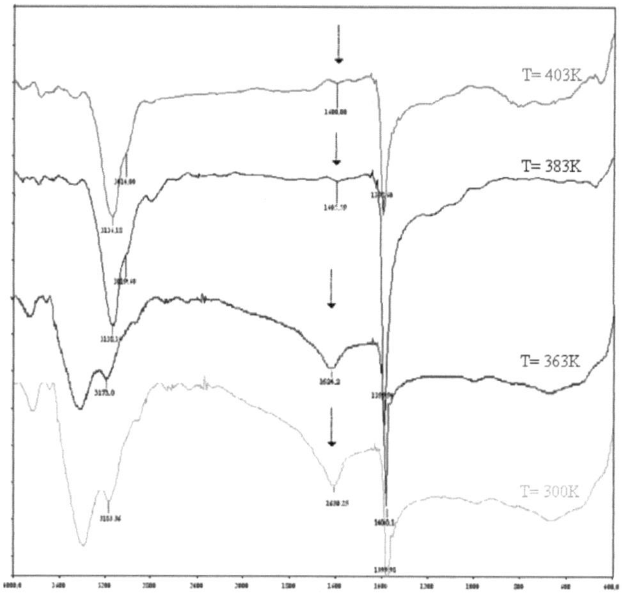

Figure 2. IR spectra at different temperatures of $[(NH_4)_{0.63}Li_{0.37}]_2TeBr_6$

At room temperature, the observed band around 3183 cm^{-1} is attributed to the symmetrical stretching vibration of ammonium groups ν_1 (NH_4^+), those around 1630 and 1400 cm^{-1} are associated with their symmetric ν_2 and antisymmetric ν_4 bending vibrations, respectively. Table 1 summarizes the observed frequencies at different temperatures for each vibration mode of the NH_4^+ tetrahedral.

From these curves, it could be seen that around 395K, the symmetric bending vibration ν_2 (NH_4^+) intensities bands decrease gradually and shift to lower frequencies. We can also note a broadening of these bands until disappearance.

Table 1. Assignment and frequencies (cm^{-1}) observed for $[(NH_4)_{0.63}Li_{0.37}]_2TeBr_6$.

T= 300K	T=363K	T=383K	T= 403K	Attribution
3183	3173	3138	3134	ν_1 (NH_4^+)
-	-	3020	3014	
1630	1624	1606	1600	ν_2 (NH_4^+)
1400	1400	1400	1399	ν_4 (NH_4^+)

The thin peak firmly resolved at 1400 cm^{-1}, assigned to the antisymmetric deformation ν_4 (NH_4^+), is almost unaffected by this transition.

The peak intensity of the symmetric stretching vibration ν_1 (NH_4^+) is slightly modified during the transition, however, there is a shift towards lower frequencies more pronounced than that seen for the symmetric bending vibration ν_2 (NH_4^+). This makes the difference between the two peaks frequencies smaller at high temperature.

A remarkable change in reorientation motion of NH_4^+ entities is found by increasing temperature. We note the appearance of a shoulder at 3119 cm^{-1}. The transition is thus accompanied by an increase in disorder on NH_4^+.

The solid solution is characterized at high temperature by a dynamic disorder of NH_4^+ highlighted by the disappearance and broadening of some bands.

This behaviour can be attributed to an "order-disorder" phase transition, which probably characterizes the motion of reorientation and H^+ diffusion.

3.3. Electrical proprieties

A combined analysis using both the complex impedance and the complex modulus formalisms were applied. Our measurements were realized in the temperature range 293 to 473 K.

Some complex impedance spectra Z" versus Z' recorded at various temperatures are presented in Fig. 3. The complex impedance plane data show a no depressed semi-circle for all frequencies and temperatures variations. This behaviour confirms that no electrode phenomenon is present. The resistance is determined by extrapolation at frequency equal to zero from the Z" vs. Z' circle arc centred under the Z' axis because at this frequency the equivalent circuit of our material presents only a resistant behaviour. The coefficient α(0.21 rad) which characterizes the deviation of the cole-cole from the Deby law is determined from the complex impedance spectrum.

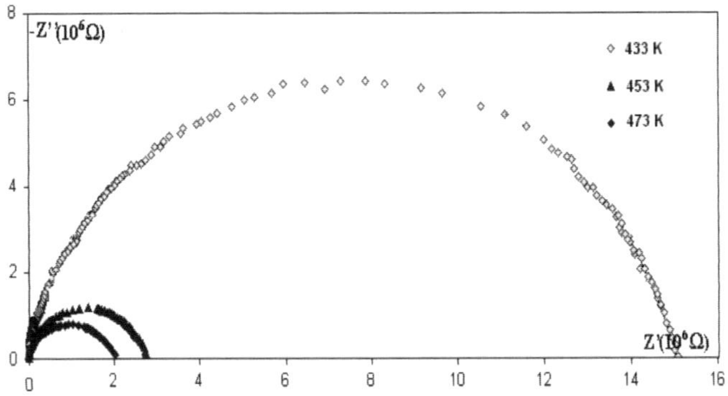

Figure 3.a Complex impedance diagrams (-Z" vs. Z') for at 293<T<453

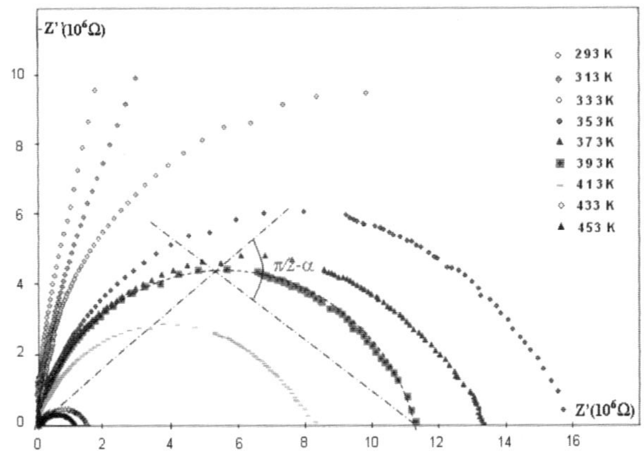

Figure 3.b Complex impedance diagrams (-Z" vs. Z') for at 433<T<473 K

Fig.4 shows the evolution of the conductivity versus inverse temperature ($\log(\sigma T)=f(1000/T)$). The conductivity plot exhibits two areas on either side of a transition temperature T = 389 K in good agreement with that determined by DSC at T = 395 K. The evolution of the conductivity in these two last regions obeys to Arrehinus law ($\sigma T = \sigma_0 \exp(-Ea/kT)$).

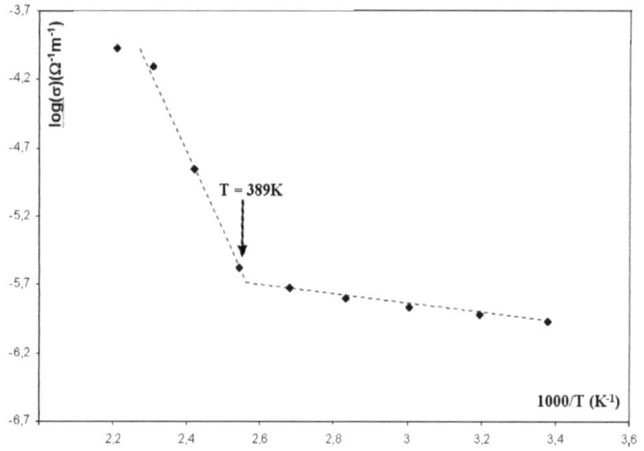

Figure 4. Temperature dependence of $\log(\sigma T)$ for $[(NH_4)_{0.63}Li_{0.37}]_2TeBr_6$

This phase transition leads to a relatively higher conductivity phase ($\sigma_{453K} = 1.1810^{-4}\ \Omega^{-1}.m^{-1}$) with a very significant increase of activation energy which varies from 0.069 eV before the transition to 1.177 eV after the transition.

This conductivity variation is due to the fact that during the transition, assumed to be of "order-disorder" based on the results obtained by the spectroscopic study, molecular motions in the potential wells weaken when the temperature decreases: the molecules or atoms tend to occupy the lowest levels of the well and to be ordered.

- At low temperature, the conductivity is mainly ensured by cations NH_4^+ and Li^+.
- At high temperatures, the conduction mechanism involves the breaking of hydrogen bonds (NH ... Br). Conduction occurs then by transport of both NH_4^+ and H^+proton.

The complex modulus formalism $M^* = j\omega C_0 Z^*$, where $j = \sqrt{-1}$, $\omega = 2\pi f$ is the angular frequency, Z^* is the complex impedance and C_0 the vacuum capacitance of the cell, has been adopted to determine the conductivity relaxation process. This formalism discriminates against electrode polarisation and other interfacial effects in solid electrolytes [11].

For a given temperature and frequency, the real part M' and imaginary part M'' of the complex modulus ($M^* = M' + jM''$) have been calculated from the complex modulus impedance data ($Z^* = Z' + jZ''$) by the following relations: $M' = \omega C_0 Z''$ and $M'' = \omega C_0 Z'$.

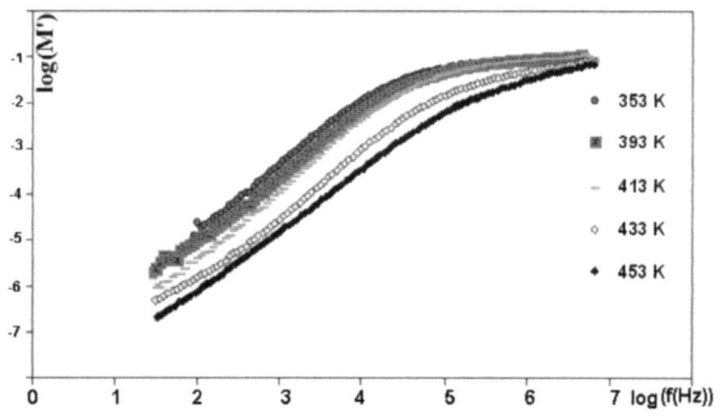

Figure 5. Frequency dependence of log(M') at various temperatures

A plot of log (M') versus log (f) at various temperatures was given in Fig. 5. Whatever temperature, M' reaches a constant value at high frequencies ($M'\infty=(1/\varepsilon'\infty)$). At low frequencies, it approaches zero, which indicates that the electrode polarisation phenomena make a negligible contribution to M^* and may be ignored when the electric data are analysed in this form [12].

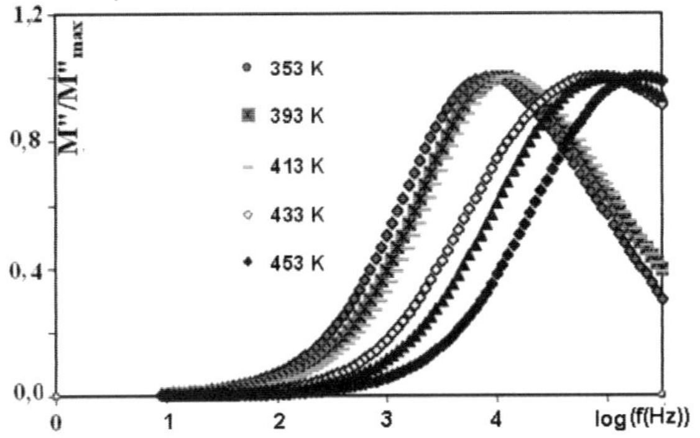

Figure 6. Plots of normalized modulus M''/M''_{max} vs. log(f) at various temperatures.

The M''/M''_{max} spectrum relative to a given temperature shows an asymmetrical peak approximately centred in the dispersion region of M' (Fig. 6). The region to the left of the peak is where the ions are mobile over long distances, the region to the right is where the ions are spatially confined to their potential wells. The frequency range where the peak occurs is indicative of the transition from short-range to long-range mobility at decreasing frequency and is defined by the condition $\omega\tau_\sigma=1$ where τ_σ is the most probable constitution proton relaxation time [13]. When the temperature increases, the modulus peak maxima shifts to higher frequencies.

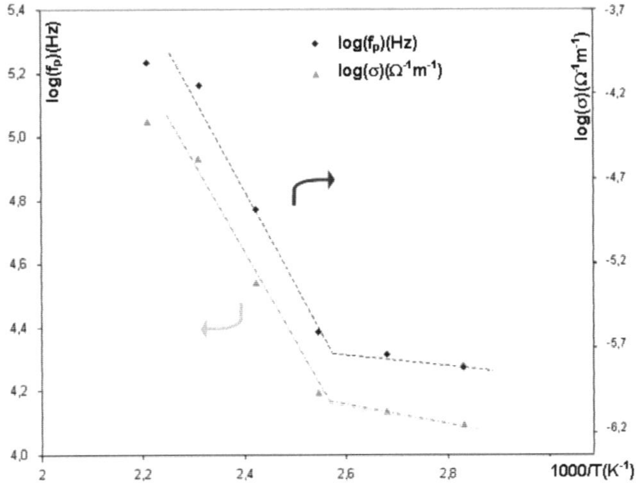

Figure 7. Temperature dependence of log(σT) and log(f_p), where f_p is the M''_{max} peak frequency, for $[(NH_4)_{0.63}Li_{0.37}]_2TeBr_6$

Fig .7 gives the temperature dependence of f_p frequency relative to M''_{max}. An Arrhenius type law is shown. The conductivity variation, the behaviour of which is also of Arrhenius-type in the temperature range 350 to 470 K, is reported in Fig. 7. Both lines observed below and above 389 K are quasiparallel, the activation energies resulting from the impedance and modulus spectra are very close(E_σ = 1177,10 ev and E_f= 1,089 ev), suggesting that the transport of Li^+ ions in the material is probably due to a hopping mechanism [14], and can be then associated with ammonium ions movement under the action of an electric field of variable frequency, the ion NH_4^+ can oscillate around its average position in a potential well or move from one to another.

3.4. Powder X-ray diffraction study at different temperatures

Diffractograms shape (Fig. 8) shows that no structural change has been detected up to 423 K, while a shift of peaks towards 2θ lowers values is observed which is due essentially to the extension of the cell.

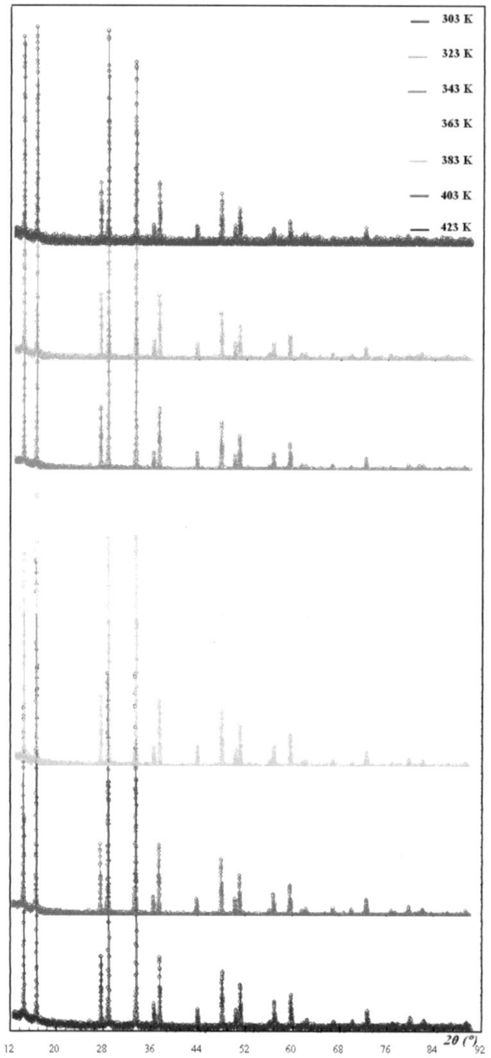

Figure 8. Powder X-ray diffractograms at different temperatures.

The indexing of each diffractogram using the DICVOL program [7] shows that the compound crystallizes in the same space group $Fm\bar{3}m$, for the different temperatures.

The temperature dependency of the unit cell parameter and volume was determined, as shown in Fig. 9, in the temperature range between 303 and 423 K. The curves variation of these two parameters is nearly linear, but deviates at higher temperatures, so that it exhibits an anomaly around 400 K, which is not very far from the endothermic peak already observed in the DSC curve.

We note an increase of the unit cell volume as well as *a* parameter with increasing temperature. This corresponds to the increase of the Te-Br distances resulting in an extension of the cations cavities which can be a source of an easier transport of the mobile ions in the tunnels.

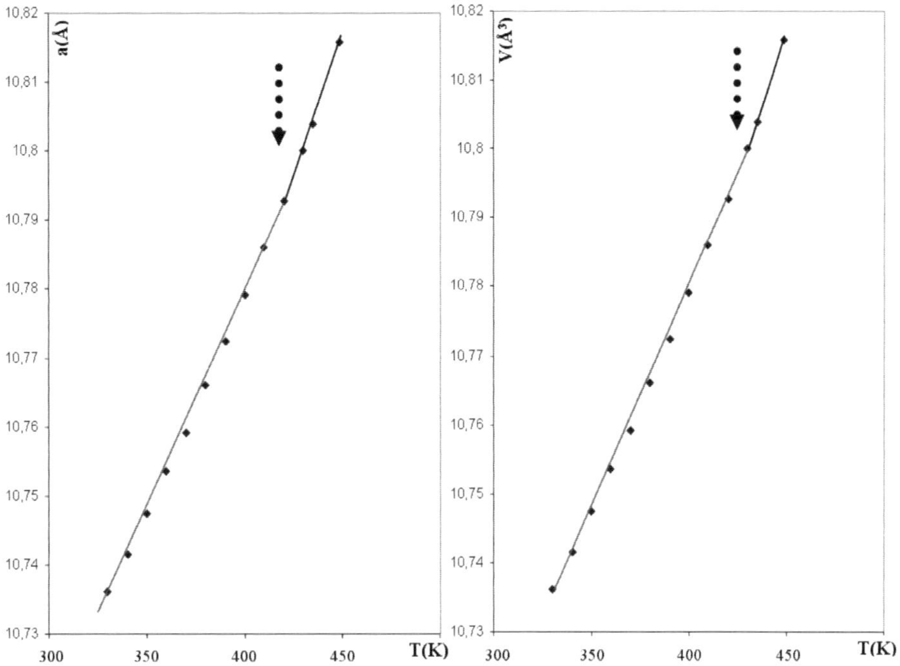

Figure 9. Temperature dependence of the unit cell parameter and volume for $[(NH_4)_{0.63}Li_{0.37}]_2TeBr_6$.

4. Conclusion

In the present investigations on the mixed compound $[(NH_4)_{0.63}Li_{0.37}]_2TeBr_6$, crystallizing in the cubic system with the space group $Fm\bar{3}m$, DSC measurements and infrared spectra show that the title compounds probably undergo an orientational "order-disorder" phase transition of NH_4^+ tetrahedra around 395K.

On the other hand, dielectric study confirms this phase transition leading to a relatively higher conductivity phase. In addition, it shows that the activation energies derived from the impedance and modulus spectra are very close, suggesting that the transport of Li^+/NH_4^+ and H^+ ions in the material investigated is probably due to a hopping mechanism.

Our studies also reveal that no change occurs in the crystal structure of the compound up to 423 K. It follows that the increase of temperature does not affect the structural arrangement, but it leads to an increase of lattice cell dimensions and consequently an extension of cations cavities which can be source of an easier transport of the mobile ions in the tunnels.

References

[1] Karray R, Kabadou A, Cisarova I, Ben Hassen R and Ben Salah A 2004 *J. Alloys Compd.* **377** 85.
[2] Karray R, Kabadou A, Loukil M and Ben Salah A 2003 *Z. Kristallogr., NCS* **218** 399.
[3] Karray R, Kabadou A, Van Der Lee A and Ben Salah A 2006 *Powder Diffraction* **21** 225.
[4] Belgaroui H, Loukil M, Karray R, Ben Salah A and Kabadou A 2010 *Journal of Structural Chemistry* **51** 4 689-695.

[5] Karray R, Kabadou A, Jarraya S and Ben Salah A 2006 *Solid State Ionics* **177** 89.
[6] Karray R, Van Der Lee A, Jarraya S, Ben Salah A and Kabadou A *Arabian journal of Chemistry (in press)*
[7] Louër D, Louër M 1972 *J. Appl. Cryst.* **5** 271.
[8] Boultif A, Louer D 2004 *J. Appl. Cryst.* **37** 724.
[9] Engel G 1935 *Krist Z.* **90** 341.
[10] Rolf Berg W, Finn Poulsen W and Niels Bjerrum J 1977 *The Journal of Chemical Physics* **67**
[11] Chowdari B. V. R, Gopalakrishnan R 1989 *Solid State Ionics* **108** 323.
[12] Howell F. S, Boje R. A, Macedo P. B and Moyinihan C. M 1974 *J. phys. Chem.* **78** 639.
[13] Patel H. K, Martin S.W 1992 *Phys. Rev. B* **45** 10292.
[14] Chowdari B. V. R, Gopalakrishnan R 1987 *Solid State Ionics* **23** 225.

IOP Conf. Series: Materials Science and Engineering **28** (2012) 012045 doi:10.1088/1757-899X/28/1/012045

Oxidations of organic matter present in the phosphoric acid 54% by the ozone: characterization of groups carbonyls upstream and downstream of the ozonation

D. Linda[a1], A. Louati[b], C. Chtara[c] and A. Kabadou[a]

[a]Laboratoire des sciences des matériaux et d'environnement, Faculté Des Sciences, Université de Sfax, 3038 Sfax, Tunisia.

[b]Laboratoire Eau-Energie-Environnement (Lab:3E), Université de Sfax, E.N.I.S, B.P.W 117, Sfax, Tunisia.

[c]Laboratoire recherche et développement, Groupe chimique Tunisien, Gabes, Tunisia.

E-mail: ahlemkabadou@yahoo.fr

Abstract. This study was focused on the oxidation of organic matter in phosphoric acid 54% by ozone. In order to understand the mechanisms involved in this process, the identification of this matter upstream and downstream of the ozonation was necessary. For the identification, after an extraction by a mixture (dichloro-methanol), the organic phase was divided into two parts: the residue and the extract:-The residue was studied by infrared spectroscopy Fourier Transform (IR-TF). It contains Kérogène which is a mixture of saturated hydrocarbons with high molecular weights. The absorption bands of the FT-IR showed that the residue contains also quantities of amino that correspond to the remains of dinoflagellate cysts, which are abundant in sediments.-The extract has been the subject of a detailed study by, chromatography on silica column, ***IR-TF spectroscopy*** and CG-SM. The passage of this extract on a silica column yielded two fractions (saturated fraction and polar fraction). Both of these fractions were analyzed by CG-SM. The yield of the reduction of the organic matter content in the phosphoric acid 54% could not exceed 29%. Therefore, we can conclude that the reduction in the rate of organic matter remains limited by the fact that some compounds are inert towards ozone.

1. Introduction

Tunisia, the fifth world producer of phosphate, transforms about seven million tons of phosphate per year in phosphoric acid. Commercially, phosphoric acid is manufactured using different processes, the most commonly used are: Thermal and wet-process. The wet-process is the process used in Tunisia to produce phosphoric acid from mineral phosphate [$Ca_3 (PO_4)_2$] by sulphuric attack as is show in eq.1.1

$$\text{eq1.1}: [Ca_3(PO_4)_2]_3 + 3H_2SO_4 + 6H_2 \rightarrow 3CaSO_4{,}2H_2O + 2H_3PO_4 \quad [1]$$

Phosphoric acid produced by this process was not pur, it contains a variety of impurities, which vary according to the origin of the minerals phosphate.

The analysis of a Tunisian phosphoric acid (54% P_2O_5) by atomic absorption spectroscopy, show that contains about 10 molar of impurities per 100 molar H_3PO_4 distributed as is shown in Table 1

[1] To whom any correspondence should be addressed.

Table 1. Wet process phosphoric acid composition.

P_2O_5 (wt.%)	F(wt.%)	H_2SO_4 (wt.%)	O.M (ppm)	Fe (ppm)	Mg (ppm)	Al (ppm)	Cr (ppm)	Zn (ppm)	Cd (ppm)	Cu (ppm)	Ni (ppm)	Mn (ppm)	V (ppm)	Ti (ppm)
54,26	0,21	2,13	687	3080	6400	3700	305	350	31,5	4	17	33	71	45

Among these impurities, organic matters are considered the most adverse. However, the presence of organic matter (O.M.) is very embarrassing on several levels, indeed it causes the clogging of the filters in the tank attack, forms foams and emulsion at the liq-liq interface during the solvent extraction process, gives a black colored phosphoric acid which penalise its commercialization and excluded it from the use in non-fertiliser applications[2].

Many research techniques have been investigated for purifying phosphoric acid, the techniques proposed were; liquid–liquid extraction, solid– liquid extraction, crystallization, concentration or addition of additives [3]. The use of these techniques was limited due to a number of disadvantages such as: -Limited efficacy.

-High costs of organic solvents and resins (more often not regenerable).

-Difficulty in recovering all the solvent from both the raffinate and the purified acid.

-Environmental pollution by some by-products. Thus, the development of new technologies for purifying phosphoric acid results is interesting. Ozone is a strong oxidizing agent, with a redox potential of 2.07 V, used frequently in drinking water and wastewater treatment. Ozone treatment is specifically suitable for partial or complete oxidation of non-biodegradable components [4, 5] and it can be used as a pre-treatment step to improve biodegradability or a post-treatment step to remove remaining O.M. This work presents the results of a study for purifying industrial phosphoric acid solutions by ozone. Some of the organics matters are not easily biodegradable. In such cases, chemical oxidation technologies help oxidizing the refractory organic compounds.

2. Materials and Methodes

2.1. Molecular ozone

The chemical properties of ozone depend on the structure of the molecule. The four extreme forms of resonance structures of ozone molecule can be expressed as follows **[1, 6]:**

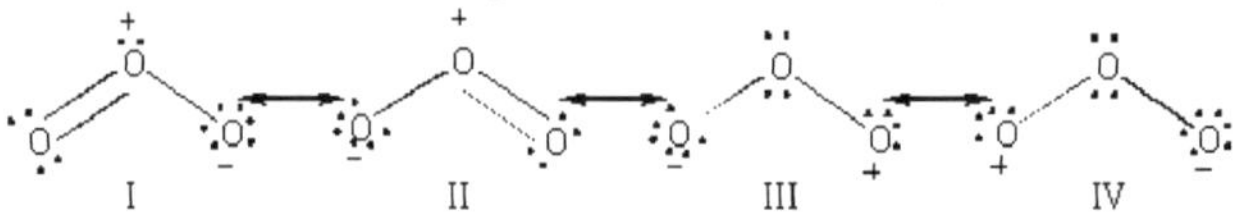

Due to its structure, molecular ozone can react as a dipole, an electrophilic or nucleophilic agent. As a result of its high reactivity, ozone is very unstable in water. The half-life time of molecular ozone varies from a few seconds up to few minutes and depends on pH, water temperature and concentration of organic and inorganic compounds in water [1, 4].

Ozone is also one of the most powerful oxidants. The ability to oxidize organic and inorganic molecules is closely related to the oxidation potential of the oxidant **[7, 8]:**

eq 2.1.1: $O_3 + 2H^+ + 2e^- \rightarrow O_2 + H_2O,\ E_0 = 2.07V$ [6]

It is used for many different purposes such as: disinfection and algae control, taste, odour and colour control, oxidation of inorganic pollutants (iron, manganese........), oxidation of organic micro- and macropollutants as well as for the improvement of coagulation [1, 9].

Ozone reacts directly with constituents found in solution and is also able to form highly reactive chemical agents that can contribute to additional oxidizing reactions. The most common of these reactive agents is the free radical hydroxyl. It can also serve as a dipole, an electrophilic or a nucleophilic agent. Its high reactivity accounts for its high unstability in water. Ozone decomposition follows the following chain reaction: The above chain reaction clearly shows that OH^-plays a significant role in initiating the ozone decomposition process in water. Also, there are several organic compounds that can initiate (OH^-, H_2O_2/HO_2^-, Fe^{2+}), propagate (R_2–CH–OH, formate, humic substances which are a complex mixtures of natural macromolecules and smaller molecules and ions which occur throughout the environment [10]) and inhibit (CH_3–COO^-, alkyl-(R), HCO^-_3 $=CO^-_3$) the chain reaction.

Summarizing Two types of reactions can be possible between ozone and organic matter, indeed it can react with organic compounds in aqueous solution either directly through selective reactions with specific functional groups (double bonds, nucleophilic positions) or through a radical reaction like hydroxyl radicals that are produced by the decomposition of ozone in water [11].

2.2. Experimental set up

The experimental set-up is shown in *figure 1.*

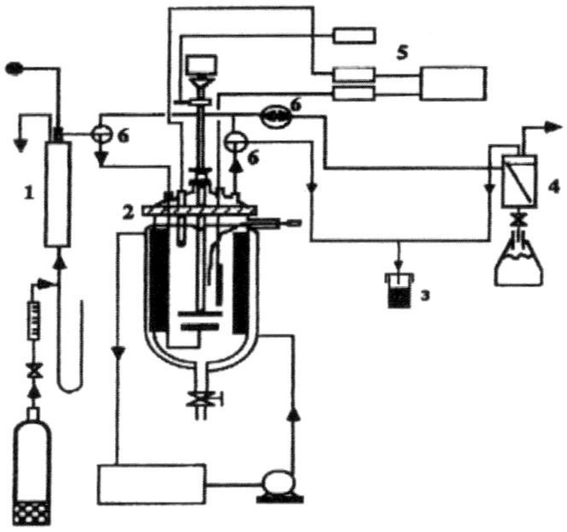

Figure 1. General Outline of an ozonolyseur; 1. Ozone generator; 2.Contactor; 3.Destroyer of residual ozone; 4.Sampler gas; 5.Measuring machine; 6. Valves.

All the experiments were carried out with a continuous mode in an reactor. Ozone was produced from pure oxygen, in an ozone generator manufactured in the faculty of sciences Gabes (Tunisia). The oxygen flow rate was controlled using a mass flow controller. Gaseous ozone leaving the generator was conducted by a system of three-way valves. A needle valve was placed before the reactor to control the ozone flow rate, the flow in excess being removed in a gas washing bottle. After the reactor, there was a three-way valve which allows the gas flow to go to the ozone analyzer , to measure the amount of ozone which was not consumed in the column),or to a gas washing bottle. Finally, gaseous ozone leaving the ozone analyzer was being distributed in a series of gas washing bottles filled with a 2% potassium iodide (KI) solution.

2.3. Experimental conditions

Ozonation experiments were carried out under continuous operation in an reactor. The effect of several parameters (inlet dye concentration, applied ozone dose, gas flow rate, pH and conductivity) in O.M removal of an phosphoric acid (54%) was investigated with the aim to optimize the operation conditions. All experiments were carried out by varying only one parameter in each situation.

The following standard operation conditions were used:

Reactor temperature = 25°C
Oxygene flow rate = 20NL/H
Sample weight = 800g
Agitation speed = 500 tours /mn

Conventional pollutant parameters for samples after ozonation were determined accordingly to predict the effect of ozonation as an advanced treatment option. ***First variable ozone contact times*** (0–360 min) were applied for all samples. The mainly observation concern the color of phosphoric acid under study.

2.4. Technical Analysis

During the ozonation, P_2O_5 and impurities amounts are determined for all samples

The P_2O_5 content was determined by volumetric titration with NaOH using bromocresel green and phenolphthalein indicators with accuracy of ± 0.2%. The water composition was determined by the Karl–Fisher method with a 4% experimental error [12]. Metallic impurities' contents were determined using a Jobin Yvon ICP 2000 spectrometer previously calibrated with adequate standards within the concentration ranges of the samples with an accuracy of ± 0.2%. The raffinate organic matter content was determined by using a redox titrimetric method using a hot mixture of $K_2Cr_2O_7$ in sulfuric medium. The excess of $Cr_2O_7^{2-}$ was back titrated with a standard solution of Fe^{2+}. The relative error was below 2.5% [13].

The organic matter is extracted in a soxhlet extractor for 1h using the mixture of dichloromethane/methanol (2:1,v/v), respecting a ratio solvent/sample of 12. The extraction yieled was 75 ±6%. This extraction was made on phosphoric acid upstream and downstream of the ozonation in order to understand the mechanisms involved during the treatment.

Upstreams the ozonation, this extraction led on a three phases: the aqueous phase, the extract and the emulsion.

The extract was separeted by chromatography column into two fractions: saturated and polar compound. The column (1.5cm i.d., 15 cm length) was filled with silica gel activated at 550°C during 4h.

The two fractions were eluted by:

-Hexane for the saturated fractions

-Methanol for the polar compounds.

Infrared absorption spectra of all fractions in KBr were recorded on a PERKIN-ELMER 1750 spectrophotometer in the 400-4000cm^{-1} range. The heating of the sample was performed in an air-atmosphere SPECTAC heating cell. A thermocouple EUROTHERM REGLER was used for the temperature measurements.

The two fractions were identified by gas chromatography (GC)/mass spectrometry (HP6890-HP5973MSD Agilent Technologies, Wilmington DE, USA). The GC was used with a 30 m fused-silica column (0.25 mm i.d.) coated with 5% phenyl methyl siloxane. Helium was used as the carrier gas at a flow rate of 1.5 ml/min. The following temperature programmer was used: 70–300 °C with ramping at 10°C/ min. The samples were injected in the split less mode with an injector temperature of 280°C. Analyses were run in the electron impact mode at 70 eV with a 2.9 s scan time over a 45–550 a.m.u. range resolution.

All solvents were glass distilled. Procedural blanks showed no interfering contribution of hydrocarbons from the freeze drying step, reagents and solvents. Replicate samples from the same samples varied ±15%. The precision of the analytical method, including all steps was ±20%.

3. Results and Discussions

The O.M removal and ozone consumption are the most important parameters to evaluate the performance of the continuous mineralization of the organic matter in phosphoric acid. The content of O.M determines the degree of purity of phosphoric acid (54%) and the ozone consumption determines the economical advantage of ozonation process. These parameters depend on: contact time, applied ozone dose, pH and conductivity. In order to study their effect, different experiments were carried out.

3.1. Chemical oxidation

3.1.1. Effect of the contact time:

Figure 2 shown the variation of the O.M amount with the contact time between ozone and O.M present in phosphoric acid. Indeed when the duration of ozonation increases, we can see a change in physical appearance of phosphoric acid, in fact the intensity of the color of acid reduces, and it turns from black to a greenish color.

To our knowledge, the color of the solution depends on the O.M content. Based in this idea and after examination of Figure.2, we can deduce that mineralization by ozonation is very low compared to color removal, which is in agreement with previous publications **[14, 15]**. This may be explained by the fast destruction of the chromospheres groups, responsible for the colour, and by the fact that the resulting sub-products are very difficult to mineralize.

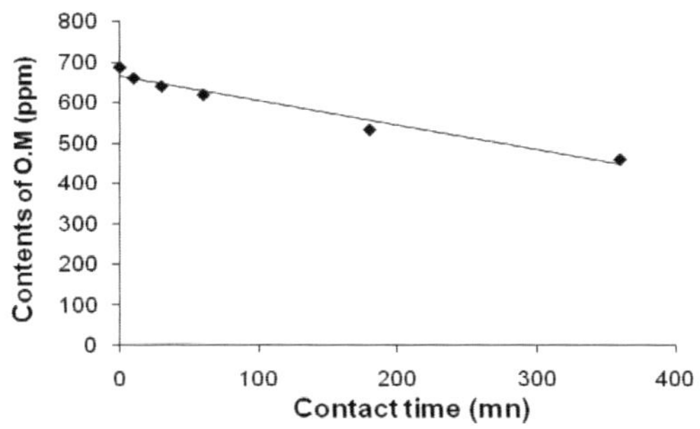

Figure 2. Effect of the contact time between ozone and organic matter in the content of TOC.

Ozonation can be effective to treat this type of acid but it would be necessary to increase the contact time to improve the decolorization and the reduction of the content of O.M in phosphoric acid.

3.1.2. Effect of excitation tension:

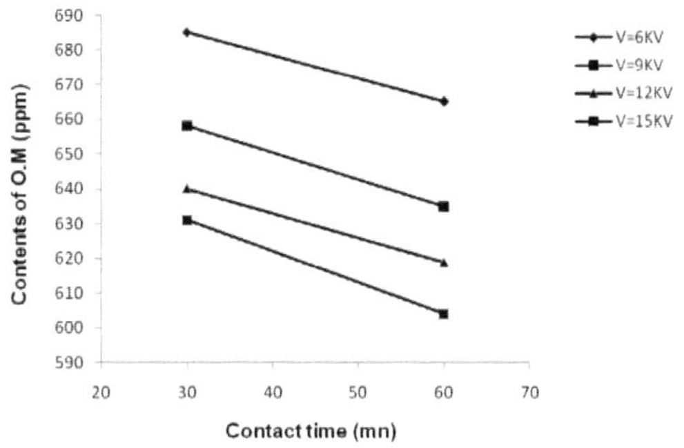

Figure 3. Variation of the TOC content with the contact ozonation for different values of tension excitation.

Ozonation of an acid was carried out in co-current by also varying a contact time between the ozone and the acid for different tension excitation. Indeed; the variation of the excitation tension from 6KV to 15KV was applied for another series of samples and presented in Figure 3. ***I***t was observed that more excitation tension is higher more the performance of ozonation is better. For excitation tension 15KV, the mineralization observed is the highest or the intensity of the color of acid is not so lower. This observation, which is in agreement with previous publications **[16, 17]**, can be explained by the fact that the ozone decomposition into hydroxyl radicals is less selective but have an oxidation potential higher than that of ozone.

Experiments show that mineralization of acid by ozonation is very low compared to color removal,

Therefore, the increase of the excitation tension is not justified because it will be increase the operation costs without the corresponding increase in the O.M content.

3.2. Mechanisms involved in an ozonation

The ozonation modify the characteristics of the molecules of organic substances present in aqueous solution.

3.2.1. GC-MS spectrum analysis

Figure 4 and 5 which present respectively the spectra of GC-MS respectively of the saturated and the polar fraction upstream the ozonation, show the presence of the following compounds: n-alkanes like n-C14, n-alkenes from n-C6:1 to n-C18:1, carboxylic diacids, alcohols, unsaturated fatty acid and carboxylic acid.

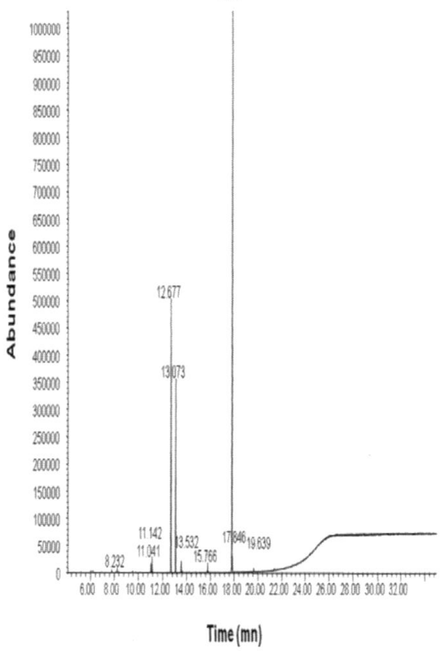

Figure 4. Gas chromatogram of saturated fraction upstream the ozonation

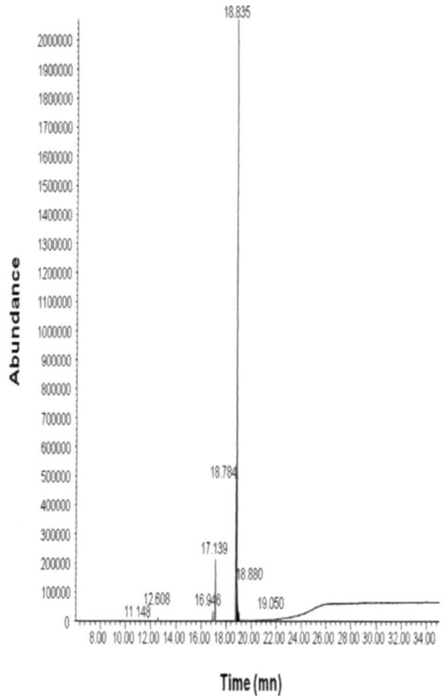

Figure 5. Gas chromatogram of polar fraction upstream the ozonation.

Figure 6 and 7 which present respectively the spectra of GC-MS respectively of the saturated and the polar fraction downstream the ozonation, shows the presence of the following compounds: n-alkane like n-C14, carboxylic diacids, alcohol, carboxylics acids and ketones

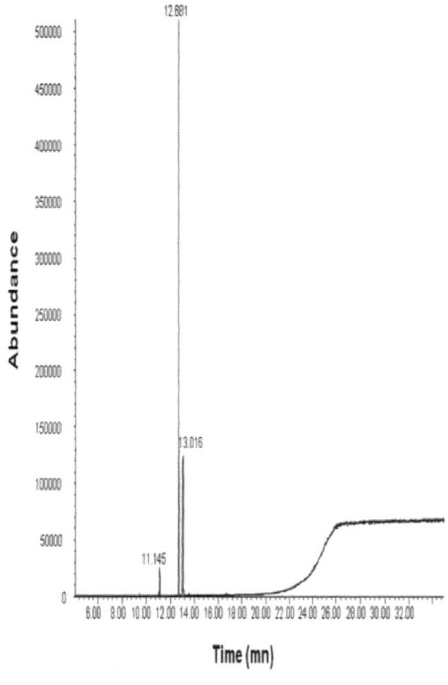

Figure 6. Gas chromatogram of saturated fraction downstream the ozonation.

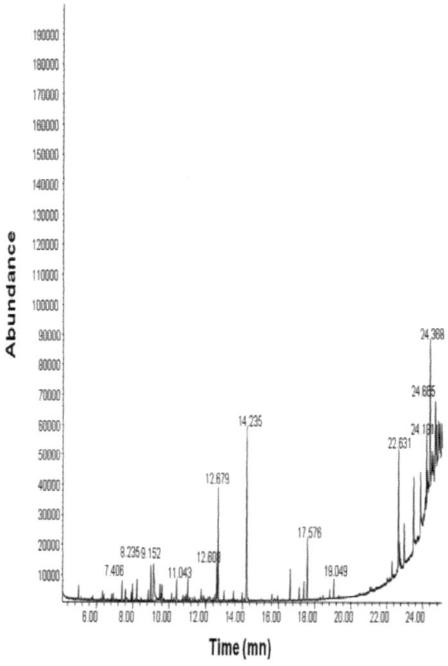

Figure 7. Gas chromatogram of polar fraction downstream the ozonation.

IOP Conf. Series: Materials Science and Engineering **28** (2012) 012045 doi:10.1088/1757-899X/28/1/012045

Table 2. Attribution of the signals to the appropriate chemical upstream ozonation of saturated fraction

chemical species	Retention time (mn)	Empirical formula	Formula developed	Molar mass g/l	% of similarity
1-Hexène	8,232	C_6H_{12}		84,16	80
1-hexadécanol	11,142	$C_{16}H_{34}O$		242,44	87
Tétradécane	11,152	$C_{14}H_{30}$		198,39	98
Phénol	12,677	C_6H_6O	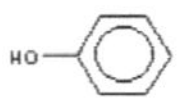	94,11	97
Acide Nonanedioique= Acide azélaique	13,073	$C_9H_{16}O_4$		188,22	87
Hexadécène	13,533	$C_{16}H_{32}$		224,43	99
1-octadécène	15,768	$C_{18}H_{36}$		252,48	93
1-Eicosene	17,793	$C_{20}H_{40}$		280,53	98
1-Nonadécène	19,639	$C_{19}H_{38}$		266,51	93

Table 3. Attribution of the signals to the appropriate chemical upstream ozonation of polar fraction

chemical species	Retention time (mn)	Empirical formula	Formula developed	Molar mass g/l	% of similarity
Tétradécane	11,143	$C_{14}H_{30}$		198,39	98
Phénol	12,608	C_6H_6O	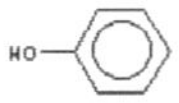	94,11	90
Acide 9-Hexadécénoique	16,946	$C_{16}H_{30}O_2$		254,41	99
Acide Hexadécanoique	17,139	$C_{16}H_{32}O_2$	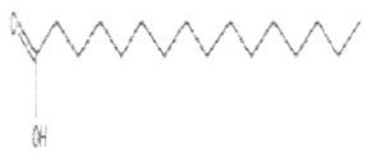	256,42	99
Acide-9,12 Octadécadienoique	18,784	$C_{18}H_{32}O_2$	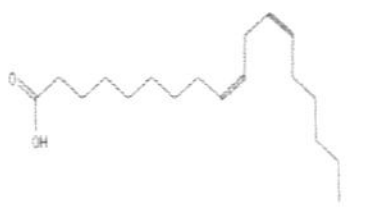	280,45	99
Acide 9-octadécénoique= acide oléique	18,835	$C_{18}H_{34}O_2$		282,48	99
Acide octadécanoique= acide stéarique	19,050	$C_{18}H_{36}O_2$	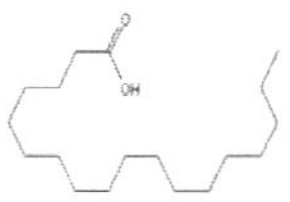	284,48	98

IOP Conf. Series: Materials Science and Engineering **28** (2012) 012045 doi:10.1088/1757-899X/28/1/012045

Table 4. Attribution of the signals to the appropriate chemical downstream ozonation of saturated fraction

chemical species	Retention time (mn)	Empirical formula	Formula developed	Molar mass g/l	% of similarity
Tétradécane	11,145	$C_{14}H_{30}$		198,39	95
Phénol	12,680	C_6H_6O		94,11	90
Acide Nonanedioique	13,016	$C_9H_{16}O_4$	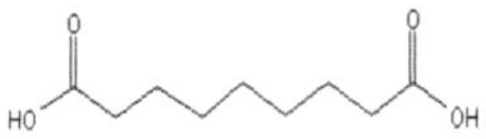	188,22	91

Table 5. Attribution of the signals to the appropriate chemical dowenstream ozonation of polar fraction

chemical species	Retention time (mn)	Empirical formula	Formula developed	Molar mass g/l	% of similarity
Acide pentanedioique	7,404	$C_5H_8O_4$	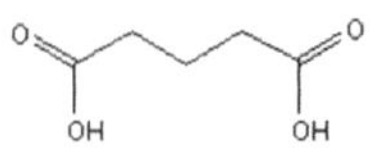	132,11	98
Acide hexanedioique	8,998	$C_6H_{10}O_4$	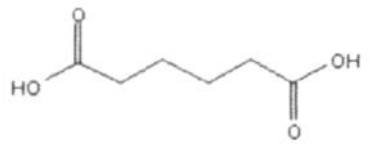	146,14	94
2H-Azepin-2-one=caprolactam	9,151	$C_6H_{11}NO$	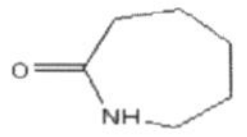	113,16	87
1-hexadécanol	11,045	$C_{16}H_{34}O$		242,44	87
Phénol	12,608	C_6H_6O	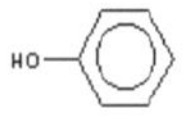	94,11	90
Acide hexadécanoique	17,576	$C_{16}H_{32}O_2$		166,13	90
Acide octadécanoique= acide stearique	19,051	$C_{18}H_{36}O_2$	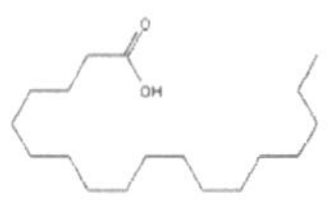	284,48	98
acide1, 2-Benzenedi-carboxylique	22,631	$C_8H_6O_4$	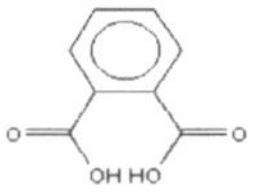	256,42	92

A tables resuming the attribution of the signals to the appropriate chemical upstream and downstream ozonation of all functions is present respectively in tables 2-5.
The study of spectra GC-MS upstream and downstream of ozonation for both saturated and polar fractions led to the following comments:

-Disappearance of alkenes and appearance of ketones. This process is explained as follows:

$H_2C{=}CH{-}CH_2{-}CH(CH_3){-}CH_3 \longrightarrow O{=}CH{-}CH_2{-}CH(CH_3){-}CH_3 + H_2C{=}O$

Without reducer can be have a pushed oxidation

-Spin-Off of the molecule at the double bond and formation of a diacid

$RCH{=}CHR'COOH + O_3 \longrightarrow RCOOH + HOOCR'COOH$

-Apparition of a new cyclic carboxylic acids, these acids come probably from the the f some molecules of kerogene.

-Clipping of phenolic groups with formation of dicarboxylic acid

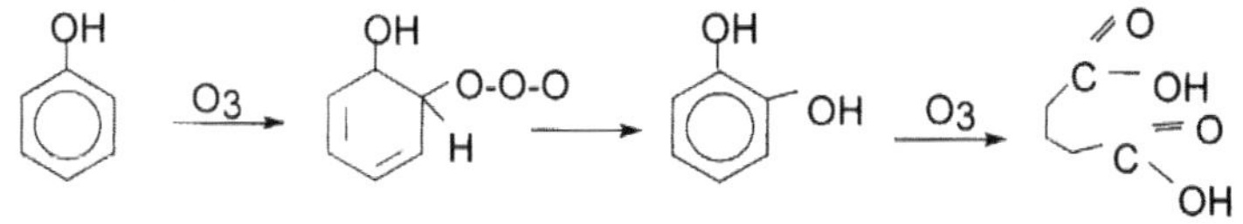

The ozone attacks the carboxylic acids with low molecular weight and converted him to carbon dioxide. This explains the absence brutal of some acids that were present and disappear as a result of the ozonation.

3.2.2. FT-IR spectrum analysis:

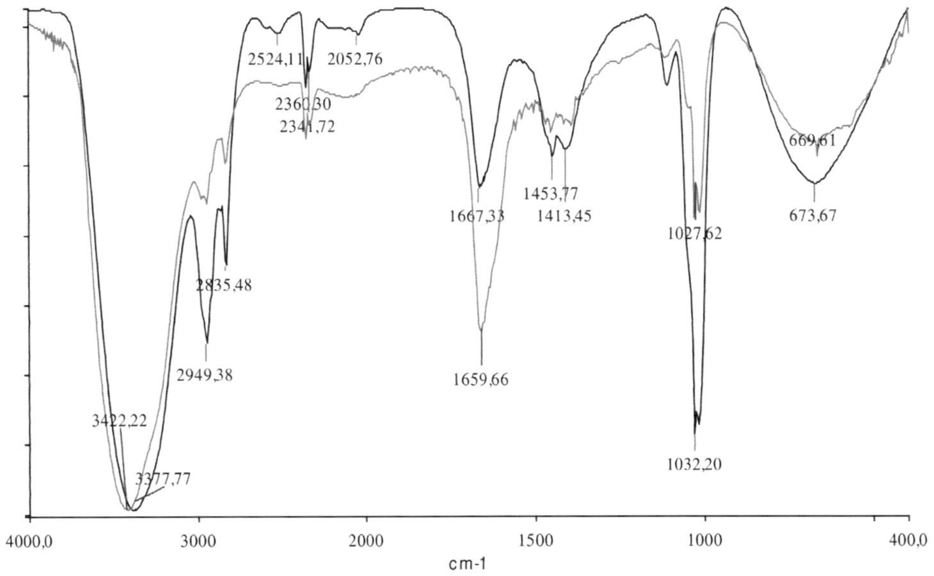

Figure 8. Supperposition of the IR spectra of the polar fraction upstream and downstream the ozonation.

After having evaporated the solvent, the extract (saturated and polar fraction) upstream (red spectrum in fig.8) and downstream the ozonation (black spectrum in fig.8) were dissolved in carbon tetrachloride and we record their IR spectrum. The superposition of the two spectra IR, see figure 8, leads to the following band presented in the table.6. The attribution of the band is based on the bibliographic study [18].

Band at 1450 cm^{-1} corresponds to the function C=C decreases in intensity for the increasing of the intensity of the band at 1660cm^{-1} which represents a ketone function

Upstream of ozonation, we notice that in addition to the saturated fraction and polar fraction there is a black layer which disappears following the action of ozone.

According to the literature this layer represents the Kérogène which is characterised by a high atomic reports (H/C) and that confirm the richness of organic matter by aliphatic compounds.

Table 6: Award of the chemical functions to frequency vibrations

Frequency vibrations (cm^{-1})	nature of the function [18]
1032	C-O-C
1450	C=C
1659	RCOR
2300	band CO_2 attributed to the environment of the machine
2835	Symmetrical stretching C-H
2949	Asymmetrical stretching C-H
3422	COOH

4. Conclusion

The continuous ozonation of organic matter in phosphoric acid was investigated. According to the experimental results, the following conclusions can be drawn:
Ozonation is highly efficient in the decolourisation of the solutions, but considerably less efficient in terms of TOC removal.

The analysis by GC-MS and FT-IR, of different extract obtained upstream and downstream of the ozonation, had identified several products (insaturated, fatty acids, ketones….).

Therefore, we can conclude that ozonation of organic compounds does not necessarily lead to a full mineralization of these compounds. The reduction in the rate of organic matter remains limited by the fact that some compounds are inert towards ozone.

References

[1] Skoromvarov J. I, Ruzin L. I and Lomonosov A.V 1988 *J. of Radioanalytical and Nuclear Chemistry*, **229** 455 467.
[2] Kotz J. C, Purcell K. F 1991 *Uses of Phosphate Containing Rock Chemistry and Chemical Reactivity, 2nd ed.*
[3] Dunn K 1985 *Phosphorus Potassium* **139** 34.
[4] Beltran F. J, Garcia-Araya J. F, Frades J, Alavarez P, Gimeno O 1999 *Water Research*, 33 723 732.
[5] Jochimsen J. C and Jekel M. R 1997 *Water Science and Technology* **35** 337 345.
[6] Nawrocki J, Biłozor S 2000 *Uzdatnianie wody, Procesy chemiczne i biologiczne, PWN, Warszawa-Poznan.*
[7] Kowal A. L, OEwiderska-Bróz M 1997 *Oczyszczanie wody, PWN, Warszawa-Wrocław.*
[8] Huang C.P, Dong C, Tang Z 1993 *Waste Manage* **13** 361.
[9] Deng Z, Frang X, Li D, Zhou S, Tao R, Dong W, Wang T, Meng G, Zhu X, 2009 *J. of* All. *and Comp.* **484** 619 621.
[10] Blinova O, Novikov A, Perminova I, Goryachenkova T, Haire R 2007 *J. of* All. *and Comp.* **444-445** 486 490.
[11] Beltran F. J, Rivas F. J, Montero-de-Espinosa R 2002 *Applied Catalysis B: Environmental* **39** 221 231.
[12] Bizot J 1967 *Bulletin* de la *Société Chimique de France 1* 151 156.
[13] Hannachi A, Habaili D, Chtara C, Ratel A 2007 *Separation and Purification Technology* **55** 212 216.
[14] Baban A, Yediler A, Lienert D, Kemerdere N, Kettrup A 2003 *Dyes Pigments* **58** 93 98.
[15] Hsu Y. C, Chen J. T, Yang H. C, Chen J. H 2001 *AIChE J.* **47** 169 176.
[16] Baban A, Yediler A, Lienert D, Kemerdere N, Kettrup A 2003 *Dyes Pigments* **58** 93 98.
[17] Hsu Y. C, Chen J. T, Yang H. C, Chen J. H 2001 *AIChE journal.* **47** 169 176.
[18] Kazuo N 1997 *Infrared and Raman Spectra of Inorganic and coordination compounds. Part B :Application in coordination, Organometallic, and Bioinorganic Chemistryp.59, 5th édition, Wiley-interscience.*

IOP Conf. Series: Materials Science and Engineering **28** (2012) 012046 doi:10.1088/1757-899X/28/1/012046

Decomposition behavior of the NaH_2 $(PO_4)_{0.48}(AsO_4)_{0.52}$ H_2O compound above room temperature shown by the study of Raman X- ray powder and ac-conductivity

Kh. Jarraya[a 1], N. Guiblin[b], N. E. Ghermani[c] and T. Mhiri[a]
[a]Laboratoire de L'Etat Solide, Faculté des Sciences BP 1171, 3000 Sfax, Tunisie.
[b]Laboratoire S.P.M.S. Ecole Centrale de Paris, 75000 Paris, France.
[c]Laboratoire de Physique Pharmaceutique UMR 8612, Université Paris XI – France.

E-mail: khaledjarraya@yahoo.fr

Abstract. The mixed compound $NaH_2(PO_4)_{0.48}(AsO_4)_{0.52}.H_2O$, which has been examined by Raman spectroscopy, XRD powder and dielectric, exhibits solid–solid phase transitions and one reversible solid–liquid phase transition. Besides, the investigation of dehydration behavior of sodium dihydrogenate phosphate-arsenate monohydrate was undertaken in the temperature range of 253 K to 333 K. The onset temperature of dehydration was determined from changes in ionic conductivity and X-ray powder on heating. It was confirmed by the Raman study not only in the disappearance of the vibrations assigned to the PO_4, AsO_4 stretching O···O bridge vibrations and the O···H–O vibrations, but also in the appearance of the new vibrations attributed to the (PO_3), which constitutes the P-O-P chain, giving a polymer (with) the formula: $NaH_2(P_2O_7)_x(As_2O_7)_{(1-x)}$. The important variation was observed in all studies at 313 K which could correspond to the transformation of the NDAP compounds of a solid crystal into a liquid form.

1. Introduction

The mixed compound $NaH_2(PO_4)_{0.48}(AsO_4)_{0.52}.H_2O$ (denoted by NDAP) [1], which crystallizes in the orthorhombic system with a space group of $P2_12_12_1$ exhibit solid–solid phase transitions and one reversible solid–liquid phase transition. This compound belongs to a group of compounds that have intermediate properties between those of a normal salt and an acid, and it undergoes a transition from a paraelectric to a superprotonic phase at room temperature [2]. This is an important discovery for the compounds derived from the KDP family (KH_2PO_4) of the general formula MH_2XO_4 (where M is a monovalent cation: K, Rb, Cs and X is P or As), whose use is possible in technological applications (fuel cells ...). The superprotonic phase of NDAP exhibits greater proton conductivity than 1.07×10^{-4} $\Omega^{-1}cm^{-1}$ to 296 K [3], and stability at room temperature. In fact, the compounds of the formulas MH_2XO_4 have all been studied excluding those in which M is sodium because the latter is not stable at room temperature except in the hydrated state. In this process, the structural phase transitions in similar compounds KDP, RDP and CDP are located at high temperature, respectively 453, 359 and 503.8 K, and the room temperature crystal structures are maintained up to the decomposition a tetragonal structure for KDP and RDP and a monoclinic symmetry for CDP. The thermal dehydration

[1] To whom any correspondence should be addressed.

takes place at the surface of the salts [4] which can be described by the formation of $K_nH_2P_nO_{3n+1}$, $RbH_2PO_4 + Rb_2H_2P_2O_7$ and $Cs_2H_2P_2O_7$ [5-7] respectively .

This work is an attempt to understand and explain the mechanisms of the transition and dehydration of the title compound above room temperature, building on studies of Raman, XRD powder and ac-conductivity. Thus, it is important to study this transition to an intermediate state described by a liquid crystal present just before the start of dehydration-decomposition.

2. Experimental

Raman spectra were measured with a powder sample between 253 K and 333 K by a HR800 micro-Raman spectrometer (Jobin Yvon Horiba, Longjumeau, France), equipped with a He–Ne laser emitting at 633 nm with a power of 20 mW in the range 50–4000 cm^{-1}. The instrument is directly controlled by a PC equipped with the software "LabSpec" (Jobin Yvon Horiba, Longjumeau, France). The measurements, whose objective has a magnifying power of 100× in an open furnace (Under the air, at an ambient pressure) on a heating from 253 to 333 K, were carried out under BX40 microscope and had a focal gradation of 1μm. It is worth noting that only the temperature range (303-323 K) would be discussed here because, with the exception of a classical narrowing and weak shift of the bands, a non spectral change occurred between 253 and 303 K. The heating rate was 5 K/min, waiting for 5 min after temperature stabilization and collecting time for each spectrum was 10 min. The measurements of the complex dielectric permittivity ($\varepsilon^* = \varepsilon' - j\varepsilon''$), were performed over a frequency interval of 10^6 –10^{-1} Hz and in a temperature range of 233 to 325 K. To cover the above frequency range, a NOVOCONTROL spectrometer integrating an ALPHA dielectric interface was employed with a precision in $\tan\delta = \varepsilon''/\varepsilon' = 10^{-5}$. In these experiments, the temperature was controlled by a nitrogen jet (QUATRO from Novocontrol) with a temperature error, during every single sweep in frequency of ± 0.1 K. Conductivity is usually associated with the generation and transport of polarization-induced charges through the compound under the action of an electric field. The X-ray powder diffraction patterns were collected from samples at various levels of dehydration. Diffraction patterns were measured between 283 and 333 K using either a Philips X'Pert Pro (Cu-Kα, 40 kV, 40 mA) at a scan rate of 0.6° 2θ min^{-1} with an angular resolution of 0.02.

3. Results and discussion

All results achieved by Raman spectroscopy at different temperatures are gathered in table 1.

3.1. Before 313 K: state solid form of NDAP

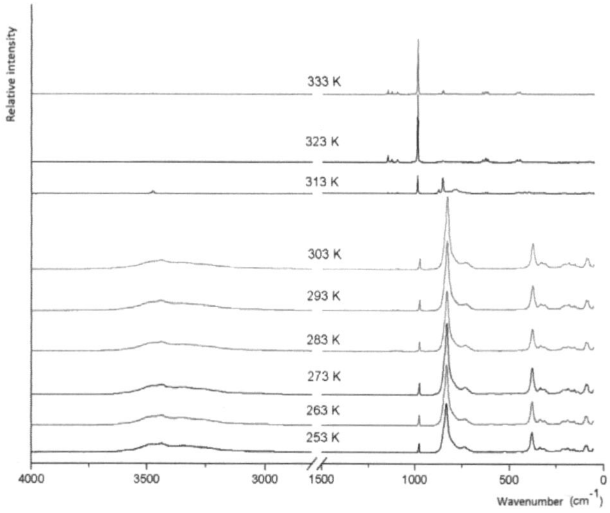

Figure 1. Raman spectra of crystalline $NaH_2(PO_4)_{0.48}(AsO_4)_{0.52}.H_2O$ between 253 and 333 K.

Table 1 Wavenumbers (cm^{-1}) and relative intensities of the bands in the Raman spectra of $NaH_2(PO_4)_{0.48}(AsO_4)_{0.52}.H_2O$.

Before 313 K	At 313 K	After 313 K	Assignments
56 w	56 vw	56 vw	Lattice vibrations
95 s 135 m 156 vw 195 w		-	δ O. . .O
209 vw 220 w		-	O...H–O vibrations
324 w 338 w	324 w	-	$\nu_2(AsO_4)$
382 vs	382 vw		$\nu_2(PO_4)$
-	-	455 w 469 w 629 m 640 m 658 m	$\delta(PO_3)$ and ρ (PO_3)
745 mb	-		$\nu_1(AsO_4)$
824 vs	794 s		ν_3 (AsO_4)
848 vs	862 s		ν (P-OH)
		862 w	ν $(As_2O_7)^{4-}$
985 s	985 s		ν_1 (PO_4)
-		991 s	$\nu_1(PO_3)$
-	-	1116 w 1145 w 1167 w	$\nu_3(PO_3)$ and (P-O-P)
1637 wb 3354 vb 3459 vb 3491 vb	1637 wb 3459 m	-	$\nu_1(H_2O)$ $\nu_3(H_2O)$

ρ: rocking *s: strong*
δ : bending *m: medium*
ν : stretching *w: weak*

The internal vibrations in Raman spectra of the PO_4 tetrahedra are presented in Fig. 1. The symmetric stretching ν_1 at 985 cm^{-1} is active with an intense peak which is a characteristic of this group. $\nu_2(PO_4)$ appears at 382 cm^{-1} with a strong peak. In the same vein, an intense peak appears at 848 cm^{-1} which corresponds to a typical vibration of (P-OH). We also detect a broad stretching vibration at 745 cm^{-1} corresponding to $\nu_1(AsO_4)$ and a vibration at 824 cm^{-1} assigned to $\nu_3(AsO_4)$. However, $\nu_3(PO_4)$ that appeared at 838 cm^{-1} is invisible because it is embedded in the strong peak measured at 848 cm^{-1}.We can observe two weak peaks at 324 and 338 cm^{-1} associated with ν_2 of AsO_4. The presence of the broad peaks detected at 3354, 3459 and 3491 cm^{-1} corresponds to ν_1 and ν_3 of water, respectively. What should be noted from the irrational spectra is that the strength and the type of O–H···O hydrogen bonds remain practically the same in the strong hydrogen bonded crystals such as KH_2PO_4, NaH_2PO_4, $CsHSO_4$ and CsH_2PO_4 [8–11]. The corresponding OH stretching vibration gives rise to broad bonds characteristics at 3467 and 1637 cm^{-1}, associated with strong hydrogen bonded systems [12]. The X-ray powder diffraction has shown the first pattern, (Fig. 2), corresponding to that of the mixed compound NDAP monohydrate, whose crystalline form has accordingly been taken to be the reference database ICSD/ Retrieve of phosphate and arsenate ions.

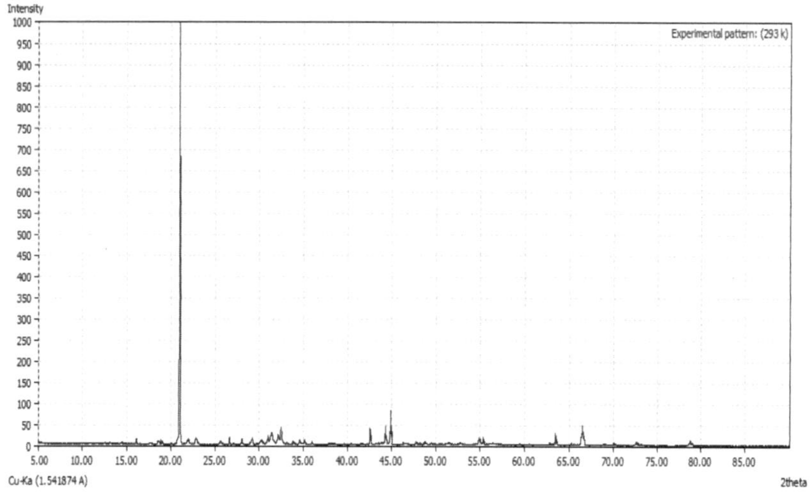

Figure 2. (a) X-ray powder diffraction for NDAP material: at 293 K.

3.2. At 313 K: liquid crystal form of NDAP

As can be seen from Fig. 1, the rise of temperature to 313 K shows modifications in the Raman spectra of this compound, these changes are detected firstly by the decrease -increase in some peaks intensity and secondly by the translation of others. In fact, the decrease of the intensity of the peaks at 382 cm^{-1} assigned to ν_2 (PO_4) and increase of the intensity of the peaks at 985 cm^{-1} are assigned to ν_1 (PO_4). Besides, the translation and decrease of the intensity of the peaks at 862 and 794 cm^{-1} are attributed to ν(P-OH) and ν_3 (AsO_4) respectively. Moreover, the peaks characteristics to the ν_2(AsO_4) at 324 cm^{-1} keep the same place. The appearance of a one medium peak at 3459 cm^{-1} instead of three broad peaks at (3354, 3459 and 3491 cm^{-1}) is related to the symmetric stretching ν_1 and ν_3 (H_2O). The disappearance of the lines at 209 a n d 220 cm^{-1} characterizes the change of the O···H–O vibrations. Indeed, this family of compounds is characterized by a large proton conduction due to the presence of hydrogen located between the tetrahedra and voids in the unit cell to facilitate the migration of the proton so as to assure the high conduction near the temperature of decomposition [2, 13]. This transformation is shown by the breaking of hydrogen bonding. What leads us to confirm this conclusion is the decrease of proton conductivity with temperature higher than that of the processing and the analysis of the variation of conductivity as a function of temperature. Fig. 3 displays the evolution of the conductivity versus inverse temperature $\log_{10}(\sigma T) = f(1000/T)$ for NDAP compound in the temperature range of 233-333 K. An Arrhenius-type $\sigma T = \sigma_0 \exp(-\Delta E_0/kT)$ characterizes the low and high temperature domain until 313 K. After this temperature, fast conductivity decreases from $13.6 \times 10^{-4}\ \Omega^{-1}cm^{-1}$ to $7.5 \times 10^{-5}\ \Omega^{-1}cm^{-1}$ at 323 K . The fast variation of the conductivity confirms the change of the structure, following the loss of hydrogen bond located between two oxygen's donor - acceptor, and certainly, for this reason, conductivity has fallen.

The X-ray powder diffraction illustrates the second pattern. (Fig. 4) corresponds to the mixed NDAP compounds, which is taken to be the reference pattern of phosphate and arsenate ions with a change in the intensity of diffraction line compared with that before 313 K. These modifications in the major intensity in the two analyses confirm the presence of the phase transition in this temperature [14].

Based on a previous thermal study made by DSC and DTA analysis [2] which has shown that no mass loss is associated with the solid – liquid phase transition at this temperature. This phase transition is also reversible of a second order type and an order-disorder character, with a gap of temperature of 0.5 K and $\Delta H = 2.02\ kJmol^{-1}$.

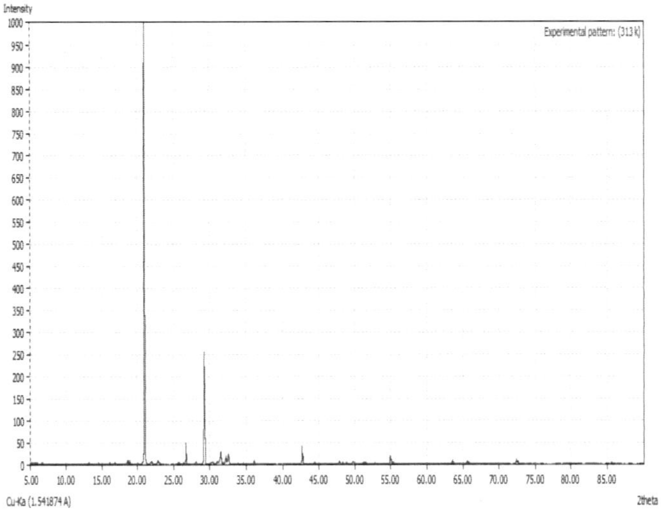

Figure 2. (b). X-ray powder diffraction for NDAP material: at 313 K.

Furthermore, from the studies of Raman, XRD powder and ac-conductivity, we are likely to conclude that the NDAP compounds were transformed from a solid crystal, before 313 K, to an intermediate state described by a liquid crystal form at 313 K. This may be due to the dynamical disorder, by elevating the temperature of the acidic protons on the hydrogen bond network linking $P(As)O_4$ groups and to the order or disorder type of the $P(As)O_4$ tetrahedra. The latter affects the motion of hydrogen atoms and increases the reorientational motion of the anions $H_2P(As)O_4^-$ tetrahedra as well as the increased thermal agitation.

3.3 At 323 K: dehydration-decomposition

The evolution of the line in Fig.1 is an important modification pertaining to the hydrogen bonds; the loss of water is very face observed with the disappearance of the vibrations near 3400 cm^{-1}. At the same time, there are significant changes of positions and intensity of some bonds corresponding to the internal vibrations of $\nu(PO_4)$ and $\nu(AsO_4)$. It is worthy to note that there is a disappearance of the peak at 985 cm^{-1} attributed to $\nu_1(PO_4)$ and substituting new strong peak at 991 cm^{-1} assigned to the stretching vibrations of $\nu_1(PO_3)$. There are also other three weak peaks that appear at 1116, 1145 and 1167cm^{-1} respectively, which are attributed to $\nu_3(PO_3)$ and (P-O-P) vibrations. The appearance of the weak lines at 455, 469 cm^{-1} and medium triobonds at 629, 640 and 658 cm^{-1} correspond to $\delta(PO_3)$ and $\rho(PO_3)$, respectively.

The important spectral alterations observed at 323 K correspond to the chemical transformation in $P(As)O_4^{3-}$ to $P_2O_7^{4-}$ and $As_2O_7^{4-}$ ions, as in the similar compounds CsH_2PO_4 [15]. Thus, when $NaH_2(PO_4)_{0.48}(AsO_4)_{0.52}.H_2O$ is heated, it tends to quickly lose its water, leading to a dehydration between 313-323 K. After 323 K, the compound is transformed into an infinite chain of polyphosphate-arsenate connected by P(As)-O-P(As) bridge. Therefore, it loses two molecule of water of its components, and constitutes a condensed phosphate-arsenate including polymer with the formula: $NaH_2(P_2O_7)_x(As_2O_7)_{(1-x)}$ according to the decomposition reaction proposed by K.-S. Lee[5]. X ray powder at 323 K has shown the pattern in (Fig. 2c), which corresponds to the dehydration-decomposition of NDAP hydrate. The latter can be shown by the DSC and ATD for previous studies [2], in which the total weight loss of this sample was 11 % corresponding well to that for the formation of $Na(P/As)O_3$. Accordingly, we take this to be the reference pattern of sodium meta-phosphate-arsenate. The transformation to $Na(P/As)O_3$, however, requires the coordinated rearrangement of all the phosphate-arsenate anions to form the metaphosphate-arsenate chain $[P(As)O_3^-]_\infty$.

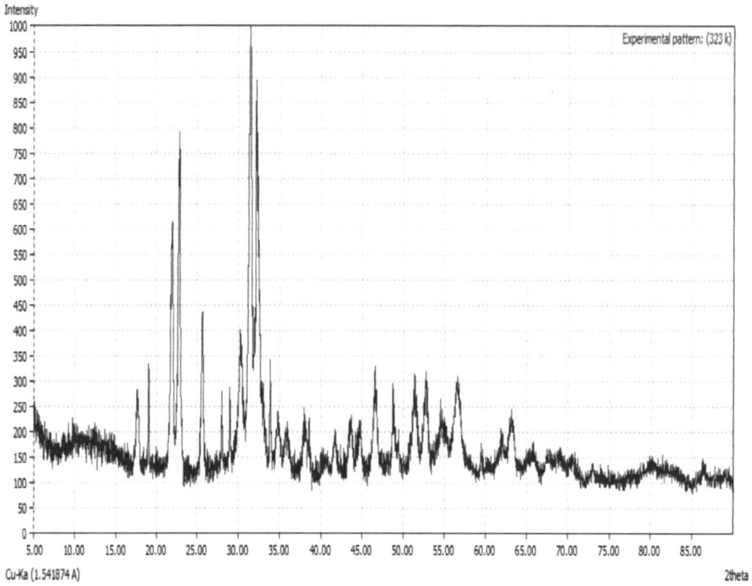

Figure 2. (c). X-ray powder diffraction for NDAP material at 323 K.

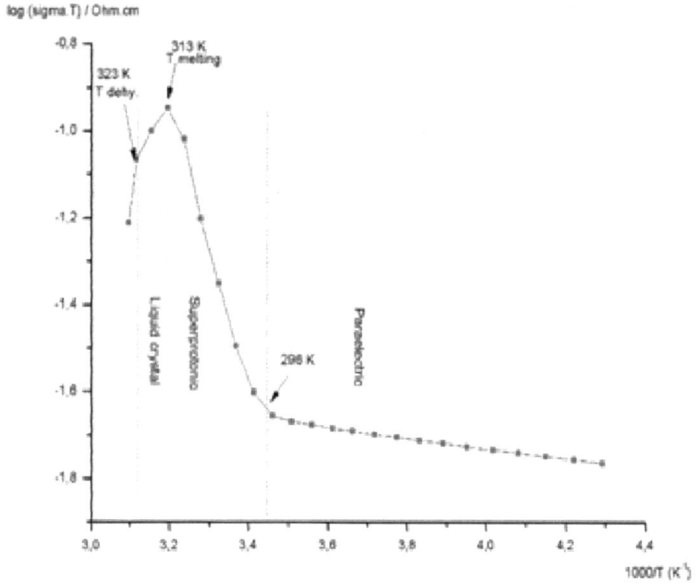

Figure 3. Temperature dependences of $\log(\sigma T) = f(1000/T)$ for NDAP.

4. Conclusion

This present study showed that the transformation of the NDAP compounds of a solid crystalline form before 313 K to a liquid crystal at 313 K is relevant. That hypothesis opens the horizon for another study on this compound at this temperature. Another significant change was observed at 323 K

corresponding to the chemical transformation in $P(As)O_4^{3-}$ to $P(As)_2O_7^{4-}$ ions. These findings have proven that the compound is transformed into an infinite chain of polyphosphate-arsenate connected by (P/As)-O-(P/As) bridge losing water of constituents.
We conclude that the dehydration-decomposition of $NaH_2(PO_4)_{0.48}(AsO_4)_{0.52}.H_2O$ begins at 323 K giving a polymer of the formula $NaH_2(P_2O_7)_x(As_2O_7)_{(1-x)}$. We can suggest the following reaction:

Reversible Phase Transition
T = 313 K

$$NaH_2(PO_4)_x(AsO_4)_{(1-x)}.H_2O \text{ (solid crystal)} \rightleftarrows NaH_2(PO_4)_x(AsO_4)_{(1-x)}.H_2O \text{ (liquid crystal)}$$

$\downarrow$ *T = 323 K : dehydration-decomposition (Polymerization)*

$$NaH_2(P_2O_7)_x(As_2O_7)_{(1-x)} \text{ (powder)} + 2H_2O$$

Finally, what differentiates NDAP of other similar compounds (RDP, CDP and KDP) is its low decomposition temperature but also its passage through an intermediate state "liquid crystal" just before its polymerization and not by the formation of liquid superficial detected in all other similar compounds. This is strongly linked to its structural properties:
- Presence of alkaline sodium
- Structural disorder by partial substitution of phosphorus by arsenic.
- Presence of a structural water molecule.

References

[1] Jarraya Kh, Ennaceur N, Mhiri T 2010 *X-ray Structure Analysis Online Japan Society of Analytical Chemistry* **26** 19.
[2] Ennaceur N, Jarraya Kh, Ledoux-Rak I, Mhiri T, submitted *in the Journal of Physics and Chemistry of Solids*.
[3] Jarraya Kh, Ennaceur N, Elmarssi M, Mhiri T 2011 *Physica B Condensed Matter* **406** 1546.
[4] Ortiza E, Vargasb R. A, Mellanderc B. E 1999 *Solid State Ionics* **125** 177.
[5] Lee K. S 1996 *J. Phys. Chem. Solids* **57** 333.
[6] Ortiz E, Vargas R. A, Mellander B. E 1998 *J. Phys. Chem. Solids* **59** 305.
[7] Ortiz E, Vargas R. A, Mellander B. E. 1999 *J. Chem. Phys.* **110** 4847.
[8] Marchon B, Novak A 1985 *J. Chem. Phys.* **78** 2105.
[9] Ohno N, Lockwood D. J 1985 *J. Chem. Phys.* **83** 4374.
[10] Choi B. K, Kim J. J 1985 *Jpn. J. Appl. Phys.* **24** 914.
[11] Baran J 1987 *J. Mol. Struct.* **162** 211.
[12] Naili H, Mhiri T, Daoud A 2001 *International Journal of Inorganic Materials* **4** 393.
[13] Naïli H, Vendier L, Jaud J, T. Mhiri 2001 *Solid State Sciences* **3** 677.
[14] Abdelhedi M, Ktari L, Dammak M, Cousson A, Kolsi A. W 2008 *J. of Alloys and Compounds* **460** (1-2) 147.
[15] Taninouchi Y, Uda T, Awakura Y, Ikeda A, Haile S. M 2007 *J. Mater. Chem.* **17** 3182.

A Theoretical study of Propagation of Rayleigh surface waves in functionally graded piezoelectric (FGPM) half-space

I. Ben Salah[a1], A. Njeh[b] et M. H. Ben Ghozlen[a]
[a]Laboratoire de Physique des Matériaux, Faculté des Sciences de Sfax, BP 1171, 3000 Sfax, Université de Sfax,Tunisie.
[b]Unité de Physique, Infomatique et Mathématique, Faculté des Sciences de Gafsa, 2112 Gafsa, Tunisie.

E-mail: bs_issam@yahoo.fr

Abstract. This work presents a theoretical study of the propagation behavior of Rayleigh waves in a functionally graded piezoelectric material (FGPM) half-space. The influence of the graded variation of FGPM coefficients on the dispersion curves of Rayleigh waves is investigated for both electrically open and short conditions. The exponential gradient may be applied separately to mechanical and electrical properties. Examples have been given illustrating the use and capabilities of the method based on the recursive stiffness matrix method, in calculating SAW coupling factor (K^2) versus crystal orientation, SAW phase velocity for particular frequency range, and field profile variations with depth. The effects of the gradient variation of material constants on the distribution of displacement, stresses and electric potential are presented and discussed in detail. It is note that this paper considers the plane problem of an anisotropic beam which the material composition varies continuously in the direction of the thickness. Some useful discussions and numerical approach are presented to show the significant influence of material inhomogeneity, and adopting a certain value of the inhomogeneity parameter α and applying suitable electric and mechanical loads can optimize the FGPM substrate structures. This will be of particular importance in modern engineering design of high-performance surface acoustic wave (SAW) devices.

1. Introduction

With the development of material technology, functionally graded piezoelectric material (FGPM) can be manufactured and used to improve the efficiency and natural life of the SAW devices [1], possibly solving not only the residual stress problem in layered structures but also the penetration depth problem in the pure piezoelectric material. Hence, the research of surface acoustic wave propagation behaviour and characteristics in FGPMs has become a topic of practical interest and importance. There is already some literature concerning the Love wave propagation in a layered structure consisting of an FGPM layer deposited on an elastic substrate [2, 3, 4] or in inhomogeneous half-space [5, 6]. However, no theoretical results have been published on the propagation of Rayleigh surface-waves in a structure that is covered by an inhomogeneous thin film of which the properties vary continuously along the depth. In reference [7], Cao and *al.* deal with dispersion relation of Rayleigh waves in a FGPM half-space by the perturbation technique. In reference [8], it has been shown that Love wave

[1] To whom any correspondence should be addressed.

can propagate in a functionally graded piezoelectric layer/substrate system by Stiffness matrix method, additionally ODE method is used to describe a state space approach which reduces the wave propagation problem to a first order matrix differential equation and which greatly simplifies the formulation of the boundary value problem for multilayer [8, 9].

In this paper, we study the propagation behavior of such Rayleigh surface waves in a functionally graded piezoelectric substrate with exponential variation when the mechanical and electrical constants vary individually in the FGPM substrate. The resolution of the problem passes by the use of the ordinary differential equation (ODE) [9] and the general stiffness matrix method (GSMM) [10, 11].

These methods are used to study multilayer structures utilizing materials of arbitrary anisotropy and piezoelectricity. We here assume that all material properties of the piezoelectric layer may have different exponential function distributions along the depth direction. The dispersion curves for the existence of the Rayleigh surface wave with respect to phase and group velocities are obtained for both the electrically open and short conditions, respectively, from which numerical matrix method are carried out to graphically show the effect of the gradient feature on the dispersion behaviour, and the electromechanical coupling factor, respectively. Results presented in this study considers the plane problem of an anisotropic beam which the dispersion curves in the main menu and shows a Cartesian plot of coupling factor versus the Euler rotation angle θ ranging from 0 to 90 degrees. The results of this study can be useful in engineering applications of the Rayleigh waves.

2. Statement of the problem

Let us consider a transversely isotropic FGPM half-space, as shown in figure 1. The coordinate system (x_1, x_2, x_3) is chosen in such way that the x_1-axis is directed along the poling direction perpendicular to x_1-x_3 plane and the x_3-axis points down into the substrate. The material parameters in the substrate change gradually along the x_3-axis direction.

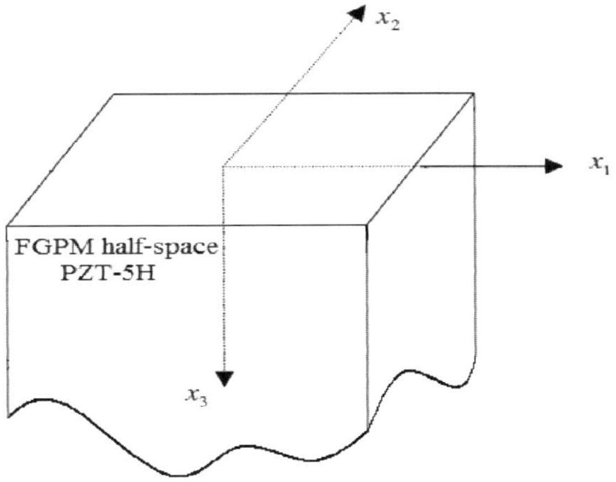

Figure 1. The functionally graded piezoelectric material half space and coordinate system

Let us consider a generally anisotropic medium (Fig.1) and a harmonic wave propagating along the x_1-axis of the form $\xi(x_3)\exp[i(k_1x_1-\omega t)]$; where k_1 is the projection of the wave vector along x_1 the guide direction. The normal stress vector t_{i3} = [t_{13}, t_{23}, t_{33}] and the particle displacement u = [u_1, u_2, u_3] are chosen as the six mechanical variables. For piezoelectric material the electric potential ϕ and the normal electric displacement component D_3, are chosen as two electric variables which must satisfy boundary relationships. This set of independent parameters is sufficient to describe the behavior of the system. The eight-component vector $\xi = [u\ \phi\ t_{i3}\ D_3]^T$ for functionally graded piezoelectric material are utilized to write constitutive equations and motion equation in the form of an ordinary differential equation system in x_3 [9, 10, 11, 12]:

$$\frac{\partial \xi}{\partial x_3} = iA\xi \tag{1}$$

Where A is the fundamental acoustic tensor [9, 10].

The study of wave propagation in inhomogeneous half-space using formalism (1) has also been used for a long time in seismology [9]. The matrix A is a function of material properties, wave number and frequency.

All material parameters in the functionally graded piezoelectric substrate, as shown in Fig. 1, are assumed to have the same exponential function variation along the x_3-axis direction according to:

$$C_{ijkl}(x_3) = C^0_{ijkl}\, e^{\alpha x_3},\ \rho(x_3) = \rho^0\, e^{\alpha x_3},\ e_{ijkl}(x_3) = e^0_{ijkl}\, e^{\alpha x_3},\ \ \varepsilon(x_3) = \varepsilon^0\, e^{\alpha x_3}, \tag{2}$$

where α = 4000 is the exponential factor characterizing the degree of the material gradient in the x_3-direction. The superscript "0" and α = 0 are obvious that corresponds to the homogenous material case. We here assume that all material properties of the piezoelectric layer as Fig. 1 have the same exponential function distribution along the x_3-axis direction. But we specify the influence of electrical and mechanical parameters are different in magnitude.

To describe the propagation behavior of the coupled Rayleigh wave in the above FGPM half-space, the following boundary and continuous conditions must be satisfied at the free surface of the material. It should be pointed out that two kinds of electrical boundary conditions, electrically open and shorted conditions, would be taken into account in this study. It must be reminded that the generalized stress vector $\boldsymbol{T}_j$ includes mechanical stress vector t_{i3} and the normal electrical displacement D_3, in the same way the generalized displacement vector $\boldsymbol{U}_j$ includes the mechanical displacement u_i and the electrical potential ϕ, with j = 1, 2, 3.

At the free surface substrate $x_3 = 0$

$$\text{Electric short case: } t_{i3} = 0 \text{ and } \phi = 0 \tag{3}$$

$$\text{Electric open case: } t_{i3} = 0 \text{ and } \sigma = 0 \tag{4}$$

σ refers to the charge density, it can be deduced from D_3 values in the neighborhood of the free surface, [9, 12] :

$$\sigma = D_3(0) - D_3(0) \tag{5}$$

Up to this stage, the problem of Rayleigh wave propagation in the FGPM half-space becomes the solution of the fundamental acoustic tensor A Eq. (1) under boundary conditions Eqs. (3), (4) and (5). To solve the Eigen value Eq. (1), we define a partial stiffness matrix $\boldsymbol{K}$ associated with a homogeneous sub-layer, which relates the general stress and displacement in plane x_3 [10, 11 12].

3. Solution of the problem

Let us consider a multilayered half-space, consisting of N arbitrarily anisotropic layers as illustrated in Fig. 2. We are dealing with the functionally graded piezoelectric half-space subdivided into N very thin layers, it is also possible to use the stiffness matrix method. This method is useful to relate the general stress and displacement in plane x_3 [10, 11, 12, 13].

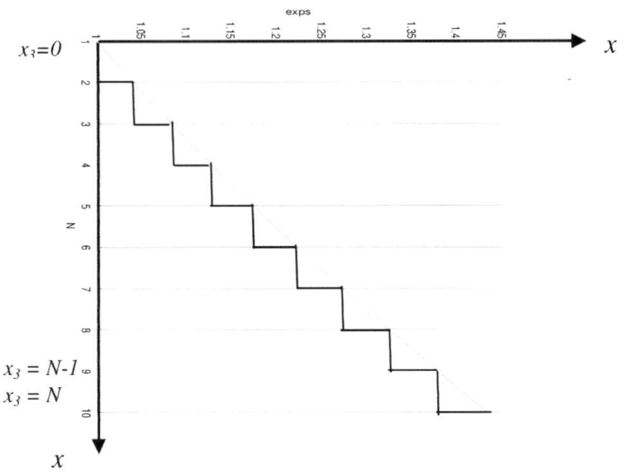

Figure 2. Geometry of a FGPM multilayered half space (PZT-5H).

Applying the recursive operation [12, 13] from the bottom layer in substrate ($x_3 = N$) to the top surface ($x_3 = 0$), therefore the amplitudes in one layer are expressed in terms of the last layer A_N which is fixed to the substrate. The total stiffness matrix relates stress to displacement generalized vectors at the film interfaces $x_3 = N$ and $x_3 = 0$:

$$\begin{bmatrix} \boldsymbol{T}_N \\ \boldsymbol{T}_0 \end{bmatrix} = \boldsymbol{K}^N \begin{bmatrix} \boldsymbol{U}_N \\ \boldsymbol{U}_0 \end{bmatrix} \tag{6}$$

For making matrix developments easier $\boldsymbol{K}^N$ is divided in four blocks: $\boldsymbol{K}_{11}^N, \boldsymbol{K}_{12}^N, \boldsymbol{K}_{21}^N, \boldsymbol{K}_{22}^N$ and the generalized stress and displacement vector at the free film surface can be deduced in term of $\boldsymbol{T}_0$ and $\boldsymbol{U}_0$. Below the results are written according to the A_N last layer amplitude basis. From boundary conditions eight dimensioned vectors $\boldsymbol{T}_0$ and $\boldsymbol{U}_0$ are replaced by $\boldsymbol{T}_N$ and $\boldsymbol{U}_N$ respectively, which are four dimensioned vectors. N sub-script refers to the last layer [12, 13].

$$\boldsymbol{T}_0 = (\boldsymbol{K}_{12}^N - \boldsymbol{K}_{11}^N (\boldsymbol{K}_{21}^N)^{-1} \boldsymbol{K}_{22}^N) \boldsymbol{U}_N + \boldsymbol{K}_{11}^N (\boldsymbol{K}_{21}^N)^{-1} \boldsymbol{T}_N \tag{7}$$

$$\boldsymbol{U}_0 = (\boldsymbol{K}_{21}^N)^{-1} (\boldsymbol{T}_N - \boldsymbol{K}_{22}^N \boldsymbol{U}_N) \tag{8}$$

Starting from the interface next to the bottom boundary, using recursively equations (7) and (8), one obtains the state vector at all interfaces. Similarly for the state vectors inside a layered system on a substrate, the surface stiffness matrix can be calculated once the state vectors at the top surface have been obtained [13].

At this level, various physical amounts can be deduced according to A_N basis, mainly stress vector t_{i3}, electrical potential ϕ and charge density σ [12, 13]. Either in the electrically shorted or open conditions, (4×4) characteristic matrix M_{sc} and M_{oc} can be constructed:

$$\begin{bmatrix} t_{i3} \\ \sigma \end{bmatrix} = M_{oc}[A_N], \tag{9}$$

$$\begin{bmatrix} t_{i3} \\ \phi \end{bmatrix} = M_{sc}[A_N] \tag{10}$$

For both cases Rayleigh velocity is the zero of $M_{oc,sc}$ matrices.

4. Numerical results

In order to demonstrate the influences of the gradient on the phase velocity V_{ph} and the coupled electromechanical factor K^2, some simulations are proposed. And some interesting phenomena are mentioned in the discussion of the numerical results. As a sample, the set of FGPM substrate system is assumed, i.e. PZT-5H. The material constants for the piezoelectric substrate are listed in Table 1.

Table 1. Elastic and electrical properties of homogeneous piezoelectric film (PZT-5H) [8]

$C_{11}=C_{22}$(GPa)	C_{12}(GPa)	$C_{13}=C_{23}$(GPa)	C_{33}(GPa)	$C_{44}=C_{55}$(GPa)
151	98	96	124	23
$e_{31}=e_{32}$ (C/m^2)	e_{33} (C/m^2)	$e_{15}=e_{24}$(C/m^2)	ε_{11}(C/Vm)	ε_{33}(C/Vm)
-5.1	27	17	15x10^{-9}	13.27x10^{-9}

The computational procedure is based both on the generalized Stiffness matrix methods and the ordinary differential equation. For the calculation of the phase velocity V_{ph} and the penetration behaviour we have used elastic and electrical properties of piezoelectric substrate (cf. Tables 1). Since the electrical and mechanical substrate characteristics are function of x_3, i.e. $C_{ijkl}=C_{ijkl}(x_3)$, $e_{ijk}=e_{ijk}(x_3)$ and $\varepsilon_{ij}=\varepsilon_{ij}(x_3)$, the substrate is divided into N=10 layers with equal thickness and different characteristics. Each layer is considered as homogenous and the ODE method can be applied [9]. The calculation of surface wave velocity is performed according to the standard procedure [10, 11], on the basis of GSMM and continuity criterion [10, 11, 12, 13]. The dispersive behavior has been completed by an investigation of the group velocity labeled V_g, which describes the rate at which energy is transported. V_g is computed on the basis of Eq. (11), where k denotes the guide component of the wave vector [14]:

$$V_g = V_{ph} + k(dV_{ph} / dk) \tag{11}$$

4.1. Effect of gradient variation on the phase velocity

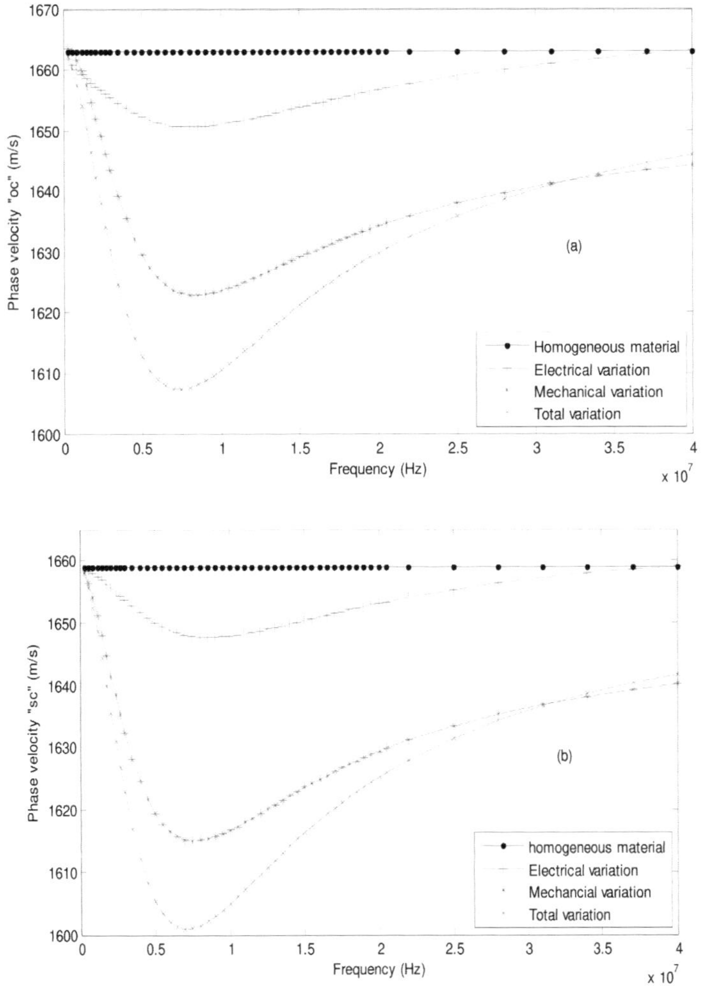

Figure 3. Phase velocity of Rayleigh waves on PZT-5H x_1-x_3 plane in "oc" (a) and "sc" (b) cases

Based on Eqs. (9) – (10), the dispersive curves for the electric open and short cases are plotted when electrical and mechanical gradient variations are applied separately and altogether, as shown in Figure It can be seen from figure 3 (a) and (b) that there is only one mode, which is the same as Rayleigh waves propagating at the free surface of the piezoelectric substrate. However the mode is changed into a dispersive one and the gradient coefficient has an obvious effect on the phase velocity at the beginning part of the mode. It can be seen that the phase velocity starts at a higher value and tends to the velocity of Rayleigh waves monotonously with an increase of the frequency. The phase velocity decreases sharply and attains the minimum when the frequency *f*, is about 7 MHz both in "oc" and "sc" cases (Fig.3a, b).

4.2. Effect of gradient variation on the coupled electromechanical factor

The coupled electromechanical factor is an important parameter for designing acoustic sensors. And the coupled electromechanical factor is defined as [15]

$$K^2 = 2(V_{ph(oc)} - V_{ph(sc)})/V_{ph(oc)} \quad (12)$$

where $V_{ph(oc)}$ and $V_{ph(sc)}$ are phase velocities, respectively, for the electric open and short cases.
Figure 4 illustrates the coupled electromechanical factors of the surface wave that propagates along the FGPM half-space. The effect of the gradient coefficient on the electromechanical coupling factor shows Fig. 4. It can be seen readily that the gradient coefficient has a strong influence on the electromechanical coupling factor, which appears more obvious with the mechanical gradient variation are applied (K^2 = 1,221% in f = 2,9MHz). If the electrical properties vary along the depth, the electromechanical factor decreases sharply and attains the minimum when the frequency, is about 2,4MHz (K^2 = 0,223%). If mechanical properties vary along the depth, the coupled electromechanical factor is larger than that in homogenous piezoelectric half-space and when the electrical properties vary along the depth.

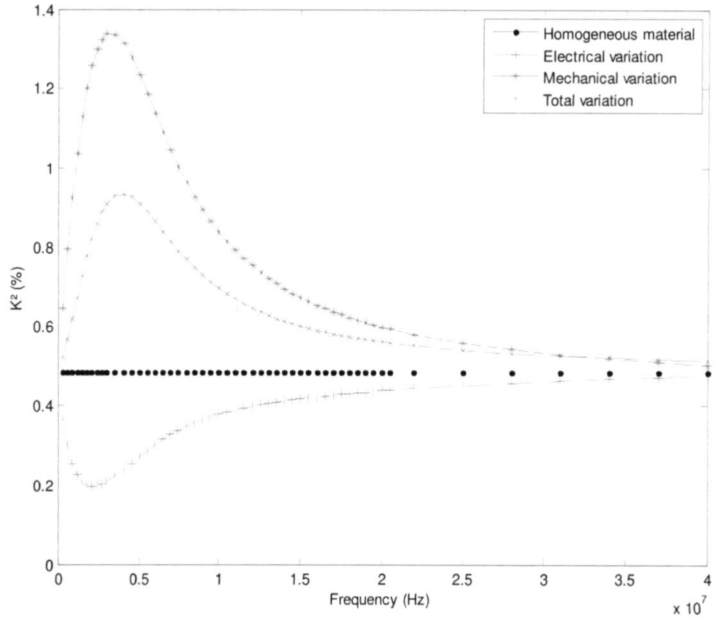

Figure 4. Variation of the electromechanical coupling factor K^2 vs. frequency for different electrical and mechanical gradient variation

4.3.Effect of gradient variation on the group velocity
Fig.5 reproduces how group velocity V_g varies with respect to frequency for "oc" (a) and "sc" (b) electrical polarizations. It is seen that for the first mode, the energy velocity is approximately the same for "oc" and "sc" electrical polarizations. The effects are plotted for the first mode when electrical and mechanical gradient variations are applied separately and altogether. It is seen that the waves exhibit normal dispersion because $dV_{ph}/dk < 0$. Over 16.5MHz, it is seen that both for the electrically open and shorted cases, the velocity of energy propagation is larger when the electrical properties vary along the depth. Less than 16.5MHz, V_g is decreased in magnitude then that when the material is ungraded.

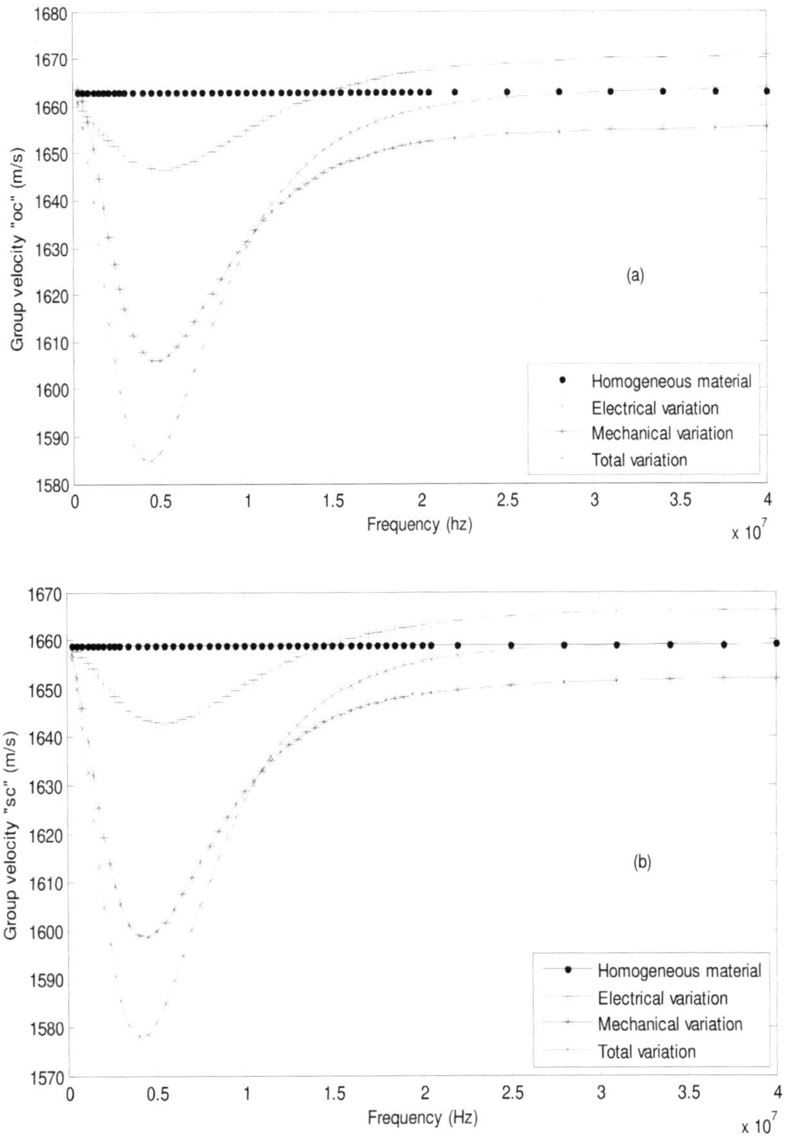

Figure 5. Group velocity of Rayleigh waves on PZT-5H x_1-x_3 plane in "oc" (a) and "sc" (b) cases

4.4. Effect of gradient variation when K^2 vary as function θ

To determine the effect of gradient variation of FGPM substrate, we have study the variation of the electromechanical coupling coefficient as function of the rotation angle of the six fold axis with respect to the normal to the free surface of the sample (x_3 axis in Fig. 1). Rayleigh velocities for different orientations of the six fold axis have been performed on the basis of preceding methods. Fig. 6 shows the variation of the electromechanical coupling coefficient in term of the rotation angle for a fixed value of frequency (5MHz). When $\theta = 90°$, the coupling factor is high when sagittal plane is in the vicinity of the symmetry axis. Particularly, it reaches a maximum of 1.265% for an angle close to 25° when the mechanical gradient varied. It means that the piezoelectricity phenomena are important when the surface acoustic wave propagate along the 6 axis. When the wave propagates perpendicular to x_3 axis (θ = 90°), the coupling factor decreases significantly and the piezoelectricity becomes inactive.

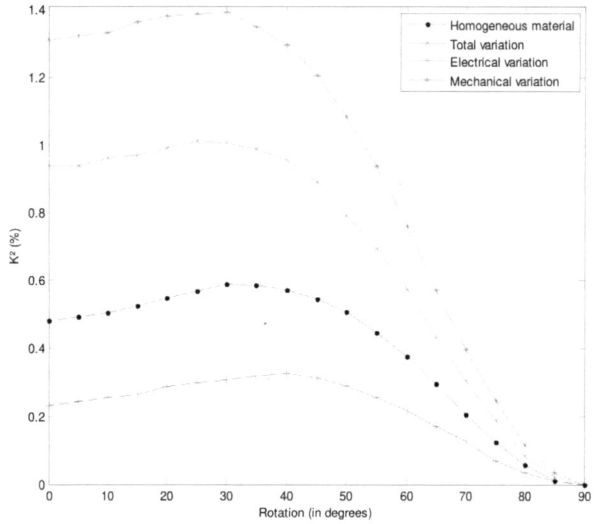

Figure 6. Variation of the electromechanical coupling factor as a function of the angle of rotation

5.Conclusions

Rayleigh surface waves can propagate at the free surface of the functionally graded piezoelectric substrate investigated in this paper by a numerical matrix method. The dispersion curves and the electromechanical coupling factor for the existence of the Rayleigh surface waves with respect to frequency are obtained for both electrically open and short conditions by means of the Stiffness matrix method. The effects of the gradient variation of material constants on the phase velocity and distribution of displacement, and electric potential are presented and discussed in detail. The electrical and mechanical gradients associated with FGPM substrate, are analyzed with respect to frequency. The obtained electromechanical coupling factor (K^2) is more affected by the mechanical gradient. The electrical gradient seems to have less influence in comparison with the mechanical gradient. The variation of K^2 shows that the conversion of mechanical energy into electrical energy is important when the surface wave propagates along the x_3 direction. The numerical method and the results can be useful for the design of the resonators and sensors.

References

[1] Qiu J, Tani J, Ueno T, Morita T, Takahashi H and Du H 2003 *Smart Mater. Struct.* **12** 115–121.
[2] Li X. Y, Wang Z. K and Huang S. H 2004 *Int. J. Solids Struct.* **41** 7309–28.
[3] Du J. K, Jin X. Y, Wang J and Xian K 2007 *Ultrasonics* **46** 13–22.
[4] Qian Z, Jin F, Wang, Z, Kishimoto K 2007 *Inter. J. of Eng. Sci.* **45** 455–466.
[5] Zheng-Hua Qian, Feng Jin, Tianjian Lu and Kikuo Kishimoto 2008 *Smart Mater. Struct.* **17** 7.
[6] Eskandari M, Shodja H. M 2008 *J Sound Vib* **313** 195–204.
[7] Cao X, Jin F, Wang Z 2008 *Acta Mechanica* **200** 247-261
[8] Ben Salah I, Wali Y, Ben Ghozlen M. H 2011 *Ultrasonics* **51** 310-316.
[9] Adler E. L 1994 *IEEE Trans. Ulfrason. Ferroelec. Freq. Contr.* **41** 876-882.
[10] Rokhlin S. I, Wang V 2004 *IEEE Trans. On Ultras., Ferro. and Freq. Cont.* **51** 1060-1071.
[11] Rokhlin S. I, Wang L 2004 *IEEE Trans. On Ultras., Ferro. and Freq. Cont.* **51** 453-463.
[12] Rokhlin S. I, Wang L 2002 *J. Acoust. Soc. Am.* **112** 822-834.

[13] Rokhlin S. I, Wang L 2002 *Int. J. Solids Struct.* **39** 5529–5545.
[14] Karl F. G 1975 Wave Motion in Elastic Solids (Oxford: Clarendon).
[15] Bernhard J, Michael J. V 1997 *Smart Materials and Structures* **6** 668–679.

IOP Conf. Series: Materials Science and Engineering **28** (2012) 012048 doi:10.1088/1757-899X/28/1/012048

Magnetic and structural properties of nanocrystalline $PrCo_3$

K. Younsi[1], L. Bessais, V. Russier, J. C. Crivello
ICMPE, UMR 7182 CNRS-Université Paris Est, 2-8 Rue Henri Dunant F-94320 Thiais, France

E-mail: younsi@icpme.cnrs.fr

Abstract. Structural and magnetic properties of nanocrystalline $PrCo_3$ obtained from high energy milling technique were presented. These properties were studied by means of X-ray diffraction (XRD), Curie temperature determination and magnetic properties measurements. The coercive field of 55 kOe measured at 10 K was obtained after annealing at 1023K for 30 mn and no saturation reached at room temperature for applied field of 90 kOe.
We have completed this experimental study by micromagnetic modeling in order to get a qualitative picture of the microstructure effect on the macroscopic magnetization curve. Density functional theory (DFT) calculations have been performed in order to obtain the electronic structure of $PrCo_x$ compounds, where x = 2, 3 and 5

1. Introduction

High performance magnetic materials are in general based on outstanding intrinsic magnetic properties and appropriate microstructure. Nanocrystalline compounds have been extensively investigated, the aim is to find the appropriate nanocrystalline state for a highest coercive field [1, 2]. A set of rare earth transition intermetallic compounds, which have quite favorable intrinsic properties as permanent magnet, have been discovered in recent years. Among nanocrystalline magnets based on rare earth and transition metal compounds, the RCo_3 compounds (R as a rare earth) have been the subject of many studies. These ones crystallize in rhombohedral $PuNi_3$-type (space group: *R-3m*) and present excellent magnets properties such as large magnetocrystalline anisotropy and important saturation magnetization. These properties are resulting from the combination of the 3*d* itinerant magnetism on Co sublattice and 4*f* localized magnetism of Pr sublattice.

In this study, we report on the elaboration of nanocrystalline $PrCo_3$ by high-energy milling, and their structural and magnetic properties in amorphous state and after recrystallization. In addition to the experiments, results from simulation and microscopic calculations are presented.

2. Experimental details

The alloys were initially prepared by co-melting the elemental metals in an induction furnace under an atmosphere of high purity Ar. The raw materials were used in the form of ingots with purities of 99.9% for Pr and for Co.

These ingots were used as pre-alloys to manufacture samples by high-energy milling. Technique and conditions of milling has been described in [3]. After milling, the as-milled amorphous powders

[1] To whom any correspondence should be addressed.

were annealed for 30 min at temperature between 1023 K and 1323 K followed by quenching in water. The crystal structure was deduced from X-ray diffraction (XRD) patterns, the grain size of the sample was determined using Rietveld method by the Scherer formula [4]. The *M−H* curves were plotted at T= 10 and 293 K with a maximum applied field of 90 kOe on sample in epoxy resin.

3. Experimental Results

Figure.1 shows, as an example, the XRD diagrams of the of $PrCo_3$ ingot using the Rietveld analysis. The results of the structure refinement performed for the alloys show that the compound is single phase with the rhombohedral $PuNi_3$-type structure (space group: $R\bar{3}m$). The Rietveld analysis leads to the unit cell parameters a= 5.069 Å and c= 24.795 Å. This ingot is an appropriate precursor for using as pre-alloys to manufacture samples by high-energy milling. In order to check the crystalline structure of $PrCo_3$ powders, the as-milled samples have been annealed in a temperature range of 1023 K to 1273 K for 30 mn.

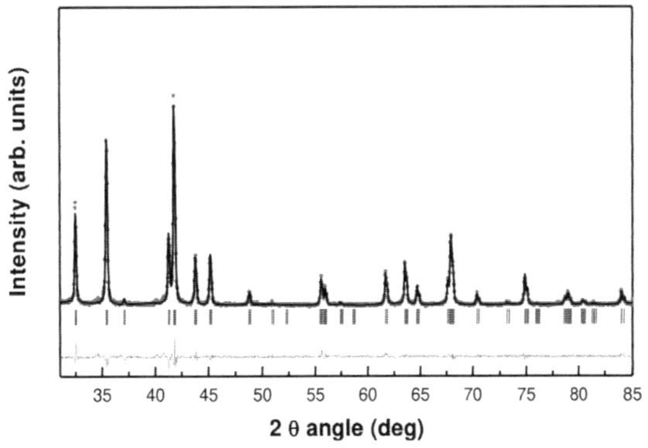

Figure 1. Rietveld analysis for $PrCo_3$ ingot.

The XRD measurement of samples annealed at distinct temperatures has been performed. The as-quenched powder is essentially quasiamorphous and the diffraction peaks become more pronounced with increasing annealing temperature. The annealed samples at 773 and 1273 K for 30 min reveal the presence of $PrCo_3$ phases crystallized. The mean diffraction crystallite size of the powder after annealing at 1023 and 1273 K for 30 min obtained by fitting the XRD diagram with Rietveld method, ranges between 31 and 45 nm respectively, this implies that synthesized materials are nonocrystalline.

The Curie temperature of $PrCo_3$ compound is 349 K, [3] this result is in agreement with the data given by Lemaire [5]. Isothermal magnetization curves *M−H* obtained at 293 K have shown that at 90kOe the saturation is not reached, which gives evidence for a magnetocrystalline anisotropy among the highest known in those kinds of systems. The highest coercivity was obtained after treatment at 1023 K for 30 min, we have obtained, H_C=55kOe and M_R=56 emu/g, inherent to nanocrystalline state. The highest coercivity confirms the large magnetocrystalline anisotropy and suggests that nanocrystalline $PrCo_3$ are interesting candidates in the field of permanent magnets.

4. Computational studies

4.1 Micromagnetic simulation

In order to get a picture of the link between the structure at the nanometer scale and the extrinsic magnetic properties we have performed a micromagnetic simulation on a model for the system. This model is treated in the framework of micromagnetism, *i.e.* through a continuous medium type of approach, and the calculations are performed by using the MAGPAR [6] code based on the finite elements method.

The picture we have in mind is that the crystallization takes place locally on dispersed seeds in the

IOP Conf. Series: Materials Science and Engineering **28** (2012) 012048 doi:10.1088/1757-899X/28/1/012048

amorphous matrix representing the as-milled sample. Hence we consider an assembly of crystallites presenting a uniaxial anisotropy comparable to what is expected in hard magnetic R-TM compounds. The saturation magnetization, J_s of the crystallites is that of the bulk $PrCo_3$, namely J_s=0.723T. We consider the simplest realization of the model, by using identical crystallites, located on the nodes of a simple cubic lattice. The crystallites are either spherical or cubic; however, the later case is likely to be closer to the experimental sample. We have chosen somewhat arbitrarily a volumic fraction occupied by the crystallites of $\varphi = 0.73$ (in the case of cubic crystallites) and the crystallite size corresponds to what has been estimated experimentally, namely, a cubic edge a = 35 *nm*. The amorphous character of the matrix is translated by a zero valued anisotropy constant, while its saturation magnetization $J_S^{(mat)}$ is a free parameter of the model. The value we get for $J_S^{(mat)}$, namely $J_S/5$ seems reasonable in our picture of an amorphous matrix. The important point is that we clearly get the signature of a two uncoupled phases behavior for $M(H)$ (Fig.2); the amplitude of the kink at zero field being determined by φ and $J_S^{(mat)}$.

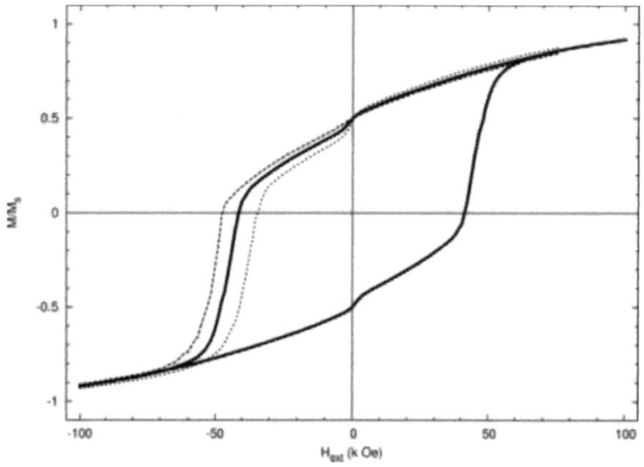

Figure 2. Calculated de-magnetization curve in terms of the external field for the systems of cubic inclusions with $\varphi = 0.73$. Randomly distributed axes and matrix saturation magnetization J_s= 0.05T, dashed line; J_s = 0.125T, solid line and J_s = 0.25T, dotted line.

4.2. DFT calculations

Band structure spin polarized calculations were carried out by using the DFT scheme implemented in the projector-augmented wave (PAW) method, performed with the VASP package [7,8]. In the present study, the generalized gradient approximation (GGA) is used for the exchange and correlation energy functionals with the Perdew-Burke-Ernzerhof (PBE) [9] functional. A plane wave basis set with a cutoff energy of 600 eV was used in all calculations converged within 0.1 eV in total energy. Brillouin zone sampling using a mesh of 0.5 k points for each primitive cell of $PrCo_x$ compounds, described respectively in *Fd-3m* space group for x=2, *R-3m* for x=3 and $P6_3$/mmm for x=5 compounds. Structural optimization were carried out under the condition that residual forces should be smaller than 0.1 eV.Å^{-1}.

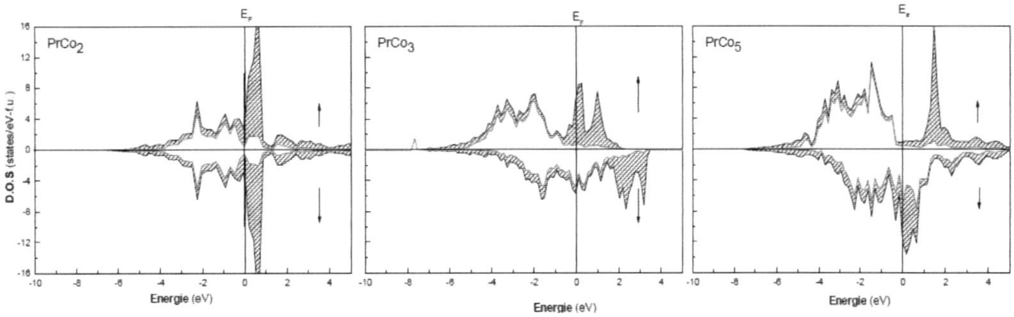

Figure 3. Total density of states (shaded region) of $PrCo_x$ (x=2, 3 and 5). The white area is corresponding to the only 3d-states of Co. The origin of the energy scale is located at the Fermi energy E_F

Figure.3 shows the density of states (DOS) for the $PrCo_x$ (x=2, 3 and 5) in the two directions of spin. The DOS show the majority contribution of orbital 3d-Co on the filled bands, with a small electronic charge transfer of Pr to Co. The increase of x leads to a distinct shift of the Co-3d states in spin majority and minority DOS as expected in the rigid-band Stoner theory. Whereas $PrCo_2$ shows the paramagnetic state as the most stable in energy, the ferromagnetic behavior is confirmed for x=3 and x=5 compounds. The magnetic anisotropy has been confirmed for $PrCo_3$ compounds, since non-collinear spin calculation along the z axis has been found $\sim 0.2\,\text{eV}\,\text{f.u}^{-1}$ more stable than calculation along x axis direction. Since the DOS at Fermi level is high enough for x=3 and 5, the itinerant magnetism properties of theses compounds can be explained by intra-atomic 3d–3d exchange. The $PrCo_x$ compounds present tiny and high density structures above the Fermi level, corresponding to localized 4f states of Pr.

5. Conclusions

The $PrCo_3$ intermetallic compound presents very high uniaxial magnetocrystalline anisotropy. The coercivity of this compound, prepared by high energy milling followed by annealing (1023 K, 30 mn), takes a very high value, up to 55 kOe at 10 K. A micromagnetic simulation and atomic scale *ab-initio* calculations have been performed to understand on the one hand the way in which the electronic structure and especially the interplay between 3d and 4f electronic states, leads to the observed anisotropy and magnetic moment, and on the other hand the relation between the nanoscale structure and the extrinsic magnetic properties.

References

[1] Bessais L and Djega-Mariadassou C 2001 *Phys. Rev. B* **63** 054412.
[2] Bessais L, Sab S, Djega-Mariadassou C and Grenèche J. M 2002 *Phys. Rev. B* **66** 054430.
[3] Younsi K, Russier V and Bessais L 2010 *J. Appl. Phys.* **107** 083916.
[4] Rietveld H 1967 *Acta Crystallogr.* **22** 151.
[5] Lemaire R 1966 *Cobalt* **33** 201.
[6] Scholz W, Schrefl T, Suess D, Dittrich R, Forster H, and Tsiantos V 2003 *Comput. Mater. Sci.*, **28** 366.
[7] Kresse G and Furthmüller J 1996 *Phys. Rev. B*, **54**(16) 11169-11186.
[8] Kresse G and Joubert D 1999 *Phys. Rev. B* **59**(3) 1758 .
[9] Perdew John P, Burke K and Ernzerhof M 1997 *Phys. Rev. Lett.* **78** (7) 1396.

Magnetic and magnetocaloric properties of monovalent substituted $La_{0.5}Ca_{0.45}A_{0.05}MnO_3$ (A=Na, Ag, K) perovskite manganites

A. Mehri [a], W. Cheikh-Rouhou Koubaa [a1], M. Koubaa [a], A. Cheikh-Rouhou [a, b]
[a] Laboratoire de Physique des Matériaux, Faculté des sciences de Sfax, Université de Sfax, B.P.1171, 3000 Sfax, Tunisia.
[b] Institute NEEL, B.P.166, 38042 Grenoble Cedex 9, France.

E-mail: wissem.koubaa@yahoo.fr

Abstract. The effects of monovalent doping in the A-site on the structural, magnetic and magnetocaloric properties in $La_{0.5}Ca_{0.45}A_{0.05}MnO_3$ (A=Na, Ag, K) powder perovskite manganites have been investigated. Our compounds have been synthesized using the conventional solid state reaction at high temperature. Rietveled refinements of X-ray diffraction patterns at room temperature show that all our powder samples are single phase and crystallize in the orthorhombic structure with *Pbnm* space group. Magnetization measurements versus temperature in a magnetic applied field of 0.05T indicate that our investigated samples display a paramagnetic- ferromagnetic transition with decreasing temperature. Moreover, in $La_{0.5}Ca_{0.45}Ag_{0.05}MnO_3$ compound, a second transition toward an antiferromagnetic state is still present around 150K. The Curie temperatures T_C are found to be 270K, 238K and 240K for A= Na, Ag and K respectively. From magnetization measurements versus magnetic applied field we have determined the magnetic entropy change as well as the relative cooling power RCP for all our samples. The maximum magnetic entropy change $\left|\Delta S_M^{Max}\right|$, in a magnetic field change of 5T is found to be 1.02, 2.43 and 1.85Jkg/K for A= Na, Ag and K respectively. The $La_{0.5}Ca_{0.45}Ag_{0.05}MnO_3$ sample exhibits the largest RCP value of 217.32 J/kg under 5T.

1. Introduction

Magnetic refrigeration technology based on the magnetocaloric effect shows advantages over traditional gas compression technology [1] and may be considered as a promising alternative to replace the conventional technique. Magnetic entropy change is known to achieve relatively high values near the para-to ferromagnetic transition [2-6]. The magnitude of the magnetic entropy change is affected by the microstructure of the material. The technical parameter called the relative cooling power (RCP) also depends on the width of the transition temperature interval [1] and determines potential applications. As the magnetocaloric properties of manganites are known to present a limited extend, a search for new refrigerant materials and new preparation methods leading to the stronger magnetocaloric effect, is still desired.

[1] To whom any correspondence should be addressed.

The substitution of trivalent element by monovalent alkali-metal in manganites induces less inhomogeneity (the substitution oxidizes two Mn ions from Mn^{3+} to Mn^{4+} state) and improves the physical properties [7-9].

The $La_{1-x}Ca_xMnO_3$ system has a rich phase diagram [10], where the doping level x and the temperature determine the presence of paramagnetism, ferromagnetism, antiferromagnetism and orbital and charge ordering. Particularly, $La_{0.5}Ca_{0.5}MnO_3$ stands as a boundary between a low temperature ground state (x<0.5) and at low temperature ferromagnetic metallic ground state (x>0.5). Close to x≈0.5 both ferromagnetism and antiferromagnetism have been observed. As the temperature decreases, it first undergoes a ferromagnetic transition at $T_C \approx$ 225K and then a charge-order antiferromagnetic transition around 155K (T_{CO}) [11]. Huy Sinh et al. [12] have widely studied the magnetocaloric properties of $La_{1-x}Ca_xMnO_3$ system where a partial calcium substitution by monovalent element is found to improve significantly the T_C value which reaches room temperature.

The authors studied the magnetocaloric properties in $La_{1-x}Ca_xMnO_3$ system and found that for x = 0.5, $|\Delta S_M^{Max}|$ reaches 3.25 J/kgK upon a magnetic applied field change of 5T [12]. Mehri et al. [13] investigated the MCE properties of $La_{0.5}Ca_{0.5-x}Na_xMnO_3$ (0.05≤x≤0.2) and found that with increasing $\langle r_A \rangle$, $|\Delta S_M^{Max}|$ increases from 1.02 J/kgK at 260 K for x=0.05 to 2.24 J/kgK at 231K for x = 0.2 under ΔH = 5T. In the present work, we synthesized $La_{0.5}Ca_{0.5-x}A_xMnO_3$ (A= Ag, K and Na) powder perovskite manganites by the conventional solid state reaction at high temperature and studied the crystallographic, magnetic and magnetocaloric effects due to 5% substitution of calcium by monovalent element.

2. Experimental section

Powder samples of $La_{0.5}Ca_{0.45}A_{0.05}MnO_3$ (A=Na, Ag, K) have been synthesized using the standard solid state reaction method at high temperature. The starting materials were intimately mixed in an agate mortar and then heated in air up to 900°C for 60h. The obtained powders were then pressed into pellets (of about 1 mm thickness and 13 mm diameter) and sintered at 1000°C in air 60h with intermediate regrinding and repelling. Finally, these pellets were rapidly quenched to room temperature in air in order to freeze the structure at the annealed temperature. Phase purity, homogeneity and cell dimensions were determined by powder X-ray diffraction at room temperature. As our samples have been synthesized in air, they are consequently stoichiometric in oxygen [14].

Structural analysis was carried out using the standard Rielveld technique [15, 16]. The amount of Mn^{4+} ions has been quantitatively checked by iodometric titration. Magnetization measurements versus temperature in the range 20 - 300K and versus magnetic applied field up to 8T were carried out using a vibrating sample magnetometer. MCE were deduced from the magnetization measurements versus magnetic applied field up to 8T at several temperatures.

3. Results and discussion

The X-Ray diffraction (XRD) spectra of our synthesized samples $La_{0.5}Ca_{0.5-x}A_xMnO_3$ (A= Ag, K and Na) have been reorded at room temperature, as shown in Figure 1. XRD patterns can be indexed n the orthorhombic system with *Pbnm* space group. As expected for silver substitution, $La_{0.5}Ca_{0.5-x}Ag_xMnO_3$ undergoes a phase mixture composed of both manganite and residual Ag metal (Fig1-c) [17-19].

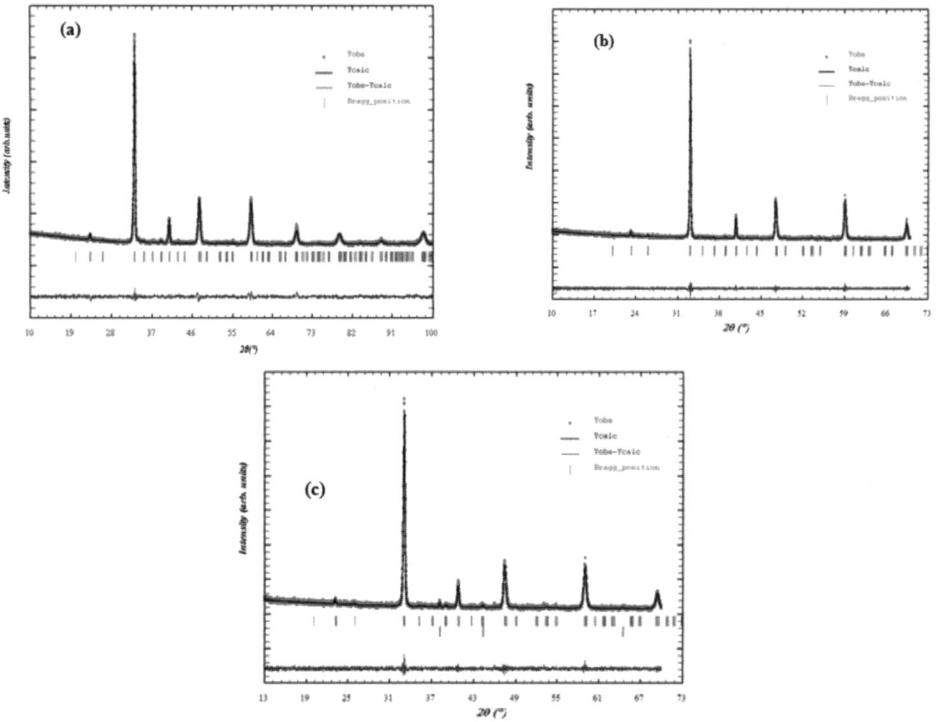

Figure 1. XRD patterns and refinement for $La_{0.5}Ca_{0.45}A_{0.05}MnO_3$ samples: (a) A = Na, (b) A = K and (c) A = Ag.

According to Tang et al. [20], this system characterized with both phases will provide interesting physical properties such as large magnetoresistance effect near room temperature.

In our compounds, the structural parameters were refined using the standard Rietveld technique based on the FULLPROF code. In the profile refinement, we began with scale and background parameters followed by the unit cell parameters. We introduced then the peak asymmetry and the preferred orientation. Finally, we treated the positional parameters and the individual isotropic parameters. The quality of refinement was evaluated through the goodness of the fit indicator χ^2. Detailed results of structural refinement as well as the average size of A-site cations ($<r_A>$) and the mismatch effect (σ^2) are listed in Table 1.

Table 1. Refined structural parameters of $La_{0.5}Ca_{0.45}A_{0.05}MnO_3$ samples at room temperature

	$La_{0.5}Ca_{0.5}MnO_3$ [18]	A = Na	A = Ag	A = K
Space group	Pnma	Pbnm	Pbnm	Pbnm
a(Å)	5.4184	5.4338	5.4371	5.4403
b(Å)	7.6418	5.4090	5.4178	5.4184
c(Å)	5.4281	7.6272	7.6427	7.6406
V(Å^3)	224.76	224.200	225.137	225.23
χ^2 (%)	2.2	1.10	1.34	1.15
$<r_A>$ (Å)	1.278	1.201	1.275	1.293
σ^2(Å^2)	0.38410^{-3}	0.38710^{-3}	0.36410^{-3}	0.99810^{-3}

Magnetization measurements recorded as a function of temperature in the range 20-300K in a magnetic applied field of 50mT show that the charge ordering is present only in $La_{0.5}Ca_{0.45}Ag_{0.05}MnO_3$ compound whereas for A = Na and K this behaviour is suppressed and we observe only a ferromagnetic transition at 270K and 240K respectively for A = Na and K (Fig.2). It should be noted that in the case of Ag substitution, we have performed iodometric titration to check the amount of Mn^{4+} ions, which is found to be 51.7% instead of 55%. This value, lower than that expected, explains the magnetic behaviour.

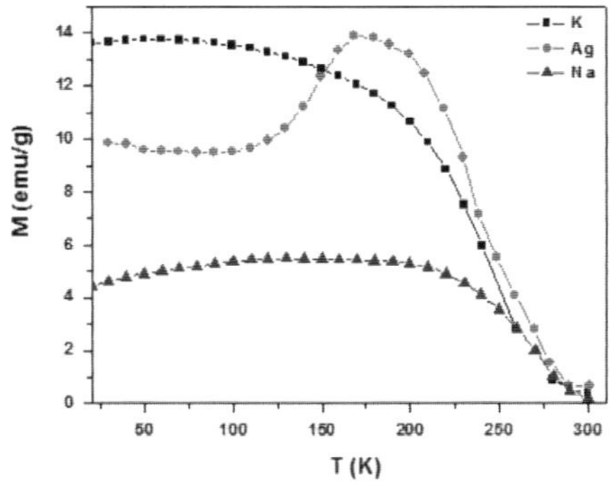

Figure 2. M (T) at 50mT for $La_{0.5}Ca_{0.45}A_{0.05}MnO_3$ samples (A = Ag, K and Na).

In order to study the magnetic behavior of samples at low temperatures, we have performed magnetization measurements versus magnetic applied field up to 8T at several temperatures. The M (H) curves for all samples are plotted in Figure 3. The magnetization evolution versus magnetic applied field for samples corresponding to A = K and Na show that below T_C, the magnetization M increases sharply with magnetic applied field for H<0.5T and then saturates above 1T. While for A=Ag the data recorded at low temperatures (Fig.3-c) show that the sample undergoes an AFM to FM transition at a certain critical field. The critical field for these transitions depends on the temperature as well as the direction of the field change. At low field the small increase in magnetization may reflect the presence of a ferromagnetic region. With the applied field increasing up to the critical field, the spins in one layer flip in unison toward the direction of the applied field, resulting in a ferromagnetic order. It is noteworthy that a relatively weak magnetic field induces a transition from an AFM state [20-170k] to a FM one [180-300k], suggesting that a large entropy can be induced by a weak field.

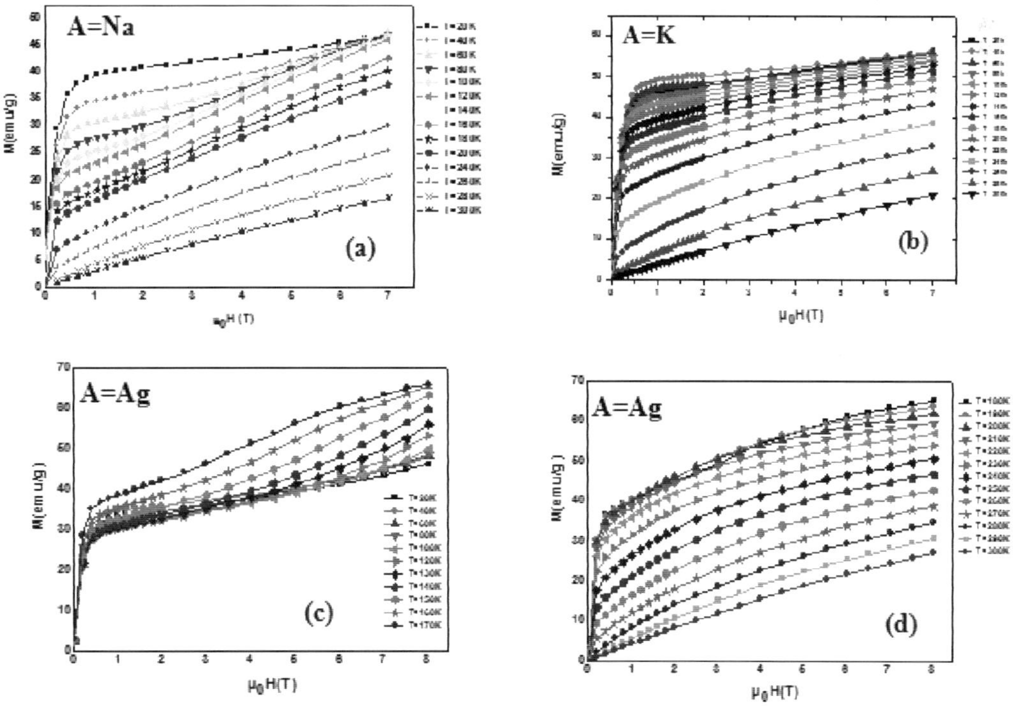

Figure 3. M (H) at several temperatures for $La_{0.5}Ca_{0.45}A_{0.05}MnO_3$ (a): A = Na, (b): A = K, (c): A = Ag at T ≤ 170K and (d): A = Ag at T ≥ 180K.

The magnetic entropy change, $\left|\Delta S_M\right|$ can be evaluated using the following equation $\left|\Delta S_M\right| = \sum_i \frac{M_i - M_{i+1}}{T_{i+1} - T_i} \Delta H_i$ where M_i and M_{i+1} are the experimental values of magnetization measured at temperatures T_i and T_{i+1} respectively, under magnetic applied field H_i [21].

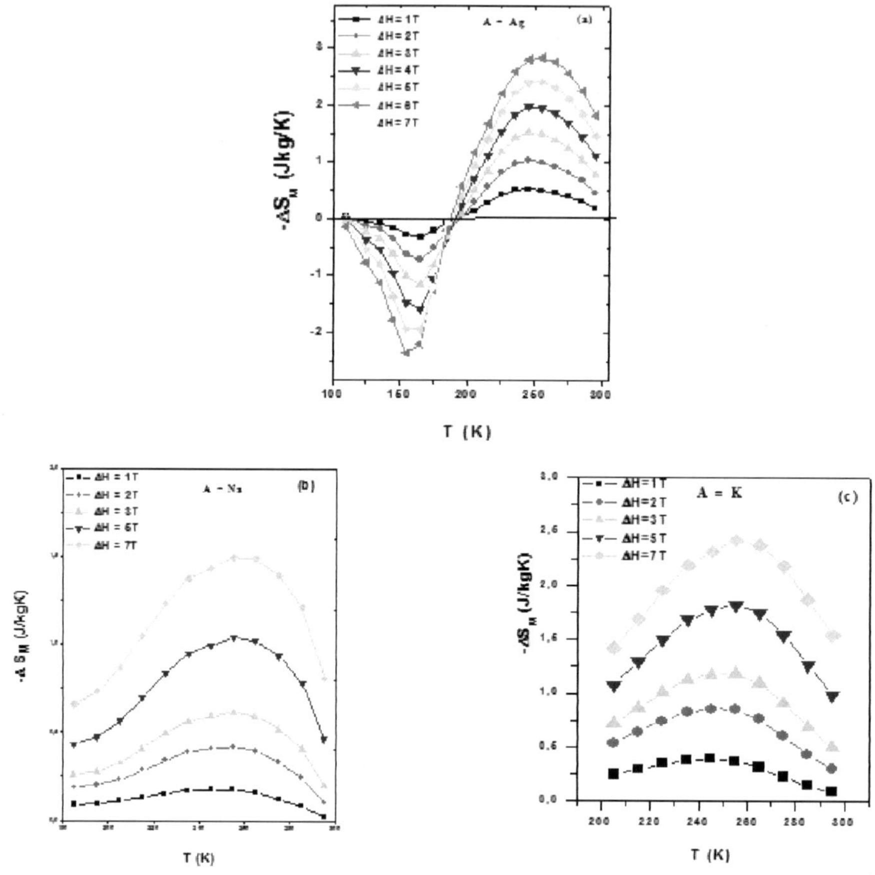

Figure 4. $\left|\Delta S_M(T)\right|$ at several magnetic applied field for $La_{0.5}Ca_{0.45}A_{0.05}MnO_3$ (a): A = Ag, (b): A = Na and (c): A = K.

Figure 4 shows the temperature dependence of the magnetic entropy change, $\left|\Delta S_M\right|$ of $La_{0.5}Ca_{0.45}A_{0.05}MnO_3$ samples (A=Ag, K, Na). For $La_{0.5}Ca_{0.45}Ag_{0.05}MnO_3$ compound still characterized by charge-order the MCE curve exhibits two peaks corresponding to T_N and Tc, the magnetocaloric effect (MCE) is positive before 190 K and becomes negative beyond that. Whereas for the compound presenting a ferromagnetic behavior, one observes only a negative MCE effect with an optimal value in the vicinity of the ambient temperature. All our samples exhibit large MCE effect around the magnetic transition temperature and the magnitude of $\left|\Delta S_M\right|$ increases with increasing the magnetic applied field. The maximum magnetic entropy change observed, $\left|\Delta S_M^{Max}\right|$ is found to be 2.43 $Jkg^{-1}K^{-1}$, 1.02 $Jkg^{-1}K^{-1}$, 1.85$Jkg^{-1}K^{-1}$ in a magnetic field change of 5T for A= Ag, Na and K respectively. The relative cooling power (RCP) is evaluated as $RCP = -\Delta S_M(T,H) \times \delta T_{FWHM}$ where δT_{FWHM} is the full-width at half-maximum of $|\Delta S_M|$ versus temperature [22]. For our samples, the RCP values are respectively 217.32 J/kg, 99.78 J/kg and 181.01 J/kg at 5T for A= Ag, Na and K (about 53.2%, 24% and 43.95% of that of pure Gd).

5. Conclusion

We investigated the effect of monovalent doping on the physical properties of $La_{0.5}Ca_{0.45}A_{0.05}MnO_3$ (A = Ag, K, Na) powder samples. The structural study shows that all our synthesized samples crystallize in the orthorhombic structure with *Pbnm* space group. 5% of partial substitution of Ca by Na and K in $La_{0.5}Ca_{0.5}MnO_3$ sample induces a paramagnetic to ferromagnetic transition near room temperature and the antiferromagnetic interaction linked to charge ordering is suppressed while for Ag compound the Charge Ordering is still present. A large magnetocaloric effect around room temperature is observed in our samples. $La_{0.5}Ca_{0.45}Ag_{0.05}MnO_3$ exhibits the highest RCP value of 217.32 J/kg in a magnetic field change of 5T.

References

[1] Giauque W. F, MacDougall D. P 1993 *Phys. Rev.***43** 786.
[2] Moritomo Y, Asamitsu A, Tokura Y 1995 *Phys. Rev. B.* **51** 16491.
[3] Medvedeva V, Bersenev Y. S, Barner K, Haupt L, Mandal R, Poddar A 1997 *Physica B* **229** 194.
[4] Xiong G. C, Li Q, Ju H. L, Greene R. L, Venkatesan T 1995 *Appl. Phys. Lett.* **66** 1689.
[5] Jin S, Tiefel T. H, McCormack M, Bryan H. M. O, Chen L. H, Ramesh R, Schurig D 1995 *Appl. Phys. Lett.* **67** 557.
[6] Satyalakshmi K. M, Fisher B, Patlagan L, Koren G, Sheriff E, Prozorov R, Yeshurun Y 1998 *Appl. Phys. Lett.* **73** 402.
[7] Tang T, Gu K. M, Cao Q. Q, Wang D. H, Zhang S. Y, Du Y. W 2000 *J. Magn. Magn. Mater.* **222** 110.
[8] Zhu H, Song H, Zhong Y. H 2002 *Appl. Phys. Lett.* **81** 3416.
[9] Das S, Dey T. K 2006 *J. Phys .Condens. Mater.* **18** 7629.
[10] Rveau B, Maignan A, Martin C, Hervieu M 1998 "Colossal Magnetoresistance and Related Properties" World Scientific (Eds), Singapore.
[11] Radaelli P, Cox D. E, Marezio M, Cheong S. W 1997 *Phys. Rev. B* **55** 3015.
[12] Sinh N. H, Tuan N. A, Mai V. T, Quang P. H, Son N. T 2005 *J. Sci. Math. Phys.* **1** 11.
[13] Mehri A, Cheikh-Rouhou Koubaa W, Koubaa M, Cheikh-Rouhou A 2009 *Phys. Proce.* **2** 975.
[14] Kuo J. H, Anderson H. U, Sparlin D. M 1989 *J. Solid State Chem.* **83** 52.
[15] Rietveld H. M 1969 *J. Appl. Cryst.* **2** 65.
[16] Roisnel T, Rodriguez-Carvajal J 2003 Computer Program FULLPROF, LLB-LCSIM.
[17] Ye S. L, Sang W. H, Dai J. M, Wang K. Y, Wang S. G, Zhang C. L, Du J. J, Sun Y. P, Fang J 2002 *J. Magn. Magn. Mater.* **248** 2633.
[18] Pal S, Banerjee A, Choudhuri B. K 2003 *J. Phys. Chem. Solids* **64** 2063.
[19] Tang T, Tien C, Hou B. Y 2008 *Physica B* **403** 2111.
[20] Tang T, Cao Q. Q, Gu K. M, Xu H. Y, Zhang S. Y, Du Y. W 2000 *Appl. Phys. Lett.* **77** 723.
[21] McMichael R. D, Ritter J. J, Shull R. D 1993 *J. Appl. Phys.* **73** 6946.
[22] Gschneider K. A, Pecharsky V. K 2000 *Annu. Rev. Mater. Sci.* **30** 387.

IOP Conf. Series: Materials Science and Engineering **28** (2012) 012050 doi:10.1088/1757-899X/28/1/012050

Effect of strontium substitution on the physical properties of $Nd_{0.5}Ca_{0.5-x}Sr_xMnO_3$ (0.0≤x≤0.5) manganites

R. M'nassri[1] [a], W. Cheikhrouhou-Koubaa[a], M. Koubaa[a] and A. Cheikhrouhou[a,b]
[a]Laboratoire de Physique des Matériaux, Faculté des Sciences de Sfax, Sfax University, B.P. 1171, 3000 Sfax, Tunisia.
[b]Institut NEEL, B.P. 166, 38042 Grenoble Cedex 9, France.

E-mail: abdcheikhrouhou@yahoo.fr

Abstract. The structural and magnetic properties of $Nd_{0.5}Ca_{0.5-x}Sr_xMnO_3$ (0.0≤*x*≤0.5) were investigated experimentally. All our samples were synthesized using the standard ceramic elaborating method at high temperatures. X-ray diffraction was systematically carried out on $Nd_{0.5}Ca_{0.5-x}Sr_xMnO_3$ polycrystalline compounds. All our samples crystallized in the orthorhombic system. The unit cell volume increased with increasing strontium amount. The substitution of Ca by Sr destroyed the charge order state observed in the parent compound and induced a ferromagnetic phase at low temperatures. The Curie temperature T_C increased from 150 K for x=0.1 to 260 K for x=0.5.

1. Introduction

Over the last fifteen years, perovskite manganese oxides, $Ln_{1-x}A_xMnO_3$ (Ln=La, Nd, Pr, Sm,... and A=Ca, Sr, Ba, Pb,...) have known considerable attention due to the wide range of crystallographic, magnetic and magneto-transport properties they exhibit [1-3]. With strategic substitution of divalent ions at the A site, some of the Mn^{3+} ions ($3d^3$, $t^3_{2g}e_g^1$) convert into Mn^{4+} ions ($3d^3$,$t^3_{2g}e_g^0$) [4], resulting in fascinating physical phenomena such as paramagnetic (PM) insulator to ferromagnetic (FM) metallic transition or various mixed magnetic phases canted antiferromagnetic (AFM)/spin glass (SG) coupled with charge/orbital ordered (CO/OO) states in a particular doping range. The physical properties in manganese perovskites are thought to arise from the strong competition among a ferromagnetic double-exchange interaction, an antiferromagnetic super-exchange interaction and the spin–phonon coupling [5-7]. These interactions are determined by intrinsic parameters such as doping level, average cationic size, cationic disorder and oxygen stoichiometry [8].

Recently, extensive studies have been focused on the charge ordering (CO) and orbital ordering (OO) phase, which is most easily observed in half-doped x = 0.5 manganites [9]. The spatial CO phase behaves as the periodic arrangement of Mn^{3+}/Mn^{4+} ions. Generally, the CO formation is accompanied by an antiferromagnetic (AFM) phase transition. The ground state, therefore, is usually assigned to be CO/AFM state. Though the origin of the CO phase is still subject of debate, it is extensively acknowledged that the mean radius of the A- cation site $\langle r_A \rangle$ plays a key role [10]. Experimentally, $\langle r_A \rangle$=1.24Å is recognized as the critical mean radius [11]. With $\langle r_A \rangle$ smaller than 1.24 Å, the CO phase exists in the system; however, with $\langle r_A \rangle$ larger than 1.24 Å, there is no CO phase. By way of

[1] To whom any correspondence should be addressed.

example, $Nd_{0.5}Ca_{0.5-x}Sr_xMnO_3$ is one of the systems which exhibit phase competition between a charge-orbital ordered antiferromagnetic state and a charge delocalized ferromagnetic state.
Recent papers aimed to study the doping effect of substituting the rare-earth element in these manganites [12-14] and many interesting effects have been shown on the magnetic properties.
In this paper, we present an experimental work on the structural and magnetic properties in the $Nd_{0.5}Ca_{0.5-x}Sr_xMnO_3$ polycrystalline samples.

2. Experimental techniques

Polycrystalline $Nd_{0.5}Ca_{0.5-x}Sr_xMnO_3$ samples were prepared by the conventional ceramic route with the following stoichiometric quantities: Nd_2O_3, $SrCO_3$, $CaCO_3$ and MnO_2 with a purity more than 99%. The stoichiometric precursors were mixed together in an agate mortar for about 30 minutes with desired proportions according to the reaction:

$$0.25Nd_2O_3 + (0.5-x)CaCO_3 + xSrCO_3 + MnO_2 \longrightarrow Nd_{0.5}Ca_{0.5-x}Sr_xMnO_3 + \delta CO_2 + \delta' O_2$$

The starting materials were intimately mixed in an agate mortar and heated in air up to 1000°C for 48 h. The obtained powders were pressed into pellets (of about 1mm thickness) and sintered at 1200°C in air for 60 h with intermediate regrinding and repelling. Finally, these pellets were cooled slowly to room temperature in air. The physical properties depend strongly on the synthesis route and also on the cooling method. Phase purity, homogeneity and cell dimensions were determined by X-ray powder diffraction (XRD) at room temperature (diffractometer using Cu Kα radiation). Structural analysis was carried out using the standard Rietveld method [15]. Magnetization (M) versus temperature (T) and versus magnetic applied field (H) were measured using a vibrating sample magnetometer equipped with a superconducting coil. M (T) data were obtained in the temperature range 20–300 K with a magnetic applied field of 5 T for x=0.0 and 0.05 T for the substituted compounds in the field cooled regime. Isothermal M (H) data were measured in the temperature range 20–300 K under a magnetic applied field ranging from 0 to 8 T.

3. Results and discussion

Rietveld analysis of the X-ray powder diffraction patterns confirms the formation of a clear single perovskite phase with an orthorhombic structure [16]. For $0.0 \leq x \leq 0.3$, the samples crystallize in the Pnma space group while the other compounds x= 0.4 and 0.5 belong to the Imma space group. All our samples are found to confirm the orthorhombic structures with tolerance factor t in the range 0.91-0.94 [4]. The various patterns are given in figure 1.

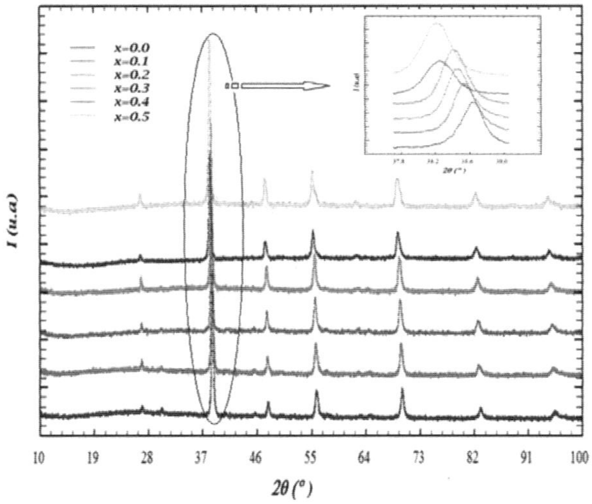

Figure 1. X-ray diffraction patterns for $Nd_{0.5}Ca_{0.5-x}Sr_xMnO_3$ ($0.0 \leq x \leq 0.5$) compounds.

A zoom on the main peak shows a shift to low angles, indicating an increase of the unit cell volume. Mean ionic radii $<r_A>$, unit cell volume, tolerance factor t and selected inter-atomic distances and angles are given in Table 1. The quality of the refinement is assessed through the goodness of the fit indicator χ^2.

A typical X-Ray diffraction pattern at 300K, including the observed and calculated profiles as well as the difference profile of $Nd_{0.5}Ca_{0.2}Sr_{0.3}MnO_3$ sample is shown in figure 2.

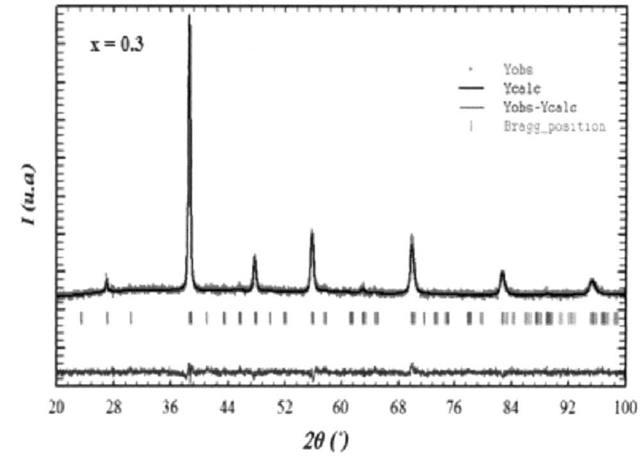

Figure 2. Observed (solid circles) and calculated (solid line) XRD patterns of the Sr-doped sample (x=0.3) at room temperature. The difference between these spectra is plotted at the bottom. Bragg reflections are indicated by ticks.

From Table 1, it can be clearly seen that the Mn-O-Mn bond angle increases with the increasing doping level x, whereas the Mn-O bond length decreases, which displays the inverse correlation to the variation in the Mn-O-Mn bond angle.

Table 1. Refined structural parameters of $Nd_{0.5}Ca_{0.5-x}Sr_xMnO_3$ samples ($0 \le x \le 0.5$) at room temperature

Parameter	x = 0.0	x = 0.1	x = 0.2	x = 0.3	x = 0.4	x = 0.5
V (Å ^{3})	221.2	222.8	223.9	225.5	226.0	226.7
Mn-O-Mn (°)	157 .78	157.96	158.09	161.12	167.09	173.66
Mn-O (Å)	1.954	1.950	1.961	1.949	1.931	1.927
W(u.a)	0.09407	0.09472	0.09296	0.09544	0.0992	0.10041
$\sigma^2(10^{-5})$	7.23	181.43	321.82	428.42	501.22	540.22
t	0.915	0.920	0.924	0.929	0.933	0.938
$<r_A>$ (Å)	1 .1715	1.1845	1.1975	1.2105	1.2235	1.2365
χ^2	1.07	1.06	1.17	1.10	1.10	1.15

It is well known that there are two possible origins of the lattice distortion of the perovskite structures: (i) the deformation of the MnO_6 octahedra originating from the Jahn–Teller (JT) effect which is decided by the concentration of Mn^{3+} ions [17], (ii) the average ionic radius $<r_A>$ of the A-cation site $<r_A>$. In the present samples, the Mn^{3+} amount is fixed and the lattice distortion is ascribed to the variation of the average ionic radius $<r_A>$ of the cation site, induced by the substitution. With increasing Sr content, the unit cell volume increases; this behavior can be explained by the ion radius of Sr^{2+} which is larger than the Ca^{2+} one. The evolutions of the average radius of the A-cation site $<r_A>$ and the unit cell volume V with Sr concentration x are plotted in figure 3.

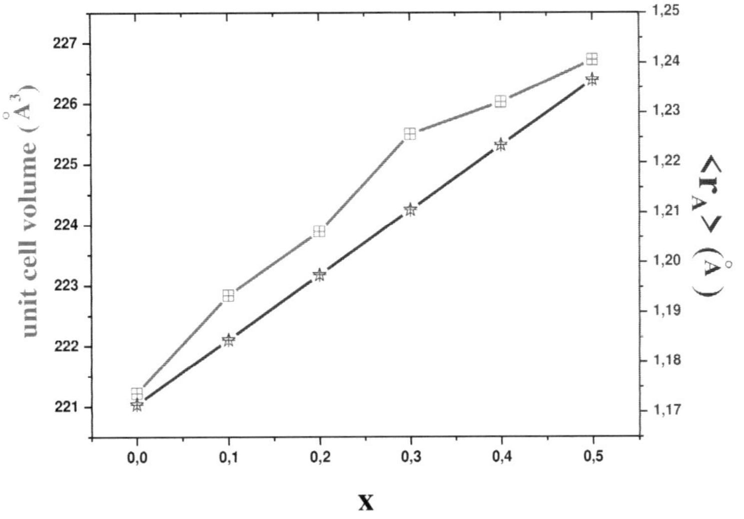

Figure 3. The unit cell volume and average radius of the A- cation site $\langle r_A \rangle$ versus x for $Nd_{0.5}Ca_{0.5-x}Sr_xMnO_3$ ($0.0 \le x \le 0.5$) compounds.

The wide bandwidth enhances the overlap between the ions, which increases the hopping rate of e_g carriers, favors the double exchange interaction (DE) between the carriers and results in on-site ferromagnetic coupling between the Mn sites and suppresses the antiferromagnetic ordering [18]. According to the DE model [5], transfer of itinerant e_g electrons between the neighboring Mn ions (local t_{2g} spins) through the O^{2-} ion results in a ferromagnetic interaction due to the on-site Hund's coupling. The strength of the DE interaction is estimated by the transfer integral, t_{eff}. The electronic bandwidth, W, is proportional to the t_{eff} and depends on Mn–O–Mn bond angles and Mn–O bond distances in MnO_6 octahedron through the overlap integrals between the Mn cation 3d orbitals and the O anion 2p orbitals. W is given by the following empirical formula [19]:

$$W \approx \frac{\cos\theta/2}{(Mn-O)^{3.5}} \ ; \ \theta = (180° - (Mn - O - Mn)) \qquad (1)$$

The wide bandwidth W plays an important role in the competition between ferromagnetic and antiferromagnetic interactions. A small change in W has a profound effect on the transport and magnetic properties of manganites [18]. The value of W is found to increase with Sr^{2+} amount and consequently the ferromagnetic DE interactions are reinforced with x (table 1).

Figure 4 shows the temperature dependence field cooled (FC) magnetization of the parent compound $Nd_{0.5}Ca_{0.5}MnO_3$ at μ_0H=5T in the temperature range 20–300 K. When the temperature is lowered from room temperature, FC magnetization increases first due to the formation of a ferromagnetic (FM) phase, but this tendency toward the FM order is destroyed by the charge ordering (CO) that favors the antiferromagnetism (AFM) phase. Two magnetic transitions can be observed from this curve: the first one is around T_{CO}=250 K and the second one appearing around the AFM Néel temperature T_N=160 K [9, 20]. In the temperature range $T_N<T<T_{CO}$, orbital order sets in progressively. Below 160 K, the manganese ions in $Nd_{0.5}Ca_{0.5}MnO_3$ form ferromagnetic zigzag chains couples antiferromagnetically at lower temperatures. The sharp rise in M (T) below 100 K is due to the ordering of the Nd ions [9].

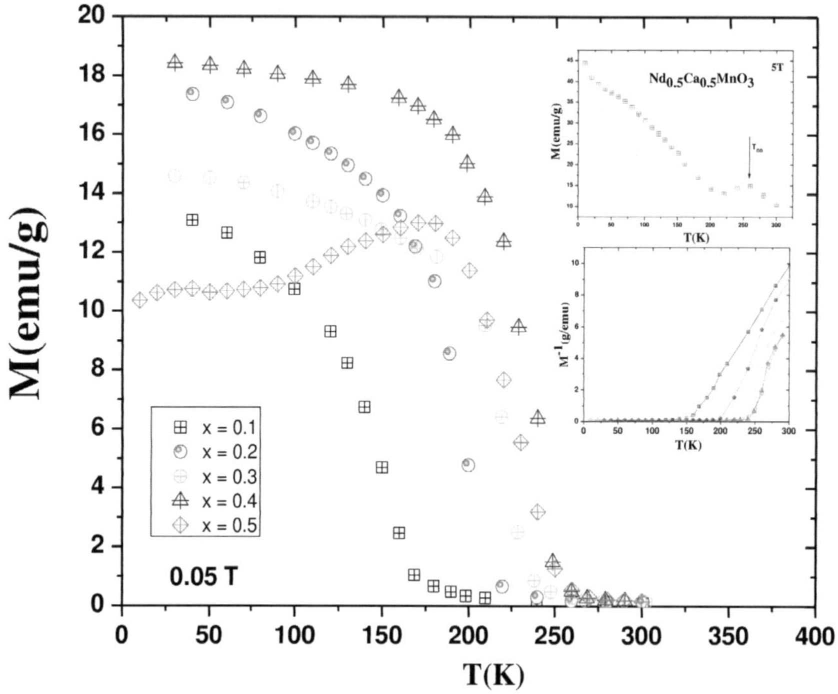

Figure 4. Magnetization measurements as a function of temperature for $Nd_{0.5}Ca_{0.5-x}Sr_xMnO_3$ ($0.1 \leq x \leq 0.5$) samples under 0.05 T. Inset $M^{-1}(T)$.

A partial strontium substitution induces an increase in the average size of the A cation site $\langle r_A \rangle$ which leads to the suppression of the CO phase and a ferromagnetic behavior is observed at low temperatures [20]. Figure 4 shows the temperature dependence of magnetization in 0.05 T field for $Nd_{0.5}Ca_{0.5-x}Sr_xMnO_3$ $(0.1 \leq x \leq 0.5)$ system. All samples exhibit a paramagnetic–ferromagnetic (PM–FM) transition. The M (T) curves do not reveal secondary magnetic phases, which confirm the good crystallization of our samples. The Curie temperature T_C, which is defined as the inflection point of M (T) curve, is found to increase from 150 K (x=0.1) to 260 K (x=0.5).

For $Nd_{0.5}Sr_{0.5}MnO_3$ sample, magnetization versus temperature in the field cooled mode exhibits an abrupt increase at 260K and tends to broaden between 200 and 150K. For T<140K, we observe a swift reduction in magnetization, which indicates the formation of long-range CO/AFM phases. With further cooling below 100K, magnetization remains unchanged.

Table 2. Physical parameters of $Nd_{0.5}Ca_{0.5-x}Sr_xMnO_3$ ($0.1 \leq x \leq 0.5$) compounds.

Parameter	x = 0.1	x = 0.2	x = 0.3	x = 0.4	x = 0.5
θ(K)	156.76	196.97	200.27	240.03	264.87
C (μ_B K /T)	14.56	10.42	8.19	8.71	9.21
μ_{eff} (μ_B)	6.6	6.82	6.04	6.23	6.41

The values of the Curie constant (C) deduced from the linear part of the $M^{-1}(T)$ curve gives an effective moment μ_{eff} as in Table 2. $Nd_{0.5}Ca_{0.5-x}Sr_xMnO_3$ contains three different magnetic ions Mn^{3+}, Mn^{4+} and Nd^{3+} and the expected effective moment μ_{eff}^{th} can be expressed as:

$$\mu_{eff}^{th} = \sqrt{0.5(\mu_{eff}^2(Nd^{3+}) + \mu_{eff}^2(Mn^{3+}) + \mu_{eff}^2(Mn^{4+}))} \quad (2)$$

The following assumptions have been made: the electronic levels of Nd^{3+} are well described by g=8/11, J=9/2, which leads to $\mu_{eff}^{th} = g\sqrt{J(J+1)}\mu_B = 3.618\mu_B$; for Mn^{4+} and Mn^{3+}, the orbital momentum is quenched so that S is the appropriate quantum number: $\mu_{eff}^{th} = g\sqrt{S(S+1)}\mu_B = 3.618\mu_B$, with g=2 and S=3/2 or [20], which leads to $\mu_{eff} = 3.87\mu_B$ for Mn^{4+} and $\mu_{eff} = 4.90\mu_B$ for Mn^{3+} [21]. This gives $\mu_{eff}^{th} = 5.10\mu_B$. The experimental values of the effective paramagnetic moments are higher than the theoretical ones in all samples (see Table 2). It is the signature of clusters of Mn^{4+} and Mn^{3+} [16, 22]. This indicates that in the paramagnetic state, the magnetic spins do not exist as individuals; they exist rather in small groups revealing the presence of ferromagnetic correlations in the paramagnetic phase.

The temperature and field dependence of the magnetization M (T, μ_0H) was measured for all samples. Figure 5 shows isothermal magnetization curves for the $Nd_{0.5}Ca_{0.5-x}Sr_xMnO_3$ samples, measured in the magnetic field range of 0–8 T and a temperature range of 20–300 K. We can observe that below T_C, magnetization increases sharply with applied field for H<0.6T and tends to saturation above 1T. At low temperature, the spontaneous magnetization M_{sp} (exp) determined by the extrapolation of the M versus H data in the high field to zero field is found to be 2.3μ_B/Mn for all samples. The magnitude of the M_{sp}(exp) is smaller than the theoretical value of 3.5μ_B/Mn calculated for full spin alignment, leading to small canted spins.

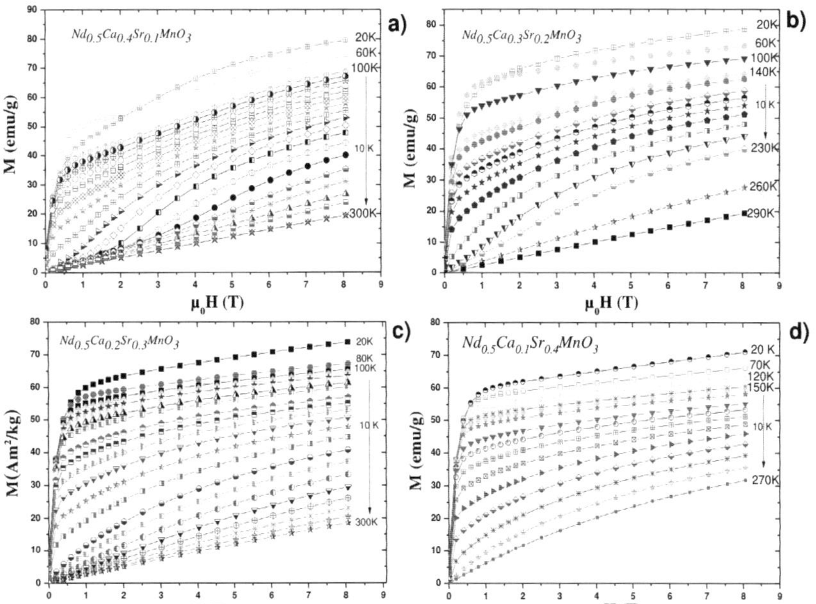

Figure 5. M (H) curves at several temperatures for $Nd_{0.5}Ca_{0.5-x}Sr_xMnO_3$ samples: (a): x=0.1; (b): x=0.2; (c): x= 0.3; (d): x=0.4

Banerjee [23] has given an experimental criterion which allows the determination of the first or second order character of the magnetic transition. This is a tool to distinguish the character of the transition by a purely magnetic method. It consists in the observation of the slope of the isotherm plots μ_0H/M versus M^2. Applying a regular approach, the straight line was constructed simply extrapolating the high magnetization parts of the curves for each studied temperature. A positive or negative slope indicates a second order or first order transition, respectively. Figure 6 shows the isotherm plots M^2 versus μ_0H/M above and below T_C for $Nd_{0.5}Ca_{0.5-x}Sr_xMnO_3$ samples (x=0.1, 0.2, 0.3, 0.4 and 0.5). The

IOP Conf. Series: Materials Science and Engineering **28** (2012) 012050 doi:10.1088/1757-899X/28/1/012050

$x = 0.3$, 0.4, 0.5 samples show positive slopes in the complete M^2 ranges, indicating that the system exhibits a second-order ferromagnetic to paramagnetic phase transition. However, the composition $x = 0.1$ and 0.2 exhibits negative slopes above 170 K and 200K respectively (where the Curie–Weiss law starts to deviate as a sharp upturn) and positive slopes in the vicinity of T_C. As reported by Mira et al. [24], samples with $t < 0.92$ exhibit a first-order magnetic transition whereas those with $t > 0.92$ do not. The significant difference in the nature of the magnetic transition of the samples is attributed to the change in the tolerance factor t.

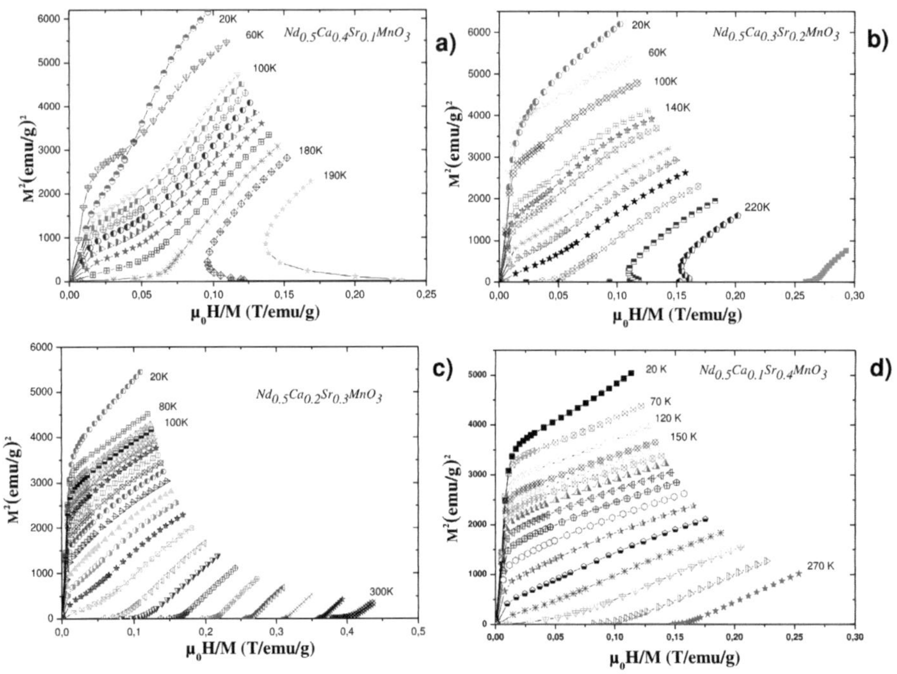

Figure 6. Arrot curves M^2 versus μ_0H/M for $Nd_{0.5}Ca_{0.5-x}Sr_xMnO_3$ samples: (a): x=0.1; (b): x=0.2; (c): x= 0.3; (d): x= 0.4

4. Conclusion

In this work, the substitution of Ca for Sr in $Nd_{0.5}Ca_{0.5-x}Sr_xMnO_3$ ($0 \leq x \leq 0.5$) suppresses the charge order state, increases the unit cell volume and the Curie temperature T_C, but reduces the amount of the lattice contraction. For x=0.3, 0.4 and 0.5, the samples exhibit a second order ferromagnetic to paramagnetic phase transition, while for x=0.1 and 0.2 the samples exhibit a first order magnetic one.

References

[1] Lin G. C, Wie Q and Zhang J. X 2006 *J. Magn. Magn. Mater.* **300** 392.
[2] Phan M. H, Tian S. B, Hoang D. Q, Yu S. C, Nguyen C and Ulyanos A. N 2005 *J. Magn. Magn. Mater.* **258** 309.
[3] Wang Z. M, Ni G, Xu Q. Y, Sang H and Du Y. W 2001 *J. Appl. Phys.* **90** 5689.
[4] Panwar N, Pandya D. K, Rao A, Wu K. K, Kaurav N, Kuo Y. K, Agarwal S. K 2008 *Eur. Phys. J. B* **65** 179.
[5] Zener C 1951 *Phys. Rev.* **82** 403.
[6] Millis A. J, Littlewood P. B, Shraiman B. I 1995 *Phys. Rev. Lett.* **74** 5144.
[7] Goodenough J. B, Wold A, Arnott R. J, Menyuk N 1961 *Phys. Rev* **124** 373.
[8] Levy P, Parisi F, Polla G, Vega D, Leyva G and Lanza H 2000 *Phys. Rev. B* **62** 6437.
[9] Millange F, De Brion S, Chouteau G 2000 *Phys. Rev. B* **62** 5619.

[10] Rivadulla F, Winkler E, Zhou J. S, Goodenough J. B 2002 *Phys. Rev. B* **66** 174432.
[11] Vanitha P. V, Santhosh P. N, Singh R. S, Rao C. N. R, Attfield J. P 1999 *Phys. Rev. B* **59** 13539.
[12] Phan M. H, Yu S. C, Hur N. H 2003 *J. Magn. Magn. Mater.* **262** 407.
[13] Gencer H, Atalay S, Adiguzel H. I, Kolat V. S 2005 *Physica B* **357** 326.
[14] Chau N, Tho N. D, Luong N. H, Giang B. H, Cong B. T 2006 *J. Magn. Magn. Mater* **303** 402.
[15] Roisnel T, Rodriguez-Carvajal J, May 2003 Computer program FULLPROF, LLB-LCSIM.
[16] Pillai S. S, Rangarajan G, Raju N. P, Epstien A. J, Santhosh P. N 2007 *J. Phys. Condens. Matter.* **19** 496221.
[17] Alonso J. A, Martı´nez-Lope M. J, Casais M. T 2000 *Inorg. Chem.* **39** 917.
[18] Moritomo Y, Kuwahara H, Tomioka Y, Tokura Y 1997 *Phys. Rev. B* **55** 7549.
[19] Medarde M, Mesot J, Lacorre P, Rosenkranz S, Fischer P and Gobrecht K 1995 *Phys. Rev. B* **52** 9248.
[20] Cheikhrouhou Koubaa W, M'nasri R, Koubaa M, Cheikhrouhou A 2010 *J. Phys. Conf. Ser* **200** 012019.
[21] Van Vleck J. H 1952 *The theory of electric and magnetic Susceptibilities* (*Oxford*) p 285.
[22] Brion S, Ciorcas F, Chouteau G , Lejay P, Radaelli P, Chaillout C 1999 *Phys. Rev. B* **59** 1304.
[23] Banerjee S. K 1964 *Phys. Lett.* **12** 16.
[24] Mira J, Riva J 2001 *Phys. Rev. B* **65** 024418.

Acknowledgments
This study is supported by the Tunisian Ministry of Higher Education and Scientific Research.

IOP Conf. Series: Materials Science and Engineering **28** (2012) 012051 doi:10.1088/1757-899X/28/1/012051

Magnetocaloric properties of $Nd_{0.5}Ca_{0.5-x}Sr_xMnO_3$ (0.0≤x≤0.5) manganites

R. M'nassri[1] [a], W. Cheikhrouhou-Koubaa[a], M. Koubaa[a] and A. Cheikhrouhou[a,b]
[a]Laboratoire de Physique des Matériaux, Faculté des Sciences de Sfax, Sfax University, B.P. 1171, 3000 Sfax, Tunisia.
[b]Institut NEEL, B.P. 166, 38042 Grenoble Cedex 9, France.

E-mail: abdcheikhrouhou@yahoo.fr

Abstract. We report in this paper the magnetic and magnetocaloric properties of polycristalline $Nd_{0.5}Ca_{0.5-x}Sr_xMnO_3$ (0.0≤x≤0.5) compounds. Our samples were synthesized using the solid-solid state reaction method at high temperatures. Structural studies showed that our samples are single phase and crystallize in the orthorhombic system with Pnma space group for 0.0≤x≤0.3 and Imma space group for x=0.4 and 0.5. The substitution of Ca by Sr destroyed the charge ordered state observed in the parent compound and induced a ferromagnetic phase at low temperatures. From isothermal magnetization versus magnetic applied field we deduced the magnetic entropy change $|\Delta S_M|$. The maximum value of $|\Delta S_M|$ reached 3.89 J/kgK for x=0.3 under $|\Delta H = 7T|$. Our results are interesting enough and suggest that these materials may be suitable candidates as working substances in magnetic refrigeration near room temperature.

1. Introduction

The mixed valence perovskites $Ln_{1-x}M_xMnO_3$ (Ln and M, respectively, being a trivalent rare earth element and a divalent alkali-earth one) have been extensively studied since the discovery in 1994 of a large magnetoresistance effect in these materials [1,2]. The substitution of the trivalent element by divalent one produces an inhomogeneous distribution of mixed valence Mn^{3+}/Mn^{4+} ions to maintain charge neutrality. The valence change of A-site in the perovskite affects significantly both magnetism and conductivity in these manganites [3]. These systems exhibit many significant properties like metal–insulator transition, ferromagnetic–paramagnetic phase change, charge and orbital ordering, etc. depending on the charge density, temperature and atomic structure.
Recently, extensive studies have been focused on the charge ordering (CO) and orbital ordering (OO) phase, which is most easily observed in half-doped x = 0.5 manganites [4]. The spatial CO phase behaves as the periodic arrangement of Mn^{3+}/Mn^{4+} ions. Generally, the CO formation is accompanied by an antiferromagnetic (AFM) phase transition. The ground state, therefore, is usually assigned to be CO/AFM state. Though the origin of the CO phase is still subject of debate, it is extensively acknowledged that the mean radius of the A- cation site $\langle r_A \rangle$ plays a key role [5]. Experimentally, $\langle r_A \rangle$ = 1.24Å is recognized to be the critical mean radius [6]. In fact, for $\langle r_A \rangle$ smaller than 1.24 Å, the CO phase exists in the system; however, no CO phase is observed when $\langle r_A \rangle$ is larger than 1.24 Å. We can, for example, consider $Nd_{0.5}Ca_{0.5-x}Sr_xMnO_3$ compound which is one of the systems exhibiting

[1] To whom any correspondence should be addressed.

phase competition between a charge-orbital ordered antiferromagnetic state and a charge delocalized ferromagnetic state.

Recent studies have been focused on the effect of rare-earth element substitution in manganites [7-9] and mainly on the magnetic properties and the magnetocaloric effect (MCE). Moreover, the possibility of using these materials, which exhibit this effect as a magnetic refrigerant, has promoted research on the magnetocaloric properties in such magnetic systems. One of the main goals of recent studies on MCE is thus to find useful magnetic materials which have a large entropy change at low magnetic field around room temperature .

In this article, we present an experimental work on the magnetic and magnetocaloric properties in the polycrystalline $Nd_{0.5}Ca_{0.5-x}Sr_xMnO_3$ samples. We are particularly interested in studying the magnetic entropy change $\Delta S(T)$ in our compounds.

2. Experimental techniques

Polycrystalline $Nd_{0.5}Ca_{0.5-x}Sr_xMnO_3$ samples were prepared by the conventional ceramic route with the following stoichiometric quantities: Nd_2O_3, $SrCO_3$, $CaCO_3$ and MnO_2 with a purity more than 99%. The stoichiometric precursors were mixed together in an agate mortar for about 30 minutes with desired proportions according to the reaction:

$$0.25Nd_2O_3 + (0.5\text{-}x)CaCO_3 + xSrCO_3 + MnO_2 \longrightarrow Nd_{0.5}Ca_{0.5-x}Sr_xMnO_3 + \delta CO_2+\delta' O_2$$

The starting materials were intimately mixed in an agate mortar and heated in air up to 1000°C for 48 h. The obtained powders were pressed into pellets (of about 1mm thickness) and sintered at 1200°C in air for 60 h with intermediate regrinding and repelling. Finally, these pellets were cooled slowly to room temperature in air. The physical properties depend strongly on the synthesis route and also on the cooling method. Phase purity, homogeneity and cell dimensions were determined by X-ray powder diffraction (XRD) at room temperature (diffractometer using Cu Kα radiation). Structural analysis was carried out using the standard Rietveld method [10]. Magnetization (M) versus temperature (T) and versus magnetic applied field (H) were measured using a vibrating sample magnetometer equipped with a superconducting coil. M (T) data were obtained in the temperature range 20–300 K with a magnetic applied field of 5 T for x=0.0 and 0.05 T for the substituted compounds in the field cooled regime. Isothermal M (H) data were measured in the temperature range 20–300 K under a magnetic applied field ranging from 0 to 8 T.

3. Results and discussion

Structural studies showed that our samples are single phase and crystallize in the orthorhombic system with Pnma space group for $0.0 \le x \le 0.3$ and Imma space group for x=0.4 and 0.5 [10].

We plot in figure 1 the temperature dependence of magnetization in 0.05 T for $Nd_{0.5}Ca_{0.5-x}Sr_xMnO_3$ $(0.1 \le x \le 0.5)$ system in the temperature range 20–300 K (inset the evolution of the field cooled (FC) magnetization of the parent compound $Nd_{0.5}Ca_{0.5}MnO_3$ at μ_0H=5T).

As we can observe, a partial strontium substitution induces an increase in the average size of the A cation site $\langle r_A \rangle$ which leads to the suppression of the CO phase in the parent compound and a ferromagnetic behavior is observed at low temperatures in the substituted samples [11]. All the substituted samples exhibit a paramagnetic–ferromagnetic (PM–FM) transition. The M(T) curves do not reveal secondary magnetic phases, which confirms the good crystallization of our samples. The Curie temperature T_C, which is defined as the inflection point of M(T) curve, is found to increase from 150 K (x=0.1) to 260 K (x=0.5).

In order to confirm the ferromagnetic behavior in our substituted samples, we plot in figure 2 the magnetization evolution versus magnetic applied field up to 8T at several temperatures in the range 20-300K for $Nd_{0.5}Ca_{0.2}Sr_{0.3}MnO_3$. Below T_C, the magnetization increases sharply with applied field for H<0.6T and tends to saturation above 1T.

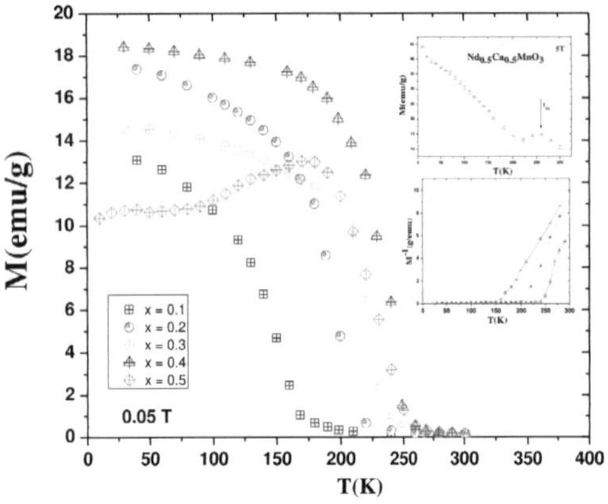

Figure 1. Magnetization measurements as a function of temperature for $Nd_{0.5}Ca_{0.5-x}Sr_xMnO_3$ ($0.1 \le x \le 0.5$) samples under 0.05 T. Inset $M^{-1}(T)$.

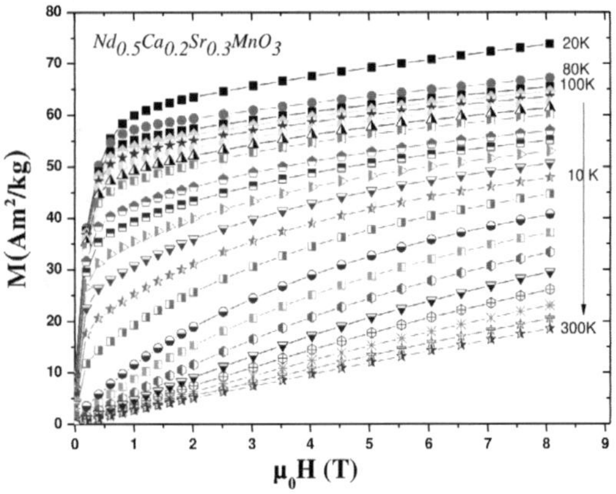

Figure 2. M (H) curves at several temperatures for $Nd_{0.5}Ca_{0.2}Sr_{0.3}MnO_3$ sample.

The large magnetocaloric effect in manganites with perovskite structure could originate from the spin-lattice coupling in the magnetic ordering process [12,13]. The strong coupling between spin and lattice has been shown by the observed lattice changes accompanying magnetic transitions in these manganites [14]. The lattice structure changes in the Mn-O bond distance as well as in the Mn-O-Mn bond angle, which would in turn favor the spin ordering. Then, a more abrupt variation of magnetization near the magnetic transition occurs and results in a large magnetic entropy change.

The isothermal magnetic entropy change $\Delta S(T)$, which is associated to the magnetocaloric effect (MCE), can be calculated from magnetization measurements as a function of the magnetic applied field at several temperatures (indirect measurement technique of the magnetocaloric effect). According to the classical thermodynamics theory, the magnetic entropy change produced by varying the magnetic field from 0 to H_{max} is given by [15]:

$$\Delta S(T,H) = \int_0^{H\max} \left(\frac{\partial M}{\partial T} \right)_H dH \quad (1)$$

The integral in equation 1 corresponds to the area enclosed between two isothermal magnetization curves M(H,T) and M(T+ΔT, H), where ΔT is the temperature difference between two isotherms. We calculated this area using a precise numerical integration. Complete $\Delta S(T)$ versus temperature curve can be derived from a series of isothermal magnetization curves obtained at discrete temperatures with an appropriate interval (relatively small) of temperatures ΔT and magnetic fields ΔH.

$$\Delta S\left(\frac{T_1 + T_2}{2} \right) = \frac{1}{\Delta T}\left[\int_0^{H\max} M(T+\Delta T, H)dH - \int_0^{H\max} M(T,H)dH \right] \quad (2)$$

In order to study the effect of the partial substitution of Ca by Sr, we have calculated the magnetic-entropy change ΔS_{max} and the relative cooling power RCP given by:

$$RCP = -\Delta S_{\max} * \Delta T_{FHWM} \quad (3)$$

Figure 3 shows the temperature dependence of ΔS under a magnetic applied field change up to 7 T in our $Nd_{0.5}Ca_{0.5-x}Sr_xMnO_3$ ($0 \le x \le 0.5$) samples. The ΔS exhibits a broad negative peak around T_C (normal MCE) for all samples and a positive peak at T_{CO} (i.e., inverse MCE) for the parent compound. It is interesting to see that $Nd_{0.5}Ca_{0.5}MnO_3$ sample exhibits a positive entropy change around T_{CO} due to the dominance of the antiferromagnetic interactions with a maximum value of 0.54 J/kgK upon a magnetic field change of 7T.

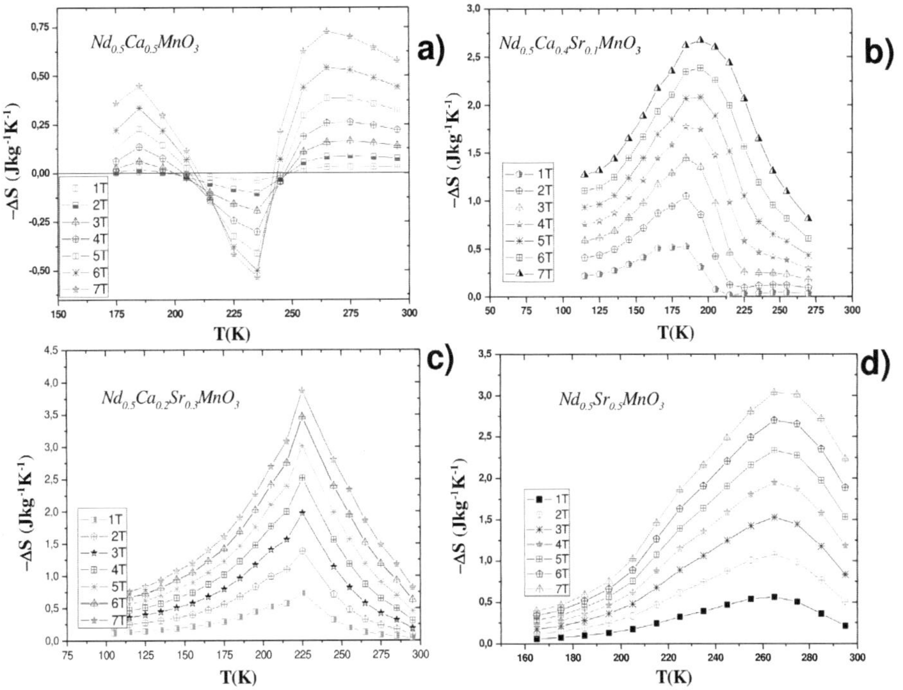

Figure 3. Magnetic entropy change versus temperature for the $Nd_{0.5}Ca_{0.5-x}Sr_xMnO_3$ samples at various magnetic applied field changes: (a): x=0.0; (b): x=0.1; (c): x= 0.3; (d): x= 0.5.

With increasing temperature and magnetic applied field, the ferromagnetic spins start to align, leading to a second negative maximum around 265 K [11].

The ΔS_M and RCP values for all our samples determined under 7T are listed in Table 1. The maximum of the isothermal entropy change for the sample with x=0.3 occurs at 221K. These values are moderate when compared to the maximum magnetic entropy change of 4.2 J/kg K^{-1} obtained in Gd for a field change of 1.5 T near its magnetic ordering temperature T_C=294K [16].

Table 1: ΔS_M, ΔT_{FHWM}(K) and RCP values occurring at the Curie temperature (T_C) and at a magnetic applied field (μ_0H = 7T) for $Nd_{0.5}Ca_{0.5-x}Sr_xMnO_3$ ($0.0 \leq x \leq 0.5$)

x	0.0	0.1	0.2	0.3	0.4	0.5
ΔS_M ($Jkg^{-1}K^{-1}$)	0.54	2.67	3.74	3.89	3.65	3.04
ΔT_{FHWM}(K)	19.15	119.51	83.66	78.23	106.88	105.58
RCP (Jkg^{-1})	10.34	319.1	312.89	304.31	390.11	320.96

4. Conclusion

In this work, the substitution of Ca for Sr in $Nd_{0.5}Ca_{0.5-x}Sr_xMnO_3$ ($0 \leq x \leq 0.5$) suppresses the charge order state, increases the unit cell volume and the Curie temperature T_C, but reduces the amount of the lattice contraction. The results of magnetic entropy change versus temperature obtained by indirect measurements are reported. The maximum value of $|\Delta S_M|$ for the x = 0.3 sample is found to be at 221 K.

References

[1] Jin S, Tiefel T. H, McCormack M, Fastnacht R. A, Ramesh R, Chen L. H 1994 *Science* **264** 413.
[2] McCormack M, Jin S, Tiefel T. H, Fleming R. M, Philips J. M, Ramesh R 1994 *Appl. Phys. Lett.* **64** 3045.
[3] Urushibara A, Moritomo Y, Arima T, Asamisu A, Kido G, Tokura Y 1995 *Phys. Rev. B* **51** 14103.
[4] Millange F, De Brion S, Chouteau G 2000 *Phys. Rev. B* **62** 5619.
[5] Rivadulla F, Winkler E, Zhou J. S, Goodenough J. B 2002 *Phys. Rev. B* **66** 174432.
[6] Vanitha P. V, Santhosh P. N, Singh R. S, Rao C. N. R, Attfield J. P 1999 *Phys. Rev. B* **59** 13539.
[7] Phan M. H, Yu S. C, Hur N. H 2003 *J. Magn. Magn. Mater* **262** 407.
[8] Gencer H, Atalay S, Adiguzel H. I, Kolat V. S 2005 *Physica B* **357** 326.
[9] Chau N, Tho N. D, Luong N. H, Giang B. H, Cong B. T 2006 *J. Magn. Magn. Mater* **303** 402.
[10] M'nassri R, Cheikhrouhou Koubaa W, Koubaa M, Cheikhrouhou A 2011 *Materials Science and Engineering* (In press).
[11] Cheikhrouhou Koubaa W, M'nasri R, Koubaa M, Cheikhrouhou A 2010 *J. Phys. Conf. Ser.* **200** 012019.
[12] Phan M. H, Yu S. C, Hur N. H, Yeong Y. H 2004 *J. Appl. Phys.* **96** 1154.
[13] Guo Z. B, Du Y. W, Zhu J. S, Huang H, Ding W. P, Feng D 1997 *Phys. Rev. Lett.* **78** 1142.
[14] Radaelli P. G, Cox D. E, Marezio M, Cheong S. W, Schiffer P. E, Ramirez A. P 1995 *Phys. Rev. Lett.* **75** 4488.
[15] Morrish A. H 1965 The Physical Principles of Magnetism, (Wiley New York) (Chapter 3).
[16] Pecharsky V. K, Gschneidner K. A 1999 *J. Magn. Magn. Mater.* **200** 44.

Acknowledgments

This study is supported by the Tunisian Ministry of Higher Education and Scientific Research.

Structure, magnetic and magnetocloric properties of $Pr_{0.7}Ca_{0.3}Mn_{1-y}Cr_yO_3$

A. Selmi[a], W. Cheikhrouhou-Koubaa[1] [a], M. Koubaa[a], A. Cheikhrouhou[a,b]
[a]Laboratoire de Physique des Matériaux, Faculté des Sciences de Sfax, Sfax University, B.P. 1171, 3000 Sfax, Tunisia.
[b]Institut NEEL, B.P. 166, 38042 Grenoble Cedex 9, France.

E-mail: wissem.koubaa@yahoo.fr

Abstract. $Pr_{0.7}Ca_{0.3}Mn_{1-y}Cr_yO_3$ compounds (with $0 \leq y \leq 0.20$) have been prepared by means of the solid-state reaction, from stoichiometric powder mixtures of binary oxides, at high temperatures in ambient air. The samples were studied by X-ray powder diffraction analysis. The structural refinement using the Rietveld method based on the FULLPROF program shows that our polycrystalline specimens are single phase and crystallize in the orthorhombic structure with Pnma space group. Magnetic measurements versus temperature and versus magnetic field were also performed in all compounds. The Cr substitution in $Pr_{0.7}Ca_{0.3}Mn_{1-y}Cr_yO_3$ is found to induce a ferromagnetic behavior at low temperatures. From isothermal magnetization curves, we can determine the magnetocaloric effect, which is of utmost important in the magnetic refrigeration technique. These results may open an important field in the search of new magnetic materials for magnetic refrigeration in order to replace the conventional expansion-compression technique.

1. Introduction

The discovery of giant magnetocaloric effect (MCE) induced by a first-order magneto structural transition in the intermetallic alloy $Gd_5Si_{1.8}Ge_{2.2}$, rejuvenated interest in search of new materials capable of showing larger MCE value (ΔS_M) [1, 2]. The MCE is the working principle of the magnetic refrigeration technology which is more energy efficient and environmental friendly compared to the conventional vapor-compression based refrigeration [3].

Manganese perovskites of general formula $Pr_{0.7}Ca_{0.3}MnO_3$ exhibit a complex phase diagram as a function of doping and temperature, characterized by a transition from a paramagnetic insulating to a charge and orbital ordered phase of the Mn^{3+} and Mn^{4+} ions in different sublattices around 215K and by an anti-ferromagnetic transition at lower temperatures (around 70 K) [4-7]. The insulating state can be transformed into a metallic one by an external magnetic field. Such a magneto-conductive transition is at the basis of the so-called colossal magnetoresistance effect (CMR) [5, 6]. Because of the crucial role of the magnetic Mn site, it is interesting to study the effects of its substitution, which may provide a tool for both exploring novel CMR materials and understanding the mechanism of CMR.

In particular, the CE-type AFM structure may be described as a combination of Mn^{3+}/Mn^{4+} CO with d^2_z orbital ordering. Cr-doping of $Pr_{0.7}Ca_{0.3}MnO_3$ induces the destruction of the charge ordered insulating state and of the associated antiferromagnetism, inducing a transition to a ferromagnetic

[1] To whom any correspondence should be addressed.

metallic state [8]. In this work we investigated the effect of Cr-doping on the structural, magnetic and magnetocaloric properties of $Pr_{0.7}Ca_{0.3}Mn_{1-y}Cr_yO_3$.

2. Experimental

Polycrystalline $Pr_{0.7}Ca_{0.3}Mn_{1-y}Cr_yO_3$ samples (with $0 \leq y \leq 0.20$) were synthesized from high purity precursors (more than 99.9%): Pr_6O_{11}, $CaCO_3$, Cr_2O_3 and MnO_2 by the conventional solid state reaction method at high temperature .The precursors were intimately mixed in the desired proportions according to the following equation:

$$0.11Pr_6O_{11} + 0.3CaCO_3 + (1-y)\, MnO_3 + 0.5yCr_2O_3 \longrightarrow Pr_{0.7}Ca_{0.3}Mn_{1-y}Cr_yO_3 + \delta CO_2 + \delta' O_2$$

The starting materials were intimately mixed in an agate mortar. The obtained powders were then pressed into pellets (of about 1mm thickness) and sintered at 800, 1000, 1100 and 1300 °C for 24h for each cycle to ensure a better crystallization with intermediate regrinding and repelling [9].

Phase identification was performed using X-ray powder diffraction analysis. Structure parameters were refined using the Rietveld method [10, 11] based on Fullprof program. Magnetization measurements versus temperature and versus magnetic applied field up to 7 T were recorded using a vibrating sample magnetometer in the temperature range 20–300 K. The magnetic entropy change was deduced from the magnetization measurements versus magnetic applied field up to 7T at several temperatures.

3. Results and discussion

The X-Ray diffraction (XRD) spectra of $Pr_{0.7}Ca_{0.3}Mn_{1-y}Cr_yO_3$ samples are characterized by the lines of perovskites structure. The compounds are single-phase and crystallize in the orthorhombic structure with Pnma space group. No structural transition has been observed. Figure 1 shows typical X-ray diffraction patterns including Bragg reflection positions as well as measured and calculated intensities at room temperature for $Pr_{0.7}Ca_{0.3}Mn_{1-y}Cr_yO_3$ samples (y = 0 and 0.2).

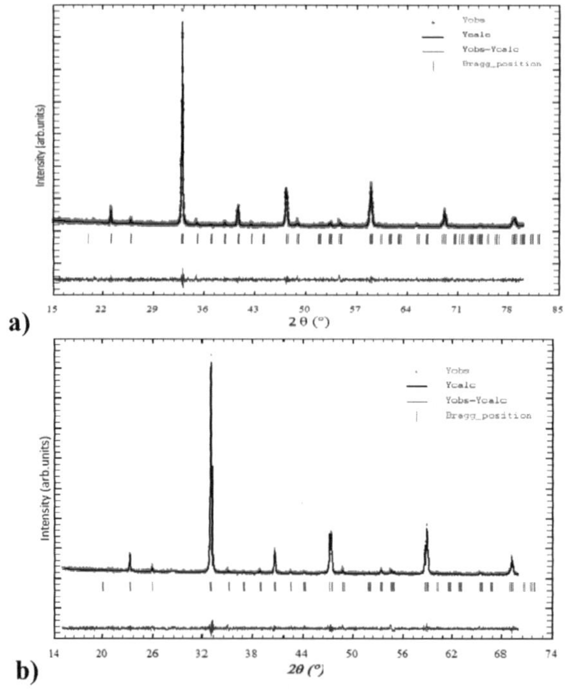

Figure 1. X-ray powder diffraction patterns and refinement at room temperature for: (a) $Pr_{0.7}Ca_{0.3}MnO_3$ and (b) $Pr_{0.7}Ca_{0.3}Mn_{0.8}Cr_{0.2}O_3$.

The structure parameters were refined using the standard Rietveld technique. In the profile refinement, we began with scale and background parameters. We then introduced the peak asymmetry and the preferred orientation. Finally, we treated the positional parameters and the individual isotropic parameters. The quality of refinement was evaluated though the goodness of the fit indicator χ^2. A good fit between the observed and the calculated profiles was obtained. The difference observed between the intensities of the measured and calculated diffraction lines can be attributed to the existence of preferential orientation of the crystallites in the samples.

In Figure 2, we present the temperature dependence of the magnetization under an applied field of 500 Oe for all our elaborated samples. The Cr doping is found to destroy the CO ordering, and to drive the system toward a ferromagnetic state at low temperatures. In fact in the Cr-doped samples, the M(T) curves show the disappearing of the CO typical bump, and a rapid increase of the magnetization is observed at temperatures below Tc. We can also notice an increase of the Curie temperature with Cr substitution from 115 for y=0 to 155 for y=0.2 as shown in figure 3.

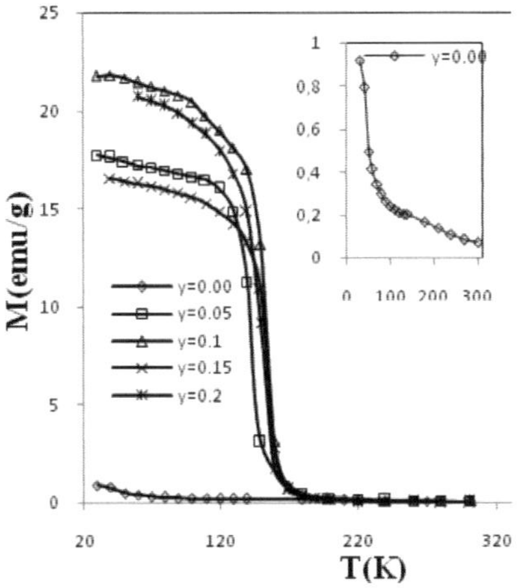

Figure 2. Temperature dependence of magnetization under 0.05T for $Pr_{0.7}Ca_{0.3}Mn_{1-y}Cr_yO_3$ ($0 \le y \le 0.15$) samples. Inset the evolution of magnetization versus temperature for the parent compound.

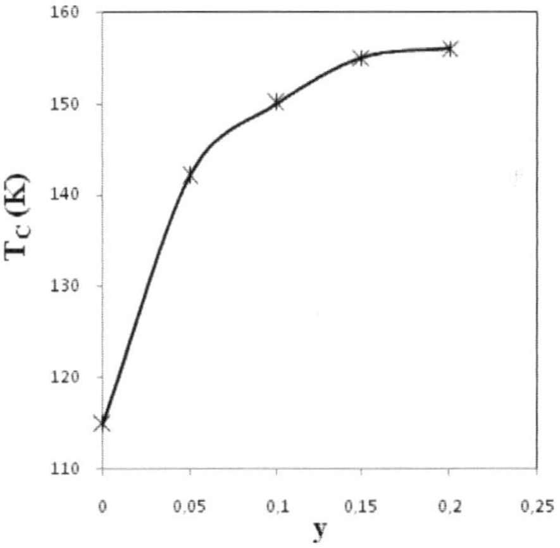

Figure 3. Curie temperature evolution versus Cr content.

We have also performed magnetization measurements versus magnetic applied field at several temperatures. Figures 4a and b show the M(H) curves for y=0 and 0.05 respectively, at various temperature. We can notice that for the substituted samples and below the Curie temperature T_C, the magnetization increases sharply for field lower than 1T and then saturates. This result confirms the ferromagnetic behavior at low temperature for these compounds.

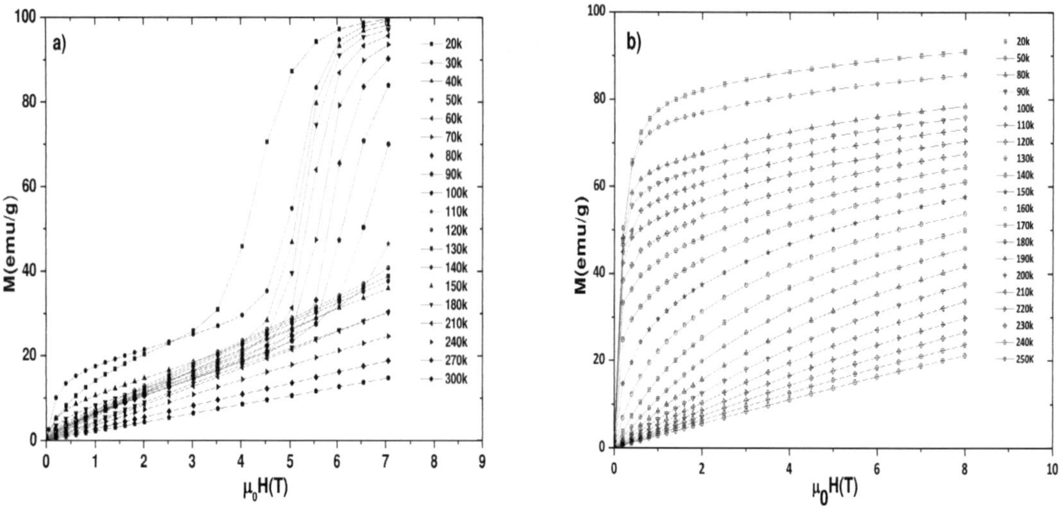

Figure 4. Isothermal magnetization curves at various temperatures for $Pr_{0.7}Ca_{0.3}Mn_{1-y}Cr_yO_3$ (a) y=0 and (b) y=0.05 samples.

In order to get deeper insight the magnetic transition, we have plotted the so-called Arrott plots (M^2versus μ_0H/M) for all our compounds. A typical curve of $Pr_{0.7}Ca_{0.3}Mn_{0.95}Cr_{0.05}O_3$ sample is shown in Figure 5. All the M^2 versus μ_0H/M curves exhibit a positive slope indicating that the ferromagnetic-paramagnetic transition is of second order, according to the Banerjee criterion [12]. The T_C values deduced from the Arrott curves are close to those deduced from M (T) measurements.

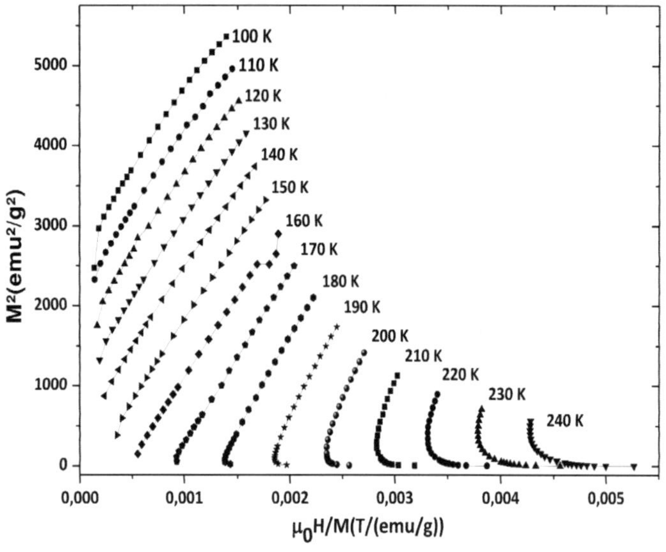

Figure 5. M^2 versus μ_0H/M isotherms for $Pr_{0.7}Ca_{0.3}Mn_{0.95}Cr_{0.05}O_3$.

The magnetocaloric effect (MCE) is expressed by means of the isothermal magnetic entropy change ΔS_m associated to the variation of magnetic field, calculated from isothermal magnetization curves. Based on thermodynamic classical theory, the magnetic entropy change in a magnetic system, resulting from ordering of spins, by the variation of external applied field is given by the following equation:

$$\Delta S_m(T,H)= S_m(T,H)-S_m(T,0)= \int \frac{dM(\cdot}{d\cdot} dH \quad (1)$$

Numerical evaluation of magnetic entropy change was carried out by approximating the above equation to the following equation:

$$\Delta S_m (T, H) = \sum \frac{(M_i(T_i,H)-M_{i+1}(T_{i+1}}{T_i - T_{i+1}} \Delta H \quad (2)$$

where M_i and M_{i+1} are the magnetization values measured at temperatures T_i and T_{i+1}, respectively. We have been essentially interested in the influence of chromium substitution on the magnetocaloric properties in our compounds since doping the B site of the perovskite strongly affect the double exchange Mn^{3+}-O-Mn^{4+} interactions and thus the magnetic part of the entropy [13].

We have reported in Figure 6 the variation of the magnetocaloric effect as a function of temperature under several magnetic field changes for all our compounds. All our samples exhibit large MCE effect around the magnetic transition temperature and the magnitude of $|\Delta S$ increases with increasing the magnetic applied field. The maximum magnetic entropy change observed, $|\Delta S_m^M$, is found to be 0.81 $Jkg^{-1}K^{-1}$, 2.76 $Jkg^{-1}K^{-1}$,2.81 $Jkg^{-1}K^{-1}$, 2.54 $Jkg^{-1}K^{-1}$ and 2.35 $Jkg^{-1}K^{-1}$ in a magnetic field change of 5T for y = 0,0.05, 0.1, 0.15 and 0.2 respectively. The relative cooling power (RCP) is evaluated as:

$$RCP = -\Delta S_m(T,H) * \delta T_{FWHM} \quad (3)$$

where δT_{FWHM} is the full-width at half-maximum of ΔS_m versus temperature [14]. For our samples, the RCP values are respectively 9.72J/kg, 405.72J/kg,269.76J/kg, 235.2J/kg and 253.2J/kg at 5T for y = 0, 0.05, 0.1, 0.15 and 0.2.

For $Pr_{0.7}Ca_{0.3}Mn_{0.95}Cr_{0.05}O_3$ compound, the magnetic entropy presents a broad and a homogeneous distribution which is a good criterion RCP of a magnetic material, it is characterized with a relative cooling power RCP= 405.72 J/kg. This value is interesting compared to either $Gd_5Si_2Ge_2$ alloy [15] or $Pr_{0.63}Sr_{0.37}MnO_3$ manganite [16].

IOP Conf. Series: Materials Science and Engineering **28** (2012) 012052 doi:10.1088/1757-899X/28/1/012052

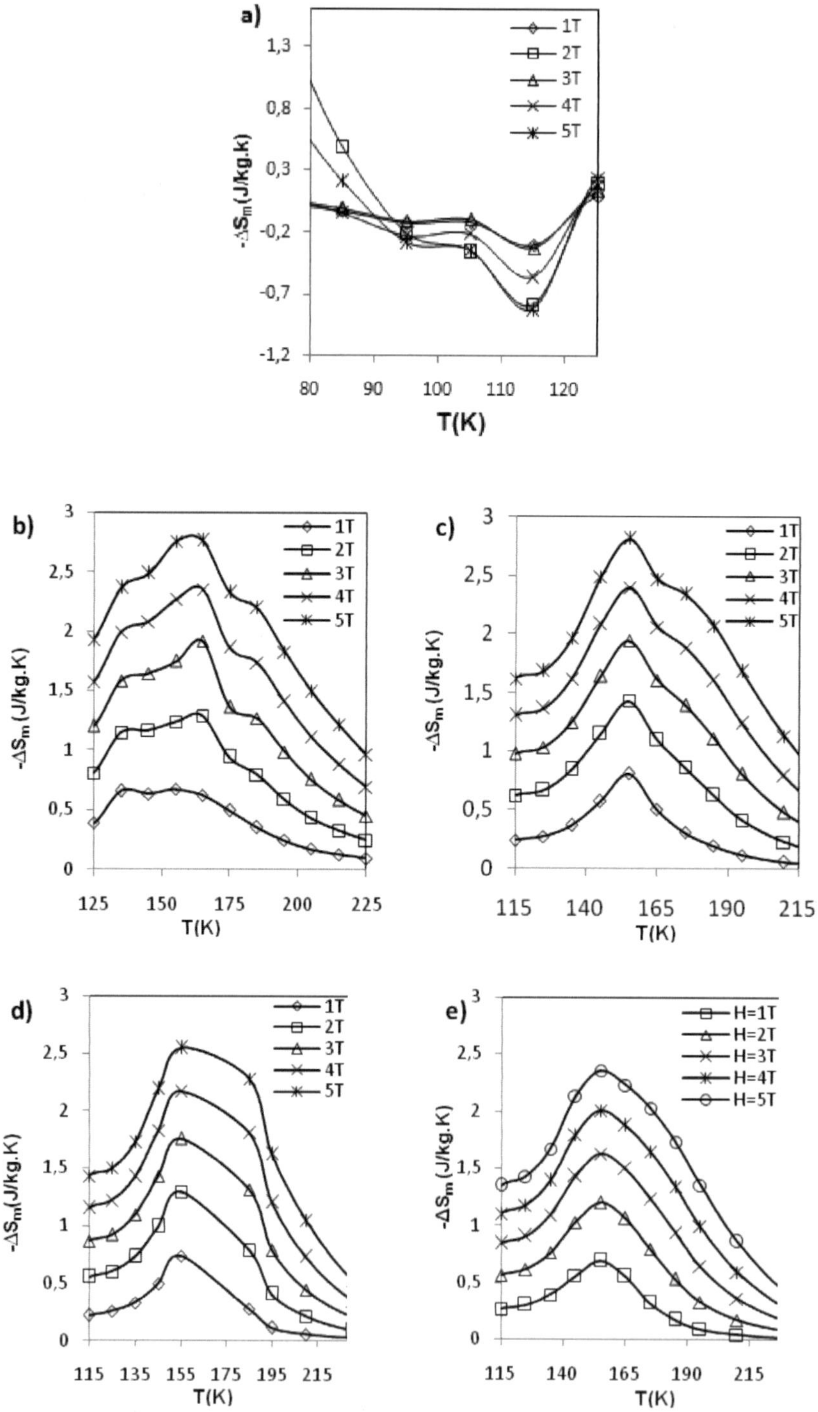

Figure 6. Magnetic entropy change for $Pr_{0.7}Ca_{0.3}Mn_{1-y}Cr_yO_3$ compounds ($0 \leq y \leq 0.2$) as a function of temperature under various magnetic fields.
a): y=0.0; **b)**: y=0.05; **c)**: y= 0.1; **d)**: y= 0.15; **e)**:y= 0.2.

4. Conclusions

In this work, we have studied the effect of chromium substitution on the B site in $Pr_{0.7}Ca_{0.3}Mn_{1-y}Cr_yO_3$ ($0\leq y\leq 0.2$) compounds. All our samples crystallize in the orthorhombic structure with Pnma space group. The chromium substitution drives the system from charge order state (antiferromagnetic) to ferromagnetic one with a Curie temperature increasing from 115K for y=0 to 155K for y=0.2. All our compounds exhibit large magnetocaloric effect around their relative T_C's.

References

[1] Gschneidner Jr. K. A, Pecharsky V. K, Tsokol A. O 2005 *Rep. Progr. Phys* **68** 1479.
[2] Morellon L, Algarabel P. A, Ibarra M. R, Blaso J, Garcai-Landa B, Arnold Z, Albertini F 1998 *Phys. Rev. B* **58** R14721.
[3] Brown G. B 1976 *J. Appl. Phys* **47** 3673.
[4] Maignan A, Varadaraju U. V, Millange F, Raveau B 1997 *J. Magn. Magn. Mater.* **186** 237.
[5] Jin S, Tiefel T. H, McCormack M, Fastnacht R. A, Ramesh R, Chen L. H 1994 *Science* **264** 413.
[6] Coey J. M. D, Viret M 1999 *Adv. Phys.* **48** 167.
[7] Martin C, Maignan A, Damay F, Hervieu M, Raveau B, Jirak Z 1999 *J. Magn. Magn. Mat.* **202** 11.
[8] Ravea B, Maigna A, Martin C 1997 *J. Solid State Chem* **130** 162.
[9] Cheikh-Rouhou Koubaa W, Koubaa M, Cheikhrouhou A 2009 *Physics Procedia* **2** 989–996.
[10] Rietveld H. M 1969 *J. Appl. Cryst* **2** 65.
[11] Roisnel T, Rodriguez-Carvajal J 2003 *Computer Program FULLPROF, LLB-LCSIM, May.*
[12] Banerjee S. K 1964 *Phys. Lett.* **12** 16.
[13] Bohigas X, Tejada J, Del Barco E, Zhang X. X, Sales M 1998 *Appl. Phys. Lett.* **73** 390.
[14] Gschneidner K. A and Pecharsky V. K 2000 *Annu. Rev. Mater.* Sci. **30** 387.
[15] Pecharsky V. K, Gschneidner K. A, Tsokol A. O 2005 *Rep. Prog. Phys* **68** 1479.
[16] Phan M. H, Peng H. X, Yu S. C 2005 *J. Appl. Phys.* **97** 10M306.

Acknowledgments

This study is supported by the Tunisian Ministry of Higher Education and Scientific Research.

IOP Conf. Series: Materials Science and Engineering **28** (2012) 012053 doi:10.1088/1757-899X/28/1/012053

Effect of Ga doping on the physical properties of $La_{0.7}(CaSr)_{0.3}Mn_{1-x}Ga_xO_3$ (x=0, 0.025, 0.05, 0.075 and 0.1)

S. Othmani[1a] and M. Balli[b]

[a]Faculté des Sciences de Sfax, Université de Sfax, Laboratoire de Physique des Matériaux, Sfax University, B.P 1171, Sfax 3000, Tunisie

[b]University of Applied Sciences of Western Switzerland, Institute of Thermal Sciences and Engineering, CH-1401 Yverdon-les-Bains, Switzerland.

E-mail: safa.othmani@gmail.com

Abstract. In this paper we report on the structural, magnetic and transport properties of $La_{0.7}(CaSr)_{0.3}Mn_{1-x}Ga_xO_3$ (x=0, 0.025, 0.05, 0.075 and 0.1) manganites elaborated by sol-gel method. The substitution of Mn by Ga affects slightly the structural parameters. A decrease of both magnetization M and Curie temperature T_C has been observed with increasing Ga content. The temperature dependence of the resistance shows a metal-insulator transition around T_C.

1. Introduction

In the past years, much attention was paid to the doped perovskite manganites $R_{1-x}A_xMnO_3$ where R is a trivalent rare-earth element and A is a divalent alkaline earth one (A=Ca, Ba, Sr, etc.) [1, 2]. The magnetic ground state of perovskite manganites is very sensitive to the doped cations [3]. It has been known that ground states of manganites, are the ferromagnetic (FM) or antiferromagnetic(AFM) structure [4, 5, 6]. In addition, some authors discovered the coexistence and the competition of the FM double exchange (DE) interaction and the AFM super-exchange (SE) in some doped perovskite manganites [7].The rich diversity of the physical properties of manganites has generated amazing interest in such materials. The metal–insulator transition, observed near the Curie temperature, results in colossal magnetoresistance (CMR) which is defined in terms of sharp decrease of the electric resistance of manganites under external magnetic field [8, 9]. The CMR materials have been focused due to many fold reasons: the comprehension of basic phenomena related to the CMR effect, the optimization of known compounds, and synthesis of new materials with the aim of enhancing and tuning their physical properties [10, 11]

The electrical resistivity of doped manganites as a function of temperature exhibits three regimes (a) low temperature metallic like conduction with an unexpectedly large absolute value of the resistivity, (b) an abrupt drop in resistivity associated with magnetic ordering, and (c) high-temperature activated conduction. Analysis of charge transport in the ferromagnetic metallic state is essential in clarifying the specific mechanisms responsible for the resistivity behavior.

This paper is devoted to the study of the effect of Ga substitution in the $La_{0.7}Ca_{0.15}Sr_{0.15}Mn_{1-x}Ga_xO_3$ manganites ($0 \leq x \leq 0.1$). The effect of Ga substitution on the structural properties is reported, than a systematic study of the transport properties of $La_{0.7}Ca_{0.15}Sr_{0.15}Mn_{1-x}Ga_xO_3$ is reported in the composition range $0 \leq x \leq 0.1$.

---

[1] To whom any correspondence should be addressed.

2. Experimental

Polycrystalline samples $La_{0.7}Ca_{0.15}Sr_{0.15}Mn_{1-x}Ga_xO_3$ with $x = 0$ to $x = 0.1$ were prepared by chemical route as described in the literature [12, 13]. The starting ingredients are La_2O_3, $SrCO_3$, $CaCO_3$, MnO_2 and Ga_2O_3. The final heat treatment was performed at 1000°C for 12h in air. The crystalline structure of the samples was checked by *X*-ray powder diffraction at room temperature using Cu/K□ radiation. *XRD* data were refined by means of the Rietveld method using the FullProf refinement program [14]. The magnetic measurements versus temperature were performed using *BS1* and *BS2* magnetometers developed in Neel Institute, Grenoble for high and low temperatures respectively. The magnetization curves were obtained under an applied magnetic field changing from 0 to 5T in the temperature range 100 - 400K. Transport properties were measured also in Neel Institute (Grenoble) by the standard four-probe method, in the temperature range 5–300K, fitted with a commercial closed cycle refrigerator operating down to 10K.

3. Results and discussions

The room temperature *X*-ray diffraction patterns of $La_{0.7}Ca_{0.15}Sr_{0.15}Mn_{1-x}Ga_xO_3$ with $0 \leq x \leq 0.1$ are shown in **figure 1.** The red points show the actual data and the calculated curve is superimposed on them.

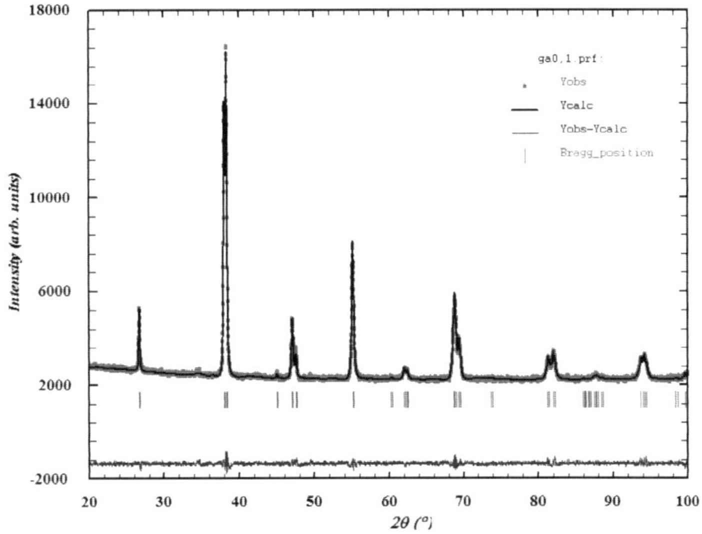

Figure 1. Observed (solid circles) and calculated (solid line) *XRD* patterns of the *Ga* doped samples (x=0.0 and 0.1) obtained at room temperature. The difference between these spectra is plotted at the bottom. Bragg reflections are indicated by ticks.

The bottom line shows the difference between the experimental and the calculated *XRD* patterns. Bragg peaks appear as vertical lines. Very good agreement between the calculated and the observed data is obtained which confirms that the samples are obtained without any other secondary or impurity phase. *XRD* patterns of the samples can be indexed in the rhombohedral system (Hexagonal setting) with *R-3C* space group. The structural parameters are listed in **table 1.** We can see that the change of lattice parameters with increasing *Ga* content is negligible [15] due to the fact that *Mn* and *Ga* have almost identical ionic radius [16].

Table 1. Lattice parameters and unit cell volume for $La_{0.7}Ca_{0.15}Sr_{0.15}Mn_{1-x}Ga_xO_3$ system.

x	0	0.025	0.05	0.075	0.1
Space group	R -3 C	R -3 C	R -3 C	R -3 C	R -3 C
a(Å)	5.487_2	5.491_3	5.495_2	5.490_9	5.495_5
a_r(Å)	3.158_5	5.463_3	5.456_9	5.466_8	5.455_1
□ (°)	60.48	60.33	60.46	60.29	60.48
V(Å^3)	346.68	348.58	348.32	348.87	348.11
Mn/Ga (site b)					
x	0.0000	0.0000	0.0000	0.0000	0.0000
y	0.0000	0.0000	0.0000	0.0000	0.0000
z	0.0000	0.0000	0.0000	0.0000	0.0000
La.Ca.Sr (site a)					
x	0.0000	0.0000	0.0000	0.0000	0.0000
y	0.0000	0.0000	0.0000	0.0000	0.0000
z	0.2500	0.2500	0.2500	0.2500	0.2500
O (site e)					
x	0.447_0	0.547_1	0.546_4	0.551_2	0.554_6
y	0.0000	0.0000	0.0000	0.0000	0.0000
z	0.2500	0.2500	0.2500	0.2500	0.2500
R_p	16.0	19.7	18.4	17.6	15.1
R_{wp}	11.2	11.3	13.1	13.3	9.48
χ^2	2.26	1.39	2.87	2.30	1.79
Mn-O	1.952_4	1.953_3	1.952_6	1.957_4	1.958_8
Mn-O-Mn	162.89	164.79	165.03	163.48	162.38
Bragg R-factor	6.00	1.43	2.64	4.57	2.91
RF-factor	6.49	1.35	2.37	5.29	1.79

The temperature dependence of the magnetization measured in magnetic applied field of 0.05T of $La_{0.7}Ca_{0.15}Sr_{0.15}Mn_{1-x}Ga_xO_3$ with $0<x<0.1$ is shown in **figure 2.** All samples exhibit a paramagnetic to ferromagnetic transition with increasing temperature. The Curie temperature T_C, defined by the minimum in *dM/dT*, has been determined to range from 336.5K for x=0 to 244.5K for x=0.1.
It is clearly seen that T_C decrease when the gallium continent increases. The decrease of T_C has been explained as a result of the dilution effect that causes a decrease of the molecular field acting on Mn ions according to the molecular field theory [17].

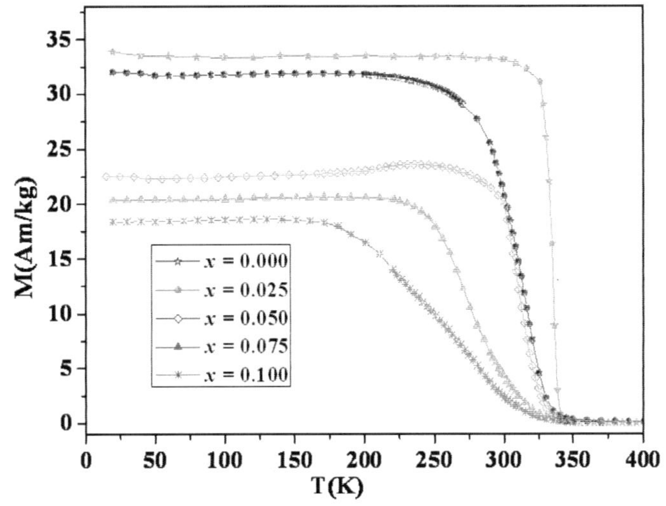

Figure 2. Variation of the magnetization M vs. temperature T for $La_{0.7}Ca_{0.15}Sr_{0.15}Mn_{1-x}Ga_xO_3$ compounds (x = 0.000, 0.025, 0.050, 0.075 and 0.100) at 0.05T.

Ferromagnetism is apparently due to the double exchange *DE* interaction between Mn^{3+} and Mn^{4+} ions. The sharp ferromagnetic-paramagnetic phase transition for x = 0, 0.025 and 0.05 indicates a first order transition and would promisingly imply a high entropy change. However the rest of the samples exhibit a second order transition.

Figure 3 shows the temperature dependence of the resistance at zero magnetic field for $La_{0.7}Ca_{0.15}Sr_{0.15}Mn_{1-x}Ga_xO_3$ for ($0\leq x\leq 0.1$). For the resistance curve, we can see that with increasing temperature an anomaly occurs in the electrical resistance. This temperature corresponds to the typical metal–insulator transition T_{mi}. The *M-I* transition temperature T_{mi} defined as the maximum in the *R(T)* curve, decreases with the *Ga* content and tends more and more to the Curie temperature (**figure 4**). We can see that the electrical transition corresponds to the magnetic transition.

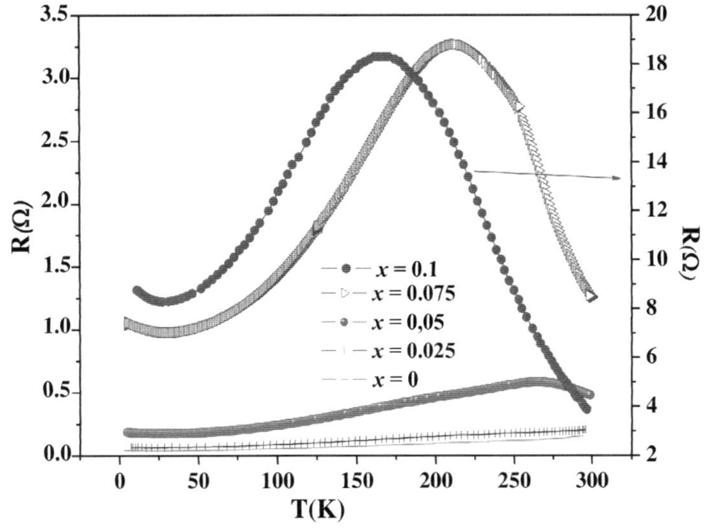

Figure 3. Temperature dependence of resistance for $La_{0.7}Ca_{0.15}Sr_{0.15}Mn_{1-x}Ga_xO_3$ ($0\leq x\leq 0.1$).

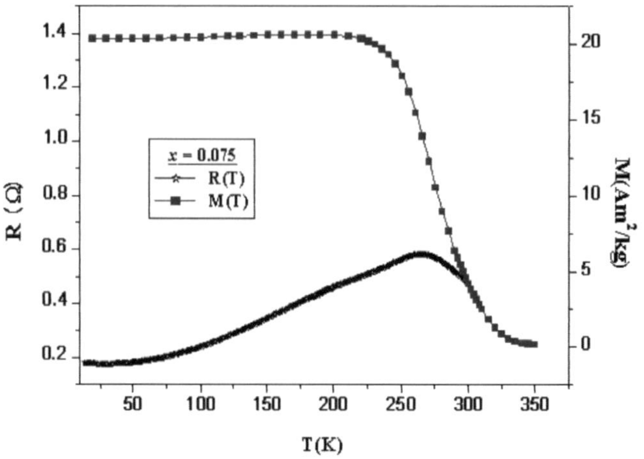

Figure 4. Temperature dependence of resistance and magnetization for $La_{0.7}Ca_{0.15}Sr_{0.15}Mn_{1-x}Ga_xO_3$ (x=0.075).

A small amount of *Ga* doping results in a large number of charge carriers in the e_g band and results in decrease in the resistivity. However, at very low temperatures (~30K), the resistivity of all the samples shows an upturn with decrease in temperature which is consistent with the earlier reports [18, 19, 20, 21]. This change of T_{mi} with doping doesn't correspond to the change with Fe doping and in A-site-doped perovskite manganites [13, 22, 23]. The metallic behavior of the samples in the temperature range ($50 \leq T \leq T_{mi}$) is retraced by using electron–phonon interaction with the model phonon spectrum comprising of an acoustic branch of Debye type and optical mode with characteristic Einstein temperature [24]. The contribution from acoustic and optical phonons together with the residual resistivity is smaller than experimental data [25, 26].

4. Conclusion

In conclusion, the relationship between compositions of $La_{0.7}Ca_{0.15}Sr_{0.15}Mn_{1-x}Ga_xO_3$ manganites for low doping *Ga* concentration ($0 \leq x \leq 0.1$) and structural as well as magnetic and transport properties has been studied. *XRD* patterns indicate that all synthesized compounds present a rhombohedral structure with space group $R\bar{3}C$. The unit-cell volume isn't affected with enhanced *Ga* doping due to the fact that Mn and Ga have almost identical ionic radius. Moreover, variation of the *Mn/Ga* ratio has the following effects: the curie temperature, which is also the working point of a magnetic refrigerant, decreases from 336.5K to 244.5K in the composition range $0<x<0.1$. Enhanced *Ga* content leaves an increased number of holes in the *Mn* e_g band, hence there are effectively more charge carriers leading to enhance hopping interaction and results in increased conductivity. The electrical resistivity of all the samples shows metal–insulator transition. It is clearly seen that T_{mi} decreases gradually with increasing *Ga* content.

References

[1] Tokura Y, Tomioka Y, Kuwahara H, Asamitsu A, Moritomo Y and Kasai M 1996 *J. Appl.Phys.* **79** 5288-92.

[2] Kuwahara H, Tomioka Y, Moritomo Y, Asamitsu A, Kasai M, Kumai R and Tokura Y 1996 *Science* **272** 80-82.

[3] Coey J. M. D, Viret M and von Molnar S 1999 *Adv. Phys.* **48(2)** 167-293.

[4] Viret M, Glattli H, Fermon C, Leon- Guevara De and Revcolevschi A 1997 *Physica B***241** 430-432.

[5] Samoilenko Z. A, Okunev V. D, Pushenko E. I, Pafomov N. N, Szymezak R, Baran M and Lewandowski S. J 2006 *Tech. Phys.* **51** 791.
[6] Phan M. H, Srinivas V, Yu S. C and Hur N. H 2003 *J. Appl. Phys* **93** 8200.
[7] De Lima O, Coaquira J. A. H, Almeida R, De Carvalho L and Malik S. K 2009 *J. Appl. Phys.* **105** 013907 - 013907-10.
[8] Sarkar T, Ghosh B, Raychaudhuri A. K 2008 *Phys. Rev.* ***B*77** 235112.
[9] Trukhanov S. V, Lobanovski L. S, Bushinsky M. V, Troyanchuk I O and Szymczak H 2003 *J. Phys. Condens. Matter* **15** 1783.
[10] Saket A, Nigam A. K, Malik S. K and Bahadur D 2008 **450** 136-141.
[11] Chen C. C, Shen C. H, Liu R. S, Lin J. G and Huang C. Y 2002 *Mater. Res. Bull.* **37** 235.
[12] Othmani S, Blel R, Bejar M, Sajieddine M, Dhahri E,. Hlil E. K 2009 *Solid. State. Commun* **149** 969-972.
[13] Othmani S, Bejar M, Dhahri E and Hlil E. K 2009 *J. Alloys Compd.* **475** 46–50.
[14] Rodriguez-Carvajal J, FULLPROF version 3.0.0, Laboratorie Leon Brillioun, CEA–CNRS, 1995.
[15] Li R. W, Sun J. R, Li Q. A, Zhang S. Y and Shen B. G 2003 *J. Magn. Magn. Mat.* **265** 248-256.
[16] Shannon R. D 1976 *Acta Cryst.* **A32** 751.
[17] Nam D. N. HN, Bau L. V, Khiem N. V, Dai N. V, Hong L. V, Phuc N. X, Newrock R. S and Nordblad P 2006 *Phys. Rev.* ***B*73** 184430.
[18] Tiwari A and Rajeev K. P 1999 *Solid State Commun.* **111** 33.
[19] Ziese M 2003 *Phys. Rev.* ***B* 68** 132411.
[20] Kumar D, Sankar J, Narayan J, Singh R. K and Majumdar A. K 2002 *Phys. Rev. B* **65** 094407.
[21] Xu Y, Zhang J, Cao G, Jing C and Cao S 2006 *Phys. Rev. B***73** 224410.
[22] Rao G. H, Sun J. R, Kattwinkel A, Haupt L, BaKrner K, Schmitt E and Gmelin E 1999 *Physica B***269** 379 – 385.
[23] Rao G. H, Sun J. R, Liang J. K and Zhou W. Y 1997 *Phys. Rev. B***55** 3742.
[24] Dinesh Varshney, Neha Dodiya, Mohammed Wasim Shaikh 2011 *J. Alloys Comp.* **509** 7447–7457.
[25] Tiwari A and Rajeev K. P 1999 *Solid State Commun.* **111** 33.
[26] Ziese M 2003 *Phys. Rev. B***68** 132411.

IOP Conf. Series: Materials Science and Engineering **28** (2012) 012054 doi:10.1088/1757-899X/28/1/012054

Effect of monovalent doping on the physical properties of $La_{0.7}Sr_{0.3}MnO_3$ compound synthesized using sol-gel technique

F. Ayadi[a], F. Saadaoui[a], W. Cheikhrouhou-Koubaa[a], M. Koubaa[a], A. Cheikhrouhou [1a,b], L. Sicard[c], S. Ammar[c]

[a]Laboratoire de Physique des Matériaux, Sfax University, Faculté des Sciences de Sfax B.P 1171, 3000 Sfax Tunisia.

[b]Neel Institute, Département Nano Science. B.P 166, 38042 Grenoble Cedex 9. France.

[c]ITODYS, Université Paris Diderot, CNRS UMR 7086, 15 rue Jean Antoine de Baïf, 75205 Paris, France.

E-mail: abdcheikhrou@yahoo.fr; ayadi.firas.1985@gmail.com

Abstract. In this work, we have studied the effect of monovalent substitution on the structural, magnetic and magnetocaloric properties of the P-doped $La_{0.7}Sr_{0.2}M_{0.1}MnO_3$ (M=Na, K) manganites synthesized by sol-gel technique. Structural investigations using X-ray diffraction show that all the elaborated compounds crystallize in the rhomboedral structure with R-3c space group. The increase in the ionic radius $\langle r_A \rangle$ induces a rotation of MnO_6 octahedra leading to a change in the Mn-O-Mn angles and the Mn-O distances. Magnetizations measurements were carried out using vibrating sample magnetometer operating above room temperature with magnetic applied field up to 5 tesla. All our samples exhibit paramagnetic to ferromagnetic transition with a Curie temperature above room temperature. From magnetic measurements versus magnetic applied field at several temperatures, we have determined the magnetocaloric effect using the classical thermodynamics based on Maxwell relations.

1. Introduction

The research in the perovskite manganites with general formula $Ln_{1-x}A_xMnO_3$ (Ln = Rare-earth cation, A= alkali-metal or alkalineearth cation) has attracted much renewed attentions after the observation of large magnetocaloric effect (MCE) in these compounds [1–3]. In fact, magnetic refrigeration is becoming a promising technology to replace the conventional gas-compression expansion technique [4,5]. For sub-room temperature magnetic refrigeration applications, gadolinium is considered as the first material known to show a large MCE, since it exhibits a maximum value of magnetic entropy change, $|\Delta S_M^{max}|$, of 5 J/kgK at 294K under magnetic applied field change of 2 T [6]. Otherwise, manganites are also interesting for application as potential candidates in magnetic refrigeration since they have low coast, are easy to elaborate and possess tunable T_C and high chemical stability. Generally, in manganites, the large magnetic entropy change could be attributed to the variation of the double exchange interaction between Mn^{3+} and Mn^{4+} ions and the strong spin lattice coupling [7–9]. A large overview of magnetocaloric effects observed in perovskite manganites has been provided by Phan and Yu [10]. The substitution of trivalent element by monovalent alkali-metal in manganites induces less inhomogeneity (the substitution oxidizes two Mn ions from Mn^{3+} to Mn^{4+} state) and improves the physical properties [11–13]. Our research group [14–16] has widely studied the magnetocaloric properties of $La_{0.65}Ca_{0.35}MnO_3$ system where a partial calcium substitution by monovalent element is found to improve significantly the T_C values which reach room temperature.

[1] To whom any correspondence should be addressed.

Zhong et al. [17] have studied the magnetocaloric properties in $La_{1-x}Na_xMnO_3$ ($0 \leq x \leq 0.2$) and found that for x = 0.2, $|\Delta S_M^{max}|$ reaches 1.96J/kgK upon a magnetic applied field change of 1 T. Hou et al. [18] investigated the MCE properties of $La_{0.8-x}Nd_xNa_{0.2}MnO_3$ and found that with decreasing $\langle r_A \rangle$, $|\Delta S_M^{max}|$ decreases from 0.43J/kgK at 334K for x = 0 to 0.36 J/kgK at 295K for x = 0.15 under ΔH = 1T. Furthermore, the study of magnetocaloric properties in $La_{1-x}K_xMnO_3$ compounds synthesized by sol gel method has shown an enhancement in both T_C and $|\Delta S_M^{max}|$ values reaching respectively 344K and 2.19 J/kgK under 1.5T for x=0.2 sample [19]. More recently, Das and Dey have observed similar behavior in $La_{1-x}K_xMnO_3$ compounds synthesized by pyrophoric method with $|\Delta S_M^{max}|$ =3J/kgK under 1T for x = 0.15 [20].

Our present study has been carried out to investigate the effect of monovalent element doping on the structural, magnetic and magneto-caloric properties of $La_{0.7}Sr_{0.2}M_{0.1}MnO_3$ (M= K and Na) compounds with fixed Mn^{4+}/Mn^{3+} ratio equal to 2/3.

2. Experimental techniques

The sol-gel process consists in dissolving the precursors (oxides and carbonates) in a concentrated nitric acid solution to transform them into nitrates. This step is generally made under the ambient temperature, however we can work up to 60C° in order to accelerate the dissolution. After reaching total dissolution, one adds the monohydrate citric acid and the ethylene glycol to complete the homogeneity and the transparency of the solution and to initialize the complexation of the atoms. This solution is slowly evaporated to 130C° until the formation of a residue of high viscosity. A transparent gel is developed during the heating process. In a later step, we raise the temperature until 300C° with a speed of climb of 10C°/minute, with the purpose of having the propagation of a combustion which transforms the freezing into fine powder (Ash).

The resulting powder is heated to 450C° during 6 hours in order to decompose the organics. After crushing, the samples are calcined to 650C° during 12 hours and then to 800C° for the same duration. Finally, the powders were then pressed into pellets (of about 1mm thickness and 13mm diameter) and sintered at 900C° and 1000C° during 24 hours. All annealing are performed in air.

The compounds $La_{0.7}Sr_{0.2}M_{0.1}MnO_3$ (M= K, Na), and the parent compound $La_{0.7}Sr_{0.3}MnO_3$ were synthesized from La_2O_3, $SrCO_3$, MnO_2, K_2CO_3 and Na_2CO_3 with high purity precursor (>99.9%) according to the following equation:

$$0{,}35\ La_2O_3 + (0{,}3-x)\ SrCO_3 + 0.5x\ M_2CO_3 + MnO_2 \longrightarrow La_{0.7}Sr_{0.3-x}M_xMnO_3 + \delta CO_2\uparrow + \delta' O_2\uparrow$$

With x=0 and 0.1 and M=K, Na

3. Characterization by scanning electronic microscopy

The morphology and the size of the crystalline grains in $La_{0.7}Sr_{0.2}M_{0.1}MnO_3$ (M= K, Na) compounds were studied using a scanning electron microscopy (SEM) (Figure 1).

We can observe that SEM images reveal a variable morphology of the grains. The characteristic parameters of the microstructure are: mean size of the grains, apparent compactness and porosity.

$La_{0.7}Sr_{0.2}K_{0.1}MnO_3$ compound is characterized by two different sizes of grain: small spherical grains of approximately 200 nm surrounded by more coarse grains (500 nm). The mean size of the grains is approximately 360 nm.

While $La_{0.7}Sr_{0.2}Na_{0.1}MnO_3$ compound also forwards a bimodal distribution of size of grains, but with a rather parallepipedic form. The mean size of the grains is more important (450 nm approximately).

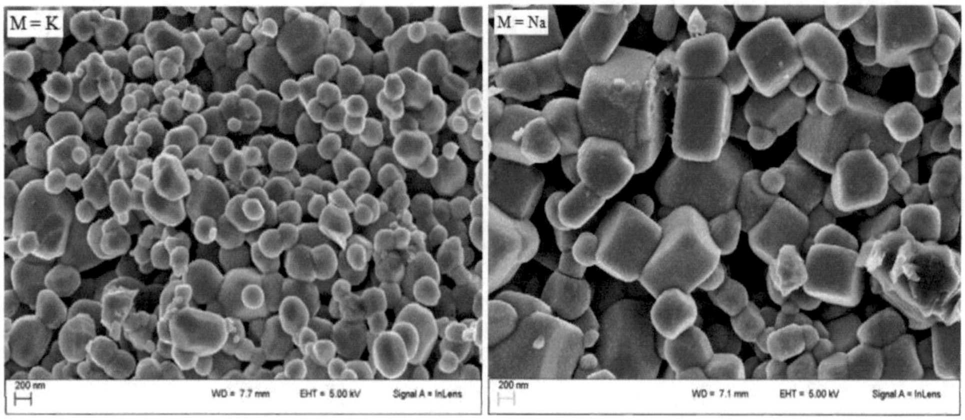

Figure. 1 Scanning electron microscopy images for $La_{0,7}Sr_{0,2}M_xMnO_3$ (M= K, Na) samples.

4. Structural characterizations

The results of the X-ray diffraction (XRD) studies at room temperature of $La_{0,7}Sr_{0,3-x}M_xMnO_3$ (x=0 and x=0,1 ; M= K, Na) compounds indicate that all our samples are single phase and present the same perovskite structure. The structural parameters of the samples are refined by the Rietveld profile-fitting method. Figure 2 shows a typical refined diagram of $La_{0,7}Sr_{0,2}K_{0.1}MnO_3$ sample.

The profile refinement is started with scale and background parameters followed by the unit cell parameters. Then, the peak asymmetry and preferred orientation corrections are applied. Finally, the positional parameters and the individual isotropic parameters are refined.

It has been found that all our samples can be indexed in the rhombohedral system with R-3c space group. The final refinement values of the structural parameters and χ^2 values of the Rietveld analysis are summarized in Table 1. We can notice an increase of the unit cell volume with the A-site ionic radius (Figure 3), as expected.

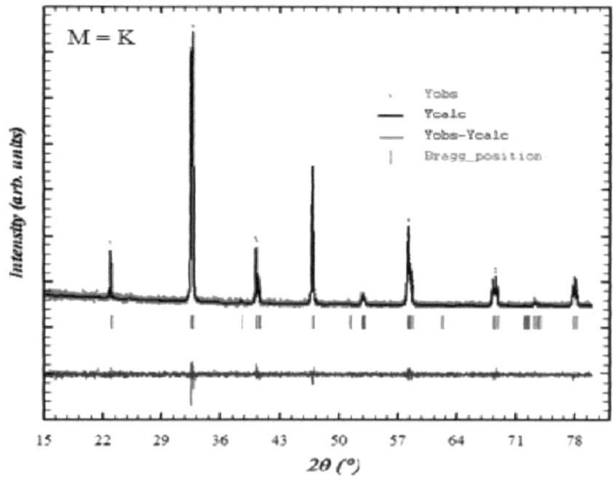

Figure. 2 X-Ray diffraction patterns at room temperature of $La_{0.7}Sr_{0.2}K_{0.1}MnO_3$ sample. Circles indicate the experimental data and the calculated data is the continuous line overlapping them. The lowest curve shows the difference between experimental and calculated patterns. The vertical bars indicate the expected reflection positions.

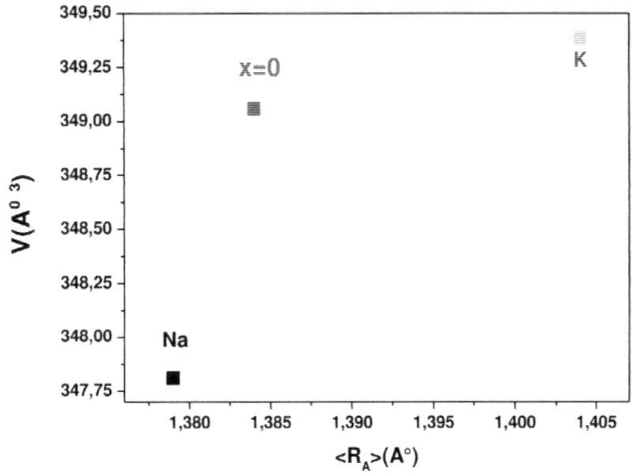

Figure 3. Unit cell volume versus $<r_A>$ for $La_{0.7}Sr_{0.3-x}M_xMnO_3$ (x=0 and x=0.1 ; M= K, Na) samples.

Table1. Refined structural parameters of $La_{0,7}Sr_{0,3-x}M_xMnO_3$ (x=0 and x=0,1 ; M= K, Na) samples at room temperature.

	$La_{0.7}Sr_{0.3}MnO_3$	$La_{0.7}Sr_{0.2}Na_{0.1}MnO_3$	$La_{0.7}Sr_{0.2}k_{0.1}MnO_3$
Groupe d'espace	R-3c	R-3c	R-3c
a (Å)	5,493(3)	5,486(5)	5,495(1)
b (Å)	5,493(3)	5,486(5)	5,495(1)
c (Å)	13,356(6)	13,342(3)	13,360(9)
V ($Å^3$)	349,058(6)	347,811(1)	349,385(8)
Mn–O (Å)	1,9501(3)	1,9498(2)	1,9500(2)
Mn-O-Mn(°)	166,9	165,9	167,3
χ^2(%)	1,274	1,192	1,254
$<r_A>$ (Å)	1,384	1,379	1,404
$<r_B>$ (Å)	0,6105	0,599	0,599
σ^2 ($Å^2$)	$1,344.10^{-3}$	$1,009.10^{-3}$	$7,184.10^{-3}$
R_P(%)	18,5	21,1	19,1
R_{WP}(%)	18,8	20,0	19,2
R_e(%)	16,7	18,3	17,1
R_F(%)	1,78	2,77	3,81
R_B(%)	2,54	2,92	2,70

5. Magnetic characterizations

Magnetization measurements as a function of temperature in the range 280–400K and in a magnetic applied field of 50mT showed that our synthesized samples exhibit a paramagnetic to ferromagnetic transition with decreasing temperature as illustrated in Figure 4 for $La_{0,7}Sr_{0,3-x}M_xMnO_3$ (x=0 and x=0,1; M= K, Na) compounds.

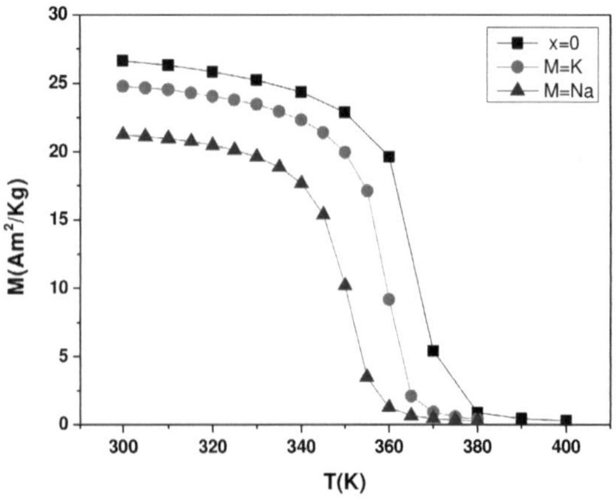

Figure. 4 Temperature dependence of magnetization at 50mT for $La_{0.7}Sr_{0.3-x}M_xMnO_3$ (x=0 and x=0.1; M= K, Na) samples.

We can also notice that M(T) the curves are characterized by a clear para-ferromagnetic transition, which confirms the good crystalline quality of our samples.

The Curie temperature (T_C) values, determined from the inflection point of the M(T) curves, are respectively 370K, 360K and 350K for x=0 and x=0.1 (M= K, Na). The highest T_C was observed in the parent compound for which the Mn^{3+}/Mn^{4+} ratio is optimum according to the theory of double exchange of Zener [21] who predicts a maximum T_C for 33% of Mn^{4+}.

Indeed, partial substitution of Sr by Na or K induces an increase in the rate of Mn^{4+} and one should consequently observe a decrease of T_C, since the antiferromagnetic super-exchange interaction Mn^{4+}-O-Mn^{4+} is strengthened with the profit of the ferro-double exchange interaction Mn^{3+}-O-Mn^{4+}. The most important decrease was observed for Na substitution although the rate of Mn^{4+} is constant, this behavior can be rather be interpreted in terms of orbital recovery since the Mn-O-Mn angle is weaker in $La_{0,7}Sr_{0,2}Na_{0.1}MnO_3$ compound. A similar behavior was observed in the same compounds elaborated with the ceramic method [22]. This decrease of magnetization was also observed by Zemni et coworkers in $La_{0,6}Sr_{0,4-x}K_xMnO_3$ compounds ($0<x<0.15$) [23].

In order to confirm the ferromagnetic behavior of our compounds to low temperatures, we investigated magnetization measurements versus magnetic field up to 5T at different temperatures in the range 280-400K. Figure 5 shows typical M(H) isotherms for $La_{0,7}Sr_{0,2}Na_{0.1}MnO_3$ compound.

For temperatures lower than T_C, the magnetization increases sharply below 0.5T and then starts to saturate for higher values, which confirms the ferromagnetic behavior of our samples at low temperatures.

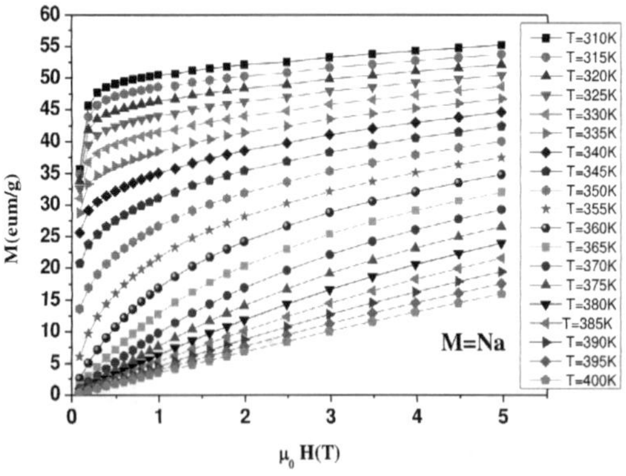

Figure. 5 Plot of magnetization M versus magnetic applied field μ_0H up to 5T at several temperatures for $La_{0.7}Sr_{0.2}Na_{0.1}MnO_3$ sample.

In figure 6, we have plotted the variations of M^2 versus μ_0H/M (Arrott curves) for $La_{0.7}Sr_{0.2}Na_{0.1}MnO_3$ compound. We can notice that for all temperatures the recorded slope is positive signature of a second order transition [24]. Furthermore, the T_C values deduced from these plots are very close to those determined from the M (T) curves.

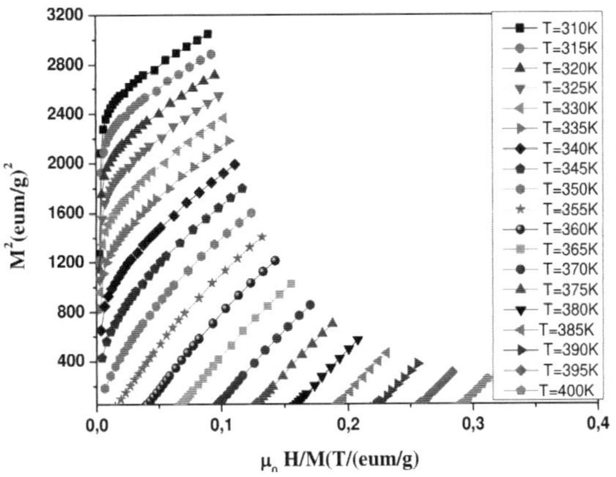

Figure. 6 M^2 versus H/M isotherms for $La_{0.7}Sr_{0.2}Na_{0.1}MnO_3$ sample.

6. Magnetocaloric characterizations

From the isothermal magnetization versus magnetic applied field, and according to the conventional thermodynamics based on Maxwell relations, we determined the variation of the magnetic entropy $|\Delta S_M^{max}|$ in our compounds as a function of temperature with various magnetic field changes.

The magnetic entropy change ΔS_M presents negative values. The maximum of $|\Delta S_M|$, observed around T_C, increases with the magnetic field change, as shown in the figure 7 for $La_{0.7}Sr_{0.2}Na_{0.1}MnO_3$ sample.

The maximum values of $|\Delta S_M|$ observed are 2.32, 2.56 and 1.93 for $La_{0.7}Sr_{0.3}MnO_3$, $La_{0.7}Sr_{0.2}K_{0.1}MnO_3$ and $La_{0.7}Sr_{0.2}Na_{0.1}MnO_3$ compounds respectively under a magnetic field change of 2T.

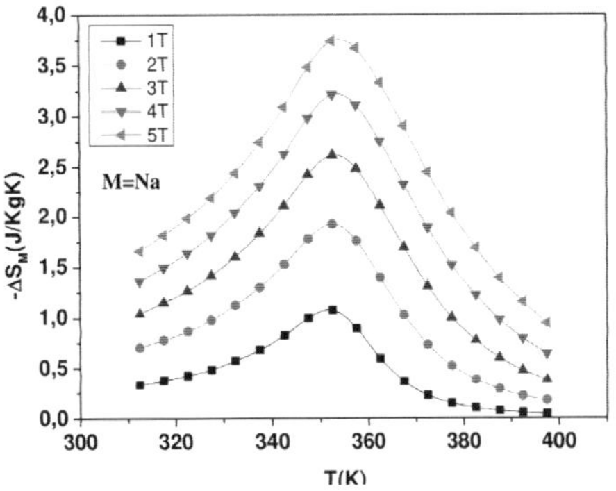

Figure. 7 Magnetic entropy change $-\Delta SM$ evolution versus temperature at several magnetic applied fields for $La_{0.7}Sr_{0.2}Na_{0.1}MnO_3$ sample.

Although the actual values of $|\Delta S_M|$ are lower than that obtained in gadolinium, which is considered as the most active material in the field of the magnetic refrigeration [25], our compounds present relatively high values with a large distribution.

In the evaluation of the magnetocaloric effect, it is important to consider not only the maximum entropy but also the relative cooling power (RCP) defined as $RCP = -\Delta S_M^{max}(T, H) \times \delta T_{FWHM}$ where δT_{FWHM} is the full-width at half-maximum of $|\Delta S_M|$ versus temperature [26]. It is then important that the magnetocaloric effect extends over a large temperature range. Figure 8 illustrates our RCP values in comparison with gadolinium under a magnetic field change of 5T.

The highest RCP value is observed in the parent compound, it reaches 60 % of that observed in Gd, which is interesting enough for technical uses.

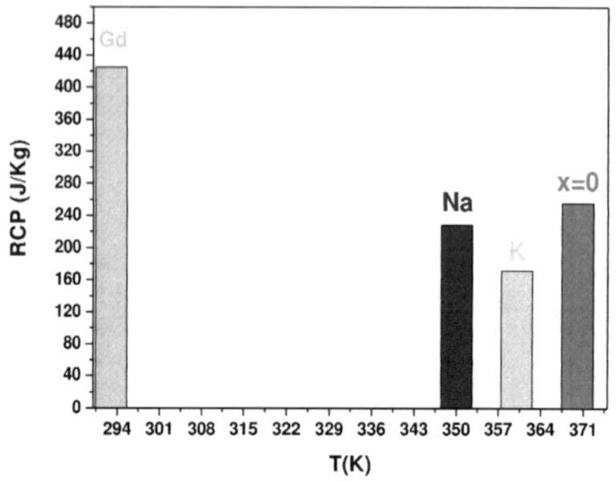

Figure. 8 RCP values versus T_C for $La_{0.7}Sr_{0.3-x}M_xMnO_3$ (x=0 and x=0.1 ; M= K, Na) and Gd.

7. Conclusions

In this work we were interested in the effect of monovalent doping in $La_{0.7}Sr_{0.2}M_{0.1}MnO_3$ (M=K, Na) compounds elaborated using sol-gel method. Our synthesized samples crystallize in the rhomboedral system with R-3c space group. All the compounds present a paramagnetic-ferromagnetic transition when the temperature decreases. The Curie temperature T_C is found to slightly decrease with partial substitution of Sr by the monovalent element. This decrease is more important in the case of sodium. Our samples present a large magnetocaloric effect and can be potential candidates in magnetic refrigeration in order to replace the conventional expansion compression technique of the gas which is very toxic for the environment.

References

[1] Guo Z. B, Du Y. W, Zhu J. S, Huang H, Ding W. P, Feng D 1997 *Phys. Rev. Lett.* **78** 1142.
[2] Wang Z. M, Ni. G, Xu Q. Y, Sang H, Du Y. W 2001 *J. Appl. Phys.* **90** 5689.
[3] Zhensheng P 2004 *J. Rare Earths* **22** 232.
[4] Shull R. D, McMichael R. D, Ritter J. J 1993 *Nanostruct. Mater.* **2** 205.
[5] Casanova F, Batlle X, Labarta A 2002 *Phys. Rev. B* **66** 212402.
[6] Gschneidner K. A, Pecharsky V. K 2000 *Annu. Rev. Mater. Sci.* **30** 387.
[7] Zener C 1951 *Phys. Rev.* **118** 440.
[8] Phan M. H, Tian S. B, Hoang D. Q, Yu S. C, Nguyen C, Ulyanov A. N 2003 *J. Magn. Magn. Mater.* **309** 258–259.
[9] Yuan S. L, Xia Z. C, Liu L, Chen W, Zhao L. F, Tang J, Zhang G. H, Zhang L. J, Cao H, Feng W, Tian Y, Niu L. Y, Liu S 2003 *Phys. Rev. B* **68** 184423.
[10] Phan M. H, Yu S. C 2007 *J. Magn. Magn. Mater.* **308** 325.
[11] Tang T, Gu K. M, Cao Q. Q, Wang D. H, Zhang S. Y, Du Y. W 2000 *J. Magn. Magn. Mater.* **222** 110.
[12] Zhu H, Song H, Zhong Y. H 2002 *Appl. Phys. Lett.* **81** 3416.
[13] Das S, Dey T. K 2006 *J. Phys.: Condens. Matter.* **18** 7629.
[14] Koubaa M, Cheikhrouhou-Koubaa W, Cheikhrouhou A 2008 *Physica B* **403** 2477.

[15] Koubaa M, Cheikhrouhou-Koubaa W, Cheikhrouhou A 2009 *J. Alloys Compd.* **473** 5.
[16] Koubaa M, Cheikhrouhou-Koubaa W, Cheikhrouhou A 2009 *J. Phys. Chem. Solids* **70** 326.
[17] Zhong W, Cheng W, Ding W. P, Zhang N, Du Y. W, Yan Q. J 1998 *Solid StateCommun.* **106** 55.
[18] Hou D, Yue C, Bai Y, Liu Q, Zhao X, Tang G 2006 *Solid State Commun.* **140** 459
[19] Zhong W, Chen W, Ding W P, Zhang N, Hu A, Du Y. W, Yan Q. J 1999 *J. Magn. Magn. Mater.* **195** 112.
[20] Das S, Dey T. K 2007 *J. Alloys Compd.* **440** 30.
[21] Zener C 1951 *Phys. Rev.* **82** 403.
[22] Cheikh-Rouhou Koubaa W, Koubaa M, Cheikh-Rouhou A 2009 *J. Physics Procedia* **2** 992
[23] Zemni S, Dhahri Ja, Cherif K, Dhahri J, Oumezzine M, Ghedvia M, Vincent H 2005 *J. Alloys comp.* **392** 55.
[24] Singh N. K, Suresh K. G, Nigam A. K 2003 *Solid State Commun.* **127** 373.
[25] Lee J. S 2004 *Phys. Status Sol. B* **241** 1765.
[26] Amaral J. S, Reis M. S, Amaral V. S, Mandonca T. M, Araujo J. P, Sa M. A, Tavares P. B, Vieira J. M 2005 *J. Magn. Magn. Mater.* **686** 290–291.

Acknowledgments
This work has been supported by the ministry of higher education and scientific research.

Effect of sodium substitution on the structural, magnetic and magnetocaloric properties of $Nd_{0.5}Sr_{0.5}MnO_3$ perovskite manganites

A. Mehri[a], W. Cheikhrouhou Koubaa[a], M. Koubaa[a], A. Cheikhrouhou[1] [a, b]
[a] Laboratoire de Physique des Matériaux, Sfax University, Faculté des sciences de Sfax, B.P.1171, 3000 Sfax, Tunisia.
[b] NEEL Institute, B.P.166, 38042 Grenoble Cedex 9, France.

E-mail: abdcheikhrouhou@yahoo.fr; afef_mehri1@yahoo.fr

Abstract. The influence of Na doping upon orbital and charge ordering in $Nd_{0.5}Sr_{0.5}MnO_3$ manganites has been investigated combining magnetic measurements with X-ray powder diffraction at room temperature. It is shown that Na doping induces or reinforces the ferromagnetic metallic state, and destroys the charge ordering one and its associated antiferromagnetism whatever is its nature. The maximum magnetic entropy change $|\Delta S_M^{Max}|$, in a magnetic field change of 4T is found to be 0.49J/kgK and 3.1J/kgK for $Nd_{0.5}Sr_{0.5}MnO_3$ and $Nd_{0.5}Sr_{0.3}Na_{0.2}MnO_3$ respectively.

1. Introduction

The manganites $Ln_{1-x}A_xMnO_3$, which exhibit colossal magnetoresistance (CMR), are generally characterized by a transition from a paramagnetic semi-conducting state to a ferromagnetic metallic state as the temperature decreases [1-6]. Moreover, $Nd_{0.5}Sr_{0.5}MnO_3$ compound exhibits a structural transition from a ferromagnetic phase with orthorhombic structure to antiferromagnetic charge ordered phase with monoclinic structure at $T_{CO} = 160K$ [7,8].

The magnetic refrigeration has attracted considerable interest due to its advantages over gas refrigeration [9]. To explore magnetic refrigerants with a large magnetic entropy change $|\Delta S_M|$ at high temperatures (T > 20 K), previous studies mainly concentrated on intermetallic compounds and rare earth alloys with high total angular momentum quantum number. These studies provided a comparatively large magnetic entropy change around the Curie temperature (T_C) [10], as observed in Gd [11], RAl_2 (R = Dy, Ho, Er) [12], etc. In particular, a giant magnetocaloric effect takes place in $Gd_5Si_{2-x}Ge_{2+x}$ [13] and $Fe_{49}Rh_{51}$ [14] at the first-order transition temperature. Recently, a large magnetic entropy change was observed in perovskite-type ferromagnetic oxides [15-18]. The intense interest in perovskite-type ferromagnetic manganese oxides $R_{1-x}B_xMnO_3$ (where R is a rare earth ion and B is a divalent alkali-ion) has been prompted by the observation of colossal magnetoresistance [3,19-21]. Experimentally the charge-ordered state (CO), which is defined as the real-space ordering of charge carriers Mn^{3+} and Mn^{4+} was observed in many of the hole-doped manganese oxides [22-25]. In particular, a drastic change of lattice constant, electron and magnetic properties take place in $Nd_{0.5}Sr_{0.5}MnO_3$ at the (CO) transition temperature (T_{CO}) [24,25]. Noting that the magnetic entropy

[1] To whom any correspondence should be addressed.

change is strongly correlated to the structural and magnetic transitions, we studied the magnetic entropy change of $Nd_{1-x}Sr_xMnO_3$ system. For $0.3 \leq x \leq 0.4$, the $Nd_{1-x}Sr_xMnO_3$ compound is known to be a ferromagnetic (FM) conductor at low temperature [24]. At x = 0.5, however, a charge-ordered antiferromagnetic (AFM) state was discovered at low temperature by neutron diffraction studies and a first-order magnetic phase transition from the charge-ordered AFM state to the charge-disordered FM state has been observed [24,25].

In the present work, we seek to explore the effect of disorder on 50-50 composition on the borderline of ferromagnetic metallic (FMM) and charge ordering insulator (COI). We elaborated using the solid state reaction at high temperature the $Nd_{0.5}Sr_{0.5-x}Na_xMnO_3$ ($0 \leq x \leq 0.2$) polycrystalline samples and studied the effect of the Na doping on the structural, magnetic and magnetocaloric properties of the charge ordering $Nd_{0.5}Sr_{0.5}MnO_3$ compound.

2. Experimental techniques

Powder samples have been elaborated using the conventional solid-state reaction at high temperature by mixing Nd_2O_3, MnO_2, $SrCO_3$ and Na_2CO_3 with a high purity up to 99.9% in the desired proportion according to the reaction:

$$\frac{1}{4}Nd_2O_3 + (0.5-x)SrCO_3 + \frac{x}{2}Na_2CO_3 + MnO_2 \rightarrow Nd_{0.5}Sr_{0.5-x}Na_xMnO_3 + \delta CO_2 + \delta' O_2$$

The starting materials were intimately mixed, ground, and fired in air at 1000°C for 4 days with intermediate regrinding and repelling. The obtained powders were then pressed under 4 tones into pellets of about 1mm thickness and 13mm diameter and sintered at 1100°C for 4 days in air with several intermediate regrinding and repressing into pellets. Finally, the pellets were rapidly quenched to room temperature. This step was done in order to keep the obtained structure at high temperature. The physical properties depend strongly on the synthesis route and also on the cooling method. Phase purity, homogeneity, and cell dimensions were determined by X-ray diffraction (XRD) at room temperature. Structural analysis was carried out using the standard Rietveld Technique [26]. Magnetization measurements versus temperature and versus magnetic applied field were carried out using a vibrating sample magnetometer in the temperature range 10-300K with an applied field up to 7T. MCE was deduced from the magnetization measurements versus magnetic applied field up to 7T at several temperatures.

3. Results and discussions

3.1. X-ray diffraction analysis

The X-ray diffraction spectra of our synthesized samples $Nd_{0.5}Sr_{0.5-x}Na_xMnO_3$ for $0 \leq x \leq 0.2$ are characterized by the lines of perovskite structure, which confirm the single orthorhombic phase with *Pbnm* space group as has been observed by Tomioka [8]. The structural parameters were refined using the standard Rietveld technique based on the FULLPROF code. In the profile refinement, we began with scale and background parameters followed by the unit cell parameters. We introduced then the peak asymmetry and the preferred orientation. Finally, we treated the positional parameters and the individual isotropic parameters. The quality of refinement was evaluated through the goodness of the fit indicator χ^2. Typical X-Ray diffraction pattern at 300K, including the observed and calculated profiles as well as the difference profile for $Nd_{0.5}Sr_{0.5}MnO_3$ and $Nd_{0.5}Sr_{0.3}Na_{0.2}MnO_3$ compounds are shown in Fig.1. Detailed results of structural refinement as well as the A-site ionic radius $\langle r_A \rangle$ and the mismatch effect σ^2 are listed in Table 1.

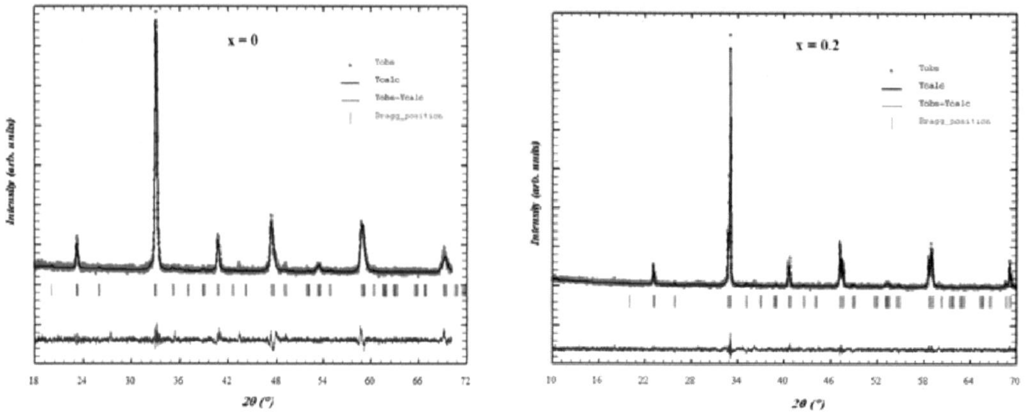

Figure 1. XRD pattern of $Nd_{0.5}Sr_{0.5}MnO_3$ and $Nd_{0.5}Sr_{0.3}Na_{0.2}MnO_3$ samples, squares indicate the experimental data and the calculated data is the continuous line overlapping them. The lowest curve shows the difference between experimental and calculated patterns. The vertical bars indicate the expected reflection positions.

Table 1. Crystallographic data of $Nd_{0.5}Sr_{0.5-x}Na_xMnO_3$ samples ($0 \leq x \leq 0.20$)

	X = 0	X = 0.05	X = 0.1	X = 0.15	X = 0.2
a(Å)	5.47	5.4642	5.4643	5.4732	5.4682
b(Å)	5.475	5.46	5.4488	5.4286	5.4292
c(Å)	7.6559	7.6479	7.643	7.6379	7.6377
V(Å^3)	229.28	228.17	227.5628	226.93	226.75
χ^2 (%)	1.33	1.86	1.86	1.5	1.23
$<r_A>$ (Å)	1.2365	1.233	1.2295	1.226	1.2225
σ^2(Å^2)	0.54010^{-3}	0.51210^{-3}	0.48110^{-3}	0.44810^{-3}	0.41310^{-3}

With increasing sodium substitution rate, $< r_A >$ and consequently the unit cell volume is found to decrease from 229.3 Å^3 for x = 0 to 226.8 Å^3 for x = 0.2 (Fig.2), which can be explained by the decrease of σ^2.

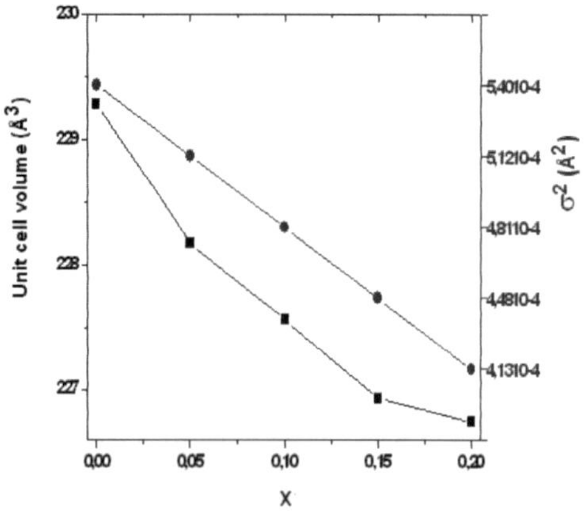

Figure 2. Evolution of the unit cell volume and σ^2 evolution versus X for $Nd_{0.5}Sr_{0.5-x}Na_xMnO_3$ compounds.

IOP Conf. Series: Materials Science and Engineering **28** (2012) 012055 doi:10.1088/1757-899X/28/1/012055

3.2. Magnetic properties

Magnetization measurements, recorded as a function of temperature in a magnetic applied field of 50mT, show that all our compounds undergo a paramagnetic to ferromagnetic transition near room temperature and the antiferromagnetic interaction linked to charge ordering is suppressed. Figure 3 illustrates the temperature dependence of the magnetization (M(T) curves) for $Nd_{0.5}Sr_{0.5-x}Na_xMnO_3$ compounds with $0 \leq x \leq 0.2$. Several authors argued that the cationic mismatch influences significantly the charge ordering in $Nd_{0.5}Sr_{0.5}MnO_3$ compounds. The M(T) curves show that for x = 0 we observe two transitions related to PM-FM at T_C = 263K and FM-AFM transitions at T_N = 115K, whereas for substituted samples, only a PM-FM transition is observed as we can see for x = 0.2 compound.

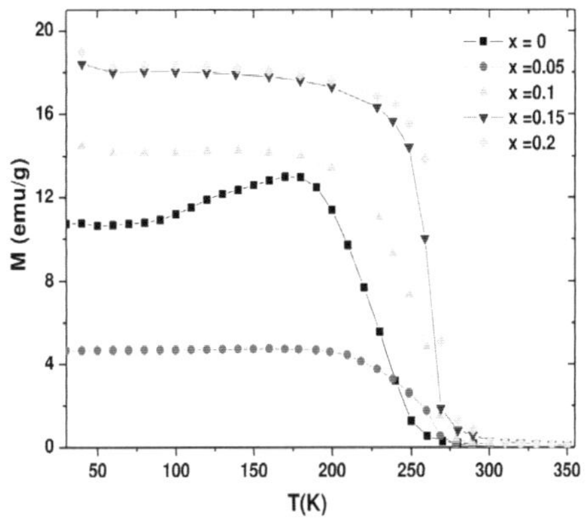

Figure 3. Evolution of the magnetization as a function of temperature at μ_0H = 50mT for $Nd_{0.5}Sr_{0.5-x}Na_xMnO_3$ samples x = 0; 0.05; 0.1; 0.15 and 0.20.

The Curie temperature TC, deduced from extrapolation of the Curie-Weiss law at 1/x =0 is found to increase with Na substitution rate from 263K for x = 0 to 292K for x = 0.2 as can be shown in Fig.4. This behavior has been previously observed in our group for monovalent substitution [27-29]
In order to confirm the ferromagnetic behavior at low temperatures, we have carried out magnetization measurements as a function of magnetic applied field μ_0H up to 7T at several temperatures.

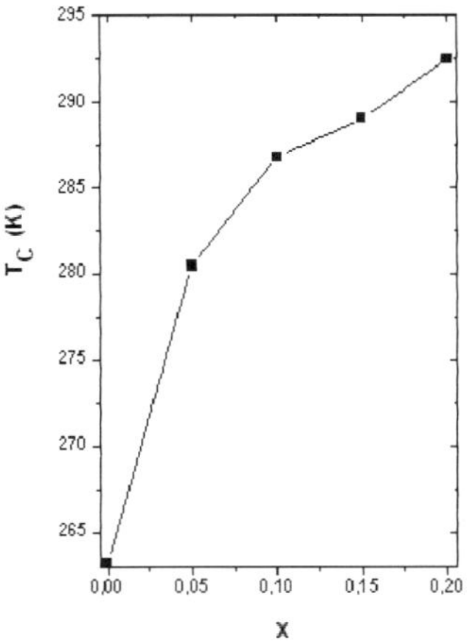

Figure 4. Curie temperature T_C evolution versus monovalent element content.

Figure 5 shows the evolution of the isothermal magnetization as a function of magnetic applied field for x=0 and 0.2. We can notice that for doped samples and below Curie temperature, the magnetization increases sharply for low field values (H<1T) and then seems to saturate, as can be observed in $Nd_{0.5}Sr_{0.3}Na_{0.2}MnO_3$.

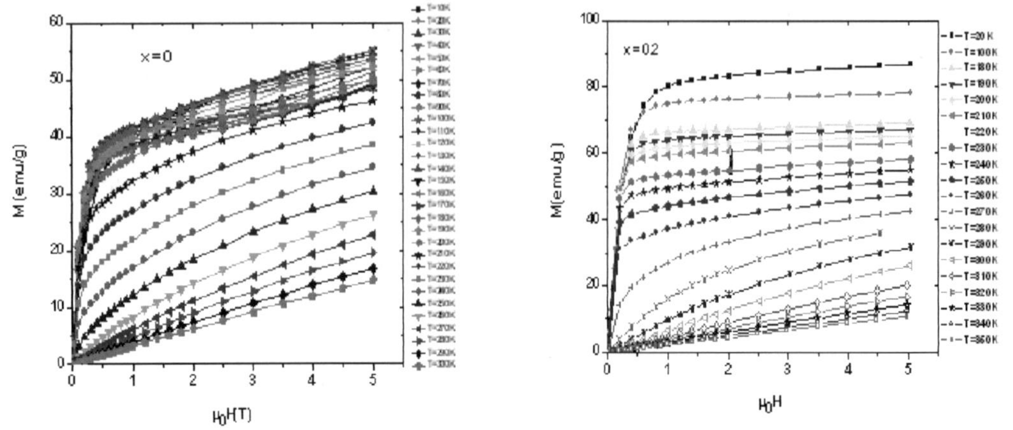

Figure 5. Magnetization evolution versus magnetic applied field up to 5T and 8T for several temperatures for $Nd_{0.5}Sr_{0.5-x}Na_xMnO_3$.

In order to get a deeper insight in the magnetic transition nature, we plot the so-called Arrott plots (M^2 versus μ_0H/M) in the temperature range 200-350K for all our samples. All the M^2 versus μ_0H/M curves exhibit a positive slope indicating that the ferromagnetic-paramagnetic transition is of second order, according to the Banerjee criterion [30]. Figure 6 shows the Arrott plots for $Nd_{0.5}Sr_{0.5}MnO_3$ and $Nd_{0.5}Sr_{0.3}Na_{0.2}MnO_3$ samples, the T_C values is about 250K for x = 0 and T_C = 275K for x = 0.2, which are close to those determined from M(T) measurements.

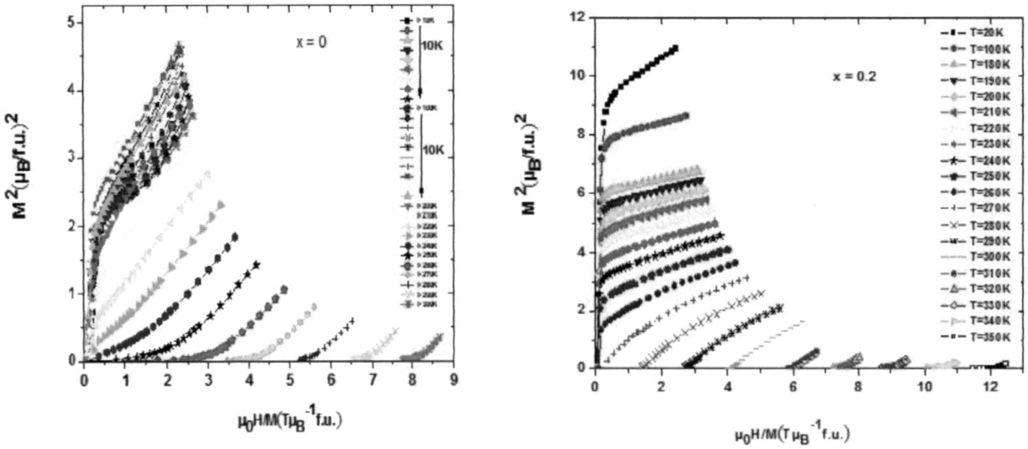

Figure 6. M^2 versus μ_0H/M isotherms for $Nd_{0.5}Sr_{0.5}MnO_3$ and $Nd_{0.5}Sr_{0.3}Na_{0.2}MnO_3$ samples.

From thermodynamical theory, the magnetic entropy change $|\Delta S_M|$ is given by [31] $|\Delta S_M| = \sum_i \frac{(M_i - M_{i+1})}{(T_{i+1} - T_i)} \Delta H_i$ where M_i and M_{i+1} are the experimental values of magnetization measured at temperatures T_i and T_{i+1}, respectively, under magnetic applied field change ΔH_i.

We plot in Figure 7 $\left|\Delta S_M\right|$ evolution as a function of temperature for different external magnetic field changes in $Nd_{0.5}Sr_{0.5}MnO_3$ and $Nd_{0.5}Sr_{0.3}Na_{0.2}MnO_3$ compounds. All our samples exhibit large MCE effect around the magnetic transition temperature and the magnitude of $\left|\Delta S_M\right|$ increases with increasing magnetic applied field.

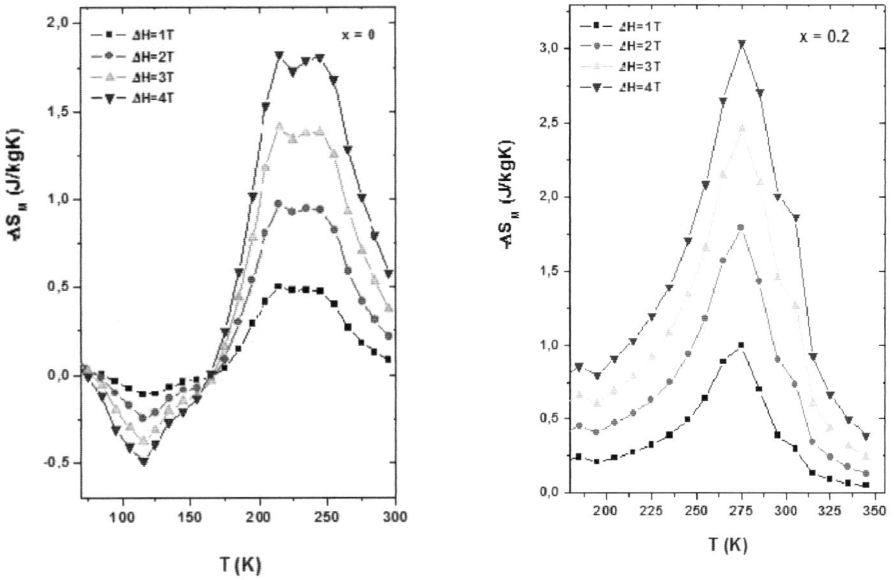

Figure 7. Magnetic entropy change $\left|\Delta S_M\right|$ versus temperature around the Curie temperature T_C for $Nd_{0.5}Sr_{0.5}MnO_3$ and $Nd_{0.5}Sr_{0.3}Na_{0.2}MnO_3$ samples.

Furthermore, for x = 0 sample, a negative magnetic entropy of -1.81J/kgK was observed at T_{CO} = 245K and a positive one of 0.49J/kgK was noticed at T_N = 115K under a magnetic field change of 4T. Whereas for x = 0.2, only negative values were recorded with $\left|\Delta S_M^{Max}\right|$=3.1J/kgK at T_C = 275K under 4T.

The evaluation of the refrigeration capacity passes trough the so-called relative cooling power RCP, $RCP = \left|\Delta S_M\right|_{\max} * \delta T_{FWHM}$, where δT_{FWHM} is the full with of half maximum of the magnetocaloric peak. It is, then, important that the magnetic entropy extend over a large temperature interval. It should be noted that a first order transition is able to concentrate the MCE in a narrow temperature range, whereas a second order transition is usually spread over a broad temperature range, which is the case in our samples and is beneficial for active magnetic refrigeration [32,33]. We have evaluated the RCP values under a magnetic field change of 4T for all samples. The RCP value is about 158J/kg for x = 0 and 209J/kg for x = 0.2. These values are somewhat smaller than those evidenced in other perovskites but high enough for technical interest.

4. Conclusion

The single-phase $Nd_{0.5}Sr_{0.5-x}Na_xMnO_3$ ($0 \le x \le 0.2$) polycrystalline samples were elaborated using the solid state reaction at high temperature. All our synthesized compounds crystallize in the orthorhombic structure with *Pbnm* space group. The Sr substitution by Na in $Nd_{0.5}Sr_{0.5}MnO_3$ sample induces a paramagnetic to ferromagnetic transition near room temperature, with a T_C slightly increasing from 263K for x = 0 to 292K for x = 0.2, and the antiferromagnetic interaction linked to charge ordering is suppressed. The largest magnetic entropy change was observed in $Nd_{0.5}Sr_{0.3}Na_{0.2}MnO_3$ sample, being 3.1J/KgK under a magnetic applied field change of 4T. The RCP values are comparable with those obtained in other perovskite-type specimens. The studied compounds may be considered as potential magnetic refrigerators operating near room temperature.

References

[1] Jonker G. H, Van Santen J. H 1950 *Physica* **16** 337.
[2] Kusters R. M, Singleton J, Keon D. A, Greedy R. M and Hayes W 1989 *Physica B* **155** 362.
[3] Jin S, Tiefel T. H, Cormack Mc, Fastnacht R. A, and Ramesh R, Chen L. H 1994 *Science* **264** 413.
[4] Maignan A, Caignaert V, Simon Ch, Hervieu M and Raveau B 1995 *J.Mat.Chem.* **5** 1089.
[5] Caignaert V, Maignan A, Raveau B 1995 *Solid State Comm.* **95**.
[6] Hwang H.Y, Cheong S.W, Radaelli P. G, Marezio M, Battlog B 1995 *Phys.Rev.Lett.* **75** 914.
[7] Jirak Z, Krupicka S, Simsa Z, Dlouha M, Vratislav Z 1985 *J. Magn. Magn. Mater.* **53** 153–166.
[8] Tomioka Y, Okuda T, Okimoto Y, Asamitsu A, Kuwahara H, Tokura Y 2001 *J. Alloys Compd.* **326** 27–35.
[9] Lacaze A. F, Beranger R, Bon Mardion G, Claudet G and Lacaze A. A 1983 *Cryogenies* **23** 427.
[10] Hashimoto T, Numasawa T, Shino M and Okada T 1981 *Cryogenies* **21** 647.
[11] Brown G. V 1976 *J. Appl. Phys.* **47** 3673.
[12] Hashimoto T, Kuzuhara T, Sahashi M, Inomata K, Tomokiyo A 1987 and Yayama H 1987 *J. Appl. Phys.* **62** 3873.
[13] Pecharsky V. K and Gschneidner K. A 1997 *Phys. Rev. Lett.* **78** 4494 Pecharsky V. K and Gschneidner K. A 1997 *Appl. Phys. Lett.* **70** 3299.
[14] Nikitin S. A, Myalikgulyev G, Annaorazov M. P, Tyurin A. L, Tishin A. M and Asatryan K. A 1990 *Phys. Lett. A* **148** 363.
[15] Guo Z. B, Du Y. W and Feng D 1997 *Phys. Rev. Lett.* **78** 1142.
[16] Guo Z. B and Du Y. W 1997 *Appl. Phys. Lett.* **70** 904.
[17] Huang H, Guo Z. B and Du Y. W 1997 *J. Magn. Magn. Mater.* **173** 302.
[18] Zhong W 1998 *Eur. Phys. J.* B3 169.
[19] Schiffer P, Ramirez Z. A. P and Cheng S. W 1995 *Phys. Rev. Lett.* **75** 3336.
[20] Ju H. L, Kwon C and Venkatesan T 1994 *Appl. Phys. Lett.* **65** 2108.
[21] Barratt J, Lees M. R and Paul D 1996 *Appl. Phys. Lett.* **68** 424.
[22] Chen C. H, Cheng S. W and Cooper A. S 1993 *Phys. Rev. Lett.* **71** 2461, Cheng S. W 1994 *Phys. Rev. B* 49 7088.
[23] Moritome Y 1995 *Phys. Rev. B* **51** 3297.
[24] Tokura Y, Tomioka Y, Kuwabara H, Asamitsu A, Moritomo Y and Kasai M 1996 *Physica C* 263 544.
[25] Kuwahara H, Tomioka Y, Asamitsu A, Moritomo Y and Tokura Y 1995 *Science* **270** 961.
[26] Rietveld M. H 1981 *J Appl. Cryst.* **2** 65, Wiles D. B, Young R. A 1981 *J.Appl.Crystallogr.***14** 149.
[27] Koubaa M, Cheikhrouhou-Koubaa W, Cheikhrouhou A 2008 *Physica B* **403** 2477.
[28] Koubaa M, Cheikhrouhou-Koubaa W, Cheikhrouhou A 2009 *J. Alloys Compd.* **473** 5.
[29] Koubaa M, Cheikhrouhou-Koubaa W, Cheikhrouhou A 2009 *J. Phys. Chem. Solids* **70** 326.
[30] Banerjee S. K 1964 *Phys. Lett.* **12** 16.
[31] McMichael R. D, Ritter J. J, Shull R. D 1993 *J. Appl. Phys.* **73** 6946.
[32] Phan M. H, Yu S. C, Hur N. H 2005 *Appl. Phys. Lett.* **86** 072504.
[33] Bruck E 2005 *J. Phys. D: Appl. Phys. R* **381**.

American College of Veterinary Internal Medicine
1997 Wadsworth Boulevard Suite A
Lakewood CO 80214-5293

CURRAN ASSOCIATES INC
proceedings.com